Springer Series on Chemical Sensors and Biosensors

Methods and Applications

O. Wolfbeis
Series Editor

Springer-Verlag Berlin Heidelberg GmbH.

Ramaier Narayanaswamy · Otto S. Wolfbeis

Optical Sensors

Industrial Environmental and Diagnostic Applications

With 197 Figures and 26 Tables

Springer

Chemical sensors and chemical biosensors are becoming more and more indispensable tools in life science, medicine, chemistry and biotechnology. The series covers exciting sensor-related aspects of chemistry, biochemistry, thin film and interface techniques, physics, including opto-electronics, measurement sciences and signal processing. The single volumes of the series focus on selected topics and will be edited by selected volume editors. The Springer Series on Chemical Sensors and Biosensors aims publish state-of-the-art articles that can serve as invaluable tools for both practitioners and researchers active in this highly interdisciplinary field. The carefully edited collection of papers ineach volume will give continuous inspiration for new research and will point to existing new trends and brand new applications.

DOI 10.1007/978-3-662-09111-1

Cataloging-in-Publication Data applied for
Bibliographic information published by Die Deutsche Bibliothek
Die Deutsche Bibliothek lists this publication in the Deutsche Nationalbibliografie; detailed bibliographic data is available in the Internet at <http:/dnb.ddb.de>.

springeronline.com

Originally published by Springer-Verlag Berlin Heidelberg New York in 2004
MyCopy version of the original edition 2004

Typesetting: medio Technologies AG, Berlin
Cover: design & production, Heidelberg

Printed on acid-free paper 52/3020 – 5 4 3 2 1 0
www.springer.com/mycopy

Series Editor

Professor Dr. Otto Wolfbeis
University of Regensburg
Institute of Analytical Chemistry, Chemo- and Biosensors
Universitätsstraße 31
93053 Regensburg
Germany
e-mail:wolfbeis@chemie.uni-regensburg.de

Volume Editor

Dr. Ramaier Narayanaswamy
Department of Instrumentation and Analytic Science
UMIST, P. O. Box 88
Manchester M 60 1QD,
England, United Kingdom
e-mail: ramaier.narayanaswamy@umist.ac.uk

Preface

The development and applications of optical chemical sensors and biosensors over the past two decades have been pursued with great interest by many researchers and have shown a tremendous growth and potential in employing such devices in many fields of application. These sensors require an interdisciplinary approach in that many scientific fields including spectroscopy, fibre optics and waveguides, optoelectronics, analytical chemistry and biochemistry, medicine and material science. The use of fibre optics and waveguides in sensors for advanced instrumentation is now well established and many novel integrated sensing systems, such as biochips, are now being developed.

In an attempt to disseminate the growth in optical sensors for many application areas such as industrial, environmental and medical diagnostics, several authors have been involved in producing this state-of-the-art type of monograph. This is the first book that is being published in a series on topics related to chemical sensors and biosensors by Springer-Verlag. The selection of topics and contributing authors for this first book on Optical Sensors have stemmed out of some of the excellent contributions made at the scientific sessions of the Europt(r)ode VI Conference that was held in April 2002 in Manchester, United Kingdom.

This book summarises various types of developments in optical sensor technology over the past two decades with an insight to future trends in this ever growing field. Chapter 1 reviews advances in this area using indicator probes and labels with a discussion of the factors that has lead to the growth of optical sensors. Chapters 2 and 3 reviews molecular recognition systems based on molecular imprints and new indicator dyes for sensing gaseous and dissolved analytes in the environment. Chapters 4 to 7 deal with aspects of optical biosensors and biochips for use in medical diagnostics and food analysis, while Chapters 8 to 15 discuss chemical transduction principles and optical sensor designs for monitoring analytes in industrial and environmental samples. These chapters illustrate a variety of transducer designs that have been investigated to date which include devices based on surface plasmon resonance, piezo-optic effects, integrated systems such as biochips, interferometric and array biosensors, and

involve analytical systems based on sensors with cross sensitivities and a biomimetic approach. Aspects of miniaturisation and automation are also discussed in some chapters as required, thus providing information on a truly integrated approach for analytical measurements. Thus the topics presented in this book are wide ranging and somewhat diverse but all contributions focus on diagnostics type of analytical systems.

This book is aimed at graduate students and researchers in academia and industry to provide knowledge on a group of devices based on a variety of optical detection techniques that are of novel trend in current measurements employed in analytical sciences. The authors, chosen from a list of multidisciplinary international team of researchers, are experts in their fields and are authoritative in respect of their contributions. The editors would like to express their sincere appreciation and gratitude to all authors for their efforts in preparing excellent contributions as chapters for this book with a common but diverse focus on the optical systems employed for environmental, industrial and medical diagnostic applications.

R. Narayanaswamy & O. Wolfbeis

Contents

Chapter 1
Optical Technology until the Year 2000: An Historical Overview
Otto S. Wolfbeis

Chapter 2
Molecularly Imprinted Polymers for Optical Sensing Devices
Marta Elena Díaz-García, Rosana Badía

Contributors

MINOO ASKARI
Oak Ridge National Laboratory
P.O. Box 2008
Oak Ridge, TN 37831-6101
USA

ROSANA BADÍA
Department of Physical and Analytical Chemistry
University of Oviedo
C/ Julian Clavería, 8
33006 Oviedo
Spain
e-mail: rbadia2@sauron.quimica.uniovi.es

KELLY R. BEARMAN
PiezOptic Ltd., Viking House
Ellingham Way, Ashford
Kent TN23 6NF
UK

MAXIMINO BEDOYA
Grupo Interlab
María Tubau 4, 3º, 28050 Madrid
Spain

DAVID C. BLACKMORE
PiezOptic Ltd., Viking House
Ellingham Way, Ashford
Kent TN23 6NF
UK

TIMOTHY J.N. CARTER
PiezOptic Ltd., Viking House
Ellingham Way, Ashford
Kent TN23 6NF
UK
e-mail: tjnc@piezoptic.com

FRANK Y. S. CHUANG
M-Division: Medical Physics and Biophysics
Lawrence Livermore National Laboratory
Livermore, California
USA
e-mail: chuangs@llnl.gov

FLORENCE COLIN
PiezOptic Ltd., Viking House
Ellingham Way, Ashford
Kent TN23 6NF
UK

BILL W. COLSTON, JR.
M-Division: Medical Physics and Biophysics
Lawrence Livermore National Laboratory
Livermore, California
USA
e-mail: colston1@llnl.gov

Marta Elena Díaz-García
Department of Physical and Analytical Chemistry, University of Oviedo
C/ Julian Clavería, 8
33006 Oviedo
Spain
e-mail: medg@sauron.quimica.uniovi.es

Tuan Vo-Dinh
Oak Ridge National Laboratory
P.O. Box 2008
Oak Ridge, TN 37831-6101
USA
e-mail: vodinht@ornl.gov

David García-Fresnadillo
Laboratory of Applied Photochemistry
Department of Organic Chemistry, Faculty of Chemistry
Complutense University
28040 Madrid
Spain

Guy Griffin
Oak Ridge National Laboratory
P.O. Box 2008
Oak Ridge, TN 37831-6101
USA

Jiří Homola, Ph.D.
Institute of Radio Engineering and Electronics
Chaberská 57, 182 51 Prague
Czech Republic
e-mail: homola@ure.cas.cz

Frank Kvasnik
Department of Instrumentation and Analytical Science, UMIST
Sackville Street, PO Box 88
Manchester M60 1QD
England, UK

L.M. Lechuga
Centro Nacional de Microelectrónica, CSIC
Madrid, Spain
e-mail: laura@imm.cnm.csic.es

Frances S. Ligler
Center for Bio/Molecular Science & Engineering, Naval Research Laboratory, Washington, DC 20375-5348, USA
e-mail: fligler@cbmse.nrl.navy.mil

Gerhard J. Mohr
Institute of Physical Chemistry
Friedrich-Schiller University
Jena, Lessingstrasse 10
07743 Jena, Germany
e-mail: gerhard.mohr@uni-jena.de

Maria C. Moreno-Bondi
Dpmts. of Analytical Chemistry
Faculty of Chemistry
Complutense University
28040 Madrid, Spain
e-mail: mcmbondi@quim.ucm.es

Ramaier Narayanaswamy
Department of Instrumentation and Analytical Science
UMIST, P.O. Box 88
Manchester M60 1QD
England, UK
e-mail: ramaier.narayanaswamy@umist.ac.uk

Guillermo Orellana
Laboratory of Applied Photochemistry
Department of Organic Chemistry, Faculty of Chemistry
Complutense University
28040 Madrid, Spain
e-mail: orellana@quim.ucm.es

Claudia Preininger
ARC Seibersdorf Research GmbH, Division Environmental and Life Sciences, Business Unit Biotechnology
2444 Seibersdorf, Austria
e-mail: claudia.preininger@arcs.ac.at

F. Prieto
National Centre of Microelectronic, CSIC
Madrid, Spain

Steven A. Ross
PiezOptic Ltd., Viking House
Ellingham Way, Ashford
Kent TN23 6NF
UK

Ursula Sauer
ARC Seibersdorf Research GmbH, Division Environmental and Life Sciences, Business Unit Biotechnology
2444 Seibersdorf, Austria

Kim E. Sapsford
Center for Bioresource Development
George Mason University
Fairfax, VA 22030-4444
USA

B. Sepúlveda
National Centre of Microelectronic, CSIC
Madrid, Spain

Peter Šimon
Department of Physical Chemistry, Faculty of Chemical and Food Technology
Slovak University of Technology
Radlinského 9
812 37 Bratislava
Slovak Republic
e-mail: simon@cvt.stuba.sk

David L. Stokes
Oak Ridge National Laboratory
P.O. Box 2008
Oak Ridge, TN 37831-6101
USA

Dimitra N. Stratis-Cullum
Oak Ridge National Laboratory
P.O. Box 2008
Oak Ridge, TN 37831-6101
USA

Ben R.Swindlehurst
Department of Instrumentation and Analytical Science
UMIST, P.O. Box 88
Manchester M60 1QD
England, UK

Hannes Voraberger
Institute of Chemical Process Development and Control
JOANNEUM RESEARCH
Forschungsgesellschaft m.b.H.
Steyrergasse 17, 8010 Graz
Austria
e-mail: hannes.voraberger@joanneum.at

Alan Wintenberg
Oak Ridge National Laboratory
P.O. Box 2008
Oak Ridge, TN 37831-6101
USA

Otto S. Wolfbeis
University of Regensburg
Institute of Analytical Chemistry,
Chemo- & Biosensors
93040 Regensburg
Germany
e-mail:otto.wolfbeis@chemie.uni-regensburg.de

John D. Wright
Centre for Materials Research
School of Physical Science
University of Kent, Canterbury
Kent CT2 7NR
UK
e-mail: j.d.wright@ukc.ac.uk

CHAPTER 1

Optical Technology until the Year 2000: An Historical Overview

Otto S. Wolfbeis

1 Introduction

This chapter reviews advances in the area of optical sensor technology using indicator probes and labels. It also presents an account of optical sensors based on the use of biomolecules and molecular receptors including molecular imprints. One may wonder why such progress has been made in the past 20 years. In the author's opinion, this is due to several factors. These include the following:

(a) New light sources have become available, including light-emitting diodes and diode lasers with their low power consumption and small size, both now spanning the entire visible range (LEDs even available with emission peaks at 370 nm);

(b) new photodiodes and CCDs have become available with hitherto unseen sensitivity for visible and near-infrared light;

(c) the advent of fiber optics and of integrated waveguide optics with excellent transmission that extends far into the near infrared;

(d) wave guides have become available that also transmit mid-infrared and infrared light;

(e) fast detectors, oscilloscopes and log-in amplifiers are now available at low cost; these allow sensitive and highly time-resolved measurements;

(f) new detection schemes including surface plasmon resonance, evanescent wave sensing have found their way into sensor technology;

(g) fluorescence spectroscopies (which are so useful for sensing purposes) have become particularly widespread and include the measurement of fluorescence (or phosphorescence) intensity, decay time, energy transfer, quenching efficiency, polarization, and allow gated measurements that can eliminate interfering background luminescence;

(h) new biomolecular receptors (synthetic, made from plastic, or genetically engineered) are available that may be used for sensing purposes;

(i) substantial progress has been made in nucleic acid chemistry and protein chemistry, including the immobilization of these biomolecules;

(j) progress in polymer chemistry, which is particularly significant since polymer chemistry plays a major role in optical sensor technology; this is particularly true for sensor matrices, biocompatible materials, and new materials including composite polymers (e.g. for molecular imprints), hydrogels, and sol-gels;

(k) powerful data processors and data loggers have become available at affordable prices; finally,

(l) it is believed that interdisciplinary teaching and interdisciplinary conferences like *Europtrode* have contributed to an improved understanding and interaction between scientists involved in optical sensor technology.

A few years ago [1] the following definition of optical chemical sensors was proposed: *Chemical sensors are miniaturized analytical devices that can deliver real-time and on-line information on the presence of specific compounds or ions in complex samples.* There are many ways to accomplish the goal of on-line and real-time sensing. However, the simplest way is to insert a sensor probe into the sample of interest so to obtain an analytical signal that can be converted into a concentration unit. Chemical sensors ideally act fully reversibly, as do physical sensors for temperature, pressure, and the like. This has been achieved in few cases only. It should also be noted at this point that the usual definitions of biosensors do not match the former definition of chemical sensors. In fact, clinicians often refer to biosensors as single shot detection elements if reagents are immobilized on a solid support. It should also be kept in mind that sensors are expected to be small and to function even in complex samples and – ideally – without sample pretreatment.

2 Very Early History

According to the Elder Pliny (23–79 AD), a papyrus test was in use in ancient Greece to detect adulteration of copper coins by iron. Coins were dissolved in acid and the presence of ferric ions was detected with a test strip impregnated with an extract of certain apples. Tachenius, in 1629, applied this method to detect increased concentrations of ferric ion in urine. Also, filter paper impregnated with sodium carbonate was used in the early 19th century to detect uric acid, and pH test strips were obtained a little later by soaking filter paper with solutions of litmus (an extract of lichens such as *Rocella tinctoria*) and subsequently gently drying them. Test strips for aluminium (consisting of filter paper with a hydroxyflavone reagent that was impregnate) were introduced and analytically characterized by Goeppelsrieder in the late 19th century, and this appears to have been the beginning of more systematic research on chemically sensitive layers. Feigl [2] summarized the art of spot testing in his famous book which describes numerous tests for ions, gases and organic compounds and is still a valuable source. These tests were later named "dry reagent chemistries" [3], and much later they were referred to by many as "optical sensors", even if good for a single shot test only. Times change. The first article on an enzymatic biosensor seems to be the one by Free et al. [4] in 1957, and the area of "immunosensors" (which are in fact irreversible) starts with the work of Ronbitaille et al. [5], if not earlier, and has had its greatest commercial success in optical pregnancy testing.

3
Early History (up to about 1985)

Most probably, the first optical chemical sensors for on-line (continuous) use where those for pH and for oxygen. It has been known for decades that cellulosic paper can be soaked with pH indicator dyes to give pH indicator strips; however, the dyes leached out and these sensors thus were of the "single-use" type. The respective research and development work is not easily traced since it is not well documented in the readily accessible literature. However, in the 1970s, indicator strips became available in which pH indicator dye was covalently linked to the cellulose matrix, usually via vinylsulfonyl groups. These "non-bleeding" test strips allowed a distinctly improved and continuous pH measurement, initially by visual inspection. In the late 1980's instruments became available that enabled the color (more precisely the reflectance) of such sensor strips to be quantified and related to pH. Respective instruments are based on the use of LEDs and are small enough to be useful for field tests in that they can even be handhelp. In 1975, the immobilization of pH indicator dyes on glass was reported by Harper [6]. Azo dyes were linked to the surface of silicate glass (said to be more stable than cellulosic supports) that was rendered reactive by treatment with the reagent aminopropyl-triethoxysilane.

The absorption (and reflectance) spectra of a common pH paper strip (e.g. the Merck Reflectoquant) displays absorption bands at about 460 nm (acid form) and 580 nm (base form), respectively. The orange band can be interrogated by a blue LED, and the blue band by a yellow LED, thus enabling 2-wavelength ratio measurements. However, in the commercial instruments, a reference signal obtained with an LED operated at about 670 nm is used to compensate for geometrical and scattering effects. This simple and low cost detection system is still superior to many of the complicated, if not expensive, pH sensors that have been described in the past 20 years. Commercial chemistries are available for a variety of pH ranges and are compatible with fiber optic systems, and an affordable fiber optic pH sensing instrument is available now from Ocean Optics Inc. (FL).

Optical sensing of oxygen started with the work of Kautsky and Hirsch in the 1930's. They reported [7, 8] that the room-temperature phosphorescence of certain dyes adsorbed on silica gel supports is strong only in the complete absence of oxygen. Traces of oxygen quench phosphorescence. Specifically, silica beads were soaked with a solution of a dye such as trypaflavine or fluorescein, dried, and placed in a flow-through cell schematically shown in Fig. 1. The room temperature phosphorescence was monitored using a fluorometer and it was found that even ppm quantities of oxygen gas (equivalent to a p_{O2} of 0.5×10^{-3} Torr) were capable of quenching phosphorescence. The effect is fully reversible within 1–2 s. The device assisted in the discovery of the Kautsky effect, i.e. the delay in the production of oxygen following illumination of a leaf.

This method was later used by Pringsheim et al. [9] and by Zakharov et al. [10] to monitor oxygen in seawater. The first complete optical sensor system was described by Bergman in 1968 [11]. His system comprised a UV light source, an oxygen-sensitive fluorescent layer comprising a porous glass or polyethylene

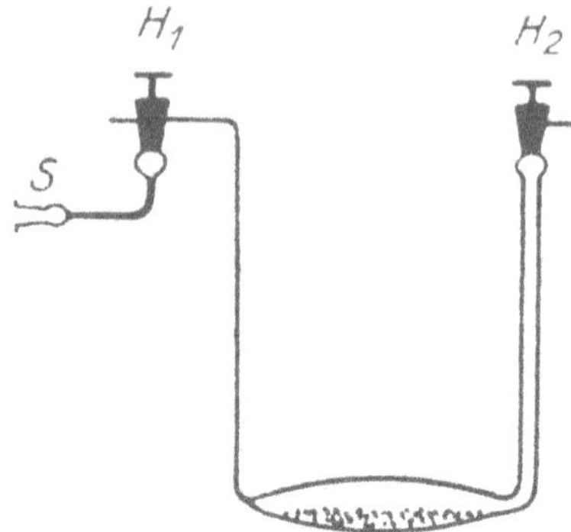

Fig. 1. Configuration of Kautsky's flow cell (1931). S, gas sample containing traces of oxygen; H_1 and H_2, valves; the sensor material is contained in the cavity on the bottom

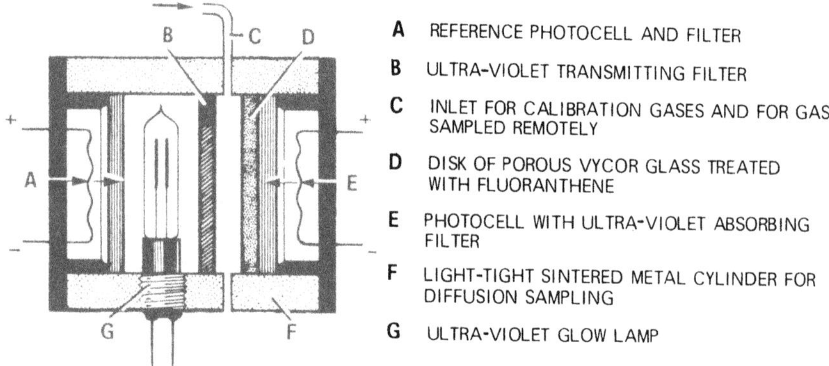

Fig. 2. Configuration of Bergman's oxygen cell (1968)

film (soaked with the oxygen-sensitive dye fluoranthene), and a photodetector. Thus, it contained all the elements of a modern optical sensor. The fluorescence of fluoranthene is strongly quenched by molecular oxygen and this served as the analytical information. The system, shown in Fig. 2, responds to oxygen at levels above 1 Torr.

A few years later Stevens described, in a patent [12], a rather similar device. He also mentions the possibility of using a radioactive source rather than a UV light source to excite fluorescence. In 1974, Hesse [13] described a fiber optic chemical sensor for oxygen and this appears to have been the very first fiber optic chemical sensor (FOCS). An oxygen-sensitive chemistry was placed in front of a fiber optic light guide through which exciting light was guided. The fluorescence emitted is guided back through either the same fiber, or through the other fibers of a bundle. The system is based on measurement of either fluorescence intensity or fluorescence decay time, both of which are affected by oxygen. A schematic of the design is shown in Fig. 3.

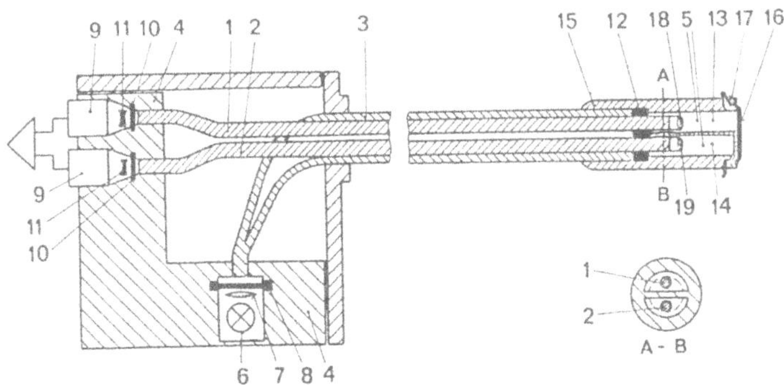

Fig. 3. First fiber optic chemical sensor (Hesse, 1974)

In 1975 and 1976, Lübbers and Opitz published three articles [14–16] to describe an instrument capable of online monitoring of oxygen or carbon dioxide (CO_2). They first [15] referred to it as an *optrode* (by linguistic analogy to the electrode), and later [16] as an *optode* (οπτιχοσ οδοσ; Greek for *the optical way*). While the detection scheme for oxygen was known, that for CO_2 was new. It resembles the Severinghaus electrode but makes use of an optical (fluorescent) pH probe contained in a bicarbonate buffer and covered with a proton-impermeable but CO_2-permeable membrane. Changes in the internal pH of a buffer as indicated by the pH probe.

Lübbers and Opitz also were the first to show that their optode was useful for monitoring CO_2 and oxygen in biological samples and in enzyme-based biosensors, for example for ethanol [17]. It was obvious, however, that the indicator probes used (pyrenebutyric acid for oxygen; 7-hydroxy-4-methylcoumarin for pH/CO_2) were not ideally suited for sensing purposes. The dyes were later replaced by decacyclene [18] and HPTS [19], respectively, both being excitable by a blue LED. Extremely fast responding ($\ll$50 ms) oxygen sensors were obtained by chemically immobilizing pyrenebutyric acid in the form of a very thin film on a glass support [20].

The late 1970s also saw initial investigations by Hirschfeld and co-workers [21–23] whose early work is difficult to trace since it is hidden in hardly accessible conference proceedings. They used fiber optic light guides in order to monitor (by either absorbance or fluorescence) the concentration of chemical species such as uranium ion which in phosphoric acid solution displays intense fluorescence if excited by the 488-nm line of the argon laser. The method was mainly applied to nuclear power stations and research stations and is paralleled by work on fiber optic sensing by Boisdé [24, 25]. The fiber optic pH sensor described by Peterson et al. [26] in 1980 was a milestone in optical fiber sensor technology. A 2-volume book that appeared in 1991 gives an account of the early work on fiber optic chemical sensors and biosensors up to about 1989 [27]. Early reviews include those by Kirkbright et al. [28], Borman [29], and Hirschfeld et al. [30].

4
Optical Sensors for Gases (Including Dissolved Gases) and Organics

Gases usually have no intrinsic absorption (except for the infrared where optical fibers have poor transmittance) so that indirect methods have to be used. Some of the early work on oxygen has been described in the previous section. Later, Hendricks [31] described a method for detection of oxygen by contacting a film comprising polyethylene naphthalenedicarboxylate with oxygen and exposing this film to ultraviolet light. On subsequent heating, the presence of oxygen is indicated by thermoluminescence. Oxygen may also be monitored via the chemiluminescence produced by oxygen on reaction with an olefin [32]. Marsoner [33] dissolved pyrenebutyric acid in poly(vinyl chloride) containing a plasticizer. The material was deposited on acrylate glass. Particles of ferric oxide added in some cases acted as optical insulators to prevent interferences by the intrinsic fluorescence of samples.

The value of phosphorescent ruthenium ligand complexes for use in optical oxygen sensing was recognized in 1986 [34]. Ruthenium tris(bipyridyl) was adsorbed onto silica gel and placed in a silicone membrane. The material can be excited with a blue LED and displays a red emission that is strongly quenched by oxygen. Similar ruthenium dyes were reported by Demas and coworkers in 1987 [35]. Since the ruthenium polypyridyl complexes have decay times in the order of a few microseconds, they are ideally suited for sensing based on measurement of decay time. The first article on the use of ruthenium complexes in luminescence decay time-based sensors appeared in 1988 [36].

Following the discovery that the fluorescence of metalloporphyrins is strongly quenched by oxygen [37], optical sensor membranes were developed that are suitable for phosphorescent sensing of oxygen [38]. In 1984, Peterson and coworkers described [39, 40] a fiber optic catheter that has been used to monitor oxygen in vivo. The sensor again exploits the quenching of fluorescence by oxygen, this time of a polycyclic aromatic hydrocarbon. Table 1 summarizes fundamental articles on optical sensors for oxygen until the year 2000.

Table 1. Fundamental contributions to optical oxygen sensor technology up to the year 2000 (only the main author/s are given)

Author	Year	Remarks
Kautsky	1931	adsorbed dyes; discovers Kautsky effect
Pringsheim	1944	sensing in biology; same as Kautsky
Hesse	1974	fiber optic sensor; also measures decay time
Lübbers	1975	quenching of PBA in teflon-covered solution; appl. to blood gas sensing
Lübbers	1978	measurement of O_2 and CO_2 in guinea pig heart
Peterson	1980	oxygen flow visualization (imaging)
Lübbers	1980	alcohol biosensor (via AlcOx)
Lübbers	1981	glucose biosensor (via GOx)
Zakharov	1981	oxygen in seawater by Kautsky method

Author	Year	Remarks
Lloyd	1981	via "residual luminescence of photobacteria" contained in silicone rubber membrane
Kroneis	1983	sterilizable oxygen probe, benzo[*ghi*]perylene as an oxygen probe
Wolfbeis	1984	glass-immobilized oxygen probes (breath gas sensor)
Peterson	1984	first fiber optic oxygen sensor
Vanderkooi, Wilson	1986,	phosphorescent sensing; imaging; Pd-coproporphyrin; decay time
Wolfbeis, Leiner	1986	ruthenium dyes in oxygen sensors
Demas	1987	ruthenium dyes in oxygen sensors
Lippitsch	1987	decay time based sensor (Ru)
Leiner	1987	plastic support for sensor mass fabrication
Wolfbeis	1988	first hybrid sensor (oxygen and CO_2)
Okazaki	1988	frequency doubled diode laser (390 nm) used to excite the oxygen probe benzoperylene
Nyberg	1988	metal oxide whose reflectivity reversibly changes with p_{O2}
Butler	1989	oxygen via reflectivity of nickel film
Gouterman	1990	barometry (pressure-sensitive paints)
Moreno-Bondi	1990	fiber optic glucose biosensor (via oxygen)
Li & Wong	1991	Pt complex as oxygen probe
Demas, DeGraff	1991	multi-site quenching model (Stern-Volmer)
Kraus, Israel et al.	1991	via phosphoresc. of pyronine in various polymers
Papkovsky	1991	phosphorescent polymer films (porphyrins)
MacCraith	1992	evanescent wave sensing using sol-gel
Lübbers	1992	glucose via GOx and oxygen
Baldini	1992	absorption-based fiber sensor (cobalt dye)
Charlesworth	1993	oxygen via RTP phosphorescence of camphorquinone
Troyanovsky, Sadovitskii	1993	pressure sensing via luminescence (pressure-sensitive paints)
Sanz-Medel, Diaz-Garcia	1994	oxygen via RTP phosphorescence quenching of ferrone chelates
Preininger	1994	bacterial sensor (for BOD)
Klimant	1995	microsensors
Holst & Lübbers	1995	oxygen flux ("FLOX") in skin
Papkovsky	1995	ketoporphyrins (diode laser-compatible probes)
Okura	1997	PtOEP in sol gel
Weldon	1997	oxygen by absorption at 761 nm (singlet-triplet absorption)
Klimant	1997	high-resolution imaging of dissolved oxygen
Hartmann	1997	oxygen flux lifetime imaging
Weldon et al.	1997	oxygen at 761 nm via its singlet-triplet transition
Campagna	1998	iridium complexes as oxygen probes
Amao	2000	europium complex as a pressure probe (in fluoropolymer)
Trettnak	2000	oxygen trace sensing (1 ppm in nitrogen) using Teflon AF polymer

Following the initial work of Lübbers and Opitz on optical sensors for carbon dioxide [14], a variety of other papers appeared. The group of Seitz [41] used a sensing material based on a solution of the indicator dye HPTS in by carbonate buffer, while Heitzmann and Kroneis [42] used polyacrylamide beads soaked with HPTS and carbonate buffer and dispersed in silicone. A fiber optic device was described by Vurek et al. in 1982 [43], and Munkholm et al. used a fiber optic system with covalently immobilized pH probes [44]. A few years later, a ruthenium complex was described whose fluorescence is quenched by carbon dioxide due to formation of protons [45].

The "plastic type" CO_2-sensor differs from the Severinghouse-type sensor in that it does not require the presence of a carbonate buffer. Kawabata et al. described the first type of such a sensor [46]. Raemer and co-workers described, in a patent [47], a similar system, this time based on an ion pair and the addition of a lipophilic base, both contained in a plastic matrix. This scheme was later widely extended by Mills et al. [48]. The first ammonia sensors based on an acid-base chemistry were described at about the same time [49, 50].

Numerous other chemistries for specific gases have been described. Butler [51] designed a hydrogen sensor based on the ability of palladium metal (Pd) to absorb hydrogen gas, thereby undergoing an expansion. If the Pd forms the coating of an optical fiber, the effect changes – through bending – the effective path length of the fiber, and this is detected by interferometry. Later, the Pd/WO_3 chemistry for hydrogen gas was applied to hydrogen sensing [52]. Exposure of this material (a palladium film on MoO_3) to hydrogen results in the appearance of a deep blue color (possibly the Mo^{5+} ion) that can be monitored at 780 nm via waveguide optics. The same effect is shown by molybdenum trioxide sputtered on glass supports and covered with a thin layer of Pd.

The poorly reproducible Fujiwara reaction may be used for quasi continuous monitoring of chlorinated hydrocarbons as they can occur in drinking water [53]. Following Hirschfeld's reservoir sensor for uranium (using phosphoric acid reagent) this appears to have been the second reservoir type of sensor. Carbon monoxide (CO) can be sensed by the effect it exerts on a complex chemistry involving certain molybdates, ferric ion, organic anions (e.g. acetate), all contained in a polymer matrix [54]. A blue color is formed which is slowly reversible. A fiber optic sensor for nitrogen oxide was reported that uses a cyanocobinamide as the optical probe which gives a coloration with NO_x (and nitrite) at 355 and 550 nm [55].

Unfortunately, no chemistries are available for less reactive gases such as nitrogen (to the dismay of the Guinness breweries!), and for many organics including aromatic hydrocarbons. Hence, hydrocarbons are usually detected by infrared spectroscopy in the gas phase, or – in the case of aromatic hydrocarbons – by direct UV absorption spectrometry. The explosive trinitrotoluene was remotely detected via fiber absorption measurements making use of its color reaction with amines [56].

5 Opt(r)odes for pH

As stated above, the beginning of optical pH sensor technology remains obscure. What is nowadays referred to as a sensor layer was formerly mostly referred to as a test strip, a dry reagent chemistry or an immobilized reagent. A general logic that is based on the immobilization chemistry of commercial reflectometric test strips was presented and extended to various pH ranges [57]. Such sensing "chemistries" are easily produced and can been coupled to fiber optics. This enabled sensing to be performed at formerly inaccessible sites. Peterson et al. were the first to report a fiber optic pH sensor [39]. The system comprised plastic fibers, a pH chemistry at their end (composed of a cellulosic dialysis tubing filled with a mixture of polystyrene particles and polyacrylamide beads dyed with phenol red), LED light sources, and photodiodes. The system is operated at two wavelengths. Fig. 4 gives a schematic of the fiber tip.

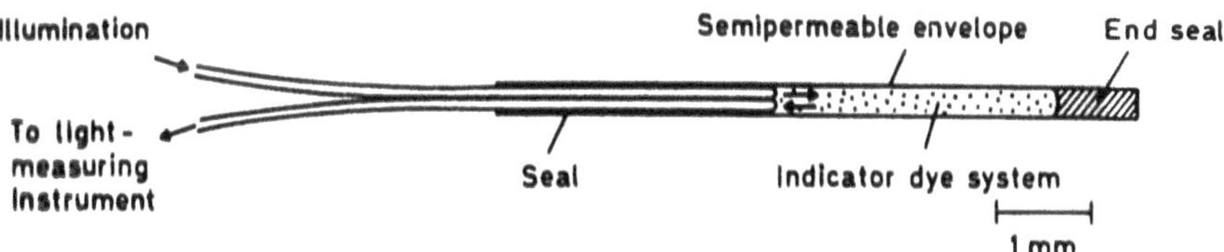

Fig. 4. Configuration of the fiber optic pH sensor of Peterson et al. (1980)

Table 2. Fundamental work on pH optical sensors until the year 2000

Author	Year	Remarks
G.B. Harper	1975	re-usable glass-bound indicators
D.W. Lübbers et al.	1977	nanoencapsulated fluorescent pH indicators
J.I. Peterson et al.	1980	first fiber optic pH probe
S. Goldstein et al.	1980	miniature fiber optic pH sensor for blood
Tait et al.	1982	fiber optic in vivo pH sensor
L.A. Saari, W.R. Seitz	1982	pH sensor based on immobilized fluorescein
N. Opitz & DW Lübbers	1983	optical sensor for pH and ionic strength
J.S. Suidan et al.	1983	fiber optic pH sensor for blood
O. S. Wolfbeis et al.	1983	evaluation of fluorescent pH indicators
J. I. Peterson & G. Vurek	1984	fiber optic sensors for biomedical applications (review)
G.F. Kirkbright et al.	1984	immobilization of pH indicators on ion exchange polymers (XAD)
Z.Zhujun, W.R. Seitz	1984	first fluorescent pH sensor (HPTS)
M.J. Goldfinch, C.R. Lowe	1984	solid phase optoelectronic pH sensor
Abraham et al.	1985 & 1986	fiber optic pH probes
D. R. Walt, F. P. Milanovich, S. M. Klainer	1986	polymer modification of a fluorescent fiber optic pH sensor

Table 2. continued

Author	Year	Remarks
A.M. Scheggi, F. Baldini	1986	comparison of pH sensing by absorption reflection, and fluorescence
B.A. Woods et al.	1986	optical pH sensing at low buffering capacity
J.L. Gehrich et al.	1986	intravasular blood pH monitoring system
J. Janata	1987	a critical review on the accuracy and precision of optical pH sensors
E. Grattan et al.	1987	2-wavelength optical fiber pH sensor
Y. Kawabata et al.	1987	fiber optic pH sensor with monolayer indicator
B.Boisdé, J.J. Pérez	1987	miniature pH sensor (1 mm)
D.M. Jordan et al.	1987	pH sensor exploiting FRET
M. Monici et al.	1987	pH sensor for seawater monitoring
O. Wolfbeis, H. Marhold	1987	pH indicators for an extended range
J.W. Attridge et al.	1987	pH sensing via refractive index
D.M. Jordan et al.	1987	pH sensor based on energy transfer
T.P. Jones, M.D. Porter	1988	immobilization of pH indicators on cellulose acetate
E.T. Knobbe et al.	1988	immobilization of pH probes in sol-gels
H.E. Posch et al.	1989	gastric pH sensor (pH 0–7)
W.P. Carey	1989	sensor for high acidities
B.H. Schaffar, O.S. Wolfbeis	1989	Langmuir-Blodgett sensing films
W. Carey, B.S. Jorgensen	1991	sensor for high acidities based on fluorescent polymers
G. Gabor, D.R. Walt	1991	pH sensing via inner filter effects
M.E. Lippitsch et al.	1992	first time domain fluorescent pH sensor
W.Tan et al.	1992	submicrometer intracellular pH sensor
T. Werner, O.S. Wolfbeis	1993	optical sensor for pH 10–13, new support material
Z. Ge et al.	1993	fiber optic evancescent wave sensor using polyaniline
J.W. Parker et al.	1993	pH sensor using the self referencing dye SNARF
G.J. Mohr, O.S. Wolfbeis	1994	azo indicators immobilized via vinylsulfonyl groups on cellulose films (on polyester support)
M.F. McCurley	1994	pH sensitive hydrogel
K.S. Bronk, D.R. Walt	1994	pH sensor array using fiber bundles
W.C. Michie et al.	1995	distributed pH sensor using fiber optics and swellable polymers
R. Koncki et al.	1995	pH sensor using near-infrared dye in PVC
S.G. Schulman et al.	1995	white range pH sensor based on photodissociation
Z. Zhang et al.	1995	pH sensor based on reflectance of swelling polymer
M.N. Taib et al.	1996	pH sensor range extended via artificial neural network
S. de Marcos, O.S. Wolfbeis	1996	polypyrrole films as pH sensor materials
D.B. Papkovsky et al.	1997	PVC sensor membranes containing porphyrins
T. Werner et al.	1997	PET pH sensor
A. Safavi, H. Abdollahi	1998	optical sensor for high pH

As the potential of optical fiber probes for pH measurements was rapidly recognized, several other articles appeared within a few years [58–66]. Most were reflectance-based, and Seitz reported the first fluorescent pH sensors [41, 61]. The article by Janata [67] on whether pH optical sensors can really measure pH is another "must" in the early literature since it points out aspects hardly addressed in pH sensor work. The dependence on ionic strength is an intrinsic limitation of pH sensors using indicator dyes. Opitz and Lübbers [68] and Offenbacher et al. [69] have presented solutions to this by making use of two indicators whose pK_a dependency on ionic strength is different, so that two independent signals are obtained from two dyes or sensors. Given the advantages of diode lasers operated at wavelengths above 600 nm, corresponding pH probes were used [70]. The fundamental work on pH sensors before the year 2000 is summarized in Table 2. A good review has recently been written by Lin [71].

6 Optical Sensors for Ions

It appears that Charlton et al. [72, 73] discovered the first methods for reversible and continuous optical measurement of clinically highly important ions. In one approach [72] they use plasticized poly(vinyl chloride) along with valinomycin as the ion carrier, and a detection scheme that was later referred to as co-extraction. In their system, potassium ion is extracted into plasticized PVC, and the same quantity of the anionic red dye erythrosine is co-extracted into it. The extracted erythrosine is quantified via absorbance or reflectance.

Charlton et al. were also the first to utilize the ion exchange principle in sensors [73]. Again they used a plasticized PVC film containing valinomycin and – in addition – a deprotonable dye (MEDPIN; a lipophilic 2,6-dichlorophenol-indophenol). On extraction of potassium into the membrane a proton is released from MEDPIN which then turns blue. The sensor layer measures potassium over the clinical range with excellent performance [74–76]. This scheme proved to be highly flexible. The dye used in the commercial system is superior (in terms of stability) to other dyes such as Nile Blue that have later been applied in the ion exchange detection scheme. Its chemical structure is shown in Fig. 5.

It has been known for many years that heavy metals can be detected qualitatively by immobilizing indicator dyes on solid supports such as cellulose. In

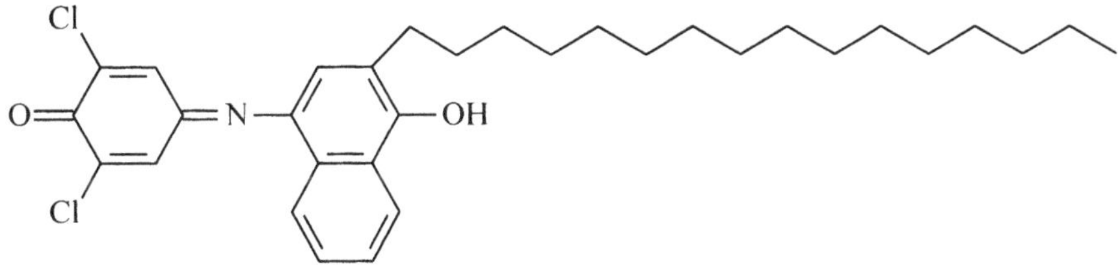

Fig. 5. Chemical structure of the deprotonable dye MEDPIN used in commercial potassium sensor strips. On deprotonation its color changes to blue (1982). It is interesting to note that this dye has been used much less often in scientific work than, for example, certain Nile Blue derivatives even though it is much more (photo)stable than the latter

fact, this approach forms the basis for the widely used test strips for heavy metals. Numerous schemes have been described for heavy metals in the past 20 years [for reviews, see 77–79]. On looking at the more recent literature one may state, however, that some of the newly described "chemistries" perform hardly any better than the rather old commercial systems based on the use of dry reagent chemistries, with the additional advantage that they are compatible with a single instrument. In fact, some of the newer systems involve rather extensive chemistry and – worst of all – seem to strongly differ in terms of spectroscopy and analytical wavelengths so that they all require their own opto-electronic platform. On the other hand, there is a pressing need for (low-cost) sensors for less common species, and those for Al^{3+} [80] and certain heavy metals [81] are typical examples.

While heavy metals can be easily detected by making use of known indicator dyes or quenchable probes, the alkali and alkaline earth elements are not easily recognized by conventional dyes at neutral pH and room temperature, and without addition of reagent. Therefore the molecular recognition properties of crown ethers and cyclic peptides have been widely used. The first studies were performed by co-extraction of ions from water into chlorinated hydrocarbons, but later on the reagents were immobilized in PVC [82]. The Amberlite ion exchanger (of the XAD type) is another widely used matrix for immobilizing indicator probes [83].

Crown ethers have excellent recognition properties and can recognize numerous ions including alkali ions. Unfortunately, those synthesizing crown ethers tend to investigate their properties in organic solutions such as acetonitrile and then claim that the findings may be useful for sensing alkali ions, thus ignoring the fact that alkali ions are usually sensed in aqueous solutions (including blood) where binding constants are often very different. An interesting scheme for an optical sensor for sodium based on ion-pair extraction and fluorescence was introduced by the Seitz group [84].

The chromo- and fluoro-ionophores form a particularly interesting class of probes since they can combine recognition properties with optical transduction such as changes in color or fluorescence. Chromo-ionophores have been used in clinical test strips for K^+ since 1985 based on the work of Voegtle [85] and others. The fluorescence of many fluorophores is particularly sensitive to perturbations of their microenvironment. For example, photo-induced electron transfer (PET) can be suppressed on binding of alkali ions. This has been demonstrated in impressive work by the groups of Valeur [86] and DeSilva [87]. The suppression of a PET effect in a naphthalimide system as described by DeSilva led to the fluoro-ionophores given in Fig. 6. Most noteworthy, the fluoro-ionophores are contained in a hydrophilic rather than hydrophobic matrix. The fluoro-ionophores are used in Roche's Opti-1 clinical electrolyte analyzer [88].

A novel approach for ion sensing based on the use of potential-sensitive or polarity-sensitive dyes (PSDs) was presented in 1987 [89]. PSDs are charge dyes and typically located at the interface between a lipophilic sensor phase and a hydrophilic sample phase. The transport of an ion into the lipophilic sensor layer causes the PSD to be displaced from the hydrophilic/hydrophobic interface into the interior of the respective phase (or vice versa), thereby inducing a sig-

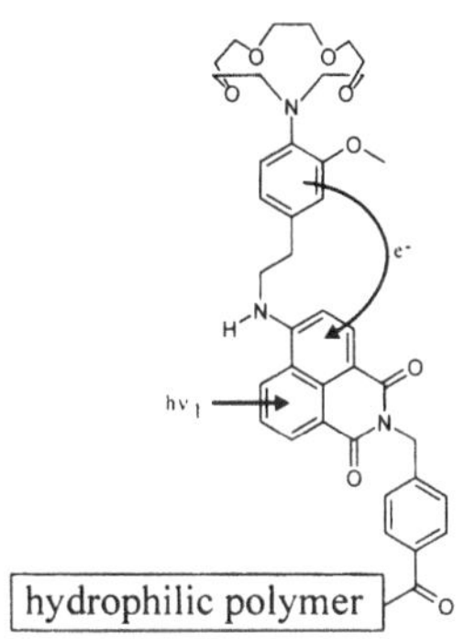

Sodium Probe (K_D 119 mmol/L at pH 7.4 and 37 °C)

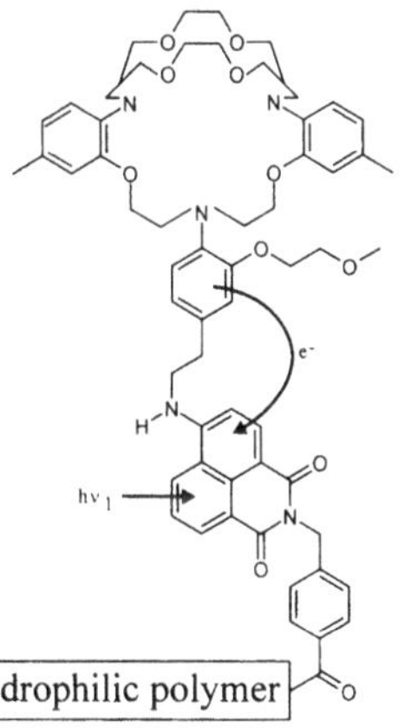

Potassium Probe (K_D 1.76 mmol/L in water at pH 7.4 and 37 °C)

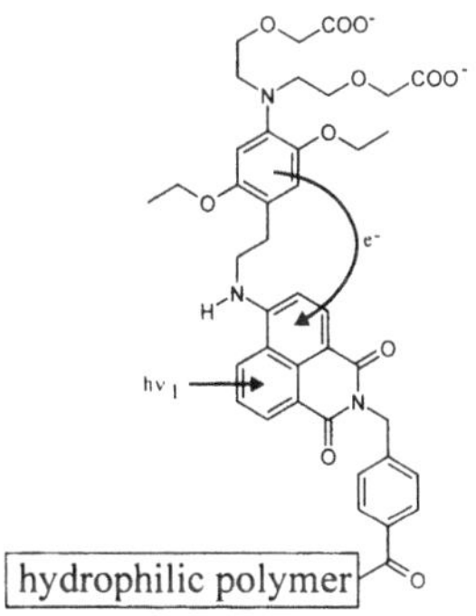

Calcium Probe (K_D 1.09 mmol/L in water at pH 7.4 and 37 °C)

Fig. 6. Chemical structures of the molecular probes for Na^+, K^+, and Ca^{2+} containing a fluorophore (a naphthalimide), a receptor, and a covalent link to the cellulosic matrix. Complexation of the alkali metal (alkaline earth) ion causes the photo-induced electron transfer (PET) to be suppressed, resulting in a strong increase in fluorescence intensity at λ_{exc} 450 nm and λ_{em} 500 nm (He, Leiner & Tusa; 1999)

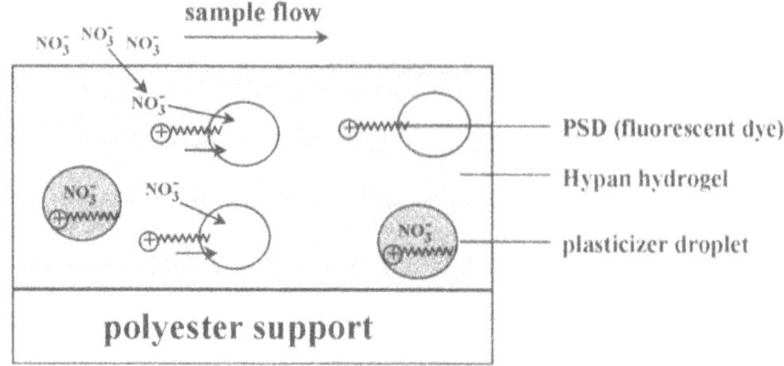

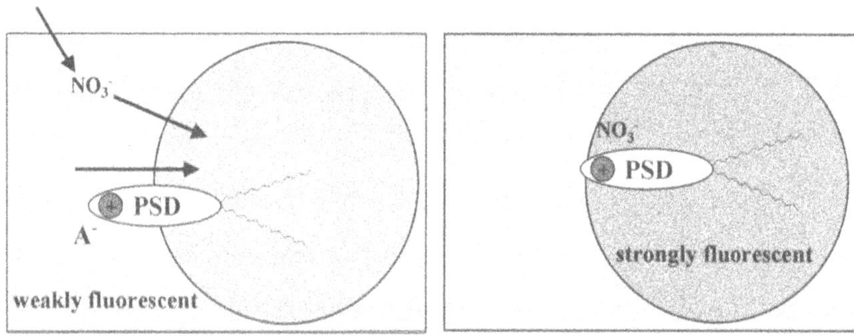

Fig. 7. Schematic representation of the function of a sensor for anions based on the use of a polarity (or potential) sensitive dye (PSD). *Top*: Cross section of a sensor membrane containing an emulsion composed of two types of polymers. The first (e.g. Hypan hydrogel) is predominantly hydrophilic, the second is a hydrophobic plasticizer that forms small droplets and contains a hydrophilic cationic dye (the PSD) . The anion (e.g. nitrate) penetrates the hydrogel layer, and is carried into the plasticizer by the ion carrier. For reasons of electrical attraction, the PSD is displaced from the interface of the droplet to its interior, thereby undergoing a large change in its fluorescence intensity and decay time. *Bottom*: Schematic representation of the micro-environment of the PSD *DiA* before (A) and after extraction of nitrate (B) from the hydrogel (mainly aqueous) phase into the plasticizer droplet. The PSD is displaced from the interface (where it is weakly fluorescent) into the interior where it is strongly fluorescent

nificant change in its fluorescence properties [90–92]. The scheme is outlined in Fig. 7. It is applicable to both cations and anions. Thus, nitrate can be sensed in concentrations such as occur in drinking water [93]. Such sensors have the unique advantage of having a virtually pH-insensitive response.

Anions were also detected by co-extraction ("anion in – proton in") using phase transfer agents such as tetraalkylammonium salts [94–97]. In contrast to extraction processes on electrodes, those occurring in optical sensor membranes require complete mass transfer. The scheme can be made pH-independent [98]. The fact that halides and pseudohalides quench the fluorescence of certain dyes as reported by Stokes in 1869 (99) was used to optically sense halides [100]. The ion exchange principle was used to design the first enantio-selective ion sensor [101]. It displays selectivity due to the use of an enantio-selective carrier.

7 Enzyme-Based Biosensors

The chemical sensors described so far, and also those for ammonium ion and ammonia, can be used to monitor enzymatic reactions. The first enzyme-based biosensors were described by the Lübbers group [102, 103] and made use of an oxygen transducer and glucose oxidase. In the following years numerous kinds of optical sensor membranes have been designed for monitoring reactions that are accompanied by the production of low molecular weight species such as H^+, CO_2, NH_3/NH_4^+, NADH, or H_2O_2, and by the early 1990's an impressive variety of detection schemes was available for enzyme substrates [104, 105]. These are summarized in Table 3. Many more have since become known but are not included in this Table.

Later it was discovered [106] that the FAD coenzymes of certain oxidases display large changes in their fluorescence if exposed to their substrates. Thus, the fluorescence of the FAD unit of lactate mono-oxygenase changes substantially on loading with lactate, and this can serve as the analytical information in an optical sensor. Subsequent work showed [107, 108] that not only enzyme substrates, but also enzyme activity may be detected via fiber optics if they contain a nonfluorescent enzyme substrate at the tip. On exposure to an enzyme, the nonfluorescent substrate is converted into the fluorescent product, and this is "seen" by the fiber optic system. Inversely, a chromogenic substrate may be assayed via immobilized alkaline phosphatase [109]. Finally, the inhibition of enzymes by pesticides and warfare agents can be detected using fiber optics [110]. In an extension of this scheme, the biochemical oxygen demand has been measured by coupling an oxygen-sensitive membrane to immobilized bacteria [111]. Goldfinch et al. [112] have exploited the effect of human serum albumin on the pK_a of immobilized pH probes to design a solid-state sensor for HSA.

8 Fiber Optic Systems

The availability of high quality fiber optic light guides facilitated the development of the respective chemical sensors. In principle, any sensing chemistry can be used to modify the surface or tip of a fiber optic system or integrated waveguide. Fiber optic sensors enable measurements to be performed in samples that cannot be placed inside an optical meter. The articles published by Hesse, Hirschfeld, Boisdé, and Peterson (as discussed in Section 2) are fundamental. A sterilizable oxygen sensor with a rod-like fiber waveguide for use in bioreactors was reported in 1983 [113]. However, such sensors (with diameters of >1 mm) are too large to be of use as non-traumatic sensors. Fiber optic microsensors for oxygen, pH and CO_2 were described first in 1984 [114, 115].

Table 3. Optical biosensors for enzyme substrates, and respective transducers (work published until 1992). Data on work by Schaffar & Wolfbeis from ref. [104]. For a recent review see ref. [105]

Analyte (substrate)	Enzyme	via	Reference
alcohols	alcohol dehydrogenase	NADH	Walters et al., 1988
alcohols	alcohol oxidase	O_2	Völkl et al., 1980; Wolfbeis & Posch, 1988
glucose	glucose dehydrogenase	NADH	Narayanaswamy & Sevilla, 1988
glucose	glucose oxidases	pH	Goldfinch & Lowe, 1984; Trettnak et al., 1989; Kulp et al., 1988
glucose	glucose oxidases	O_2	Völkl et al., 1980; Trettnak et al., 1988; Kroneis et al., 1987; Schaffar & Wolfbeis, 1990; Dremel et al., 1989; Moreno-Bondi et al., 1990
lactate, pyruvate	lactate dehydrogenase	NADH	Wangsa & Arnold, 1988
lactate	lactate oxygenase	O_2	Dremel et al., 1989
lactate	lactate mono-oxygenase	O_2, CO_2	Lübbers et al., 1981; Trettnak & Wolfbeis, 1989
creatinine	creatinine iminohydrolase	NH_4^+	Wolfbeis & Li, 1991
carboxy esters	esterases	pH	Luo & Walt, 1989
urea	urease	NH_3/NH_4^+	Rhines & Arnold, 1989; Wolfbeis & Li, 1991
urea	urease	pH	Goldfinch & Lowe, 1984; Luo & Walt, 1989 Yerian et al., 1986
glutamate	glutamate oxidase	O_2	Dremel et al., 1991
glutamate	glutamate ecarboxylase	CO_2	Dremel et al., 1991
oxalate	oxalate decarboxylase	CO_2	Schaffar & Wolfbeis, 1991
phenols	phenolase	O_2	Schaffar & Wolfbeis, 1991
sulfite	sulfite oxidase	O_2	Schaffar & Wolfbeis, 1991
penicillin	penicillinase	pH	Goldfinch & Lowe, 1984; Kulp et al., 1987
bilirubin	bilirubin oxidase	O_2	Schaffar, 1988; Trettnak, 1989
ascorbate	ascorbate oxidase	O_2	Schaffar, 1988
uric acid	uricase	O_2	Schaffar & Wolfbeis, 1991
xanthine	xanthine oxidase	O_2	Völkl et al., 1980
cholesterol	cholesterol oxidase	O_2	Trettnak & Wolfbeis, 1990

True microsensors, with diameters of <50 μm were first reported in 1995 [116, 117] and enabled submillimeter resolution studies in marine microbiology. This kind of oxygen sensor is commercially available now (Fig. 8) and has found widespread application in biological and marine research.

The first nanosensors were reported soon after by Kopelman et al. (118), including one for NO [119]. It needs to be stated, however, that in these nano-

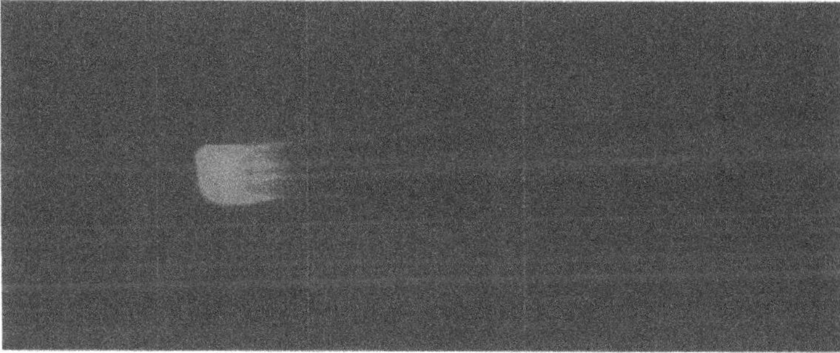

Fig. 8. Fluorescence photograph of the 20-μm fiber tip of the Microx™ fiber optic oxygen sensor displaying the strong fluorescence of the oxygen probe immobilized at the tip of the fiber

sensors it is only the exciting light that is guided via fibers to the area to be analyzed, but not the emitted light.

9
Signal Referencing

Practically all early optical sensors measured just the intensity of absorbed, reflected, or emitted (fluorescent) light. Unfortunately, the intensity of the light detected by the photodetector also depends on factors other than the concentration of an analyte. These include the intensity of the light source and the sensitivity of the photodetector, and also interfering effects such as stray light, temperature, background luminescence (from both the sample and the materials employed), and fiber bending. Therefore, referenced detection schemes are preferred. Among those, the measurement of fluorescence decay time is a superb parameter but can be applied to fluorescent systems only.

Two-wavelength referencing is among the most effective methods and can be applied both to absorbing/reflecting and fluorescent/phosphorescent chemistries, provided two bands appear in either absorption, excitation, or emission. It is fairly simple and can account for many drifts including photobleaching. While pH probes usually display 2 absorption or emission bands which enables 2-wavelength referencing (see above), this is not the case for oxygen probes. Seitz was the first to introduce ratiometric (2-wavelength) oxygen indicators [120]. In the absence of a second band, a reference dye may be added [115] which, however, needs to display the same leaching and bleaching features as the indicator dye. Fluorescence polarization is another intrinsically self-referenced parameter.

A highly attractive referencing scheme was introduced in 1998 by Klimant et al. [121]. It was termed *dual luminophore referencing* (DLR) and makes use of two dyes, one being the indicator, the other being inert and usually contained in polymer nanoparticles. Unlike in the 2-wavelength ratiometric approach, the DLR scheme is not based on the measurement of intensity but rather of a phase shift that is modulated by the varying contributions of the amplitude of the

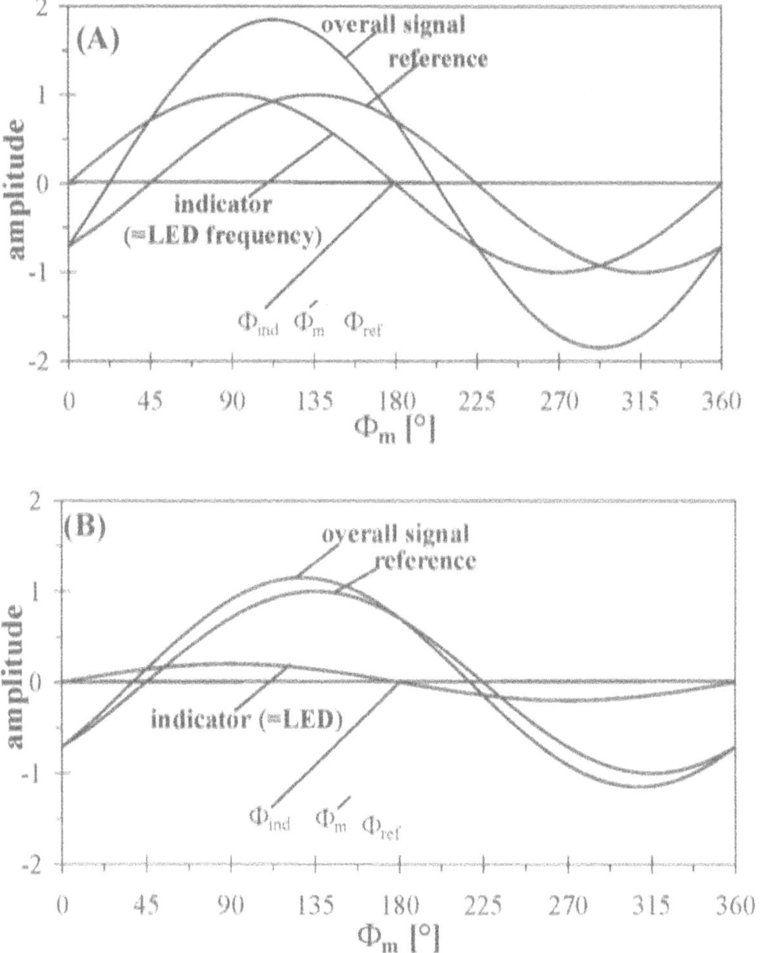

Fig. 9. Dual Lifetime Referencing (DLR). *Top:* phase signal of the overall (measured) luminescence (λ_m) composed of the signal from amplitude reference dye (λ_{ref}) and that from the indicator dye (λ_{ind}); since the fluorescence of the indicator dye is relatively strong, the phase shift between λ_{ref} and λ_m is relatively large. *Bottom:* phase signal of the overall (measured) luminescence (λ_m) composed of the signal from amplitude reference dye (λ_{ref}) and that from the indicator dye (λ_{ind}); since the fluorescence of the indicator dye is relatively weak, the phase shift is small (Klimant et al., 1999)

indicator dye to the total signal amplitude. Fig. 9 shows a schematic of the DLR method. In practice, the reference dye is added to the sensor chemistry in the form of a dye having a long decay time (e.g. a ruthenium-ligand complex) and incorporated into a polymer (e. g. polyacrylonitrile) that shields the dye from interferences such as quenching by other chemicals.

An attractive feature of this scheme is its applicability to any indicator-based sensing scheme, independent of whether it is based on absorption or fluores-

Table 4. Methods for referencing fluorescence signals, and their respective advantages. Symbols: (++), well compensated; (+), partially compensated; (-), not compensated for

A: Interferences caused by the instrumental system

Interference caused by	two wave-length (2-λ)	τ (time domain)	τ (frequency domain)	Aniso-tropy	DLR (time domain)	DLR (frequency domain)
optical components	-	++	+	-	+	+
electronics	++	-	+	++	-	+
instrumental drift	+	++	++	+	++	++
misalignment of fibers	++	++	++	++	++	++

B: Interferences caused by the sample

Interference caused by	Two wave-length (2-λ)	τ (time domain)	τ (frequency domain)	Aniso-tropy	DLR (time domain)	DLR (frequency domain)
background fluorescence (sample and optical system)	-	+	-	-	-	-
light scatter by turbid samples	+	++	++	-	++	++
intrinsic color	-	++	++	+	+	+

C: Effects caused by the sensor chemistry

Interference caused by	Two wave-length (2-λ)	τ (time domain)	τ (frequency domain)	Aniso-tropy	DLR (time domain)	DLR (frequency domain)
dye leaching & bleaching	-	+	+	+	-	-
inhomogeneous dye loading	+	++	++	+	+	+

cence. It is mandatory, though, that the two dyes used have at least partially overlapping absorption and emission bands, and that the concentration ratio of inert (reference) dye to total indicator dye remains constant. An additional requirement for DLR to work properly is that the two dyes have to have widely different decay times, typically by a factor of >300. The DLR scheme can be applied in the frequency domain, the time domain, and in polarization [122]. It also is applicable to imaging, as will be shown later. Several gases and ions, but also seawater salinity have been sensed with superb precision and sensor stability. Table 4 summarizes the methods often used for referencing optical signals, and their respective merits.

10 Optical Sensing Schemes

All early tests were read visually. Often the color was (and is) compared with a color chart. Results can be surprisingly precise, occasionally better than ±10%. Instrumental quantitation of test strips, dry reagent chemistries, or sensor strips is frequently performed by reflectometry, less so by fluorescence, and hardly by absorption, simply because all commercial supports are nontransparent. Fluorescence is much less easily read-out than color. The law of diffuse reflectance was established by Kubelka and Munk [123] in 1931 and forms the basis for data processing in most instrumental readers. Unlike in absorption or fluorescence, diffuse reflectance involves interaction of light with both the reagent and the matrix (or particles), so that both the physical form of the support and the color of the reagent determine signal intensity.

Most sensors are based on layers composed of a support material and a more or less selective "chemistry" deposited thereon or therein. Sensor layers can be deposited in various ways. Often they have been placed directly in the (sterile)

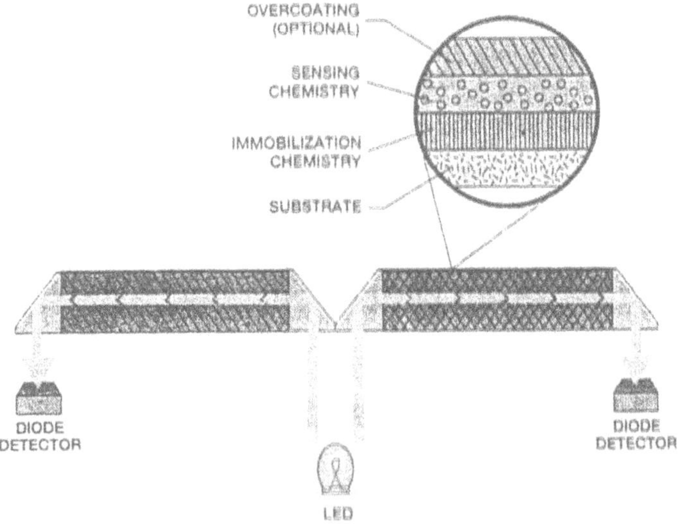

Fig. 10. Referenced waveguide chip package containing an LED light source, a sensor "chemistry" and 2 photodiodes. Light is split into two paths that pass through the two identical chemistries, one exposed to the analyte, the other not. The chemistry is immobilized on a solid substrate via an immobilization chemistry and, in one case, made insensitive by an overcoat (1997)

sample compartment, for example in the form of a disposable kit that can hold a blood sample. They also have been placed in flow-through cells that act as sample compartments [124, 125], in disposable cuvettes with integrated sensor layers [126], in microtiter plates to form a sensor layer on the bottom of a well [127, 128], and inside hollow waveguides [129]. A smart sensor chip was presented by Texas Instruments [130]. It includes an LED light source, a photodiode, a chemically sensitive waveguide and an inert reference waveguide. It is schematically shown in Fig. 10.

The most common way for sensor materials to be deposited is in the form of a planar layer, mostly inside a flow-through cell or a sample compartment. The sensor is then interrogated (often via optical fiber cables) from outside through a transparent window. Figure 11 shows the instrumental arrangement of the commercially most successful optical chemical sensor between 1984 and 2000. It is widely used to monitor pH, p_{CO2} and p_{O2} during cardiopulmonary bypass operations [124] and contains 3 fluorescent spots, each sensitive for one parameter, in contact with blood. Fluorescence intensity is measured at two wavelengths and the signals are then submitted to internal referencing and data processing.

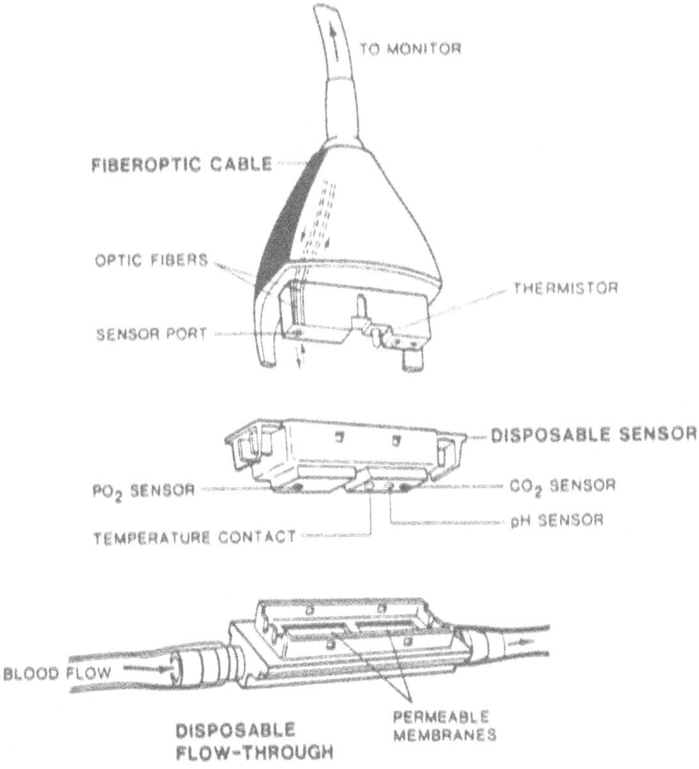

Fig. 11. Schematic of the GasStat™ system for continuous fluorescent monitoring of blood gases and pH in blood during cardiopulmonary bypass (Gerhrich et al., 1994)

Sensor layers may, however, also be deposited in (and on) waveguides and even form the cladding of a fiber core. In the late 1980s, a sensing method known as distributed sensing became known [131–133]. It makes use of the fact that instruments are available that allow for highly time resolved measurements – for example via optical time domain reflectometry (OTDR). In this scheme, the sensing chemistry is deposited (in the form of small spots) on the core of fiber optics at certain distances. The echo from these sensor spots following a pulse of light from a central station is monitored via OTDR or OTD fluorescence.

It became obvious in the late 1980s that optical sensors may just as well be arranged in the form of bundles or arrays. On the one hand, they can yield multiple information which is needed in many situations. On the other, they can help to overcome non-selectivities. Thirdly, arrays are needed in proteomics and genomics. In 1988, Smardzewski [134] introduced a multi-element optical waveguide sensor as a general format for multi-analyte sensing, and also recognized the potential of applying pattern recognition methods. Gehrich et al. reported triple sensors for blood gases [115], and Walt's group reported on fiber arrays for genomic screening [135].

The problem of the diversity of optical schemes and analytical wavelengths has limited the applicability of chemical sensors in arrays where only a single type of light source is normally available. Instruments that are designed to report more than one parameter therefore have to have spectrally "matched" chemistries. The blue LED is most often used in blood gas analyzers, and a consistent set of luminescence decay time-based chemical sensor materials for clinical applications has been presented [136] for a variety of parameters, all based on the measurement of the luminescence of ruthenium probes partially linked to other sensing schemes, e.g. via resonance energy transfer.

Sensing two species at the same time became of interest in view of the possibility of spectral or frequency multiplexing. The first sensors to measure two species simultaneously with one sensor layer were reported for the case of CO_2/O_2 [137], for halothane/oxygen [18], glucose/oxygen [138, 139], and different halides [140]. The first scheme was based on the finding that two indicator chemistries can be placed in one sensor layer, both being excitable by the same LED but displaying highly different (green and red) emissions. The second approach makes use of permeation-selective layers. Teflon, for example, is permeable to oxygen but not to halothane. Two sensors have to be employed, one covered with teflon, the other not, and two signals are obtained that can be processed to give information on the concentration of either species.

Catalytic sensors make use of a catalytic reaction that leads to products that are more easily detected than the analyte itself. A sensor containing an inorganic catalyst (MnO_2) on the surface of an oxygen sensor was used to monitor hydrogen peroxide [141]. The metal oxide decomposes hydrogen peroxide, and the increase in oxygen partial pressure can be related to the H_2O_2 concentration, provided oxygen partial pressure is kept constant or its variation is compensated for. Obviously, enzyme-based biosensors for substrates (see Section 6) also fall into the category of catalytic sensors.

Chemically sensitive coatings have been placed inside capillaries that can serve both as light guides and as flow-through cells, the inner sensing layer being excited by the evanescent field of a light beam propagating through the walls of the capillary [142]. Lieberman and Brown [143] have presented intrinsic fiber optic sensors based on two-stage fluorescence coupling, a generic technique for chemical sensing and leading to a more than 100-fold increase in coupling of photons from a chemically sensitive coating into the guided mode of a fiber optic waveguide.

Another interesting form of spectroscopy for use in optical sensor technology was presented by Okura and Amao [144]. It is based on the measurement of the reflectance of an indicator whose triplet state is easily populated and whose triplet-absorption can measured as a function of quenching efficiency of oxygen. It enables the monitoring of any quenching process by measuring the reflectivity of this triplet state. In addition it enables the use of non-luminescent indicator dyes for the determination of species like oxygen which so far was mostly detected via quenching of luminescence. The finding that oxygen quenches the excited state complex (exciplex) formed between pyrene and perylene can increase the efficacy of quenching by oxygen (and thus the sensitivity of respective sensors) but the effect still awaits explanation [145].

The effect of photo-induced electron transfer (PET) has attracted considerable attention since it is often observed in supramolecular systems as they are applied in sensor technology. PET usually occurs intramolecularly and involves the transfer of an electron from one moiety in a supramolecular system to another, typically over distances of maximally 3 nm. It is accompanied by the quenching of the fluorescence of the acceptor moiety. Such systems are sometimes referred to as switches, but in fact they are not since there is no strict transition between fully "on" and fully "off" as in true switches. The PET scheme has been applied to sense ions (pH, metal ions), and carbohydrates. Pioneering work has been presented by several groups including those of Czarnik [146], De Silva [147], Diaz-Garcia [148], Fabrizzi [149], Shinkai [150], Tsien [151], and Valeur [86].

11
Materials for Optical Chemical Sensors and Biosensors

Material science plays a major role in the design and fabrication of optical chemical sensors. Materials include polymers (including sol-gels), plasticizers, composite materials, plastic supports, molecular imprints, and of course ion carriers and indicator probes. Hydrophobic polymeric matrices have been used to sense gases such as oxygen and CO_2, while hydrophilic matrices were used to sense pH, ions, and water-soluble enzyme substrates such as glucose or lactate. Among the hydrophilic materials, the polyurethane hydrogels [137, 152, 153] and the Hypan hydrogels (polyacrylonitrile-co-polyacrylamide) [154] possess unique properties and have often been used in our research.

Two types of immobilized reagents need to be distinguished. The first type consists of physically immobilized (or entrapped) components, dissolved, for example, in silicone rubber [18, 68, 113], polyethylene [11], ethyl cellulose [48; also used in Boehringer's ions sensors], plasticized poly(vinyl chloride) [95, 96], fluoropolymers [144], or adsorbed on silica particles [34]. The other type is based on chemical (covalent) immobilization of dyes and additives, e.g. on glasses and sol-gels, cellulose, hydrogels, poly(acrylamides), metal surfaces such as aluminum, and the like. Most of these materials were known by the year 1990 [27] but were often re-invented later.

Optical sensor layers are rather thin and usually are not easy to handle. Hence, solid but inert supports were sought. The first supports were usually glasses and the like. Leiner et al. have shown that for many applications poly(ethylene terephthalate) supports (Mylar) such as are used for making photographic films were found to be ideally suited [34, 50, 111, 137, 138, 154]. Commercial polyester layers with gelatine or cellulose coatings (as used for ink jet printers) are well suited for covalent immobilization of dyes and reagents [57, 155], as are certain preactivated membranes (as used for immunoblotting) for immobilization of enzymes. Also, polyurethanes were found to be most suitable polymers to host dyed cellulose beads [155]. An interesting approach was reported by Lübber's group [156] who immobilized indicators for oxygen and pH in nanocapsules which retain the probes but are permeable to the analyte.

The optical signal of a sensor layer is subject to interference from the intrinsic color or fluorescence of the sample. One elegant way to overcome interference by the sample is to provide a so-called optical isolation. It consists of a thin (≈5 μm) and optically non-transparent layer that is permeable to the analyte and placed between sample and the sensor material. Depending on the analyte, the layer is hydrophobic (in case of gases and vapors) or hydrophilic (in case of charged or highly water-soluble analytes). Typical examples that were used in the 1980s include carbon black in silicone, black teflon, ferric oxide in ethyl cellulose or silicone, titanium dioxide or barium sulfate (also in hydrogels). White optical isolations also act as reflecting areas and usually increase the signals of reflectometric and fluorescent sensors.

Langmuir-Blodgett films are materials useful for making sensor layers of nanometer dimensions. Such films were used to sense oxygen [157, 158] and pH, or chloride, oxygen, or glucose [158]. Other less conventional materials for sensor matrices include the sol-gels [159–161] and zeolites [162]. Sol-gels have been modified in many ways, e.g. by making composite materials [163], since conventional sol-gels are less suited for chemical sensing purposes. For this reason, modified sol-gels rather than "pure" sol-gels are used [164] in the commercial fiber optic microsensor shown in Fig. 8.

Plasticized poly(vinyl chloride), a material widely used in ion-selective electrodes, has been employed in academic research for oxygen sensing [33, 165)] and ion sensing via co-extraction [95–97] but hardly in commercial optical systems because the slope of the response strongly depends on the fraction of plasticizer which tends to leach, thus seriously affecting long-term stability. Polystyrene, silicone, ethyl cellulose, cellulose, or two-domain polymers such as certain polyurethanes or poly(acrylamide-co-acrylonitrile) are by far preferred over PVC [166].

Other novel materials for use in optical sensors or probes consist of polymers that have an inherent optical response, and these were introduced after 1992. They include polypyrroles [167], polyanilines [168], composites composed of polypyrrole and Prussian Blue [169], and polyanilines with receptor sites for

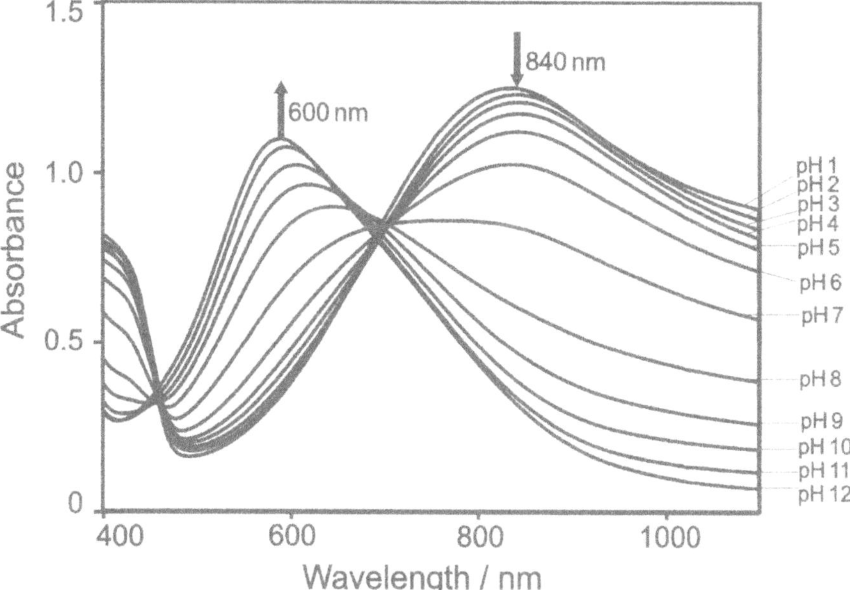

Fig. 12. pH dependence of the UV-VIS spectra of a 60-nm film of polyaniline deposited on polystyrene. Note the strong and broad absorption that extends far into the near infrared

saccharides [170]. Their absorption spectra extend far into the near infrared, as can be seen in Fig. 12.

Molecular imprints (MIPs) found their first application in sensor technology in the 1990s [171, 172]. MIPs are ideal means to recognize molecular species, in particular those of rigid structure. Binding is usually irreversible, so that probes rather than continuous sensors are obtained in most cases. In certain – though probably rare – cases, they may replace proteinic recognition elements such as antibodies [173].

The traditional plasticizer for PVC is *ortho*-nitrophenyl octyl ether (NPOE) but it can act as a quencher for many indicator probes and was therefore replaced [174] by the non-quenching plasticizer *ortho*-cyanophenyl dodecyl ether (CPDDE). Various other polymers and composites have been proposed including polymers with fillers and unusual speciality polymers [175]. A polymer with extraordinarily high (so far unsurpassed) permeation for oxygen was reported by the groups of Asai, Amao and Osuka [176].

Many indicator probes have poor solubility in certain polymers. The method of ion pairing can substantially improve the solubility of charged indicator dyes. In essence, the inorganic counter ion of a dye (which limits its solubility in hydrophobic polymers) is replaced by an organic counter ion [165, 177]. The ion pair may even consist of a dye cation and a dye anion, thereby also enabling resonance energy transfer [136, 178–179] and thus the possibility of converting intensity-based sensors into decay time-based sensors [180, 181].

Photobleaching of indicator dyes has always been a nuisance in optical sensor technology, but tertiary amines such as diaza-bicyclooctane (DABCO) can greatly reduce this [182]. Carotene has a similar effect. Both are capable of quenching reactive (singlet) states of oxygen. So-called antifading agents are commercially available.

12 Imaging and Pressure-Sensitive Paints

Peterson and Fitzgerald in 1980 reported a method for air flow visualization that is based on the quenching of the fluorescence of a dye by oxygen present in air [183]. Wilson, Vanderkooi and coworkers demonstrated imaging of oxygen on the surface of organs at various rates of perfusion using phosphorescent porphyrins [184]. The first imaging experiments were based on measurement of fluorescence intensity.

Given the advantages of decay time-based sensing, it was soon extended to imaging. The first decay time-based images for a variety of parameters were demonstrated by Lakowicz [185]. All these images were obtained by adding a fluorescent probe to the sample to be analyzed. Hartmann et al. [186] used, for the first time, a chemical sensor layer (sensitive both to oxygen monitored surface oxygen pressure and also to the flux of oxygen through skin by a phase-fluorimetric method). The sensor layer containing a ruthenium-based oxygen probe was excited by a frequency-modulated blue LED. Liebsch et al. [187] imaged pH, oxygen and temperature using sensor membranes placed in microtiter plates and employing decay time-based data acquisition. In subsequent work, pH was imaged using the DLR scheme [188]. Decay time-based imaging is well suited for the detection of the underoxygenation of tumorous skin tissue and to monitor the growth of engineered tissue. Figure 13 shows the oxy-

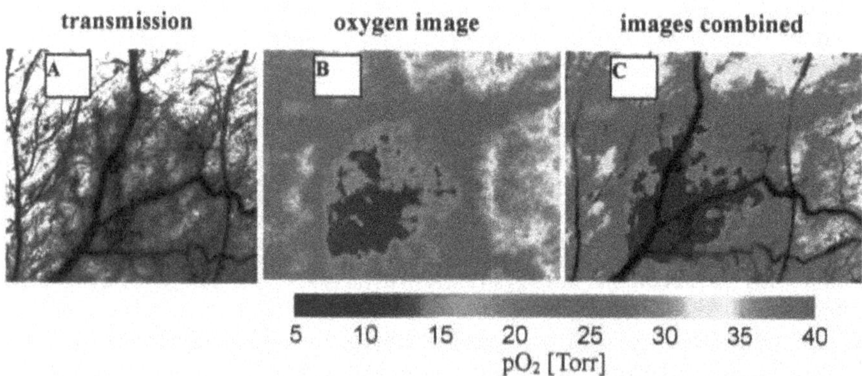

Fig. 13. *Left*: transmission photograph of tumorous skin of a hamster; *center*: distribution of the oxygen partial pressure, demonstrating the heavy under-oxygenation of the tumorous area. *right*: the two pictures combined. The data of the "oxygen image" were obtained by placing an oxygen-sensitive sensor film on the skin and imaging the film via fluorescence lifetime imaging (≈1560 sensels)

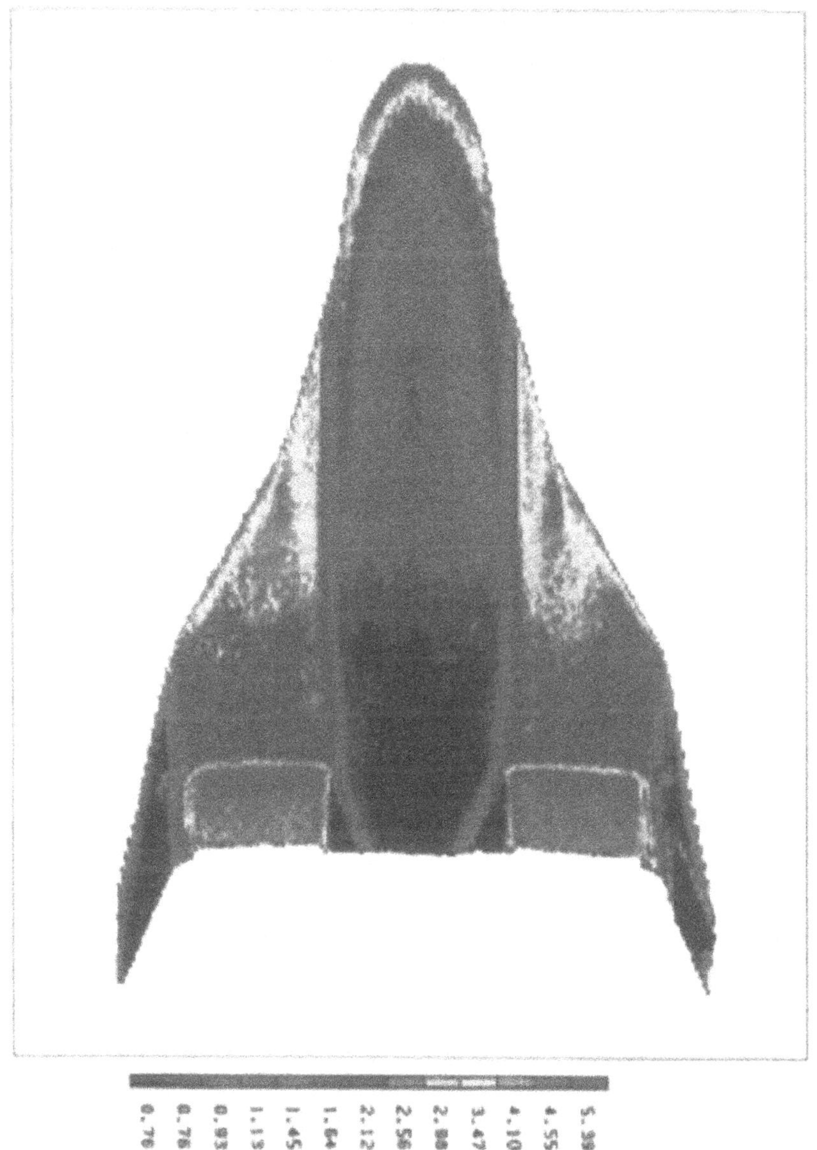

Fig. 14. Pressure distribution in a model of an airplane as visualized via the fluorescence of a so-called pressure-sensitve (but in fact oxygen-sensitive) "paint" (PSP)

gen image of normal skin and of tumorous skin tissue, demonstrating the large underoxygenation of the latter.

A fascinating application of oxygen imaging is seen in so-called pressure-sensitive paints (PSPs) which in fact are oxygen-sensitive paints. Such PSPs were first described in the late 1980s by Russian, US, and Canadian researchers [189–191]. The function of PSPs is based on the quenching by air (oxygen) of the luminescence of quenchable dyes such as pyrenes, platinum porphyrins, or ruthenium bipyridyl complexes. Figure 14 shows a typical image as obtained with a pressure-sensitive paint deposited on a model of an airplane placed in a wind tunnel. Pressure-sensitive paints are being marketed by various companies now. More recently, europium complexes have been proposed for PSPs [192] along with polymers of unsurpassed oxygen-permeability, while Engler and co-workers use pyrenebutyric acid [193] which requires UV excitation but has a fast response.

13 Commercial Instrumentation Using Opt(r)odes

Many of those working on optical sensors since 1970 have been overoptimistic. While many chemical sensors and biosensor have found applications in the laboratory and in research, they are by far less often used than physical sensors, e.g. those for temperature, pressure, strain. This may be due to several factors, of which the following are considered to be the most significant:

(a) Most chemical sensors suffer from poor selectivity; therefore, they can only be applied to rather specific matrices;

(b) most chemical sensors and biosensors suffer from inadequate stability (both storage stability and operational stability);

(c) most sensors are too expensive; this is true not only for the costs for product development, but also for the instrument itself and the sensor materials used; the market for such devices is too small. It is quite unrealistic to assume that an investment in the development of a commercial sensor for, e.g. copper(II) ion, may ever produce an adequate return, unless the same instrument can sense several other species as well.

The commercialization of optical sensor technology therefore started slowly, initially with various test strips, first for pH, then for numerous ions including heavy metal ions and certain anions. The Merck Reflotron instrument can quantify over 30 environmental parameters. Numerous sticks are available for clinical parameters that can be optically read, at least in principle. These include first of all the common parameters tested for in blood and urine including erythrocytes (introduced in 1974), pH (1964), glucose (1964), blood urea nitrogen (BUN), drugs, total protein, but also pregnancy tests, all of which are optical. Boehringer (now Roche) has offered quantitative optical tests for the non-proteinic parameters K^+, urea, bilirubin, glucose, cholesterol, creatinine since 1988, and these appear to be as good as some of the tests ("sensors") reported in recent years, if not simpler. The article by Free et al. [194] was a milestone paper and describes a triple test strip for urinary glucose, protein, and pH.

In 1984, CDI (later 3M) introduced their critical care monitoring system (GasStat 300) for oxygen, pH and CO_2 for cardiopulmonary monitoring (see Fig. 11). The sensors employed are based on the work of Lübbers et al., Wolfbeis et al., Yafuso, Miller and Tusa. The sensor head consists of a flow-through cell containing sensor spots and fiber optic cables attached to the cell in order to optically interrogate the sensor spots. The sensor exhibits excellent performance if properly calibrated [195]. Other products based on optical sensor technology include those of Cardiomed (system 4000), Puritan-Bennett, and more recently of Radiometer (Copenhagen).

AVL (now Roche) in 1994 introduced its Opti-1 system, a portable (and near patient) station for measurement of blood gases and blood electrolytes. It is based on a disposable cassette containing planar fluorescent sensors of 3 mm i.d. but also acting as a sample compartment. Its analytical scope has recently been extended to enzyme substrates such as glucose and urea. In 1986, Gehrich et al. [107] described an optical fluorescence sensing system for intravascular blood gas monitoring, and Miller et al. [106] reported an in-vivo catheter for blood gas monitoring. Although announced, the systems were never commercialized.

Optical sensors for CO_2 act as transducers in the blood bacteria detection systems of Organon BV and of Becton Dickinson [196]. Mycobacteria, in contrast to "conventional" bacteria, are more specifically detected via an oxygen transducer, and a simple and affordable version of a microbacteria growth indicator tube has been developed whose fluorescence can be inspected visually. Some of the commercial optical (non-fiber) chemical sensors and biosensors are compiled in Table 5, and some fiber optic sensors in Table 6, thus demonstrating the significance of optical (fiber) technology for environmental, industrial, and clinical sensing.

Table 5. Selection of commercial chemical sensors and biosensors

Product	Manufacturer	Analytes	Remarks
RQ-Flex	Merck	pH, heavy metals, disinfectants	affordable, fast, 2 LEDs & photodiodes, hand-held
Reflotron	Roche (Boehringer)	many clincal parameters including glucose and lactate	green and red LED, dry reagent chemistry (H_2O_2)
AccuSport	Roche	lactate	LOx, used in sports medicine, hand-held
AccuTrend	Roche α	glucose	hand-held, GOx and phospho-molybdate reagent
Opti-1 and Opti R	Roche (AVL)	O_2, pH, CO_2, K, Na, Ca, Cl, glucose etc.	portable, fluorescence based, versatile; the Opti-R is re-usable
CDI 400	3M (Cardiovasc. Dev. Inc.)	blood gases and pH	fluorescent sensor spots in contact with blood in an extracorporeal loop; highly successful
BacT Alert	Organon Teknika	bacteria in blood	via CO_2 (reflectometry)
BacTec	Becton Dickinson	bacteria in blood	via CO_2 or O_2 (fluorescence)
MGET	Becton Dickinson	M. *tuberculosis*	via O_2; disposable tube, visual inspection of fluorescence intensity

Table 6. Selection of commercial fiber optic chemical sensors

Product	Manufacturer	Analytes	Remarks
GeoSensor	GeoCenters (1991)	pH 3–8	± 0.1 units; robust
PetroSense	FiberChem (1992)	hydrocarbons; gasoline	1-mm fibers, robust
MicrOx	Presens	O_2	30 μm tip; via fluorescence decay time
FibOx	Presens	O_2	2-mm fibers, DLR based
YSI 8500	Yellow Springs	CO_2	for bioprocess control; autoclavable
FiberDek	Soundek	oil and gasoline on water	reflectance based; 2 LEDs, hand-held
TP300	Ocean optics	pH (various ranges)	reflectance based, fiber bundles
FOXY	Ocean optics	O_2	1-mm fibers, intensity based

In conclusion it can be said that the past 20 years have seen tremendous steps forward in optical sensor technology. Sensing schemes have become reliable enough to be of practial utility, instrumentation has become available at costs that make this sensor technology competitive with established sensors, and numerous sensors have been presented that are of interest more from an academic (analytical) rather than from an industrial point of view at present, but conceivably will find their way into practical applications as well.

References

1. Cammann K, Hall EAH, Kellner R, Schmidt HL, Wolfbeis OS (1995) The Cambridge Definition of Chemical Sensors
2. Feigl F (1949) Specific, Selective and Sensitive Reactions. Academic Press, New York
3. Walter B (1983) Anal Chem 55: 499A
4. Free AH, Adams EC, Kercher ML (1957) Clin Chem 3: 163
5. Roubitaille D, Rousseau F, Audovin F, Forest JC (1994) Clin Chem 41: 320
6. Harper GB (1975) Anal Chem 47: 348
7. Kautsky H, Hirsch A, Chem Ber (1931) 64: 2677
8. Kautsky H (1939) Trans Faraday Soc 35: 216
9. Pollack M, Pringsheim P, Terwood D (1944) J Chem Phys 12: 295
10. Zakharov IA, Grishaeva TI (1980) Zh Anal Khim 35: 481 and refs cited
11. Bergman I (1968) Nature 218: 396
12. Stevens PR (1982) US Pat 3.612.866
13. Hesse HH (1974) East Ger Pat 106086
14. Lübbers DW, Opitz N (1975) Z Naturforsch 30C: 532
15. Opitz N, Lübbers DW (1975) Eur J Physiol 355: R120
16. Lübbers DW, Opitz N (1976) Adv Exptl Biol 75: 65
17. Völkl KP, Opitz N, Lübbers DW (1980) Fresenius J Anal Chem 301: 162
18. Wolfbeis OS, Posch HE, Kroneis H (1985)Anal Chem 57: 2556
19. Wolfbeis OS, Fürlinger E, Kroneis H, Marsoner H (1983)Fresenius Z Anal Chem 314: 319
20. Wolfbeis OS, Offenbacher H, Kroneis H, Marsoner H (1984) Mikrochim Acta I: 153
21. Hirschfeld T, Deaton T, Milanovich FP, Klainer SM (1983) Opt Eng 22: 27
22. Milanovich FP, Hirschfeld T (1983) Adv Instrum 38: 407

23. Miller HH, Hirschfeld TB (1987) Proc SPIE-Int Soc Opt Eng 718: 39
24. Boisdé G, Perez JJ (1975) Fr Pat 2.317.638
25. Perez JJ, Boisdé G (1980) Analusis 8: 344
26. Peterson J I, Goldstein SR, Fitzgerald RV, RV Buckhold (1980) Anal Chem 52: 864
27. Wolfbeis OS (ed) (1991) Fiber Optic Chemical Sensors and Biosensors, CRC Press, Boca Raton, Florida
28. Kirkbright GF, Narayanaswamy R, Welti NA (1984) Analyst 109: 15
29. Borman S (1981) Anal Chem 53: 1616A
30. Hirschfeld T, Callis JB, Kowalski BR (1984) Science 226: 312
31. Hendricks HD (1971) Mol Phys 20: 189; US Pat 3.709.663
32. Freeman TF, Seitz WR (1981) Anal Chem 53: 98
33. Marsoner H, Kroneis H (1984) Eur Pat Appl 109958 & 109959; US Pat 4.587.101
34. Wolfbeis OS, Leiner MJP, Posch HE (1986) Mikrochim Acta (Vienna) III: 359
35. Bacon JR, Demas JN (1987) Anal Chem 59: 2780
36. Lippitsch ME, Pusterhofer J, Leiner MJP, Wolfbeis OS (1988) Anal Chim Acta 205: 1
37. Vanderkooi JM, Maniara G, Green TJ, Wilson DF (1987) J Biol Chem 262: 5476
38. Kahil HE, Gouterman MP, Green E (1987)PCT Appl. WO 87.00023
39. Peterson JI, Fitzgerald RV, Buckold DK (1984) Anal Chem 56: 62
40. Peterson JI, Vurek GG (1984) Science 224: 123
41. Zhujun Z, Seitz WR (1984) Anal Chim Acta 160: 305
42. Heitzmann H, Kroneis N (1985) US Pat 4.557.900
43. Vurek GG, Feustel PJ, Severinghaus JW (1983) Ann Biomed Eng 11: 499
44. Munkholm C, Walt DR, Milanovich FP (1988) Talanta 35: 109
45. Moreno-Bondi MC, Orellana G, Camara C, Wolfbeis OS (1990) Proc SPIE-Int Soc Opt Eng 1368: 157
46. Kawabata Y, Kamichika T, Imasaka T, Ishibashi N (1989) Anal Chim Acta 219: 223
47. Raemer DB, Walt DR, Munkholm C (1991) US Pat 5.005.572
48. Mills A, Chang Q, McMurray N (1992) Anal Chem 64: 1383
49. Arnold MA (1986) Anal Chem 58: 1137
50. Wolfbeis OS, Posch HE (1986) Anal Chim Acta 185: 321
51. Butler MA (1984) Appl Phys Lett 45: 1007
52. Nishizawa K, Yamazaki T (1986)Jap Kokai 60.209.149
53. Klainer SM, Goswami K, Nelson NR, Simon SJ, Eccles LA (1990) US Pat 4.892.383
54. Goswami K, Saini DPS, Klainer SM, Ejiofor CH (1994) US-Pat 5.302.350
55. Freeman MK, Bachas LG (1992) Anal Chim Acta 256: 269
56. Zhang Y, Seitz WR (1989) Anal Chim Acta 221: 1
57. Mohr GJ, Wolfbeis OS (1994) Anal Chim Acta 292: 41
58. Edmonds TE, Ross ID (1985) Anal Proc 22: 206
59. Kirkbright GF, Narayanaswamy R, Welti NA (1984) Analyst 109: 1052
60. Boisde G, Perez JJ (1987) Proc SPIE-Int Soc Opt Eng 798: 238
61. Saari LA, Seitz WR (1982) Anal Chim Acta 54: 821
62. Bacci M, Baldini F, Scheggi AM (1988) Anal Chim Acta 207: 343
63. Tait GA, Young RB, Wilson GJ, Steward DJ, MacGregor DC (1984) Am J Physiol Heart Circ Physiol 15: H232
64. Suidan JS, Young BK, Hetzel FW, Seal HR (1983) Clin Chem 29: 1566
65. Grattan KTV, Mouaziz Z, Palmer AW (1987/88) Biosens 3: 17
66. Offenbacher H, Wolfbeis OS, Fürlinger E (1986) Sens Actuators 9: 73
67. Janata J (1987) Anal Chem 59: 1351
68. Opitz N, Lübbers DW (1983) Sens Actuat 4: 473
69. Wolfbeis OS, Offenbacher H (1986) Sens Actuators 9: 85
70. Wolfbeis OS, Werner T, Rodriguez NV, Kessler MA (1992) Mikrochim Acta (Vienna) 108: 133
71. Lin J (2000)Trends Anal Chem 19: 541Ð551
72. Charlton SC, Fleming RL, Zipp A (1982) Clin Chem 28: 1857

73. Charlton SC (1987)US Pat 4.645.744
74. Gibb I (1987) J Clin Pathol 40: 298
75. Ng RH, Sparks KM, Statland BE (1992) Clin Chem 38: 1371
76. Zipp A, Hornby WB (1984) Talanta 31: 863
77. Narayanaswamy R, Sevilla F (1988) J Phys E: Sci Instrum 21: 10
78. Seitz WR (1988) CRC Crit Rev Anal Chem 19: 135
79. Oehme I, Wolfbeis OS (1997) Mikrochim Acta 126: 177
80. Saari LA, Seitz WR (1983) Anal Chem 55: 667
81. Zhujun Z, Seitz WR (1985) Anal Chim Acta 171: 251
82. Nakamura H, Takagi M, Ueno K (1980) Anal Chem 52: 1668
83. Alder JF, Ashworth DC, Narayanaswamy R, Moss RE, Sutherland IO (1987) Analyst 112: 1191
84. Zhujun Z, JL Mullin, WR Seitz (1986) Anal Chim Acta 184: 251
85. Voegtle, F (ed) (1996) Comprehensive Supramolecular Chemistry, Volume 2: Molecular Recognition: Receptors for Molecular Guests,), Pergamon Press, Oxford, UK
86. Valeur B, Leray I (2000) Coord Chem Rev 205: 3
87. Prasanna de Silva A, Gunaratne HQN, Gunnlaugsson T, Huxley AJM, McCoy CP, Rademacher JT, Rice TE (1997) Chem Rev 97: 1515 (review)
88. He H, Leiner MJP, Tusa J, Proc Europt(r)ode 2000 (Lyon, Fr), Book of Abstracts, Coulet P (ed), p 49Ð50
89. Wolfbeis OS, Schaffar BPH (1987) Anal Chim Acta 198: 1
90. Wolfbeis OS (1995) Sens Actuators 29B: 140
91. Krause Ch, Werner T, Huber Ch, Wolfbeis OS (1999) Anal Chem 71: 5304
92. Kawabata Y, Tahara R, Imasaka T, Ishibashi N (1990) Anal Chem 62: 1528
93. Mohr GJ, Wolfbeis OS (1997) Sens Actuat 37: 103
94. Tan SSS, Hauser PC, Chaniotakis NA, Suter G, Simon W (1989) Chimia 43: 257
95. Morf WE, Seiler K, Lehmann B, Behringer C, Hartman K, Simon W (1989) Pure Appl Chem 61: 1613
96. Seiler K, Simon W (1992) Anal Chim Acta 266: 73
97. Wang E, Meyerhoff M E (1993) Anal Chim Acta 283: 673
98. Huber Ch, Werner T, Krause Ch, Wolfbeis OS, Leiner MJP (1999) Anal Chim Acta 398: 137
99. Stokes GG (1869) J Chem Soc 22: 174
100. Urbano E, Offenbacher H, OS Wolfbeis (1984) Anal Chem 56: 427
101. He H, Uray G, Wolfbeis OS (1990) Proc SPIE-Int Soc Opt Eng vol 1368: 175
102. Völkl KP, Grossmann U, Opitz N, Lübbers DW (1981) Adv Physiol Sci 25: 99
103. Lübbers DW, Opitz N (1983) Sens Actuators 3: 641
104. Schaffar BPH, Wolfbeis OS (1991) Chemically Mediated Fiber Optic Biosensors, chapter 8 in: Biosensors Principles and Applications, Blum LJ, Coulet PR (eds), M Dekker, New York, chapter 8: pp 163Ð194
105. Marazuela MD, Moreno-Bondi MC (2002) Anal Bioanal Chem 372: 664
106. Trettnak W, Wolfbeis OS (1989) Fresenius Z Anal Chem 334: 427
107. Wolfbeis OS (1986) Anal Chem 58: 2874
108. Zhang Z, Seitz WR, O'Connell K (1990) Anal Chim Acta 236: 251
109. Freeman MK, Bachas L (1992) Biosens Bioelectron 7: 49
110. Trettnak W, Reininger F, Zinterl E, Wolfbeis OS (1993) Sens Actuators B11: 87
111. Preininger C, Klimant I, Wolfbeis OS (1994) Anal Chem 66: 1841
112. Goldfinch MJ, Lowe CR (1984) Anal Biochem 138: 430
113. Kroneis HW, Marsoner HJ (1983) Sens Actuat 4: 587
114. Miller WW, Yafuso M, Yan Ch, Hui HK, Arick S (1987) Clin Chem 33: 1538
115. Gehrich JL, Lübbers DW, Opitz N, Hansmann DR, Miller WW, Tusa JK, Yafuso M (1986) IEEE Trans Biomed Eng 33: 117
116. Klimant I, Meyer V, Kuhl M (1995) Limnol Oceanography 40: 1159
117. Holst G, Glud RN, Kuehl M, Klimant I (1997) Sens Actuators B38: 122, and refs cited

118. Song A, Parus S, Kopelman R (1997) Anal Chem 69: 863
119. Barker SL Kopelman R, Meyer TE, Cusanovich MA (1998) Anal Chem 70: 971
120. Lee ED, Werner TC, Seitz WR (1987) Anal Chem 59: 279
121. Review: Klimant I, Huber Ch, Liebsch G, Neurauter G, Stangelmayer A, Wolfbeis OS (2001) in New Trends in Fluorescence Spectroscopy: Application to Chemical and Life Sciences, B Valeur, J C Brochon (eds), Springer Verlag, Berlin, chap 13, pp 257-274
122. Lakowicz JR, Gryczynski I, Gryczynski Z, Dattelbaum JD (1999) Anal Biochem 267: 397
123. Kubelka P, Munk F (1931) Z Techn Phys 12: 593
124. Weigl BH, Holobar A, Trettnak W, Klimant I, Kraus H, O'Leary P, Wolfbeis OS (1994) J Biotechnol 32: 127
125. Hill AG, Groom RC, Vinanasky RP, Lefrak EA (1985) Proc Am Acad Cardiovasc Perfusion 6: 148
126. Preininger C, Wolfbeis OS (1996) Biosens Bioelectron 11: 981
127. Koenig B (1996) Oxygen Sensors in Microtiterplates, Diploma Thesis, Univ of Regensburg
128. Piletsky SA, Panasyuk TL, Piletskaya EV, Sergeeva TA, Elkaya AV, Pringsheim E, Wolfbeis OS (2000) Fresenius J Anal Chem 366: 807
129. Weigl BH, Wolfbeis OS(1994) Anal Chem, 66: 3323
130. Polina RJ, Klainer SM (1997) Proc SPIE-Int Soc Opt Eng 3105: 71
131. Rogers AJ (1986) J Phys D: Appl Phys 19: 2237
132. Blyler LL, Jr, Lieberman RA, Cohen LG, Ferrara JA, Macchesney JB (1989) Polym Eng Sci 29: 1215
133. Kvasnik F, McGrath AD (1990) Proc SPIE-Int Soc Opt Eng 1172: 75
134. Smardzewski RR (1988) Talanta 35: 95
135. Steemers FJ, Ferguson JA, Walt DR (2000) Nature Biotechnol 18: 91
136. Wolfbeis OS, Klimant I, Werner T, Huber Ch, Kosch U, Krause Ch, Neurauter G, Duerkop A (1998) Sens Actuat B51: 17
137. Wolfbeis OS, Weis L, Leiner MJP, Ziegler WE (1988) Anal Chem 60: 2028
138. Trettnak W, Leiner MJP, Wolfbeis OS (1988) Analyst 113: 1519
139. Wolfbeis OS, Oehme I, Papkovskaya N, Klimant I (2000) Biosens Bioelectron 15: 69
140. Wolfbeis OS, Urbano E (1983) Anal Chem 55: 1904
141. Posch HE, Wolfbeis OS (1989) Mikrochim Acta 41
142. Wolfbeis OS (1996) Trends Anal Chem 15: 225
143. Lieberman RA, Brown KE (1989) Proc SPIE-Int Soc Opt Eng 990: 104
144. Amao Y, Asai K, Okura I (1999) Analusis 27: 885
145. Sharma A, OS Wolfbeis (1988) Appl Spectrosc 42: 1009
146. Czarnik AW (1993) Fluorescent Chemosensors for Ion and Molecule Recognition, Am Chem Soc Publ, Wash, DC
147. Daffy LM, Prasanna de Silva A, Gunaratne HQN, Huber Ch, Lynch PLM, Werner T, Wolfbeis OS (1998) Chem Eur J 4: 1810
148. Granda-Valdes M, Badia R, Pina-Luis G, Diaz-Garcia ME (2000) Review on PET in sensors, Quim Anal 19: 38
149. Fabrizzi L, Poggi A (1995) Chem Soc Rev 197
150. Shinkai S, Takeuchi M (1996) Trends Anal Chem 15: 188
151. Tsien RY (1980) Biochemistry 19: 239
152. Oehme I, Prattes S, Wolfbeis OS, Mohr GJ (1998) Talanta 47: 597
153. Oehme I, Prokes B, Murkovic I, Werner T, Klimant I, Wolfbeis OS (1994) Fresenius J Anal Chem 350: 563
154. Leiner MJP (1995) Sens Actuators B29: 169
155. Werner T, Wolfbeis OS (1993) Fresenius J Anal Chem 346: 564
156. Lübbers DW, Opitz N, Speiser PP, Bisson HJ (1977) Z Naturforsch 32C: 133
157. Beswick RB, Pitt CW (1988) Chem Phys Lett 143: 589
158. Schaffar BPH, Wolfbeis OS (1989) Proc SPIE-Int Soc Opt Eng 990: 122

159. Knobbe ET, Dunn B, Gold M (1988) Proc SPIE 906: 39
160. MacCraith BD, Ruddy V, Potter C, O'Kelly B, McGilp, JF (1991) Electron Lett 27: 1247
161. Review: Wolfbeis OS, Reisfeld R, Oehme I (1996) Sol-Gels and Chemical Sensors, in: Structure Bonding, vol 85, Springer, pp 51Ð98
162. Meier B, Werner T, Klimant I, Wolfbeis OS (1995) Sens Actuators B29: 240
163. Lobnik A, Oehme I, Wolfbeis OS (1998) Anal Chim Acta 367: 159
164. Klimant I, Ruckruh F, Liebsch G, Stangelmayer A, Wolfbeis OS (1999) Mikrochim Acta 131: 35
165. Klimant I, Wolfbeis OS (1995) Anal Chem 67: 3160
166. Wolfbeis OS (1997) Chemical Sensing Using Indicator Dyes, in: Optical Fiber Sensors, B Culshaw, J Dakin (eds), Artech House, Boston-London, vol IV, chap 8, pp 53Ð107
167. de Marcos S, Wolfbeis OS (1996) Anal Chim Acta 334: 149
168. Pringsheim E, Terpetschnig E, Wolfbeis OS (1997)Anal Chim Acta 357: 247
169. Koncki R, Wolfbeis OS (1998) Anal Chem 70: 2544
170. Pringsheim E, Terpetschnig E, Piletsky SA, Wolfbeis OS (1999) Adv Mater 10: 865
171. Kriz D, Kempe M, Mosbach K (1996) Sens Actuators B33: 178
172. Kriz D, Ramstrom O, Mosbach K (1997) Anal Chem 69: 345A
173. Yano K, Karube I (1999) Trends Anal Chem 18: 199
174. Papkovsky DB, Mohr GJ, Wolfbeis OS (1997) Anal Chim Acta 337: 201Ð205
175. Cox ME, Dunn B (1986) J Polym Sci 24A: 621; 24A: 2395
176. Asai K, Amao Y, Ishimi Y, Okura I, Nishide H (2002) J Thermodyn. Heat Transfer 16: 109
177. Mohr GJ, Werner T, Oehme I, Preininger C, Klimant I, Kovacs B, Wolfbeis OS (1997) Adv Mater 14:1108
178. Kosch U, Klimant I, Werner T, Wolfbeis OS (1998) Anal Chem 70: 3892
179. Neurauter G, Klimant I, Wolfbeis OS (1999) Anal Chim Acta 382: 67
180. Wolfbeis OS, Klimant I, Werner T, Huber Ch, Kosch U, Krause Ch, Neurauter G, Duerkop A (1998) Sens Actuators B51: 17
181. Werner T, Klimant I, Huber Ch, Krause Ch, Wolfbeis OS (1999) Mikrochim Acta 131: 25
182. Atkinson RS, Brimage DR, Davidson RS (1973) J Chem Soc Perkin I 960
183. Peterson JI, Fitzgerald RV (1980) Rev Sci Instrum 51: 670
184. Rumsey WL, Vanderkooi JM, Wilson DF (1988) Science 241: 1649
185. Lakowicz JR, Szmacinski H, Nowaczyk K, Johnson ML (1992) Cell Calcium 13: 131
186. Hartmann P, Ziegler W, Holst G, Lübbers DW (1997) Sens Actuat B38: 110 and refs cited
187. Liebsch G, Klimant I, Frank B, Holst G, Wolfbeis OS (2000) Appl Spectrosc 54: 548
188. Liebsch G, Klimant I, Krause Ch, Wolfbeis OS (2001) Anal Chem 73: 4354
189. For a review, see; Gouterman M (1997) J Chem Educ 74: 697
190. Baron AE, Danielson JDS, Gouterman M, Wan JR, Callis JB (1993) Rev Sci Instrum 64: 3394
191. McLachlan BG, Kavandi JL, Callis JB, Gouterman M, Green E, Khalil G, Burns D (1993) Exp Fluids 14: 33
192. Amao Y, Ishikawa Y, Okura I, Miyashita T, Bull Chem Soc Jpn (2001) 74: 2445
193. Engler RH, Klein C, Trinks O (2002) Meas Sci Technol 11: 1077
194. Free HM, Collins GF, Free AH (1960) Clin Chem 6: 352
195. Gothgen IH, Siggaard-Andersen O, Rasmussen JP, Wimberley PD, Fogh-Andersen N (1987) Scand J Clin Lab Invest, Suppl 188: 27
196. Swenson FJ (1993) Sens Actuators B11: 315

CHAPTER 2

Molecularly Imprinted Polymers for Optical Sensing Devices

Marta Elena Díaz-García, Rosana Badía

1 Introduction

A problem of paramount importance in analytical chemistry is selectivity, particularly at low analyte concentrations in the presence of interfering substances. The sensitive and selective determination of a large number of trace compounds in complex samples is of great relevance in many fields such as biotechnology, the environment, food and pharmacentrical industries and health care for diagnosis or treatment of diseases.

Living beings depend on molecular recognition, a process by which one molecule binds specifically to its target. These interactions rely on selective binding events between complex biological molecules and their target ligands. For example, the specific interaction of an enzyme with its substrate, that of a hormone and its cellular receptor, the ability of an antibody to attach to a foreign invader or even that of a virus to attach to a host cell, are all subtle and specific interactions that depend on molecular recognition.

The exquisite specificity and selectivity of biorecognition by these biological receptors make them highly attractive for the development of sensing devices, as a result of which biosensors have attracted much attention. However, sensors that rely on biological recognition elements often lack storage life and operational stability and are accompanied by difficulties in generating the affinity sensing phase and the specificity towards the desired target molecule. Due to these limitations biosensors have not become quite the commercial success expected in the early euphoric development phase.

During the last decade, studies on the complex interactions between molecular species, the molecular recognition process, and the ability to mimic natural binding phenomena have led to the development of abiotic materials which combine the properties of both the natural and the synthetic components [1-3]. As the structure, properties and mechanisms of biological systems become understood scientists are attempting to transfer this knowledge to synthetically designed materials. The goal in the design of these artificial receptors is the construction of model systems capable of mimicking the molecular level selectivities observed in nature by developing host systems with cavities possessing complementary fuctionalities to that of the guest molecule or ion. Extremely high selectivity is obtained if, as in nature, a cavity exits with both an appropriate shape to match that of the ligand and with binding sites in a topochemical arrangement, so that the ligand is bound inside the cavity. In this context, receptor molecules such as cyclodextrins, cyclophanes, crown ethers and calixarenes [4–6] have been actively studied because of their three-dimensional structure, associated with versatile recognition properties related to neutral molecules as well as metal or organic ions.

Another approach to create receptor-like binding sites is molecular imprinting. The concept of molecular imprinting emerged from Pauling's theory of the formation of antibodies, according to which antigen molecules are the templates around which serum proteins assemble to form antibodies [7,8]. Although Pauling's speculation was later discarded in favour of the more appropriate "clonal selective" theory of antibody formation, chemists found it useful to adopt the concept of forming a three dimensional structure around a template to create nanostructured synthetic materials (imprinted polymers) that can recognise target molecules with selectivities similar to their biological counterparts. These models laid the foundations for Biomimetic Chemistry.

In this chapter an overview of imprinted polymers (MIPs) is given and their potential in optical sensing approaches for environmental and industrial applications is discussed. Emphasis will be placed on the recent developments in this field and the challenges that lie ahead.

2 Molecular Imprinting Process

Molecular imprinting is a general strategy to produce tailor-made cross-linked polymer materials that "remember" a particular molecule. This "memory" is created in the presence of a template via close interactions with functional monomers. The process involves three main key steps (Fig. 1):

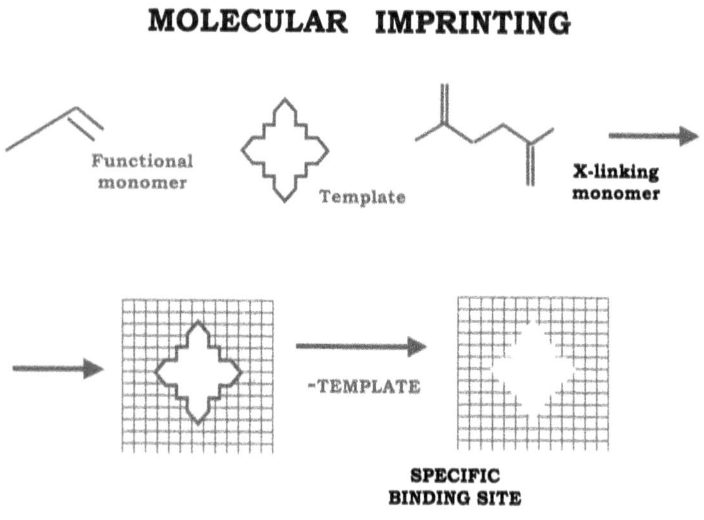

Fig. 1. Schematic diagram of the molecular imprinting principle

i) Association of the template (print molecule) with the functional monomers (pre-arrangement step).

ii) Co-polymerisation of the resulting printing assembly with an excess of a cross-linking agent in an inert solvent to form a highly cross-linked matrix.

iii) Template extraction by washing in a suitable solvent or by acid or basic hydrolysis.

Template removal creates a specifically imprinted groove in the polymer in the form of well-defined cavities possessing a shape and stereochemical arrangement of functional groups complementary to that of the template. Molecular recognition in MIPs is a direct consequence of their rigidly cross-linked structure and the process is similar to a "lock-and-key" fit. Two general imprinting strategies have been employed to design the positioning of the polymer functional groups for optimum recognition efficiency: the pre-organised (covalent) imprinting method and the self-assembly (non-covalent) imprinting approach [9].

2.1 Covalent Molecular Imprinting

Covalent imprinting, as suggested by Wulff and co-workers [10, 11], involves the synthesis of a polymerizable derivative of the template. This derivative is a composite of the imprint molecule and the functional monomer linked by strong reversible covalent bonds such as boronate esters [12], Schiff bases [13], or ketal [14]. Covalent imprinting has been highly successful in the cases of molecules bearing 1,2- and 1,3-diol groups, e.g. various sugar derivatives. An advantage of this approach is that the resulting imprinted polymers show a weak selectivity dependence on the solvent used during the polymerisation process. This

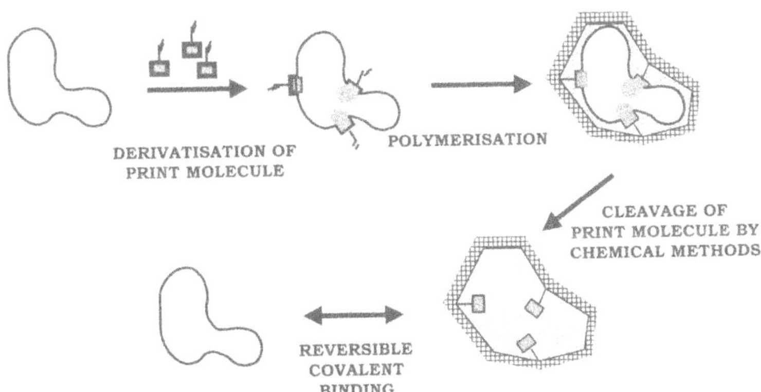

Fig. 2. Covalent molecular imprinting process

phenomenon is ascribed to the limited binding to the imprint derivative by the solvent and even other non-specific molecules present in the solution prior to polymerisation. Template cleavage by chemical methods (acid/basic hydrolysis) from the imprinted polymer yield the moieties of the functional monomers in positions determined by the imprint molecule (Fig. 2). Covalent imprinting is limited by several disadvantages such as the necessity of first preparing a monomer-template conjugate, chemically cleaving the template from the polymer and slow binding kinetics.

2.2 Self-assembly Molecular Imprinting

The concept of the self-assembly method of imprinted polymer formation is shown in Fig. 3. This relies on the self-assembly of functional monomers around the template in the pre-polymerization mixture in a way that maximizes the binding interactions between the two species. The binding interactions can be single or multiple point in nature and involve hydrogen bonding, electrostatic, hydrophobic, dipole-dipole and/or π-π interactions. After polymerisation, the template is extracted from the solid polymer using a solvent capable of disrupting the specific interactions. The topochemistry of the imprinted binding cavities is complementary in size, shape and functionality to the target molecule.

Non-covalent imprinting is more flexible in terms of choice of functional monomers to complement the template molecule's functional groups. In some cases, the monomers can even be combined to increase the strength of binding and to achieve greater functional complementarity to the template. This, in turn, results in better selectivity in template recognition.

Polymers prepared using a self-assembly imprinting protocol are characterised by non-uniform distribution of binding sites, as a result of which part of the

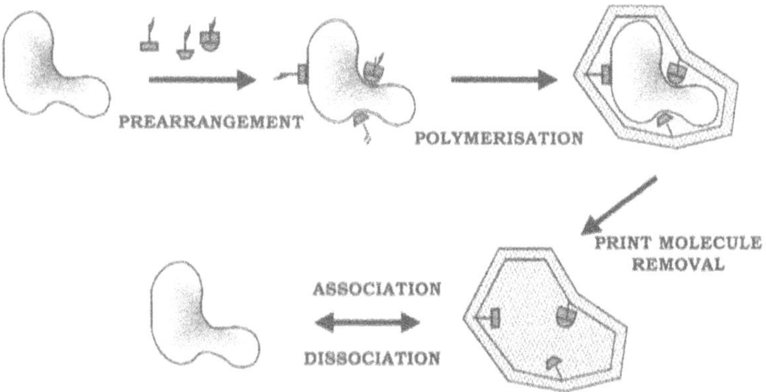

Fig. 3. Principle of non-covalent imprinting

imprinted sites may react favourably with the template, whereas other imprinted sites may interact in a less favourable or even non-specific manner [15]. This is the result of unavoidable random incorporation of functional monomers, which in turn depends on the specifics of the polymerisation conditions such as temperature and the particular solvent [16, 17]. However, it is not at all obvious that the covalent approach is, in practical terms, superior to the non-covalent one as rebinding heterogeneity has also been observed with polymers prepared using non-covalent interactions [18]. It is probable that molecular imprinting results in materials with a broad distribution of binding sites, only a fraction of which exhibit template selectivity. Non-covalent imprinting is the simplest way to prepare recognition materials for biomimetic sensing.

3 Polymer Composition

3.1 Templates

Molecular imprints have been developed for many classes of molecules, including drugs, antibiotics, pesticides, dyes, metal ions and others [9, 19, 20]. Templates offering multiple sites for interaction with the functional monomer are likely to yield binding sites of higher specificity and affinity for the template. A quantitative structure binding relationship was demonstrated by imprinting a number of structurally related basic N-heterocycles by co-polymerisation of methacrylic acid (MAA) and ethylene glycol dimethacrylate (EDMA) [21,22]. High affinity and selectivity were reported when 9-ethyladenine was used as the template [23, 24]. This was thought to be due to a combination of Watson and Crick- and Hoogsten-type hydrogen bonds between the base and carboxylic acids. Sellergren and Dauwe observed similar effects [25] in their studies on imprinting a series of structurally similar triazine templates. Thermodynamic considerations indicate that non-rigid templates, such as proteins or polypeptides, yield imprinted polymers with a large excess of poorly selective sites [26]. Templates with structural rigidity that can fit into the binding site of the polymer with minimal change in its conformation will increase the affinity and selectivity in the recognition. This is the result of a minimal loss in entropy due to conformational changes in the recognition binding site as well as in the template upon binding [9]. Based on these considerations, the "epitope approach" to molecular imprinting has been recently described by N. Minoura et al. [27] for developing MIPs selective towards peptides and proteins. The approach is based on using a short peptide chain as the template, which represents part of a larger peptide or protein. The resultant MIP should recognize the whole protein in a way similar to that in which an antibody operates in the presence of an antigen; only a small part of it, the epitope, is recognized.

Besides the type of template, the ratio of the template to the functional monomer plays a key role in the selectivity of the imprinted polymers when the possible interactions involved inside the matrix are taken into account. The optimum ratio has to be determined for each individual template. Investigations on the effect of stoichiometry have been reported in several imprinting studies [9]. In most cases an excess molar ratio of the monomer to the template produces more favourable results.

3.2 Type of Monomer and Crosslinker

The type of monomers used for producing an analytically useful MIP is very important, since they are components that are vital for successful templating of synthetic polymers. The functional binding monomer is used to create the template central structure about which a rigid matrix is formed by co-polymerization with added crosslinking agents. In general, the structure and the chemical characteristics of the analyte are used as the starting point to select binding functional monomers.

Acrylic (AA) and methacrylic acid (MAA) (Fig. 4) are the most widely used monomers if basic groups are present in the template. They interact ionically with amine functional groups and via hydrogen bonding with a variety of polar functionalities such as carboxylic acids, carbamates and carboxylic esters [28]. The former interaction is stronger than the latter. The introduction of 4-vinylpyridine (4-Vpy) as a functional monomer in non-covalent molecular

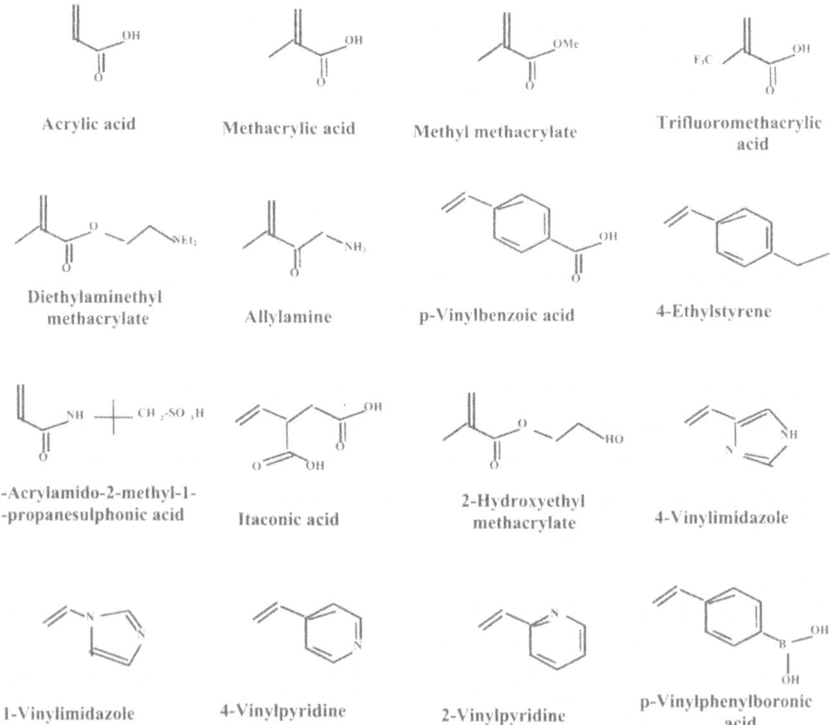

Fig. 4. Common monomers for MIP production

imprinting [29,30] produces polymers with better selectivity for carboxylic acid templates as compared to polymers prepared with MAA [31]. This is explained as being due to the formation of ionic bonds between the recognition sites of the polymers and templates containing carboxylic groups. Metal chelating functional monomers are suitable for print molecules capable of forming complexes with certain metal ions [32].

Recently, 2-(trifluoromethyl) acrylic acid (TFMAA) was introduced as a functional monomer for basic templates. Enhanced selectivity and affinity were obtained for such polymers when compared to those prepared using MAA [33]. Better recognition ability has been achieved in some cases by combinations of two or more functional monomers provided that the different monomers do not interact with one another more strongly than with the template [34–36]. Some of the most widely used functional monomers for molecular imprinting are shown in Fig.4.

In the imprinting process, a very high degree of cross-linking is used in order to achieve a high specificity and selectivity for the polymer. This enhanced selectivity is mainly due to the rigidity that the imprint cavity brings about by the preservation of monomer-template assemblies during polymerisation, which allows functional groups to be fixed in a stable arrangement complementary to the template.

An example of such a cross-linking monomer is ethylene glycol dimethacrylate (EDMA) (Fig. 5) that has been extensively used as cross-linker in non-covalent molecular imprinting. New cross-linkers, which are tri- or tetrafunctional, have been used by different investigators [37]. These include 2,2-bis

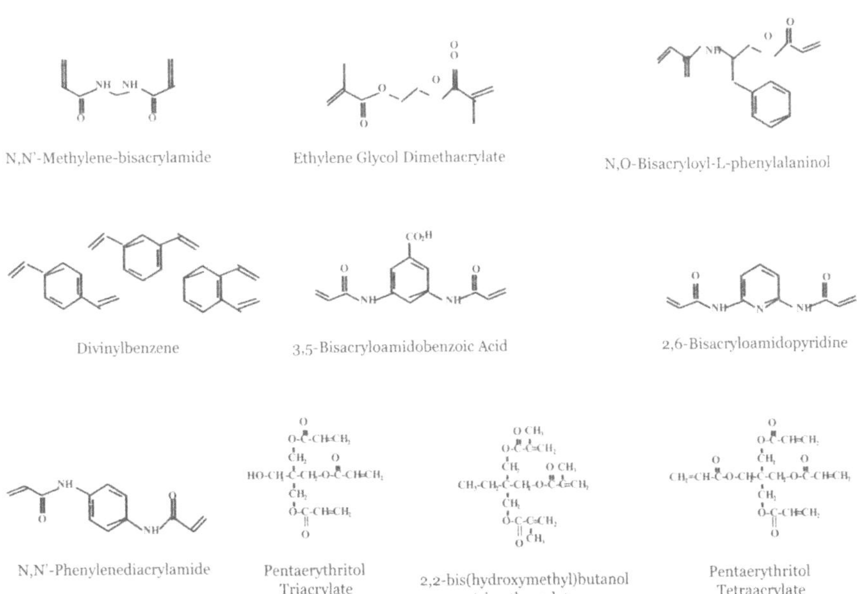

Fig. 5. Cross-linkers for MIP synthesis

(hydroxymethyl)butanol trimethacrylate (TRIM), pentaerythritol tetraacrylate (PETEA) and pentaerythritol triacrylate (PETRA). Molecular imprinted polymers with high load capacity and excellent resolving capability were obtained using TRIM as compared with EDMA polymers [37].

The selectivity is greatly influenced by the kind and the amount of the linking monomer used in the imprinted polymer synthesis. In fact, because of the large proportion of cross-linking involved, it determines to a large extent the hydrophobicity of the polymer which, in turn, affects the extent of non-specific binding.

3.3 Porogenic Solvents

The choice of the porogenic solvent is critical in most molecular imprinting procedures as porogens not only govern the strength of non-covalent interactions but also influence the polymer morphology (inner surface area and average pore size). The solvent should not interfere with the interactions between the template and the monomers during polymer growth. Thus, in the non-covalent imprinting approach, the solvent used should be as non-polar as possible in order to maximise ionic interactions between the print molecule and the monomer. Moreever, solvents with low hydrogen-bond acidity/basicity should be used when dealing with hydrogen bond interactions in order to achieve better imprinting.

The use of solvents that weaken or disrupt the interactive forces between the print species and the functional monomers should be adequate to extract the template. In the rebinding step, as a rule of thumb, the solvent of choice should be the same as (or similar to) that used during the polymer synthesis.

Porogenic solvents affect the physical characteristics of the imprinted material. It was found [30] that acetonitrile –a fairly polar solvent – leads to macroporous polymers with higher surface areas than chloroform. On the other hand, as swelling is dependent on the surrounding medium, it was observed that swelling of imprinted polymers was most pronounced in chlorinated solvents (chloroform, dichloromethane) as compared to acetonitrile or tetrahydrofuran. Although swelling behaviour may result in changes in the 3D-configuration of the recognition sites, Sellergren et al. [16] concluded that polymer morphology is not critical for the selectivity and the strength of substrate rebinding.

3.4 Radical Initiators

In molecular imprinting free radicals may be generated either by thermal or by photochemical means at different temperatures. For the commonly used azo-initiators such as 2,2′-azo-bis-isobutyronitrile, radicals are generated by heating the reaction mixture at 50–60°C in order to provide a suitable rate of decomposition. For thermally unstable templates, free radicals may be generated by UV radiation at low temperatures (e.g. 4°C). In this case, the free-radical propagation rate is slow due to the low temperature of polymerisation. Depending on the radical initiation mode, materials with different selectivity at room temperature may result. Photochemically prepared polymers are associated with good recognition properties at room temperature, while thermally initiated polymers work better at higher temperatures [9, 38].

4 MIP Optical Sensing Applications

4.1 Optical Sensing Approaches for Metals of Environmental Concern

The basic principles of metal ion recognition based on imprinted polymers were first applied in the solid phase extraction and recovery of precious metals such as gold and silver from aqueous solutions [39], for clean-up of waters polluted with cadmium, lead, mercury and other toxic metals or for preconcentration of low contents of heavy/toxic metals prior to their determination in order to improve detection limits and reduce interferences from macrocomponents [40, 41].

Selective recognition arises from a combination of binding of the template functional moieties into the binding sites, ion interaction and van der Waals shape complementarity. This purification/extraction strategy takes advantage of the fact that other metal ions with distinct chemical reactivity should not interact with the binding sites. The chemical nature of the metal ion recognition process permits the use of these imprinted materials in the development of optical chemical sensors [41].

4.1.1 Imprinted Metal Ion Sensors Based on Polymerizable Metal Chelates (Covalent Imprinting)

As the first example of this idea, a fluorescent lead templated chemosensor was recently developed by Güney et al. [42]. The imprinted polymer was prepared by free radical polymerization of 9-vinylcarbazole and lead methacrylate with ethylene glycol methacrylate as cross-linker. Dioxane was used as the porogenic solvent. This polymer, prepared with a functional monomer containing carbazole as a fluorescent tag, allows the detection of lead ions at very low concentrations through a binding-dependent change (quenching) of the carbazole fluorescence. The sensing scheme described is based on an imprinted film in contact with the metal ion solution; lead is detected and quantified during the binding. No data about selectivity and/or non-specific binding are reported in this study, but as noted by the authors the recognition material could be used in a remote detection configuration to detect very low concentrations of lead.

In a similar manner, R.H. Fish et al. [43] have developed imprinted polymers that can be used to selectively trap Cu^{2+} ions in the presence of Fe^{3+}, Co^{2+}, Ni^{2+}, Zn^{2+} and Mn^{2+}. To prepare the polymer, the target metal ion is sandwiched between *N*-[4-vinylbenzyl]-1,4,7-triazacyclononanes (TACN) ligands. After cross-linking with divinylbenzene, the metal ion is washed away with a strong acid (6M HCl). Interaction of copper ions with the imprinted polymer turns the slightly off-white material to a green colour. Recently, and using a similar protocol, R.H. Fish et al. [44] have developed imprinted polymers with TACN ligands with high selectivity for Hg^{2+} in competition with Cd^{2+}, Ag^{+}, Pb^{2+}, Cu^{2+} and Fe^{3+}. Fish claims that imprinted polymers using TACN monomers containing attached fluorescent groups could be polymerized on the tip of an optical fiber in order to develop real-time optical probes for a variety of metal ions of environmental concern [45].

4.1.2
Optical Sensors Based on Non-covalent Imprinting of Fluorescent Metal Chelates

The use of fluorescent metal-chelates as templates offers a potentially simple method for developing non-covalent imprinted polymers for metal ion recognition/sensing. In this approach the fluorescent metal chelate is allowed to self-assemble in the pre-polymerization mixture, thus maximising the binding interactions with functional monomers. After polymerisation, the metal ion is extracted from the polymerised material, leaving a nanopore that is supposed to be selective for binding of the metal (due to the complementary positioning

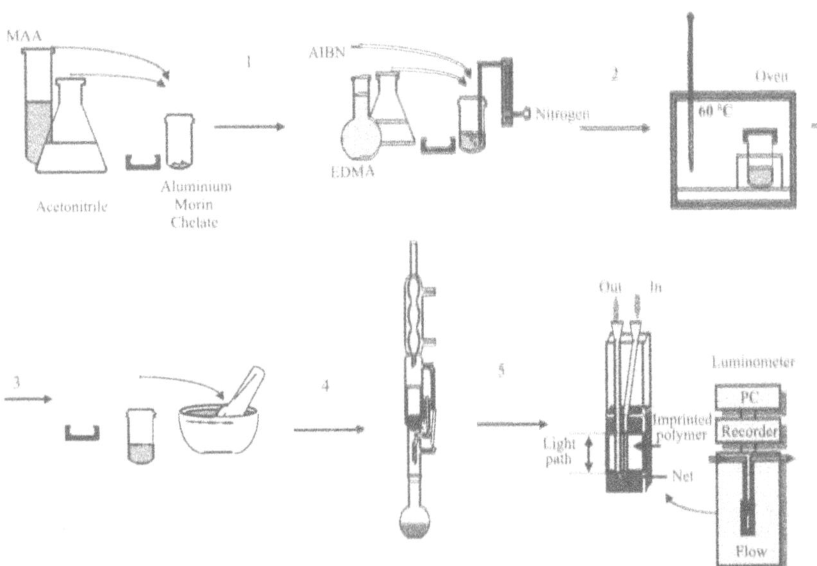

Fig. 6. Schematic of molecular imprinting using a metal chelate as the template. 1) Dissolution of the metal chelate (aluminium-morin) and functional monomers in a solvent of low polarity. 2) Free radical polymerisation initiated with an azo initiator and a crosslinking agent, thermochemically at 60 °C. 3) Griding of the polymer chunk (150–250 µm). 4) Removal of the metal ion from the polymer matrix. 5) Packing the flow-cell with the MIP for its evaluation as sensing material in a flow regime. (MAA: methacrylic acid, EDMA: ethylene glycol dimethacrylate, AIBN: 2,2'-azo-bis-isobutyronitrile)

of the fluorescent chelator functional groups). A schematic of this type of preparation is illustrated in Fig. 6.

This polymer enables the detection and measurement of aluminium in the presence of other metal ions such as Mg^{2+}, Ca^{2+}, Be^{2+} [46]. The aluminium recognition factor of the imprinted polymer was found to be dependent on the porogenic solvent used for making the polymer. The chemical sensing principle behind this sensing material is based on the fluorescence changes observed upon rebinding of aluminium. The application in flow-through sensing applications seems very promising as a low detection limit (0.01 μg mL^{-1} aluminium) and a reversible response for aluminium were obtained, while no leaching of the morin was observed.

From the aforementioned discussions, it is clear that each approach contains advantages and disadvantages when considering their use as optical sensing materials for metal ions. For example, it has been observed that formation of polymerizable derivatives of aluminium- and beryllium-morin [47] complexes for use as templates, resulted in a loss of the fluorescent properties of the recognition material, which rendered the polymers unsuitable for use as optical sensing phases. On the other hand, non-covalent imprinting can yield sensing materials with analytical potential for optical sensing, but non-uniform dispersion of the template in the polymer matrix during preparation and washing off of the chelating ligands during use limit their practical application.

4.2 Optical Sensing Approaches for Environmentally Harmful Compounds

In recent years, the potential of MIPs has been exploited in optical sensing technology to monitor toxic and carcinogenic compounds in water. Thus, MIP sensor membranes for the detection of polycyclic aromatic hydrocarbons in waters have been reported recently [48, 49]. Implementation of this concept involved the synthesis of polyurethane imprinted polymers with PAHs of different structure (naphthalene, pyrene, anthracene, perylene, benzoperylene, etc). The interaction driving forces during polymer synthesis are mainly non-covalent π-π and van der Waals interactions. The selective molecular recognition and rebinding processes are determined by the size and the shape of the cavities; larger molecules are size-excluded and smaller ones are washed out due to the lack of tight fit to the imprinted cavity. The imprinted polymers were prepared as layers of up to several micrometers thickness and were operated within a flow regime. The native fluorescence of the templates was exploited as the optical transduction concept. The analytical performance characteristics of these recognition materials are very promising for the potential use of these materials for field monitoring. For pyrene, as model analyte, the detection limit is in the parts per trillion range, with a IUPAC *S/N* ratio of 3:1. The calibration curve shows linearity with the pyrene concentration up to ca. 40 μg/L, a linear range that can be extended by raising the amount of imprint molecules during the polymer synthesis [49]. The equilibrium constant of a sensor layer imprinted with 3% of pyrene toward pyrene was found to be 9.4×10^7. The response time of the sensor was a few minutes and the effect fully reversible. These imprinted polyurethane layers offer the possibility of being combined with fiber-optic systems or with mass-sensitive universal transducers such as quartz microbalance or surface acoustic wave devices.

New generations of optical MIP-based sensors are emerging. Mizaikoff and co-workers [50] describe an interesting approach to detect the herbicide 2,4-dichlorophenoxyacetic acid (2,4-D) in water. Their method relies on the synthesis of an imprinted polymer using 4-vinylpyridine as the functional monomer in the presence of the polar protic solvent methanol and water, which is unusual for non-covalent MIPs. Layers of the imprinted polymer were prepared on the surface of zinc selenide attenuated total reflection elements and infrared evanescent wave spectroscopy was used as the transduction principle. This method provides good detectability for 2,4-D and reversible response by rinsing with a phosphate buffer. The discrimination between the structurally similar analytes, 2,4-D and POAc, was enhanced due to the inherent selectivity of the transducer. This effective combination allows the scope of MIP-based sensors to be extended to applications in complex matrices, where non-specific interactions or cross reactivities may be a problem.

Compounds that possess the ability to bind metal ions to form complexes in solution are amenable to molecular imprinting with metal ion complexation. Thus, with a judicious choice of metal ion with useful spectroscopic properties, selective and sensitive devices may be produced. Using this approach, Murray et al. [51] have developed a fluorescent fiber optic MIP-based sensor for the hydrolysis product of the nerve gas soman. Soman is an organophosphorus compound that resembles diisopropylfluorophosphate, a drug that blocks the action of acetylcholinesterase. This blockage results in contraction of smooth muscles and eventual asphyxiation. In the imprinted polymer, cavities are formed with coordinated Eu^{3+} as part of the binding site. The sensor is free from interferences from organophosphorus pesticides and herbicides, the compounds most chemically similar to the nerve agent. The sensor is extremely sensitive with a limit of detection of 600 parts per quadrillion in an alkaline solution.

A recent study [52] demonstrated the feasibility of a photoinduced electron transfer (PET) mechanism as a means of signal transduction for MIP-based luminescent sensing of 2,4-D. A sol-gel derived molecular imprinted luminescent sensing material was developed by a conventional sol-gel process using anthryl fluorescent tailor-made organosilane functional monomer. The resulting fluorescent imprinted material was found to selectively respond to 2,4-D in aqueous media. Upon rebinding, the fluorescent properties of the functional monomer were perturbed through acid-base ion-pair formation with 2,4-D, which led to suppression of PET quenching of its anthryl fluorophore. Although the background emission of the imprinted material is a shortcoming for sensitivity, the approach presented in this work is the first attempt to use PET mechanisms in MIP optical sensing (see Fig. 7). There is still much work to be done in order to lower background emission, to optimise the MIP composition by exploring other sensor monomers and to further understand the factors affecting the behaviour of these MIPs.

When the target analyte does not exhibit an optical property that can be used for transduction it is possible to design competitive binding assays using MIPs

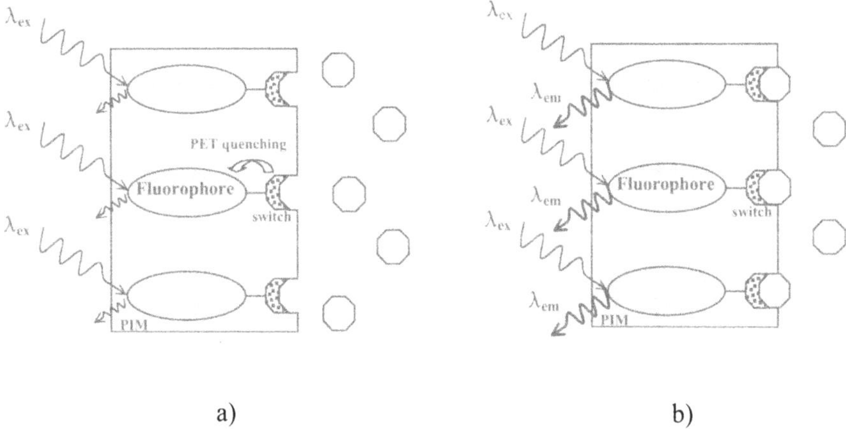

Fig. 7. PET-MIP-based switch a) Fluorescence is off due to the PET process, b) Fluorescence is on due to PET suppression

as antibody mimics. Detection is then based on displacement of a labelled analyte by the unlabelled one. These systems can use either a fluorescent or a coloured marker. This idea was first proposed by Mosbach et al. in 1995 [53] using a fiber-optic device and an MIP phase sensitive to dansyl-phenylalanine. Piletsky et al. [54] developed a competitive sensing assay for the pesticide triazine using fluorescein-labeled triazine as the competitive probe for the binding points in an acrylic MIP against triazine. The system was evaluated by sorption analysis, measuring the concentration of free fluorescent triazine probe in the supernatant after 3.5 hours incubation. Attractive features of this assay are its high selectivity for triazine over structurally related herbicides (atrazine and simazine); sensitivity for triazine is in the 0.01–100mM range and testing time lower than conventional ELISA assays.

A similar fluorescent ligand displacement assay was developed by Haupt et al. [55] for the herbicide 2,4-dichlorophenoxyacetic acid based on an MIP and using a coumarin derivative as a non-related fluorescent probe. The specificity and sensitivity of this MIP biomimetic assay are on a par with a radioligand displacement format using the same MIP and radiolabeled 2,4-dichlorophenoxyacetic acid. An attractive feature of the assay is that it can be used both in aqueous buffer and in organic solvents, with a detection limit of about 100 nM.

Although the use of MIP in immunoassay-type applications is unlikely to be universally applicable, the above results demonstrate that optical competitive MIP-based formats provide useful systems in many instances for use in environmental analysis. These assays offer the advantage of safer experimental use than radioactive materials. Improvements in sensitivity and detection limits of optical MIP competitive sensing assays are still required for them to compete with affinity sensors. Progress in this area could be achieved by the use of chemiluminescent or long-lived luminescent labels to increase the signal-to-noise ratio and eliminate background signal problems.

4.3
MIP Optical Sensing Materials for Organic Volatile Compounds

Many MIP-based sensor layers developed for applications involving optical detection have the property of changing colour or fluorescence upon interaction with the analyte. Dickert et al. [56] have prepared polyurethane polymers imprinted with several kinds of volatile organic solvents. This kind of volatile analyte creates cavities by evaporation. Furthermore, introduction of chromogenic or fluorescent dyes into the polymer network results in sensitive materials for detection of organic solvent vapours in air via changes in optical properties (solvatochromic changes). According to Dickert et al. [56] two principles of selectivity are combined in these materials, e.g. "concave chemistry" due to molecular cavities and "convex chemistry" realised by donor-acceptor (solvent-dye) interactions. No mention is made of the analytical performance characteristics of these materials, but the authors drew conclusions about the potential for sensor applications. Little work has so far been done in the development of selective MIP-based gas sensors for environmental monitoring. Research in this direction is a primary goal.

5
Conclusions and Outlook

The results given in this chapter illustrate the possibilities that exist with respect to the nature of the templates (metal ions, organic compounds, solvent vapours), the polymer synthesis conditions and the optical transduction concepts that can be used to develop MIP-based optical sensing schemes for industrial/environmental applications. The field is still in its infancy and is currently receiving significant attention. Besides the many advantages of the technique various problems remain open. Development of new approaches for MIP synthesis, improvement of the MIP specificity, lowering nonspecific binding or use of different optical transduction schemes (e.g. luminescence with time discrimination, energy transfer, etc.) are some guidelines for future efforts in the development of novel biomimetic sensors or ligand-binding assays, applicable to a wide range of analytes of environmental/industrial concern.

Acknowledgements. Financial support by the Ministerio de Ciencia y Tecnología (Proj.#MCT-00-MAT-0600) is most gratefully acknowledged.

References

1. Sellergren B, Angew Chem Int Ed (2000) 39: 1031
2. Haupt K, Mosbach K, Chem Rev (2000) 100: 2495
3. Al-Kindy S, Badía R, Suárez Rodríguez JL, Díaz-García ME (2000) Crit Rev Anal Chem 30: 291
4. Lehn JM (1988) Angew Chem Int Ed Eng 27: 89
5. Rebk J Jr (1990) Angew Chem Int Ed Eng 29: 245
6. Cram DJ (1992) Nature 356: 29

7. Burnt FM (1959) The Clonal Selection Theory of Acquired Immunity. Cambridge University Press. Cambridge
8. Silverstein AM (1989) A History of Immunology. Acad Press, San Diego
9. Sellergren B (Ed) (2001) Molecularly Imprinted Polymers. Man-made mimics of antibodies and their applications in analytical chemistry. Elsevier. Techniques and Instrumentation in Analytical Chemistry. Vol. 23
10. Wulff G, Vesper W (1978) J Chromatogr 167: 171
11. Wulff G, Schauhoff S (1991) J Org Chem 56: 395
12. Wulff G (1995) Angew Chem Int Ed Eng 34: 1812
13. Wulff G, Vietmeir J (1989) Makromol Chem 1901: 1727
14. Shea KJ Sasaki DY (1991) J Am Chem Soc 113: 4109
15. Vincent T, Rhemcho Z, Tan J (1999) Anal Chem 71: 248A
16. Sellergren B, Shea KJ (1993) J Chromatogr A 635: 31
17. Sellergren B, Shea KJ, J Chromatogr (1995) 690: 29
18. Shea KJ, Sasaki DY (1989) J Am Chem Soc 111: 3442
19. Suárez Rodríguez JL, Díaz-García ME (2000) Anal Chim Acta 405: 67
20. Suárez Rodríguez JL, Díaz-García ME (2001) Biosens Bioelectron 16: 955
21. Vlatakis G, Andersson LI, Muller R, Mosbach K (1993) Nature 361: 645
22. Shea KJ, Spivak DA, Sellergren B (1993) J Am Chem Soc 115: 3368
23. Spivak DA, Gilmore MA, Shea KJ (1997) J Am Chem Soc 119: 4388
24. Yu C, Mosbach K (1997) J Org Chem 62: 4057
25. Dauwe C, Sellergren B (1996) J Chromatogr A 735: 191
26. Nicholls IA (1995) Chem Lett 1035
27. Rachkov A, Minoura N (2001) Biochim Biophys Acta 1544: 255
28. Kempe M, Mosbach K (1995) J Chromatogr A 694: 3
29. Kempe M, Fischer L, Mosbach K (1993) J Mol Recog 6: 25
30. Kempe M, Mosbach K (1994) J Chromatogr A 664: 276
31. Andersson LI, Mosbach K (1990) J Chromatogr 516: 313
32. Chen G, Guan Z, Chen C-T, Fu L, Sundaresan V, Arnold FH (1997) Nat Biotechnol 15: 354
33. Matsui J, Nicholls IA, Takeuchi T (1998) Anal Chim Acta 365: 89
34. Ramström O, Andersson LJ, Mosbach K (1993) J Org Chem 58: 7652
35. Hosaya K, Shirasu Y, Kimata K, Tanaka N (1998) Anal Chem 70: 943
36. Tukerkewitsch P, Wandelt B, Darling GD, Powell WS (1998) Anal Chem 70: 2035
37. Kempe M (1996) Anal Chem 68: 1948
38. Nicholls IA, Andersson LI, Mosbach K, Ekberg B (1995) Trends Biotechnol 13: 47
39. Nishide H, Tsuchida E (1976) Makromol Chem 177: 2295
40. Kabanow VA, Efendiev AA, Orujev DD (1979) J Appl Polym Sci 24: 259
41. Bartsch RA, Maeda M (Eds) (1998) Recognition with imprinted polymers. American Chemical Society Washington DC
42. Güney O, Yilmaz Y, Pekcan Ö (2002) Sens Actuators B 1–4: 4261
43. Chen H, Olmstead MM, Albright RL, Devenyi J, Fish RH (1997) Angew Chem Int Ed Engl 36: 642
44. Lo HC, Chen H, Fish RH (2001) Eur J Inorg Chem 9: 2217
45. Preuss P (1997) Symposium Series Book Recognition with Imprinted Polymers Science Articles Arch. Berkeley Lab California
46. Al-Kindy S, Badía R, Díaz-García ME (2002) Anal Lett, 35: 1763
47. Carvalho Torres AL, Díaz-García ME (unpublished results)
48. Dickert FL, Besenböck H, Tortscjanoff M (1998) Adv Mater 10: 149
49. Dickert FL, Tortscjanoff M, Bulst WE, Fischerauer G (1999) Anal Chem 71: 4559
50. Jakusch M, Janotta M, Mizaikoff B, Mosbach K, Haupt K (1999) Anal Chem 71: 4786
51. Jenkins AL, Uy OM, Murray GM (1999) Anal Chem 71: 373
52. Leung MK-P, Chow C-F, Lam MH-W (2001) J Mater Chem 11: 2985

53. Kriz D, Ramström O, Svensson A, Mosbach K (1995) Anal Chem 67: 2142
54. Piletsky SA, Piltskaya EV, El'skaya AV (1997) Anal Lett 30: 445
55. Haupt K, Mayes AG, Mosbach K (1998) Anal Chem 70: 3936

Chromogenic and Fluorogenic Reactands: New Indicator Dyes for Monitoring Amines, Alcohols and Aldehydes

GERHARD J. MOHR

1 Introduction

In the past few years a wide range of optical sensors for ions has been presented. Sensors for pH are based on the protonation/deprotonation of pH indicator dyes [1] and sensors for cations and anions use a combination of pH indicator dyes with selective ionophores (*via* the mechanisms of coextraction or ion-exchange) [2]. In the case of coextraction and ion-exchange, the uncoloured ionophore recognizes the analyte while the pH indicator dye changes its colour [2]. This approach is highly cross-sensitive to pH and has not found practical application so far. A more sophisticated and synthetically challenging approach is to use selective fluoro- and chromoionophores [1,3,4]. They are advantageous because the dyes both selectively recognize the analyte and simultaneously

Ligands for ionic analytes
(analyte recognition via complexation)

Dye — Na^+ K^+ Li^+ ⇌ *Dye* — K^+

Reactands for neutral analytes
(analyte recognition via formation of a covalent bond)

Fig. 1 Principle of analyte recognition using fluorogenic ligands and reactands. Apart from an increase in fluorescence upon interaction with the analyte, changes in absorbance can be observed as well

change their colour. Fluoroionophores for sodium and potassium with very low cross-sensitivity to pH can be found in the AVL OPTI devices [5].

The situation is more complex when electrically neutral analytes have to be detected. First, the interaction between indicator dyes and neutral analytes usually is rather weak (Van der Waals interactions, hydrogen bonding, hydrophobic interactions) [6]. Secondly, complexation of a neutral analyte has a much weaker effect on the electron delocalisation of an indicator dye than the complexation of an ion, thus causing only small changes in colour.

In order to provide sufficient signal changes upon exposure to neutral analytes, chemical reactions have been introduced into optical sensing (Fig. 1). Narayanaswamy et al. have used pararosaniline immobilized on ion-exchange resin for the detection of formaldehyde and acrolein in aqueous solution [7]. A sensor for the detection of hydrazine made use of the reaction of hydrazine with *p*-dimethylaminobenzaldehyde to form a coloured benzalazine in sol-gel glass [8]. A sensor for glucose has been developed using 3-aminophenyl boronic acid copolymerised with aniline to form a polyaniline layer with analyte-dependent absorbance changes in the near infrared spectral range [9]. Turner et al. investigated the ability of a hemithioacetal-based polymer to react with primary amines and to form a fluorescent isoindole complex [10]. Pretsch et al. used the bisulphite addition to a lipophilic aldehyde for optical detection of sulphur dioxide [11]. A method to detect alcohols was based on trifluoroacetophenone derivatives that gave signal changes at around 305 nm [12].

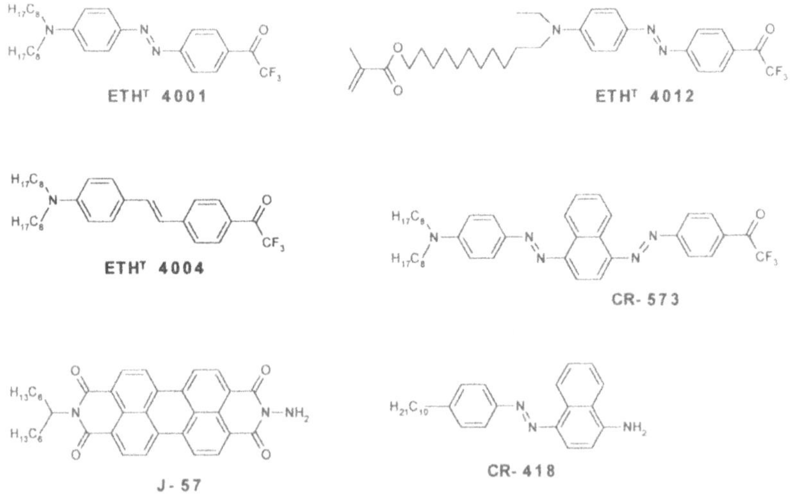

Fig. 2 Chemical structures of 4-*N*,*N*-dioctylamino-4′-trifluoroacetylazobenzene (ETH^{T} 4001), 4-trifluoroacetyl-4'-[*N*-(11-methacryloxyundecyl)-*N*-ethylamino]-azobenzene (ETH^{T} 4012), 4-*N*,*N*-dioctylamino-4′-trifluoroacetylstilbene (ETH^{T} 4004), 4-[4-(4-trifluoroacetylphenylazo)-1-naphthylazo]-*N*,*N*-dioctylaniline (CR-573), *N*-amino-*N*′-(1-hexylheptyl) perylene-3,4:9,10-tetracarboxylbisimide (J-57), and 1-amino-4-(4-decylphenylazo)-naphthalene (CR-418)

The approach of using chemical reactions for detecting neutral analytes has been improved by synthesizing new long-wavelength absorbing and fluorescing indicator dyes (Fig. 2). Since they combine both the properties of a chemical reagent with those of a selective ligand, these indicator dyes are termed chromo- and fluororeactands. Here, reactands for analytes such as amines, alcohols and aldehydes are presented and reaction mechanism and parameters such as selectivity, sensitivity, and (of special importance in the case of chemical reactions) reversibility and response time are discussed.

2 Sensing Amines

2.1 Trifluoroacetylazobenzene Dyes

A wide range of amines are pollutants in industrial and manufacturing areas because they are extensively used in the preparation of fertilizers, pharmaceuticals, surfactants, biological buffer substances and colorants. Volatile amines can be found in agricultural areas, and their presence may be an indicator of food quality. The chromoreactand 4-*N*,*N*-dioctylamino-4′-trifluoroacetylazobenzene **ETH^T 4001** is capable of performing a reversible chemical reaction with amines. The conversion of the trifluoroacetyl group into a hemiaminal causes a

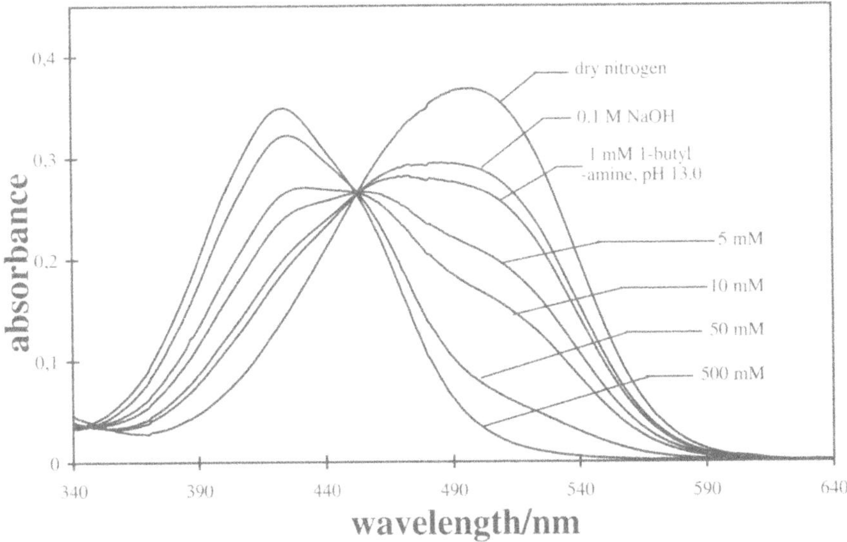

Fig. 3 Absorbance spectra of a sensor layer composed of ETH^T 4001, PVC and bis-(2-ethylhexyl)-sebacate in contact with dry nitrogen and different concentrations of aqueous 1-butylamine at pH 13.0. On changing from nitrogen to 0.1 M sodium hydroxide solution, the diol is formed, whereas, on changing from plain buffer to aqueous 1-butylamine, the hemiaminal is formed. Both types of reaction are fully reversible

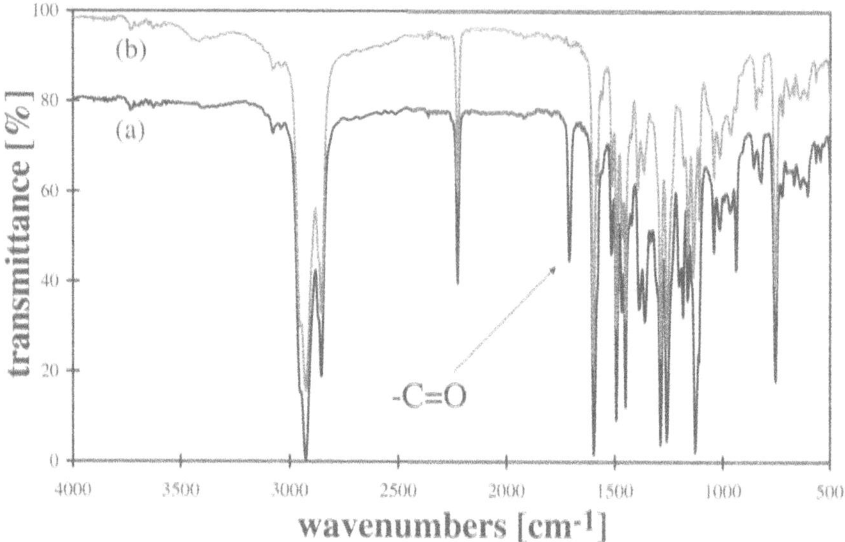

Fig. 4 Infrared spectra of the amine sensor membrane (composed of 16% ETH^T 4001, 28% PVC and 56% 2-octyloxybenzonitrile) upon exposure to air (a) and gaseous 1-hexylamine (b). The carbonyl band of the trifluoroacetyl form of the reactand at 1709 cm^{-1} disappears upon conversion to the hemiaminal. Also shown are the nitrile group of the plasticizer at 2227 cm^{-1} and the CH-vibrations of PVC and the plasticizer at 2856 and 2927 cm^{-1}

change in electron delocalisation resulting in a colour change from red to yellow [13].

Figure 3 shows the spectral behaviour of **ETH^T 4001** in plasticised PVC upon exposure to aqueous 1-butylamine. First, the formation of a diol upon interaction with water is observed, and then, due to the higher nucleophilicity of 1-butylamine, the hemiaminal formation. These two chemical reactions cause the absorbance of **ETH^T 4001** to decrease at a wavelength of around 490 nm (representing the trifluoroacetyl form of the dye) and to increase at a wavelength of around 420 nm (corresponding to the diol and hemiaminal forms). The sensor layer exhibits a sensitivity range of 1 mM–100 mM for aqueous 1-butylamine and a detection limit of 0.3 mM. The conversion of the trifluoroacetyl form of the reactand is also visible in the infrared spectral range and confirms that the sensor membrane responds via a chemical reaction (Fig. 4).

The mathematical description of the interaction of **ETH^T 4001** with amines in sensor membranes includes two processes, namely the extraction of the amine from the aqueous into the polymer phase (governed by the lipophilicity of the analyte) and the chemical reaction of the analyte with the reactive group (governed by nucleophilicity). The overall reaction of the reactand R in the membrane phase with the amine A in the aqueous sample to form the hemiaminal AR in the membrane is described by:

$$R_{\text{membrane}} + A_{\text{aqueous}} \Leftrightarrow AR_{\text{membrane}} \qquad (1)$$

and consequently:

$$K = \frac{[AR]}{[R] \cdot a_A} \tag{2}$$

where $[R]$ is the concentration of the free reactand, a_A the amine activity in the aqueous phase, $[AR]$ the concentration of the hemiaminal in the membrane phase, and K the combined equilibrium constant for extraction of the amine from the aqueous into the membrane phase and the subsequent nucleophilic addition reaction. The reactivity, φ, is defined as the ratio between the concentration of the hemiaminal in the membrane and the total reactand concentration, and is described in (3):

$$\varphi = \frac{[AR]}{[AR]+[R]} = \frac{(S_X - S_R)}{(S_{AR} - S_R)} \tag{3}$$

where S_X is the absorbance or fluorescence signal at a defined amine concentration, S_{AR} the signal of the hemiaminal and S_R the signal of the reactand. Combining Eqs 2 and 3 gives the dependence of absorbance (or fluorescence) of the sensor membranes and allows the calculation of the sensitivity of optode layers towards the investigated amines:

$$S_X = \frac{(K \cdot S_{AR} \cdot a_A + S_R)}{(1 + K \cdot a_A)} \tag{4}$$

The selectivity of chemical reactions is different from the selectivity encountered with complexing agents. Whereas Coulomb, Van der Waals and hydrophobic interactions are responsible for the selective recognition of the guest by a host, a reactand such as **ETH^T 4001** provides selectivity via its different chemical reactivity towards interfering species. As mentioned above, the first parameter for the encountered selectivity of **ETH^T 4001** in a polymer layer towards amines is their different lipophilicity. More lipophilic amines are more efficiently extracted from the aqueous phase into the lipophilic organic membrane phase. This phenomenon is well described by the *n*-octanol-water partition coefficient [14]. The different nucleophilicity of amines additionally contributes to the selectivity pattern. The selectivity for primary, secondary, and tertiary amines of a sensor layer composed of 1% of **ETH^T 4001**, 33% of PVC and 66% of DOS is shown in Table 1. The selectivity coefficients $\log K^{opt}$ (calculated according to Eq. 4) and the logarithms of the *n*-octanol-water partition coefficient ($\log K_{OW}$) indicate that the sensor exhibits highest sensitivity for lipophilic primary amines. Although the lipophilicity of secondary and tertiary amines lies within the same range, the response is significantly lower. This difference is attributed to the fact that secondary and tertiary amines are sterically hindered in approaching the trifluoroacetyl group by their bulky groups near the ami-

Table 1. Selectivity coefficients $\log K^{opt}$ of the sensor layer based on ETH[T] 4001/PVC/DOS for amines and alcohols relative to 1-butylamine calculated according to equation 4, and $\log K_{OW}$ values of the investigated amines and alcohols

Analyte	$\log K^{opt}$	$\log K_{OW}$	Analyte	$\log K^{opt}$	$\log K_{OW}$
ammonia	-2.3	N/A	*tert*-butylamine	-1.8	0.40
methylamine	-1.5	-0.57	1-hexylamine	0.9	2.06
ethylamine	-1.2	-0.13	pyridine	-1.7	0.65
diethylamine	-1.6	0.58	aniline	-1.5	0.90
triethylamine	-1.1	1.45	ethanol	-2.8	-0.30
1-propylamine	-0.5	0.48	1-propanol	-2.3	0.25
2-propylamine	-1.8	0.26	amphetamine	0.0	1.76
1-butylamine	0.0	0.86	methamphetamine	0.2	2.07

no moiety. A similar discrimination is observed for primary amines with bulky substituents such as *tert*-butylamine or 2-propylamine.

The response time of the plasticised PVC layer is in the range of 5–15 min when investigated in flow-through cells. However, these values represent both the time for an exchange of sample solutions in the flow cell and the true response of the reactand in the layer. Gas-phase measurements with sensor layers using a CMOS-based calorimetric transduction of the chemical reaction have shown that the reaction of **ETH[T] 4001** with dry 1-butylamine in plasticised PVC can be as fast as 3 s for forward and 15 s for reverse response [15].

When investigating different polymer matrices for **ETH[T] 4001** it was found that all the sensor membranes showed very similar selectivity and response time, independently of whether plasticised or plasticiser-free polymers were used [16]. Consequently, it is not possible to tailor the selectivity pattern by changing the matrix but rather by developing reactands with bulky substituents for steric discrimination.

The reactand exhibits a pronounced positive solvatochromism, i.e. the absorbance maximum shifts to longer wavelengths with increasing polarity of the solvent. Thus, it is possible to correlate the reactand's absorbance maximum in the polymer with the polymer's dielectric constant [16]. In addition, the extent of diol formation by **ETH[T] 4001** correlates with the water uptake of the respective polymer matrix [17]. Finally, the sensitivity to 1-butylamine indicates the lipophilicity of the polymer material [16]. Consequently, a single dye yields a significant amount of information about the polymer's physical properties (Table 2).

Real measurements have shown that the detection of absorbance changes at around 500 nm can be affected by sample colour, and furthermore, for the development of small sensing devices compatibility with cheap light sources is required. Recently, a chromoreactand has been synthesized whose absorbance changes are greatest at around 630 nm rather than at 490 nm. The dye was obtained by increasing the length of the chromophore via the insertion of a naphthylazo moiety into **ETH[T] 4001**, giving the bisazo dye 4-[4-(4-trifluoroacetylphenylazo)-1-naphthylazo]-*N*,*N*-dioctylaniline (**CR-573**). Additional-

Table 2. Correlation between diol formation of ETH^T 4001 and water uptake of the polymer layers upon conditioning, and correlation between the relative dielectric constant ε_r of layers and the absorbance maximum λ_{max} of ETH^T 4001 in dry layers. The composition of the membranes is 1% ETH^T 4001, 33 % PVC and 66% plasticizer (or 1% ETH^T 4001 and 99% of plasticiser-free polymers)

Polymer	Extent of diol formation [%]	Water uptake [%]	λ_{max} [nm]	ε_r
PVC/2-octyloxybenzonitrile	12	N/A	510	13.9
PVC/bis(2-ethylhexyl)sebacate	24	0.48	497	4.8
PVC/tris(2-ethylhexyl)phosphate	90	1.15	501	7.0
PVC/tris(2-ethylhexyl)trimelitate	9	0.32	500	5.5
Poly(hexyl methacrylate)	9	N/A	494	N/A
Poly(ethylene vinylacetate)	31	N/A	494	N/A
Polybutadiene	1	< 0.1	483	2.3
Polydimethylsiloxane	1	< 0.1	462	3.2

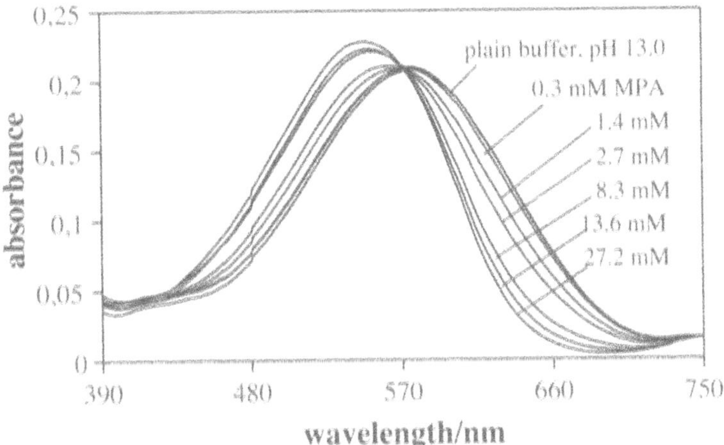

Fig. 5 Change in absorbance of chromoreactand CR-573 as a function of amphetamine (MPA) concentration. All measurements were performed at pH 13.0 to provide amphetamine in its chemically reactive neutral form

ly, this reactand exhibited a threefold increase in sensitivity compared to ETH^T 4001 and has been used to detect amphetamines (Fig. 5).

2.2 Trifluoroacetylazobenzene Copolymers

The basic principle of optical sensors is the fast and reversible interaction of indicator dyes with the analyte, resulting in changes in absorbance or luminescence. The dyes are usually incorporated into polymer layers which are permeable for the analyte present in the sample solution or in the gas phase. Such polymer-based optical sensor layers are simple to prepare because all components (polymers, dyes, additives) are dissolved in one common solvent and spread on various types of optical waveguides. In most cases, plasticisers are required in order to facilitate analyte diffusion and to speed up response times. The inherent problem, however, is that the sensor components have to be highly lipophilic or they are prone to leaching, especially when exposed to samples such as blood or serum. Furthermore, plasticisers tend to evaporate, causing migration, crystallization or decomposition of dyes and additives [18].

An alternative is to develop materials where all components (polymer, plasticiser and indicator dye) are covalently linked to each other. Such materials, namely copolymers made of acrylates and ionophores containing acrylate groups, have already been published for use in ion-selective electrodes [19,20] and optodes [21]. So-called self-plasticised copolymers with good flexibility and low glass transition temperature, T_g, are obtained by using methyl methacrylate and adding different amounts of alkyl acrylates [19,20]. The latter lowers the T_g of the resulting copolymers and increases the flexibility of the material (see Table 3).

A trifluoroacetylazobenzene reactand with a methacrylate group attached to the *N*-alkyl chain was synthesised (**ETH^T 4012**) and copolymerised with methyl methacrylate and butyl acrylate [22]. The response behaviour of membranes using this reactand linked to the polymer is comparable to that of the structurally related reactand **ETH^T 4001** dissolved in plasticised PVC. However, covalently immobilizing the reactands brings about response times which are 2 – 3 times slower. A similar decrease in response time was described for a calcium sensor where the indicator (Nile Blue) was linked to carboxy-PVC via an alkyl spacer [23]. Certainly there is a difference in chemical reactivity and diffusion, dependent on whether the indicator is dissolved in a solvent polymeric membrane or covalently immobilized via a spacer to a polymer chain.

The selectivity pattern of copolymer membranes is similar to that of membranes based on plasticised PVC, with a preference for primary amines over secondary and tertiary amines [22,24]. However, the copolymer-based sensor layers exhibit several advantages over PVC-based membranes, namely: (a) all components are copolymerised and thus no leaching is observed, (b) due to the covalent linkage, no crystallization, migration or aggregation of the dye

Table 3. Compositions of the chromogenic copolymers (MMA, methyl methacrylate; BA, butyl acrylate), and corresponding physical properties

Polymer	Ratio ETH^T 4012/MMA/BA [mol%][a]	T_g [°C]	M_n[b]	Forward response	Reverse response
MB10	0.4/99.6/0.0	121	16400	3–5 h	> 5 h
MB32	0.4/61.5/38.1	27	46000	10–15 min	20–30 min
MB23	0.4/39.6/60.0	−10	73900	5–10 min	15–20 min

[a] molar ratio of chromoreactand, methyl methacrylate and butyl acrylate; [b] M_n, number average molecular weight

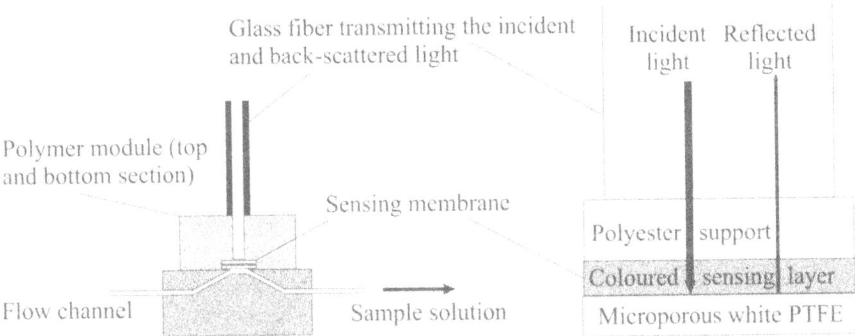

Fig. 6 Set-up of the flow-through cell connected to the optical fibre. The sensor layer is placed between top and bottom section of the module. Light is guided via the fibre to (and through) the sensor layer and is then reflected by the PTFE layer on top of the sensor layer. The reflected light is guided back to the detector via the fibre

is found, a behaviour often observed with optical sensor membranes based on plasticised PVC, and (c) the copolymers do not require plasticisers, thus enhancing the operational and shelf lifetime. Copolymer membranes composed of alkyl acrylates are generally rather unpolar but the use of more polar monomers such as (meth)acrylates with 2-cyanoethyl, *N*,*N*-diethylaminoethyl or 2-alkoxyethyl groups for copolymer preparation is possible in order to facilitate the extraction of polar analytes into the sensor layer. The mechanical stability of membranes based on copolymers and of those based on plasticised PVC is comparable.

In order to develop a device for simplified and rugged sensing of amines, a copolymer layer based on **ETH^{T} 4012** has been coated with microporous white PTFE serving both as a protection against mechanical stress and as a reflector of light directed onto the sensor layer. Additionally, the microporous PTFE layer is only permeable for neutral analytes and not for ionic species. Reflectance measurements have then been performed in a specifically designed flow cell connected via optical fibres to a diode array spectrophotometer (Fig. 6). This approach resulted in a sensor with operational and shelf life in the range of months [25]. In later experiments, the photometer was replaced by single LEDs and photodiodes in order to miniaturize the device.

3 Sensing Alcohols

3.1 Trifluoroacetylstilbenes

Ethanol is the main analyte in the wine- and brewing industries. Since various strains of yeast have become popular and are widely used in bio-transformations, there is also a need for monitoring ethanol as a parameter in flavour synthesis and other areas of biotechnology. The reactand 4-*N*,*N*-dioctylamino-

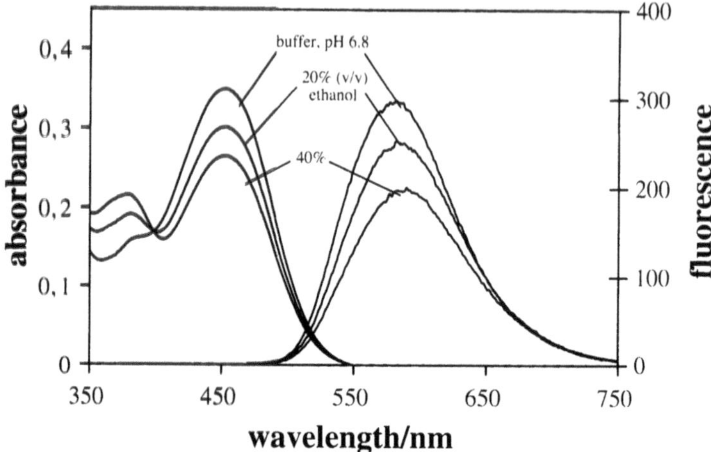

Fig. 7 Absorbance and fluorescence spectra of the ethanol sensor layer in contact with plain buffer and aqueous ethanol, showing the decrease in both absorbance and fluorescence of the trifluoroacetyl form and the increase in absorbance of the hemiacetal

4′-trifluoroacetylstilbene (**ETH^{T} 4004**) is a luminescent analogue of the amine-sensitive trifluoroacetylazobenzene derivative **ETH^{T} 4001** [26,27]. **ETH^{T} 4004** is considered to be advantageous over absorbance-based dyes because it is more easily adapted to fibre optics and it can be combined with self-referencing methods such as dual luminophore referencing in order to yield stable sensor signals [28–30]. On dissolving **ETH^{T} 4004** together with the catalyst tridodecylmethylammonium chloride (TDMACl) in plasticised PVC, a sensor layer for alcohols is obtained. A response of the trifluoroacetyl group to alcohols is also observed without TDMACl, however, with a reaction rate in the range of hours. The absorbance maximum of **ETH^{T} 4004** is located at 453 nm. Upon interaction with aqueous ethanol, the absorbance of the trifluoroacetyl form at 453 nm is decreased and a new absorbance of the hemiacetal at 373 nm is observed (Fig. 7).

A similar behaviour is found on measuring the fluorescence of the sensor layers (now coated with perm-selective microporous PTFE in order to avoid possible cross-sensitivity to pH). The fluorescence excitation maximum of the reactand in plasticised PVC is located at 452 nm, and the emission maximum at 576 nm (Fig. 7). Exposure to alcohol results in a decrease in fluorescence intensity at 576 nm, which correlates with the decrease in absorbance of the membrane at 453 nm. At the same time the conversion of the reactand results in an increase in the fluorescence of the hemiacetal form at 420 nm, which correlates with the increase in absorbance at 373 nm. The response is fully reversed on exposure to plain buffer. When measuring the luminescence of the trifluoroacetyl form at 576 nm, the relative signal change on going from 0 vol% to 40 vol% ethanol is as high as −35%, and response times are in the range of 5–10 min (Fig. 8). The layer exhibits comparable sensitivity to methanol while it is more responsive to the more lipophilic alcohols such as propanol and butanol.

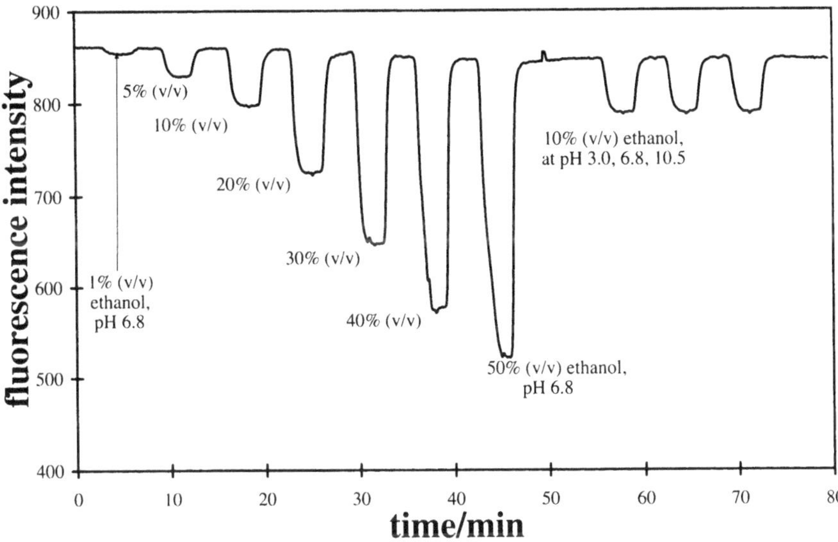

Fig. 8 Response of the ethanol sensor layer to aqueous ethanol solutions. The excitation and the emission wavelengths were set to 450 and 560 nm, respectively

Table 4. Measurement of alcohol content in beverages using the ethanol sensor layer based on ETH^T 4004

Sample	Measured using ETH^T 4004 in PVC/DOS [% (*v/v*)]	Reported [% (*v/v*)]
Chasselas de Geneve	10.7 ± 0.4	11.2
Chai de Bordes-Quancard	11.6 ± 0.2	11.5
Rioja Glorioso (red wine)	17.4 ± 0.4	12.5
Rioja Glorioso decolourised	12.9 ± 0.3	12.5
Baselbieter Kirsch	39.0 ± 0.2	37.5
Vodka Moskovskaya	40.4 ± 0.5	40.0

A sensor layer (coated with microporous PTFE) was used to measure the alcohol content of different wine and vodka samples. The results are shown in Table 4 and mostly correlate with the reported data. However, the measured ethanol concentration for red wine is significantly higher than the reported value. Therefore, red wine was investigated after decolourisation with activated carbon. Then, the ethanol concentration was found to agree with the reported value. This clearly shows that sample colours can cause problems. It seems that the coloured components of red wine diffused through the PTFE layer and quenched the fluorescence of the reactand. As a consequence, reactands (such as **CR-573**) with both absorbance and emission at longer wavelengths are required for real measurements.

The use of **ETH**^{T} **4001** and **ETH**^{T} **4004** for amine and ethanol sensing is based on the chemical reactivity of trifluoroacetyl derivatives with nucleophilic analytes. At first sight, such a general reactivity with nucleophilic species might limit practical application because of a low selectivity. However, there are several parameters that allow tailoring of the selectivity of the above sensors. Alcohols, for example, are usually encountered in bioreactors or during fermentation at neutral or slightly acidic pH. Under such experimental conditions, possibly interfering aliphatic amines are not reactive because they are protonated. In contrast, the measurement of amines is less affected by the presence of alcohols because alcohols are significantly less nucleophilic. Thus, the sensitivity to alcohols is usually smaller by a factor of thousand and the response to alcohols is slow without catalyst. Amines can be differentiated by their pK_a because aromatic amines are usually chemically reactive (unprotonated) at neutral pH whereas aliphatic amines have to be measured at pH values of above 10. Finally, the reactivity of trifluoroacetyl derivatives towards water (diol formation) can be restricted by the use of hydrophobic polymers such as poly(hexylmethacrylate) or a combination of PVC with specific plasticisers [16].

4 Sensing Aldehydes

4.1 Perylene Tetracarboxylbisimides

Aldehydes are present in disinfectants, are widely used for the preparation of resins and may be found in indoor air, in auto exhaust gases or rain water. The fluororeactand *N*-amino-*N*′-(1-hexylheptyl)perylene-3,4:9,10-tetracarboxyl bisimide can be used to detect aldehydes and ketones because it exhibits a reactive amino group. This amino group quenches the fluorescence of the perylene dye but if the amino group is converted into a Schiff base (via heating in dry organic solvents), dequenching of fluorescence occurs and an increase in fluorescence upon chemical reaction with aldehydes/ketones is observed [31]. The dye proved to be unsuitable for labelling aldehydes because the stability of the condensation products in organic solvents was low. However, the chemical instability of these products indicated a possible reversibility in polymer layers and consequently, the perylene dye was used for the preparation of optical sensors.

Sensor layers composed of the fluororeactand, PVC and tris(2-ethylhexyl)-phosphate were investigated for their response to buffered solutions of aqueous propionaldehyde. The sensor membrane exhibits only weak fluorescence upon exposure to plain buffer of pH 2.5. When the layer is in contact with aqueous buffer solutions containing propionaldehyde, a strong increase in fluorescence at 534 nm, 576 nm, and 613 nm is observed (Fig. 9). The fluorescence intensity of the reactand is dependent on the concentration of the aldehyde, and the response is fully reversible. The sensor membrane shows highest sensitivity to propionaldehyde in the 5 mM–100 mM range, with a LOD of 0.3 mM. A significant effect of sample pH on the response time is observed, indicating acid catal-

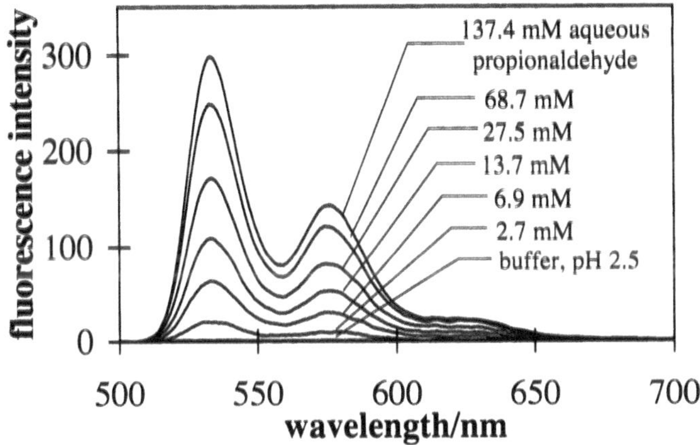

Fig. 9 Fluorescence emission spectra of a sensor layer composed of plasticised PVC and *N*-amino-*N'*-(1-hexylheptyl)perylene-3,4:9,10-tetracarboxylbisimide in contact with plain buffer and buffered solutions of propionaldehyde, all at pH 2.5. The excitation wavelength was set to 485 nm

Table 5. Selectivity coefficients $\log K^{opt}$ of the sensor membrane containing the fluororeactand J-57 towards aldehydes and ketones relative to propionaldehyde, and $\log K_{ow}$ of the respective aldehydes and ketones

Analyte	$\log K^{opt}$	$\log K_{OW}$	Analyte	$\log K^{opt}$	$\log K_{OW}$
formaldehyde	-1.8	0.35	glutaraldehyde	-0.9	N/A
acetaldehyde	-0.4	0.45	acetone	-1.5	-0.24
propionaldehyde	0.0	0.59	ethyl methyl ketone	-1.3	0.29
butyraldehyde	1.0	0.88	diethyl ketone	-1.6	0.82
glyoxal	<-3.0	N/A	ethanol	<-3.0	-0.30

ysis for the chemical interaction of the reactand with aldehydes in the membrane. The response time on changing from plain buffer to 137.4 mM propionaldehyde and vice versa is in the range of 2–6 min at pH 2.0, 15–45 min at pH 4.0 and 50–180 min at pH 6.0. However, the magnitude of the signal changes is not affected by pH.

While a Schiff base is formed on heating the perylene reactand with aldehydes in dry chloroform [31], it is thought that a hemiaminal rather than a Schiff base is formed in the sensor membrane because of the high water content under experimental conditions and because the reaction proceeds reversibly at room temperature. Unfortunately, due to rather low solubility of the reactand in the polymer layer, an IR-spectroscopic evaluation of the mechanism was not possible.

The selectivity pattern is comparable to that already observed with sensors for amines and alcohols based on **ETH^{T} 4001** and **4004**. Again a dependence of the response on the lipophilicity of the respective aldehyde is found in that highest sensitivity is observed for butyraldehyde, followed by propionaldehyde, acetaldehyde, glutaraldehyde, formaldehyde and glyoxal [32]. The sensitivity to all investigated ketones is relatively small (Table 5). This difference is due to the fact that ketones are less electrophilic than aldehydes and have a sterically demanding structure. Recently, a new chromoreactand for aldehydes (1-amino-4-(4-decylphenylazo)-naphthalene, **CR-418**) has been synthesized that deviates from this selectivity pattern in that it exhibits higher sensitivity to formaldehyde than to acetaldehyde or propionaldehyde. The resulting sensor membrane responds to formaldehyde in the 1 – 100 mM range with a limit of detection of 0.2 mM.

5
Conclusions and Outlook

Chromo- and fluororeactands represent a new approach to the detection of electrically neutral analytes with advantages such as absorbance and fluorescence in the visible spectral range, compatibility with cheap light sources and detectors, and large signal changes upon exposure to the analyte. The sensor layers exhibit good operational and shelf life, especially in the case of copolymer materials. Considering that chemical reactions are taking place rather than complexation processes, the response is usually quite fast and can be improved by appropriate catalysts.

Nevertheless, sensor characteristics such as selectivity and sensitivity still demand improvement. In this context, the preparation of molecularly imprinted polymers (MIPs) is one interesting approach [33]. Trifluoroacetylazobenzene dyes with methacrylate groups are promising candidates because the trifluoroacetyl group will be responsible for forming a covalent bond during the imprinting process while the comonomers record the analyte's shape during polymerisation. Releasing and binding of the analyte will then cause analyte-selective colour changes.

Molecular imprinting is thought to provide an enhancement in selectivity and sensitivity for analytes such as sugars, amino acid derivatives, drugs or neurotransmitters. Using more than one reactand in the imprinted polymer is expected to give multiple interactions with the analyte and colour changes at different wavelengths corresponding to these different interactions (to be evaluated by chemometrics).

Reactands may be used in optical artificial noses [34] but the development of more selective calorimetric and capacitive microsensors for electronic noses is promising as well, because chemical reactions yield changes in both the reaction enthalpy and the dipole moment [15]. The covalent immobilisation of reactands on AFM or SNOM sensor surfaces may be beneficial for a selective recognition of functional groups with a high local resolution.

References

1. Wolfbeis OS (ed.) (1991) Fiber Optic Chemical Sensors and Biosensors, CRC Press, Boca Raton, FL
2. Bakker E, Simon W (1992) Selectivity of ion-sensitive bulk optodes. Anal Chem 64: 1805-1812
3. de Silva AP, Gunaratne HQN, Gunnlaugsson T, Huxley AJM, McCoy CP, Rademacher JT, Rice TE (1997) Signalling recognition events with fluorescent sensors and switches. Chem Rev 97:1515
4. Wolfbeis OS (2000) Fiber optic chemical sensors and biosensors. Anal Chem 72:81R
5. He HR, Mortellaro MA, Leiner MJP, Fraatz RJ (1999) Optical sensor for determination of blood sodium based on a new fluoroionophore. Abs Pap Am Chem Soc 218:239; http: //www.roche.com/diagnostics/
6. Israelachvili J (1992) Intermolecular & Surface Forces, Academic Press Limited, London
7. Baker MEJ, Narayanaswamy R (1994) Development of an optical formaldehyde sensor based on the use of immobilized pararosaniline. Analyst 119:959
8. Gojon C, Dureault B, Hovnanian N, Guizard C (1997) A comparison of immobilization sol-gel methods for an optical chemical hydrazine sensor, Sens Actuators B-Chem 38: 154
9. Pringsheim E, Terpetschnig E, Piletsky SA, Wolfbeis OS (1999) A polyaniline with near-infrared optical response to saccharides. Adv Mater 11:865
10. Subrahmanyam S, Piletsky SA, Piletska EV, Chen B, Day R, Turner APF (2000) Bite-and-switch approach to creatine recognition by use of molecularly imprinted polymers. Adv Mater 12:722
11. Kuratli M, Pretsch E (1994) Sulfur dioxide-selective optodes. Anal Chem 66:85
12. Seiler K, Wang KM, Kuratli M, Simon W (1991) Development of an ethanol-selective optode membrane based on a reversible chemical recognition process. Anal Chim Acta 244:151
13. Mohr GJ, Demuth C, Spichiger-Keller UE (1998) Application of chromogenic and fluorogenic reactands in the optical sensing of dissolved aliphatic amines, Anal Chem 70: 3868
14. Sangster J (1997) Octanol-Water Partition Coefficients: Fundamentals and Physical Chemistry, Wiley-VCH, West Sussex
15. Mohr GJ, Zhylyak G, Nezel T, Spichiger UE, Kerness N, Brand O, Baltes H, Grummt UW (2002) Using reactands in CMOS-based calorimetric sensors: New functional materials for electronic noses. Anal Sci 18:109
16. Mohr GJ, Nezel T, Spichiger-Keller UE (2000) Effect of the polymer matrix on the response of optical sensors for dissolved aliphatic amines based on the chromoreactand ETH^T 4001. Anal Chim Acta 414:181
17. Dürselen LFJ (1989) Beitrag zur Untersuchung asymmetrischer Eigenschaften von PVC-Flüssigmembranen, PhD Thesis ETH No. 8927, ETH-Zurich, Zurich
18. Eugster R, Rosatzin T, Rusterholz B, Aebersold B, Pedrazza U, Ruegg D, Schmid A, Spichiger UE, Simon W (1994) Plasticizers for liquid polymeric membranes of ion-selective chemical sensors. Anal Chim Acta 289: 1
19. Heng LY, Hall EAH (1996) Methacrylate-acrylate based polymers of low plasticiser content for potassium ion-selective membranes. Anal Chim Acta 324:47
20. Heng LY, Hall EAH (2000) Producing self-plasticizing ion-selective membranes. Anal Chem 72:42
21. Peper S, Tsagkatakis I, Bakker E (2001) Cross-linked dodecyl acrylate microspheres: novel matrices for plasticizer-free optical ion sensing. Anal Chim Acta 442:25
22. Mohr GJ, Tirelli N, Lohse C, Spichiger-Keller UE (1998) Development of chromogenic copolymers for optical detection of amines. Adv Mater 10:1353

23. Rosatzin T, Holy P, Seiler K, Rusterholz B, Simon W (1992) Immobilization of components in polymer membrane-based calcium-selective bulk optodes. Anal Chem 64: 2029
24. Mohr GJ, Tirelli N, Spichiger-Keller UE (1999) Plasticizer-free optode membranes for dissolved amines based on copolymers from alkyl methacrylates and the fluoro reactand ETH^T 4014. Anal Chem 71:1534
25. Moradian A, Mohr GJ, Linnhoff M, Fehlmann M, Spichiger UE (2000) Continuous optical monitoring of aqueous amines in transflectance mode. Sens Actuators B-Chem 62: 154
26. Mohr GJ, Lehmann F, Grummt UW, Spichiger Keller UE (1997) Fluorescent ligands for optical sensing of alcohols: Synthesis and characterisation of p-*N,N*-dialkylamino-trifluoroacetylstilbenes. Anal Chim Acta 344:215
27. Mohr GJ, Spichiger-Keller UE (1997) Novel fluorescent sensor membranes for alcohols based on p-*N,N*-dioctylamino-4'-trifluoroacetylstilbene. Anal Chim Acta, 351:189
28. Huber C, Klimant I, Krause C, Wolfbeis OS (2001) Dual lifetime referencing as applied to a chloride optical sensor. Anal Chem 73:2097
29. Liebsch G, Klimant I, Krause C, Wolfbeis OS (2001) Fluorescent imaging of pH with optical sensors using time domain dual lifetime referencing. Anal Chem 73:4354
30. Mohr GJ, Klimant I, Spichiger UE, Wolfbeis OS (2001) Fluoro reactands and dual luminophore referencing: A technique to optically measure amines. Anal Chem 73:1053
31. Langhals H, Jona W (1998) The identification of carbonyl compounds by fluorescence: A novel carbonyl-derivatizing reagent. Chem Eur J 4:2110
32. Mohr GJ, Spichiger UE, Jona W, Langhals H (2000) Using *N*-aminoperylene-3,4:9,10-tetracarboxylbisimide as a fluorogenic reactand in the optical sensing of aqueous propionaldehyde. Anal Chem 72:1084
33. Haupt K (2001) Molecularly imprinted polymers in analytical chemistry. Analyst 126: 747
34. Dickinson TA, White J, Kauer JS, Walt DR (1996) A chemical-detecting system based on a cross-reactive optical sensor array. Nature 382:697

CHAPTER 4

Design, Quality Control and Normalization of Biosensor Chips

CLAUDIA PREININGER, URSULA SAUER

1 Introduction

With the completion of the human genome project biochip technologies have boosted and revolutionized automated genomic and proteomic analysis (www.microarrays.org; www.gene-chips.com) [1–7]. Based on conventional biomolecular techniques such as Southern and Northern blotting, sample preparation and assay was miniaturized by micromachining and microbiochemistry implemented efficiently by automated processes. To use biochip technologies for high throughput applications the system was adapted for high levels of parallelization. The potential of biochips lies in the parallel analysis of a huge number of probes, measured at once instead of one probe after the other. Such a technique speeds up biomolecular analysis tremendously. DNA chips have been widely used for gene expression, functional analysis, gene mapping and genotyping. Measuring RNA levels, however, might not give a complete or accurate description of a biological system. Because proteins mediate nearly all cellular activities, biochips have also been applied at the protein level ("proteomics") [8, 9].

The term "biochip" is derived from the term designating microelectronic chips produced by the photolithographic industry. While in the early days of biochip technology biochips were defined as devices made of semiconductor materials forming miniaturized electronic circuits, the term biochip has taken on several meanings over the years and has become more generic. Nowadays, a biochip is defined as a material or device that consists of a solid substrate providing reactive test sites and containing highly ordered grids of biomolecular probes for parallel, high throughput analysis. In the literature, various terms have been used to refer to biochips: biochip, DNA chip, protein chip, DNA microarray, protein microarray, biosensor array etc.

There are two categories of biochips:

- Biochips consisting of a solid or gel-type substrate: the biological reaction takes place on the chip; sampling, washing and detection are performed separately.
- Integrated biochips [10] or lab-on-chips, which have all the necessary elements built into one chip. Microfluidics, which can be described in the present context as fluid-based biochemistry reduced to very small volumes in order to facilitate large scale analysis on a chip, is the process implemented

in order to maintain adequate access to the solutions needed in each element of the chip.

There is a strong industry commitment to the development of biochip technologies for faster and more effective food control, water analysis and environmental restoration. Biochips can be used for the identification of toxic bacteria and organic pollutants in water, and as a preselection and screening method for genes that enable natural enzymes to metabolize and detoxify chemicals. Such genes can be transferred into common bacteria for remediation of contaminated land or water. In agriculture biochips are applied for disease and mutation detection to identify genes that increase the plants' resistance against e.g. mold.

In the last twenty years questions of environmental and industrial interest have been coming to the fore of research. The requirements for testing threats in complex samples are high sensitivity and specificity, together with a high throughput in a short time. Small sample volumes and low reagent demand call for the miniaturization of the analytical devices. The development of biochips takes these specifications into account; size reduction and the integration of several sensors in one device. Technologies which allow on-site use, additionally have to be easy to apply and inexpensive, and sample preparation should not be time-consuming.

The main field of application of biochips is gene expression profiling. As the detection of pollutants, toxicants and biowarfare agents in water, soil, and food products is of rising concern, the potential of biochip technology is explored for these applications [11]. In contrast to conventional techniques (Northern blotting, differential display, SAGE, and dotblot analysis), new technologies allow screening of high numbers of mRNA under numerous different conditions. The production of series of biochips facilitates comparative studies of a large number of samples [12].

Up to now, assessing the risk of transgenic agricultural products has been limited to comparisions of a small number of nutrients and known toxins, between the transformed line and the starting plant. Analysis of altered gene expression will give a broader insight into the effect of food components on the human intestine [12]. Another application of biochips is in the rapid identification of pathogens and food-spoilage bacteria and subsequently the development of new antimicrobial strategies. The mechanisms by which bacteria become resistant against antimicrobial substances can be analyzed by the use of biochips with sensitve and resistant strains [13]. Wu et al. [14] demonstrate the use of microarrays for the investigation of selected genes of microbial communities in marine sediments and soil samples. Diagnostic oligonucleotide microarrays, using variable regions of microbial 16S rRNA as probes, were employed for the detection and quantification of microbial populations in the natural environment [15, 16]. In the field of toxicogenomics, a discipline dealing with the question of how genes respond to a toxicant, chip technology provides faster, cheaper and more accurate results than animal testing [17].

Miniaturized immunoassays and enzymatic reactions were, for instance, developed for water pollution control [18, 19] and for the detection of dioxins, polychlorinated biphenyls and atrazine [20]. Rowe-Taitt et al. [21] developed a biochip for detecting hazardous bacterial strains and toxins. Antibodies specific for *Bacillus anthracis*, *Francisella tularensis* and *Brucella abortus* were immobilized on the surface of a planar waveguide, to capture antigen present in the sample. An array biosensor consisting of sandwich immunoassays was employed for the identification of bacterial, viral and protein analytes [22].

Taylor and Walt [23] developed a cell based biosensing array for measuring the response of cells to changes in the environment. By etching wells into the distal end of an optical immaging fiber, compartments for individual cells were produced. The location of the cells was determined by labels, each cell population was stained with a non-toxic, live-cell encoding dye.

2
Principle

A fluorescently labelled target is hybridized to the chip, consisting of a set of regularly arranged spots of biomolecular probes such as oligonucleotides, PCR-amplified cDNA or proteins. At positions on the array where the immobilized probe recognizes a complementary target, binding occurs which is then detected by a fluorescence scanner. The data output consists of a list of binding events,

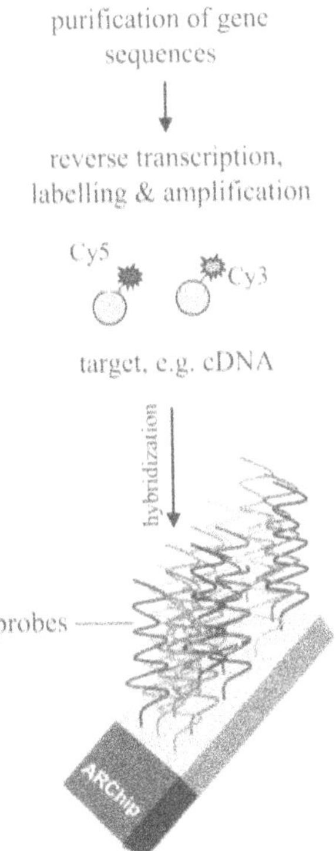

Fig. 1. Principle of biochip technology: sequencing-by-hybridization

indicating the presence or the relative abundance of specific targets in the sample. In Fig. 1. the principle of on-chip hybridization is shown.

Biochips can be used either for measuring differential expression between two populations [12] or for testing for the presence of a DNA sequence (resequencing) [24, 25]. Protein chips have been applied in expression profiling and antibody detection [26], binding specificities of a protein expression library and protein-protein interactions [27, 28].

3 Biochip Fabrication

3.1 Biomolecular Probes

Modified oligonucleotides, cDNA, gene fragments and PCR products are used as probes in DNA chips. Targets come either from genomic DNA, cloned DNA, cDNA or RNA. Fluorescently labelled targets are prepared by appropriate methods, for instance isolated DNA or RNA is amplified by PCR using fluorescent primer pairs. For gene expression studies, the fluorescent probes are usually produced from RNA by incorporating fluorescent nucleotides into complementary DNA (cDNA). 20 μM oligonucleotide probe and 50–250 ng/μL PCR product are recommended for chip immobilization. Lower concentrations result in hybridization signals that are too weak and thus in low sensitivity, whereas an overly high concentration sometimes causes comet tails produced by DNA that do not bind to the chip surface. For transcriptional profiling, oligos have to be designed to minimize cross-hybridization with any other DNA sequences that may be present in the sample. On the other hand, oligos for detection purposes have to hybridize to each member within the targeted group of sequences, thus these probes consist of regions of high sequence conservation [16]. For probe design software programmes are used, such as the ARB phylogenetic software, developed by Strunk and Ludwig (www.arb-home.de). As an alternative to oligonucleotides, peptide nucleic acids (PNAs) are used for diagnostics. PNAs exhibit greater binding affinity and specificity, and are capable of discriminating among single base mismatches [29].

Monoclonal and polyclonal antibodies serve as specific affinity ligands in protein chips. Though monoclonal/polyclonal pairs are more readily available than monoclonal/monoclonal pairs, polyclonal antibodies often cause higher background and lower sensitivity and specificity.

3.2 Array Manufacture

A chip array can be formed by
- *in situ* light-directed combinatorial on-chip synthesis
- arraying pre-synthesized biomolecular elements on pre-activated chip surfaces
- electronic addressing

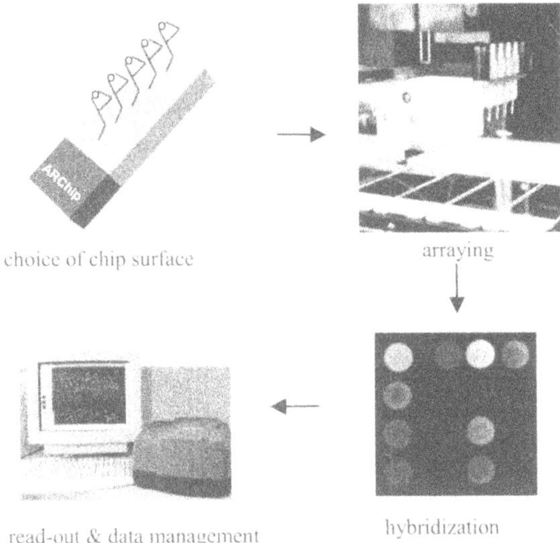

Fig. 2. Processes involved in a biochip experiment

Processes involved in chip manufacture and analysis are outlined schematically in Fig. 2.

Light-directed chemical synthesis [30, 31] (see Fig. 3.) makes use of photolithographic masks to define the chip exposure sites. In this process, the chip surface containing light-protective groups is selectively illuminated by light passing through a photolithographic mask. Deprotection of the illuminated sites leads to activation and chemical coupling of nucleosides at the specific test sites. The process is repeated several times in order to grow base chains and complete the multiple probe array. *In situ* light-directed synthesis can produce the highest packing density of all spotting techniques (at least one order of magnitude higher than conventional biochip arrayers). However, errors occurring in the synthesis process cannot be corrected, and, compared to printing techniques, the photolithographic method is rather expensive, since a different photolithographic mask is required for every DNA letter. By using 5′[2-(2-nitrophenyl)-propyloxycarbonyl]-2′-deoxynucleoside phosphoramidites instead of the Affymetrix chemistry, the yield during oligo synthesis on the chip surface can be improved by at least 12% per condensate reaction [32].

Non-contact biochip arrayers, commonly based on the piezoelectric effect, can apply controlled sub-nanoliter probe volumes to pre-specified locations on the chip surface. Due to the fact that the dispenser does not touch the surface, a non-contact arrayer provides low risk of contamination and is most suitable for printing on soft materials such as hydrogels. Contact arrayers make use of pins or capillaries to transfer probes to specific test sites on the chip. The Pin-and-Ring^{TM} technology by Affymetrix consists of a horizontal open ring and a vertical pin. During arraying, the ring is dipped into the probe – the probe is held within the ring by means of surface tension – and moved to the desired position on the chip. In order to make a spot, the pin is driven down through the ring and

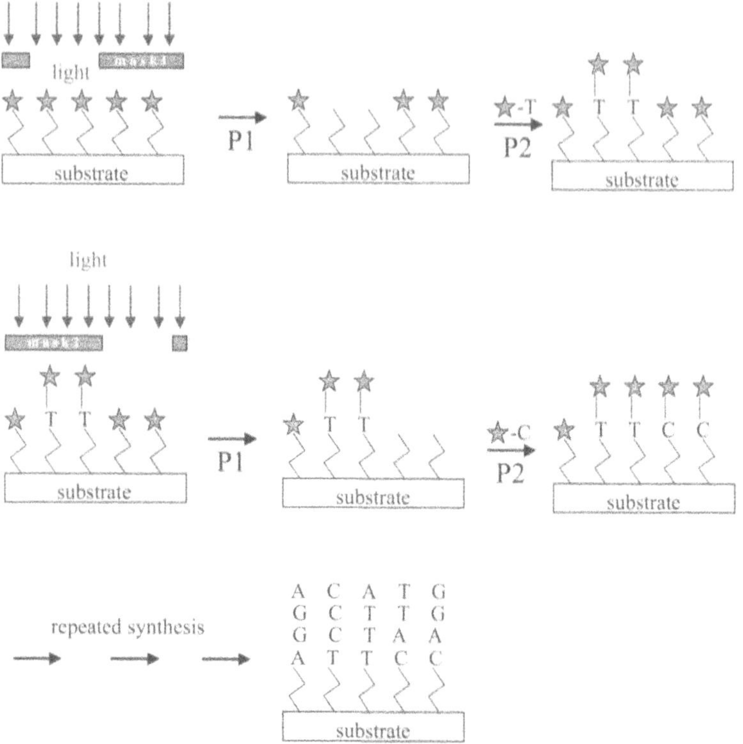

Fig. 3. Principle of light-directed *in situ* synthesis; P1: deprotection of reactive groups, P2: chemical coupling of DNA letter

a portion of the probe is transferred to the pin's end and deposited at the desired position on the surface.

The degree of uniformity among printed spots strongly affects the quality of data obtained from chip analysis. The spots must be of the same shape and size throughout a slide and from one slide to another. The uniformity of an array's grid depends to a critical extent on the arrayer's ability to precisely and accurately move to the desired location. The resolution of the *xy*-movement is about 10 μm. The *z*-axis placement, i.e. the distance between the pins and the chip surface, influences the spreading of the droplet. Another limiting factor of the array system is the set of pins and dispensers. It is truly difficult to get a matched set of printing tips that are uniform in size, shape and height. Some arrayers are equipped with enclosed filter, source and slide-plate cooling, as well as humidity and temperature control. In deciding what sort of arrayer is most suitable, reseachers are recommended to refer to several reviews on different arraying technologies and instruments:

www.biorobotics.com/; www.genemachines.com/;
www.lab-on-a-chip.com;
www.xenopore.com/products.htm.

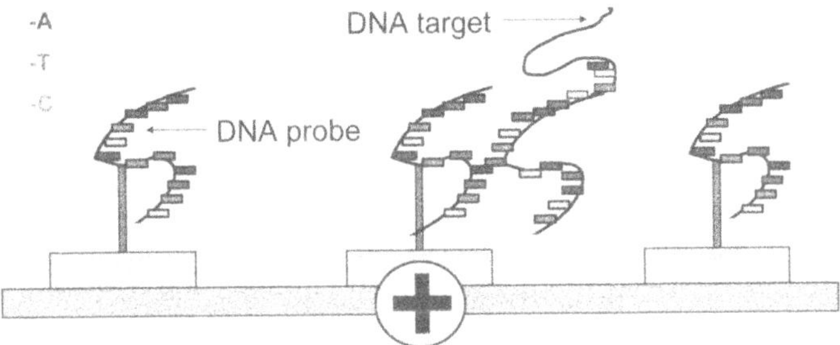

Fig. 4. Placement of negatively charged DNA to positvely activated test sites

Electronic addressing (see Fig. 4.), introduced by Nanogen (www.nanogen.com), is based on the fact that, negatively charged DNA moves toward specific test sites when positive voltage is applied. In contrast to conventional biochip analysis, in this case hybridization does not take place under conditions involving reaction rates controlled by temperature and salt concentrations in solutions, rather it is accelerated by applying an electrical field, resulting in a significant improvement in performance.

3.3
Slides and Immobilization

Several solid substrates (glass [33], porous membranes [34], polypropylene [35]) have been utilized for application in biochips. Glass offers a number of practical advantages, such as mechanical stability and low autofluorescence. Due to the non-porous character of glass chips, the volume of the hybridization solution can be kept to a minimum and probe-target interaction is not limited by diffusion into pores. However, three-dimensional microporous surfaces such as nitrocellulose [34], polyacrylamide [36] and poly(vinyl alcohol) [37] yield stronger signals and thus more consistent data than do two-dimensional glass surfaces.

Important criteria for biomolecule immobilization are high functionality of the chip surface, high immobilization capacity and density of the attachment, stable linkage between the biomolecule and the solid support, good accessibility for the interacting molecules, and low non-specific binding. Depending on the size of the probe, either covalent or electrostatic immobilization may be preferred. In general, oligonucleotides and DNA fragments of approximately 20 to 70 bases are amino-modified and bound covalently to the chip surface. Complete or partially complementary DNA of up to 5000 nucleotide bases is bound to the chip by electrostatic adsorption.

Commercially available glass chips provide reactive aldehyde, amino, mercapto or epoxy groups for covalent binding of DNA. To covalently bind oligonucleotides to reactive chip surfaces, the oligonucleotides are modified at their 5′end, most often with a primary amino group. However, thiol- and allyl-modified [36], silanized and acrylic-labelled [38], acrylamide-modified [39], hydrazide [40] and biotinylated oligonucleotides have also been synthesized. Synthetic oligonucleotides that provide an acrylamide group can be utilized for attachment to a thiol-functionalized surface, forming a stable thioether linkage. Zhao et al. [41] described the efficient binding of oligodeoxyribonucleotides modified with multiple phosphorothioate moieties when binding to bromoacetamidosilane-coated slides. Strategies for the covalent binding of pre-synthesized oligonucleotides to chip surfaces such as reactive glass slides, gold films, polyacrylamide gel pads, and polypyrrole films [42] are reviewed in [43]. In [44] plasma polymerization was used to form an anchorage layer for streptavidin immobilization. Using biotinylated DNA, a biochip was created that showed improved non-specific binding compared with poly-L-lysine coated glass chips. Zammatteo et al. [45] compared different immobilization chemistries on glass surfaces optimizing the coupling procedure and the hybridization efficiency. As a result of these studies, aldehyde-modified glass chips have come to be considered the reactive surface of choice.

The covalent attachment of disulfide-modified oligonucleotides to 3-mercaptopropyl silane-modified surfaces via a thiol/disulfide exchange reaction is an alternative method, allowing array densities of about 20000 spots/cm^2. The results of Rogers et al. [46] show that the hybridization efficiency was directly related to probe attachment density. In reference [47] the covalent and directed immobilization of DNA on glass-type oxide surfaces is described. A characteristic feature of the protocol is the deposition of DNA-droplets on a heated surface, which results in a more efficient coupling reaction (150–300 fmol/mm^2). Lindroos et al. [48] compared six different commercially available slides with respect to their fluorescence background, the efficiency of the attachment reaction, and signal-to-noise ratio. A seven times stronger hybridization signal was found on epoxy-slides than on aldehyde-, isothiocyanate-, mercapto-, or unmodified glass slides. Dolan et al. [49] reported on diazotized chip surfaces for the immobilization of unmodified oligonucleotides. They showed that the *p*-aminophenyltrimethoxysilane (ATMS)/diazotization chemistry developed was superior to commercial poly-L-lysine and silylated slides with respect to probe concentration and fluorescence background. However, in order to keep the diazonium salt stable, all processes involved in chip manufacturing have to be performed at 4 °C. A highly functional chip was constructed by using polyamidoamine (PAMAM) dendrimers containing 64 primary amino groups modified with a glutaric anhydride and *N*-hydroxysuccinimide layer for immobilization of amino-modified oligonucleotides. Compared with planar, amino- and epoxy-silanized surfaces, the hybridization signal was 8 to 10 times higher [50]. Sung [51] compared flat aminosilane layers with aziridine-polymerized layers. As expected, hybridization fluorescence was enhanced on aziridine surfaces due to the higher density of reactive groups. However, surface density was not found to have any effect on the discrimination of terminal and internal mismatched oligonucleotides. In general, surface density affects the extent of interaction between neighbouring probes as well as interaction between the immobilized probes and the substrate surface. High immobilization capacity results in a strong fluorescence signal, yet a reduced dynamic range.

Chips for non-covalent immobilization use nylon, poly-L-lysine and nitrocellulose as immobilization matrices, interfacial streptavidin-biotin layers or 3D link hydrogels, such as polyacrylamide. Cationic polyelectrolytes such as poly-

(diallyldimethylammonium) (PDDA) can be applied in order to electrostatically bind negatively charged DNA as well [52]. It is obvious from literature cited that the sensitivity and fluorescence background of biochips critically depend on the effective immobilization of the biomolecular probes on the chip. As functions of varying surface chemistries, probe concentration, probe length and print buffer need to be optimized. Slide autofluorescence, reproducibility of arraying, spot morphology, binding efficiency and probe purification are crucial for the accuracy and reliability of chip analysis data. The signal intensity depends mainly on the functionality of the chip surface, the immobilization capacity and the density of attachment as well as accessibility for the interacting molecules. Southern et al. [53] reported that the orientation and the packing density of the immobilized probes have a direct bearing on the hybridization process. In general, the immobilization of longer oligonucleotides (up to 70 bases) leads to a stronger hybridization fluorescence. Oligonucleotides on long spacers are lifted away from the surface and their neighbors and are therefore more accessible for interacting with targets. Clearly, the effect of spacer length on the hybridization signal is a result of the surface chemistry used for probe immobilization [54]. In most cases the optimal spacer length is a trade-off between the optimal signal intensity and the risk of cross-hybridizations. Long target sequences are likely to fold in on themselves and their bulk hinders them from approaching the surface. Short targets can better interact with an immobilzed probe and do not form duplexes as a result of intramolecular base pairing. In the ideal case, probe and target will have the same length.

For the immobilization of proteins [55–57], hydrogels which provide a three-dimensional, solution-like environment are needed in order to keep the protein spots hydrated at all times and to prevent the proteins from denaturing. Immobilized proteins may be denatured by hydrophobic or ionic interaction with the chip surface or by the potential energy stored at the air-liquid interface. In order to avoid denaturation and to bind proteins site-specifically to the surface without provoking steric hindrance, linker molecules, such as biotin, protein G or protein A or Fc-fusion products are used. Stabilizers such as glycerol are added to the print buffer in order to prevent the protein spots from drying during array fabrication and to maintain the protein's structure and activity. Stabilizers, however, can cause blocking of pins due to differences in viscosity and wettability. Humidity control during arraying is also a crucial parameter. Affinity chemistry methods, based on the reversible complexation of phenyl(di)boronic acid (P(D)BA) with salicylhydroxamic acid (SHA), were exclusively developed for protein immobilization (www.prolinx.com). Proteins can also be immobilized on poly(ethylene glycol)-modified surfaces via biotin-streptavidin interaction [58, 59]. A number of other approaches include hydrogels [36, 37] and dextran-based platforms [60] (www.biacore.com). Attempts at reducing non-specific binding involve the use of poly(ethylene glycol) [61] or blocking agents. When using recombinant proteins, amino- or carboxy-terminal tags can be introduced. His-tags, amino-terminal serine or threonine residues can be used for oriented protein immobilization, so that the biologically active site is lifted away from the surface, rendering it easily accessible for the target.

Eickhoff et al. [62] have described 2D/3D biochips which consist of a highly ordered grid of drops on a lipophilic chip surface. The drops serve as anchors for the incorporation and analysis of liquid samples and represent a virtually barrier-free three-dimensional reaction chamber for biochemical reactions.

4 Optical Read-out

In conventional chip experiments, fluorescence scanners are used for chip read-out. In the case of laser scanners, HeNe lasers are used as excitation sources and photomultiplier tubes as detectors, whereas CCD-based scanners use white light sources. The optical system can be confocal or non-confocal. There has been controversy among researchers as to which technology is superior. Cheung et al. [63] have shown that a non-confocal approach leads to the best signal quality as well as a good signal-to-noise ratio; in this procedure all of the photons are captured and all of the fluorescence reflected from the surface is detected. In the case of confocal fluorescence scanners, fluorescence occurs in the plane illuminated by the excitation light cone. The correct plane of focus being critical, the confocal method clearly reduces out-of-focus fluorescent light. Standard biochip experiments are performed using two fluorescent labels as reporter molecules. The most widely used fluorescent labels are Cy3 (λ_{ex} = 532 nm) and Cy5 (λ_{ex} = 635 nm) (Amersham Biosciences). Alternative indicators are the Alexa Fluor dyes from Molecular Probes and a series of bridged hemicyanine dyes, e.g. Dy-630-NHS from Dyomics. A third fluorescent label might be attached to the biomolecular probes and used to check the spot quality. Multiple laser excitation becomes necessary with simultaneous detection of single nucleotide polymorphisms (SNPs).

The background problem can be further overcome when using a surface-confined fluorescence excitation and detection scheme: at a certain angle of incident light, total internal reflection (TIR) occurs at the interface of a dense (e.g. quartz) and less dense (e.g. water) medium. An evanescent wave is generated which penetrates into the less dense medium and decays exponentially. Optical detection of the binding event is restricted to the penetration depth of the evanescent field and thus to the surface-bound molecules. Fluorescence from unbound molecules in the bulk solution is not detected. In contrast to standard fluorescence scanners, which detect the fluorescence after hybridization, evanescent wave technology allows the measurement of real-time kinetics (www.zeptosens.com, www.affinity-sensors.com).

Alternative optical methods not requiring any label are based on either surface plasmon resonance (SPR) (www.biacore.com) or reflectometric interference spectroscopy (RifS) (www.analytik-jena.de). SPR relies on the excitation of surface plasmons at a metal/liquid interface, reducing the reflected light intensity at a certain angle and wavelength. Biomolecular binding events cause changes in the refractive index at the surface layer and these are detected as changes in the SPR signal. Fluorescent targets can also be detected in a surface plasmon field if they are at a certain distance (a few nm) from the metal surface. Liebermann et al. [64] used surface-plasmon-field-enhanced fluorescence spectroscopy to detect the hybridization of a fluorophor-labelled oligonucleotide target. In order to keep losses due to energy transfer to the metal small, an interfacial layer of beyond 2 Förster radii based on the biotin/streptavidin chemistry is built up. The hybridization event could be quantitatively measured with a detection limit as low as 100 target molecules/μm^2. Care must be taken that the fluorophores attached to the target are placed within the exponentially decaying optical field of surface plasmons after binding and that the fluorophore is not in too close proximity to the metal surface in order not to quench a substantial part of the light. RifS is based on the interference of light beams reflected at interfaces with different refractive indices. Upon illumination by white light, the intensity of the reflected light is detected in dependence on the wavelength and an interference pattern is obtained. When binding occurs, the optical thickness changes. These changes lead to changes in the interference spectrum which can be detected [65, 66].

Though biochip technology is advancing and gaining increasing importance in life sciences, biochips suffer from insufficient sensitivity at low RNA concentrations and from a sometimes poor signal-to-noise ratio. By applying surface-enhanced fluorescence techniques (SEF) using metal nanofilms and nanoclusters, this problem can be overcome. The principle is based on the enhanced absorption and emission of a fluorophore when bound at a certain distance to a resonant layer of a metal or a semiconductor or both. On excitation the fluorescence of the fluorophor changes due to radiative losses of the molecular field to nonradiating plasmons in the metal. At zero distance from the metal, the fluorescence emission is completely quenched.

Mayer et al. [67] reported on a high throughput chip for detection of structural and conformational changes in DNA and proteins based on metal nanocluster resonance transducers. The chip consists of a reflecting mirror layer preferably made of an electron-conducting metal or cluster layer, a resonance layer made of a lipophilic polymer or an inorganic glass, a biointeraction layer, e.g. crosslinked proteins about 2–100 molecules thick (10–300 nm), and a layer of metal nanoclusters sputtered onto the top of the chip or adsorbed from aqueous solutions. Optimal resonance is achieved within 10–400 nm distance between cluster and mirror. Surface-enhanced light adsorption was measured as a shift of the maxima or as an increase or decrease of the reflectance at a defined wavelength.

In the place of fluorescent indicators, nanoparticles such as latex fluorescent nanospheres (commercially available), luminescent quantum dots and optically active metal nanoparticles are used for labelling DNA and protein probes. Compared to organic fluorophores, nanoparticle probes are more photostable and provide bright and steady fluorescence. However, 20-nm sized nanoparticles are much larger than fluorescent indicators, a fact that could cause problems due to binding kinetics and steric hindrance. Because each nanoparticle contains hundreds of dye molecules, nanoparticle probes are not suitable for chip detection using fluorescence resonance energy transfer (FRET). To overcome this problem 2–5 nm sized quantum dots can be applied, offering the advantages of size-tuneable emission, symmetrical spectra, and simultaneous excitation of different quantum dots at a single wavelength. Taylor et al. [68] reported on bioconjugated fluorescent nanoparticles crosslinked to the restriction enzyme EcoR1 via 1-ethyl-3-(3-dimethyl-aminopropyl)carbodiimide (EDAC) for targeting specific sequences of single DNA molecules. Köhler et al. [69] described oligonucleotide-modified gold nanoparticles with diameters of 15 to 60 nm as novel labels in biochip technology. Chip read-out was performed by detection of transmitted and reflected light. With this method 4-μm spots could easily be imaged.

DNA probes are attached to gold nanoparticles which bind to the target present in the sample solution. Upon binding the nanoparticle probe changes color. No amplification step is required. Taton et al. [70, 71] note that this type of technology could eliminate the need for PCR as a diagnostic tool. Using gold nanoprobe technology, the chips can be developed like photographs and thus do not require expensive optical set-ups. The silver in the photographic developing solution reacts with gold and amplifies the probe signal by as much as 100 000 times. Hybridized spots appear as grey spots in the scan. The darker the spot, the more target DNA is present.

Size-selected CdSe nanocrystals (18–70 Å) wrapped in several monolayers of ZnS show promise as new, bright labels in genomics and proteomics. Han et al. [72] reported the incorporation of semiconductor dots in polymer beads in controlled ratios and demonstrated the use of beads with three colors of quantum dots in a DNA hybridization assay. Semiconductor quantum dots (QDs) such as ZnS-capped CdSe nanocrystals allow multicolor optical coding of biochips by means of conjugating a biomolecular probe with the surface of a polymer bead containing an identification code in its interior. In contrast to previous studies using water-soluble quantum dots [73–75], hydrophobic quantum dots were incorporated into crosslinked beads formed by emulsion polymerization of styrene, divinylbenzene, and acrylic acid and applied to multiplex spectral coding. Quantum dots act as ideal fluorophores because their fluorescence emission wavelength can continuously be tuned by changing the particle size, and a single wavelength can be used for simultaneous excitation of different-sized QDs. In general, n intensity levels with m colors generate (n^m-1) unique codes. Han et al. [72] have shown that embedded QDs have optical properties similar to free QDs and that the ratio of the two fluorescence intensities equals the number of QDs per bead. The linear relation between the fluorescence intensity and the number of incorporated QDs confirms the lack of FRET among the QDs in the bead, which is a key requirement for multiplexed optical coding.

Quantum dots are nanometer scale particles containing nanocrystallites of zinc and cadmium sulfides and selenides that provide extraordinary optical properties. Quantum dots are photostable and emit absorbed light in multiple, resolvable colors varying according to size. Thus, the wavelength of both the incident and the emitted light can be tuned by the particle size, producing a whole set of colors ranging from ultraviolet to infrared. In contrast to fluorophores, a number of quantum dots can be excited simultaneously at a single wavelength using only one light source. Suitable light sources are lamps, lasers and light emitting diodes (LEDs). Because each quantum dot has the same excitation wavelength regardless of size, yet an emission wavelength that is size-dependent, these dots eliminate the need for multiple laser sources in microscopy set-ups. Furthermore, quantum dots show narrow, symmetrical emission spectra with minimal spectral overlap, which allow multiple colors to be detected with simple optics, resulting in turn in a significant decrease in cost and tremendous increase in sensitivity.

An increase in sensitivity and reliability of chip analysis can also be achieved by using fluorescence resonance energy transfer (FRET). For this purpose both the probe and the target are labelled with a fluorophore. The spectra of both fluorophores, the donor and the acceptor, are overlapping. When the emission spectrum of the donor, e.g. Cy5, overlaps with the absorption spectrum of the acceptor, e.g. Cy5.5, and the donor and the acceptor are at a certain distance from each other, energy is transferred from the donor to the acceptor on excitation of the donor fluorophor.

5 Quality Control

Variability in chip data can be caused by irreproducible probe arraying, uneven slides and inhomogeneous coating, sample processing, probe concentration, hybridization conditions, insufficient post-hybridization washing, poor incorporation and amplification of fluorescent labels, and by the scanning process. Being aware of the fact that there are numerous sources of variance, we will focus mainly on variances caused by arraying, quality of support and layer, hybridization conditions, and fluorescent labels.

5.1 Autofluorescence

The autofluorescence of the glass substrate at $\lambda_{ex} = 532$ nm and $\lambda_{ex} = 635$ nm, the reproducibility of the coating procedure and the manufacturing process, and the fluorescence background caused by experimental conditions such as blocking and washing [76] are crucial for the accuracy of chip analysis. Fig. 5 compares the autofluorescence of various commercial slides at $\lambda_{ex} = 635$ nm. As is obvious, porous matrices such as nitrocellulose (FAST), Nylon, and poly(vinyl alcohol) (ARChip Gel) cause the highest autofluorescence.

5.2 Arraying

Uniform chip surfaces are essential for printing high quality arrays with improved sensitivity. Dust particles and scratches on the chip surface, bent or broken arrayer tips or contaminants in the probe can cause irregular spot morphology which in turn can strongly affect chip analysis. In order to control spot quality, some arrayers are equipped with a CCD camera which continuously monitors and ensures the correct deposition of the probe on the chip. Due to the fact that laser light is scattered by DNA molecules, spots can also be visualized and checked, when scanning the unprocessed slide after printing. In a controlled experiment, outlying and poor quality spots along with problematic slides are filtered out.

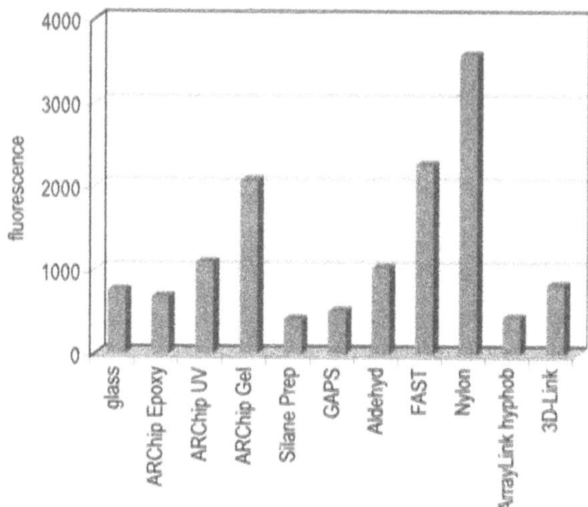

Fig. 5. Autofluorescence of bare glass (Sigma), ARChips-Epoxy, -UV, and -Gel, Silane Prep (Sigma), GAPS (Corning), Aldehyde (Telechem), FAST (Schleicher & Schuell), Nylon (Amersham), ArrayLink hyphob (Genescan) and 3D-Link (Amersham Biosciences) at λ_{ex} = 635 nm and 1000 V laser power

5.3 Print buffer

The choice of print buffer can drastically affect the hybridization efficiency and spot morphology. As a matter of different surface chemistries the optimum print buffer has to be determined for each kind of chip. Fig. 6 shows a 3D-view of a spot of dTAlf1b, a 16S rRNA sequence, printed in 3 × SSC, 0.1 N phosphate buffer, and 50% DMSO onto a nitrocellulose slide, and hybridized with *Rhizobium freddi*. The most homogeneous spots were obtained in 3 × SSC. Due to the destruction of the nitrocellulose, donut-like spots were observed in 50% DMSO.

5.4 Immobilization

The fact that spots of biomolecular probes are present on the chip does not necessarily mean that probe immobilization has been successful. The quality of probe immobilization can be tested, for example, by unspecific hybridization with fluorescently labelled DNA or by staining the array with nucleic acid binding fluorescent indicators. Several nucleic acid stains matching the commonly used filter sets are available.

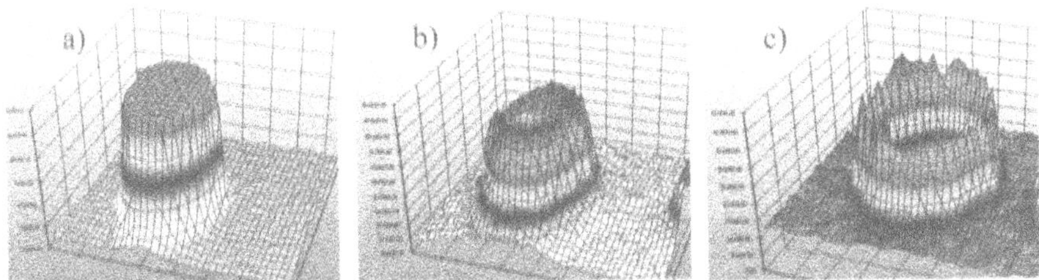

Fig. 6. 3D-view (Iconoclust™) of dTAlf1b-spots in a) 3 × SSC, b) 0.1 N phosphate buffer, and c) 50% DMSO

5.5 Fluorescent label

In order to make available a highly fluorescent target leading to a strong hybridization signal, an efficient and uniform incorporation and amplification of the fluorescent label is required. The most widely used fluorescent indicators in chip analysis are Cy3 (λ_{ex} = 532 nm) and Cy5 (λ_{ex} = 635 nm) (Amersham Biosciences). Alternative fluorophores are the Alexa Fluor dyes, developed by Molecular Probes, and a new series of bridged hemicyanine dyes developed by Dyomics. In [76] the extent of incorporation of Cy5 and Dy-630-NHS was compared by using 0.3, 0.5, and 1 μL of extracted *Rhizobium fredii* DNA as a template for the PCR reaction. As reported by the authors, Dy630 led to a higher yield of labelled DNA. Thus, beginning from the same amount of DNA, amplified DNA with a higher level of label incorporation and therefore higher fluorescence can be produced. As a result, less DNA is needed for chip analysis, which is of great importance, especially in medical diagnostics and cancer research, where probe material is very limited.

5.6 Validation

Due to the fact that biochip technologies are a new, not yet standardized high throughput technique, it is important to validate the obtained results. To distinguish between closely related sequences on expression arrays and encode protein isoforms quantitative RT-PCR can be used. Using RT-PCR the presence of mutants can be confirmed and the level of DNA expression can be determined. Relative expression levels of a gene can be compared by Northern blotting. To confirm, whether the expressed RNA reflects the protein level achieved in the cells, Western blotting is the method of choice [24]. Fig. 7 shows the hybridization profiles of various 16S rRNA probes hybridized with a fluorescent target on a) ARChip Epoxy, b) ARChip UV, c) nitrocellulose, d) aldehyde, e) aminosilane, f) epoxy slides. Signal intensities were computed as mean fluorescence of three replicate spots minus background. On nitrocellulose and ARChip Epoxy slides weak signals were also obtained for less specific probes indicating an enhanced sensitivity on these materials.

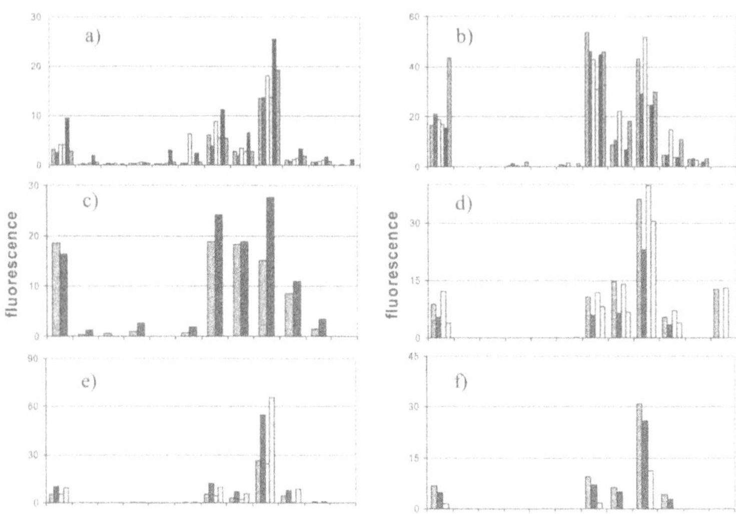

Fig. 7. Hybridization profiles on various slides: a) ARChip Epoxy, b) ARChip UV, c) nitrocellulose (FAST, Schleicher &Schuell), d) aldehyde (Telechem), e) Silane prep (Sigma), f) 3D-Link (Amersham Biosciences)

6 Data Collection and Analysis

Biochips produce huge data sets. Analysis of these data is a quickly evolving field, and there is still no consensus on methods and algorithms [77]. Data collected from microarray experiments are random snapshots with errors, inherently noisy and incomplete. Extracting meaningful information from thousands of data points by means of bioinformatics and statistical analysis is sophisticated and calls for collaboration among researchers from different disciplines. A well thought-out experimental design is the first step toward successful data mining. An increasing number of image and data analysis tools, in part freely accessible in the web, is available. Some examples are found in Tables 1 and 2.

6.1 Imaging

Microarray slides are imaged at one, two or more wavelengths with a fluorescence microscope, a charge-coupled device (CCD) camera or a laser scanner. The detector (e.g. a photo-multiplier tube, PMT) converts the incident photons into electrical current. Tagged Image File Format (TIFF) images are produced of the fluorescence intensities of each pixel for each fluorescence channel, converting the electrons into a sequence of digital signals. Each spot consists of a number of pixels. The signal correlates with the area density of dye molecules. The level of PMT voltage has to be adjusted so as to maximize the dynamic range of the instrument, but without the brightest pixels reaching saturation, indicating that

Table 1. Image analysis software

Product	Source	Web address
Tigr Spot-finder*	The Institute for Genomic Research	www.tigr.org/software/
ScanAlyze*	University of California Berkeley, Eisen Lab	http://rana.lbl.gov/EisenSoftware.htm
F-scan*	National Institutes of Health, Bethesda	http://abs.cit.nih.gov/fscan/
P-scan*	National Institutes of Health, Bethesda	http://abs.cit.nih.gov/pscan
UCSF Spot* [100]	UCSF, Jain Lab	http://jainlab.ucsf.edu/Downloads.html
GenePix Pro	Axon Instruments	www.axon.com
Iconoclust	Clondiag Chip Technologies	www.clondiag.com
ArrayPro	Media Cybernetic	www.mediacy.com
Spot	CSIRO Mathematical & Information Science	www.cmis.csiro.au/iap/spot.htm
QuantArray	Packard Bioscience	www.perkinelmer.com
ArrayVision	Imaging Research Inc.	www.imagingresearch.com/products/ARV.asp
ImaGene	BioDiscovery	www.biodiscovery.com/imagene.asp
Gene Traffic	Iobion Informatics	www.iobion.com/products/products.html

*Freely available to academic researchers and non-profit institutions

the photodetection device is being overloaded. In dual-color experiments the two pseudo-color images are merged for visualization purposes.

6.2 Image Analysis

Image analysis is a crucial step on the way to meaningful data. There are a numerous software packages available using different algorithms for spot characterization, most of them providing basic data mining tools as well (Table 1). A high quality spot is characterized by a high signal-to-noise ratio, stable spot size and regular shape. Intensity variations within a spot and spot homogeneity can be checked with the standard deviation of the mean pixel intensities and visualized by 3D views showing the intensity values of each pixel (IconoClust^{TM}). A regularly spaced mask (grid) is placed over the array and aligned with the features automatically or manually, though the latter method is prone to errors. In order to facilitate finding the position of the grid on the chip, guide dots may be added. The guide dots are located at a certain position and contain labelled material. The algorithms for spot image analysis include matched filtering, hierarchical methods and thresholding (e.g. TIGR_Spotfinder), seeded region grow-

Table 2. Data analysis software

Product	Source	Web address
Cyber T [92]*	University of California, Irvine	http://visitor.ics.uci.edu/genex/cybert
SNOMAD* [78]	Pevsner Lab, Johns Hopkins University	http://pevsnerlab.kennedykrieger.org/snomadinput.html
GeneViz*	Contentsoft AG	www.contentsoft.de/index.htm?geneviz.htm
MA-ANOVA* [88]	The Jackson Laboratory	www.jax.org/staff/churchill/labsite/software/anova/index.html
BASE*	Lund University	http://base.thep.lu.se/
AMIADA*	Hong Kong University	http://web.hku.hk/~xxia/software/amiada.html
DNA-Chip Analyser*	Harvard University, Wong Lab	www.dchip.org/
Genesis*	IBMT Graz University of Technology	http://genome.tugraz.at/Software/GenesisCenter.html
Cleaver*	Stanford Biomedical Informatics	http://classify.stanford.edu/
Data-Machine*	Dep. of Medicine, Boston University	http://people.bu.edu/strehlow/
SOTA*	alma Bioinformatica	www.almabioinfo.com/
CTWC* [101]	Weizmann Institute of Science	http://ctwc.weizmann.ac.il/
FASTA*	University of Virginia	ftp.virginia.edu/pub/fasta
GeneSpring	Silicon Genetics	www.silicongenetics.com/cgi/SiG.cgi/company/index.smf
Spotfire	Spotfire	www.spotfire.com/products/comp.asp
Acuity	Axon Instruments	www.axon.com/GN_Acuity.html

*Freely available to academic researchers and non-profit institutions

ing [78], and other methods. These should be robust against outliers and meet the problems of overlapping spots, contamination, strong background and varying spot size. The software should check the image automatically with regard to its quality: flagging, that is the marking of spots deemed bad, is a critical step in establishing data reliability. Reasons for flagging may be saturated measurement, dust, scratches, donuts, heavy background, or spots that are not at the expected location. The software has to be able to associate a feature on the array with a probe identity, and possibly with its sequence in a database.

6.3 Background

Estimating background intensities is necessary for the purpose of correcting non-specific hybridization. Reactive groups on the slide surface at sites where no probe is immobilized have to be blocked before hybridization. Otherwise, labelled targets and impurities will bind with these regions and result in a strong background. The composition of the blocking buffer, which effectively reacts with residual binding sites, has to be optimized for the surface chemistry used [79]. Autofluorescence of the support (nylon or glass) and of the layer has to be considered as well. Therefore, a quality control step is recommended: a subset of slides from each production batch should be scanned prior to printing and improperly coated slides should be excluded. Low uniform background is important for good signal-to-noise ratios. Several calculation methods are implemented in image analysis software packages for estimating background data. These are usually obtained from selected regions surrounding the spots, called local background, in contrast to a global background estimate for the entire array. Local background data may be biased positively by bright pixels belonging to a neighboring spot. Alternatively, background can be estimated from signals emitted by foreign array elements included for this purpose [80].

6.4 Quantification

The analysis tool computes the arithmetic mean or the median of the pixel intensities for each spot in both color channels. Median intensities are less susceptible to extreme values, whereas variability of the data can be estimated from mean intensities. Local sampling of background or spots from only buffer (negative controls) can be used to establish a threshold which a true signal must exceed, e.g. two standard deviations above background. Raw intensities or background-substracted intensities may be used for further analysis. Tran et al. [81] found that a threshold based on mean to median correlation of pixel intensities allows determination of more reliable signals. In dual-color microarray experiments, signal intensity ratios are calculated. Usually these ratios are $\log_2$ transformed, since this operation facilitates comparisons of levels of expression. For instance, a $\log_2$ ratio of +2 indicates a fourfold increase in expression, a log ratio of -2 corresponds to a four fold decrease.

A large amount of probe DNA per spot is a first essential toward establishing a linear relationship between measured fluorescence and the concentration of the target. The amount of DNA material ranges from 50 ng for nylon macroarrays to 1 ng or less for glass arrays [79]. The optimum probe concentration is a trade-off between optimal signal intensity and low availability of probes. Printing too highly concentrated probes causes comets and smeared spots.

Kane et al. [82] report a reproducible minimum detection limit of ≈10 mRNA copies/cell for both oligonucleotide and PCR probes: 50mer oligonucleotides and PCR probes, 322–393 bases in length with respect to sensitivity and specificity. Sensitivity was defined as the ability to detect a 3-fold change in RNA. The probes were designed to detect a subset of procaryotic transcripts spiked to eucaryotic total RNA at increasing levels, from 1 compared to 3 copies per cell to 10 000 compared to 30 000 copies per cell. Oligonucleotides and PCR probes did not differ in sensitivity. Data from microarray experiments is evaluated by comparing results with those derived using standard techniques, e.g. Northern hybridization [83].

6.5 Normalization

Several technical factors cause variance, disguising actual differences in signal intensities. Such sources of variation include measurement errors and print-tip effects. Control spots may be added to normalize print-tip effects caused by systematic variation in pin geometry: pins may print different amounts of a probe. Furthermore, the amount of a probe bound to the surface is unknown. Probe, target and array preparation and processing have an impact on data reproducibility. First of all, oligonucleotide arrays depend on the specificity of base pairing, thus hybridization conditions are critical [84]. High-quality hybridization should be reproducible and specific, leading to strong signals and minimum background. The efficiency of the hybridization reaction is affected by temperature, time, buffers and target concentration. Surface inhomogeneities may be a source of noise as well [85]. Optimizing factors involved in chip manufacturing, array design and hybridization, and elaborating reproducible protocols are important steps toward gaining more reliable data.

To account for differences in labelling and for quantum yield of the fluorescence labels, as well as for differences in the quantities of targets, normalization of the fluorescence intensities in each channel is necessary in dual color experiments [86]. Furthermore, normalization is needed for making comparisons across arrays and experiments. Methods for normalization have to be adapted to experimental designs and issues. At least the majority of systematic and random fluctuation should be eliminated by the normalization procedure. Global normalization is widely used in gene expression experiments, although it turns out to be the weakest method. This strategy assumes that the total mass of RNA labelled with the two dyes is equal, and subsequently the average ratio of an array should be set to one, thereby introducing a normalization factor, or the median of the distribution of log ratios is set to zero [87]. Alternatively, a subset of genes with constant expression, so-called house-keeping genes, may be used for normalization. In toxicological studies this method is limited due to the lack of genes not somehow regulated by a toxic compound. When employing the spiked controls method, synthetic DNA sequences or DNA sequences from another organism are spotted on the array and included in the two different samples in equal amounts. Thus, they should have equal red and green intensities and can be used for normalization. Another approach uses a function of intensities of added controls in increasing concentrations for normalization.

7 Statistical Analysis

Statistical analysis faces a number of problems, in particular, analysis of gene expression data is a big challenge. It must be born in mind that data gained from microarray experiments are estimates with a margin of error, in contrast to precisely known quantities, and that errors in these estimates have to be assessed through replication [88]. Perusal of the literature shows that levels of replication are usually low due to limited availability of targets. The

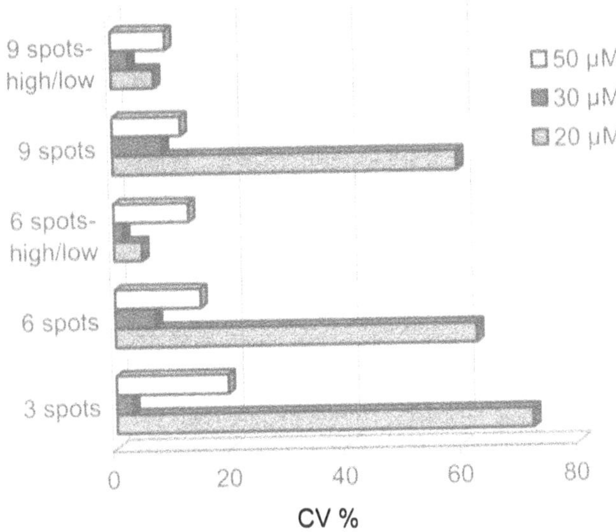

Fig. 8. Coefficients of variation of mean hybridization fluorescence of replicate spots of 20, 30 and 50 μM Alf 1b probe

number of replicate microarrays for each experiment should first be determined by statistical methods. Reproducibility of replicated experiments can be estimated by correlation analysis and represented graphically by plotting two data sets against each other. Generally speaking, correlation coefficients in the range of 0.8 and above are considered to be fairly successful. Another approach for extracting more robust data involves replicate spots for each sample. We recommend printing spots in replicates of three. Increasing the number of replicates improves reproducibility only, when the highest and lowest signal intensities are discarded (Fig. 8).

In statistical analysis, usually means or medians of the replicate spots and log-transformed signal intensity ratios are computed and normalized. Coefficients of variation or standard deviations serve as measures of the uniformity of individual pixel intensities. Standard errors of replicate experiments and spot morphologies allow spots worthy of further investigation to be filtered out. Normalized data may be filtered and grouped in order to reduce features before further analysis. So-called supervised methods use information (e.g. functional classes of genes, type of tissue, treatments) gained from outside the microarray experiment, while unsupervised techniques do not require additional information and therefore are used for more exploratory tasks [89]. Supervised analysis is employed for constructing classifiers which assign predefined classes to a given expression profile [90]. Statistically significant differences between treatments are frequently evaluated by means of t-statistics or analysis of variance (ANOVA).

In basic analysis of differential gene expression the observed fold change in expression is deemed significant if it is above an arbitrary threshold. More sophisticated approaches have recently been introduced. In a comparative study of the capability of means, t-statistics and an empirical Bayes method for revealing differential expression, the latter procedure resulted in the lowest error rates [91]. Long et al. [92] describe statistical tests based on analysis of variance and a Bayesian prior, which are implemented in a web interface called Cyber-T.

Clustering algorithms are a useful tool for identifying groups of related objects which may be genes as well as samples and combinations of both [89, 93, 94]. Hierarchical clustering [95], a procedure based on self-organising maps (SOMs) for the discovery and prediction of cancer classes [96], K-means [97], where the desired number of clusters has to be chosen by the user, and growing self-organising trees [98] are frequently used.

In another microarray design, namely high-density oligonucleotide arrays (GeneChip, Affimetrix), each perfect match (PM) probe is accompanied by a mismatched probe (MM) with a single base change right in the centre of the oligonucleotide as a control for non-specific hybridization. The array consists of subsets of 16-25 oligos for each gene; the intensities of these features have to be combined to make for a single value, e.g. by averaging the difference between PM and MM. For a stepwise description of statistical analysis of GeneChip data refer to [99].

References

1. Schena M (ed) (1999) DNA Microarrays: a practical approach. Oxford University Press
2. Case-Green SC, Mir KU, Pritchard CE, Southern EM (1998) Analysing genetic information with DNA arrays. Curr Opin Chem Biol 2:404-410
3. Wang J (2000) From DNA biosensors to gene chips. Nucl Acid Res 28:3011-3016
4. Sanders GHW, Manz A (2000) Chip-based microsystems for genomic and proteomic analysis. Trends Anal Chem 19:364-378
5. Lockhart DJ, Winzeler EA (2000) Genomics, gene expression and DNA-arrays. Nature 405:827-836
6. Kurian KM, Watson CJ, Wyllie AH (1999) DNA chip technology. J Pathol 187:267-271
7. Vo-Dinh T, Cullum BM, Stokes DL (2001) Nanosensors and biochips: frontiers in biomolecular diagnostic. Sens Actuators B 74:2-11
8. Weinberger SR, Morris TS, Pawlak M (2000) Recent trends in protein biochip technology. Pharmacogenomics 1:395-416
9. Mirzabekov A, Kolchinsky A (2001) Emerging array-based technologies in proteomics. Curr Opin Biotechnol 6:70-75
10. Vo-Dinh T, Alarie JP, Isola N, Landis D, Winterberg AL, Ericson MN (1999) DNA Biochip Using a Phototransistor Integrated Circuit. Anal Chem 71:358-363
11. Iqbal SS, Mayo MW, Bruno JG, Bronk BV, Batt CA, Chambers JP (2000) A review of molecular technologies for detection of biological threat agents. Biosens Bioelectron 15:549-578
12. van Hal NLW, Vorst O, van Houwelingen AMML, Kok EJ, Peijnenburg A, Aharoni A, van Tunen AJ, Keijer J (2000) The application of DNA microarrays in gene Expression analysis. J Biotechnol 78:271-280
13. Kuipers OP (1999) Genomics for food biotechnology: prospects of the use of high-throughput technologies for the improvement of food microorganisms. Curr Opin Biotech 10:511-516

14. Wu L, Thompson DK, Guangshan L, Hurt RA, Tiedje JM, Zhou J (2001) Development and Evaluation of Functional Gene Arrays for Detection of Selected Genes in the Environment. App Environ Microbiol 67:5780-5790
15. Guschin DY, Mobarry BK, Proudnikov D, Stahl DA, Rittmann BE, Mirzabekov AD (1997) Oligonucleotide Microchips as Genosensors for Determinative and Environmental Studies in Microbiology. App Environ Microbiol 63:2397-2402
16. Bodrossy L (2003) Diagnostic oligonucleotide microarrays for microbiology. In: Blalock E (ed): Microarrays and Bioinformatics for Beginners. Kluwer Academic Publishers, New York. In Press.
17. Nuwaysir EF, Bittner M, Trent J, Barrett JC, Afshari CA (1999) Microarrays and toxicology: the advent of toxicogenomics. Mol Carcinog 24:153-159
18. Barzen B, Brecht A, Gauglitz G (2002) Optical multiple-analyte immunosensor for water pollution control. Biosens Bioelectron 17:289-295
19. Weller MG, Schuetz AJ, Winklmair M, Niessner R (1999) Highly parallel affinity sensor for the detection of environmental contaminants in water. Anal Chim Acta 393:29-41
20. Shimomura M, Nomura Y, Zhang W, Sakino M, Lee K-H, Ikebukuro K, Karube I (2001) Simple and rapid detection method using surface plasmon resonance for dioxins, polychlorinated biphenyls and atrazine. Anal Chim Acta 434:223-230
21. Rowe-Taitt CA, Golden JP, Feldstein MJ, Cras JJ, Hoffman KE, Ligler FS (2000) Array biosensor for detection of biohazards. Biosens Bioelectron 14:785-794
22. Rowe CA, Tender LM, Feldstein MJ, Golden JP, Scruggs SB, MacCraith BD, Cras JJ, Ligler FS (1999) Array Biosensor for Simultaneous Identification of Bacterial, Viral, and Protein Analytes. Anal Chem 71:3846-3852
23. Taylor LC, Walt DR (2000) Application of High-Density Optical Microwell arrays in a Live-Cell Biosensing System. Anal Biochem 278:132-142
24. Christensen CBV (2002) Arrays in biological and chemical analysis Talanta 56:289-299
25. Hacia JG (1999) Resequencing and mutational analysis using oligonucleotide microarrays. Nature Genetics Suppl 21:42-47
26. Leuking A, Horn M, Eickhoff H, Buessow K, Lehrach H and Walter (1999) G Protein microarrays for gene expression and antibody screening. Anal Biochem 270:103-111
27. MacBeath G, Schreiber SL (2000) Printing proteins as microarrays for high-throughput function determination. Science 289:1760-1763
28. Schweitzer B, Kingsmore SF, Measuring proteins on microarrays (2002) Curr Opin Biotechnol 13:14-19
29. Lowe CR, Chemoselective biosensors. (1999) Curr Opin Chem Biol 3:106-111
30. Fodor SPA, Read JL, Pirrung MC, Stryer L, Lu AT, Solas D (1991) Light-directed, spatially addressable parallel chemical synthesis. Science 251:767-773
31. Southern EM, Maskos U, Elder JK (1992) Analyzing and comparing nucelic acid sequences by hybridization to arrays of oligonucleotides: evaluation using experimental models. Genomics 13:1008-1017
32. Beier M, Hoheisel JD (2000) Production by quantitative photolithographic synthesis of individually quality checked DNA microarrays. Nucl Acid Res 28:e11
33. Zhai JY, Wang C. Making DNA microarrays on glass slides, Axon Instr, www.axon.com
34. Stillman BA, Tonkinson JL (2000) FAST Slides: a novel surface for microarrays. BioTechniques 29:630-35
35. Beier M, Hoheisel JD (1999) Versatile derivatization of solid support media for covalent bonding on DNA-microchips. Nucl Acid Res 27:1970-77
36. Vassiliskov AV, Timofeev EN, Surzhikov SA, Drobyshev AL, Shick VV, Mirzabekov AD (1999) Fabrication of microarray of gel-immobilized compounds on a chip by copolymerization. BioTechniques 27:592-606
37. Preininger C, Chiarelli P (2001) Immobilization of Oligonucleotides on Crosslinked Poly(vinyl alcohol) for Application in DNA chips. Talanta 55:973-980

38. Kumar A, Larsson O, Parodi D, Liang Z (2000) Silanized nucleic acids: a general platform for DNA immobilization. Nucl Acid Res 28:e71
39. Rehman FN, Audeh M, Abrams ES, Hammond PhW, Kenney M, Boles TC (1999) Immobilzation of acrylamide-modified oligonucleotides by co-polymerization. Nucl Acid Res 27:649-655
40. Raddatz S, Mueller-Ibeler J, Kluge J, Wäß L, Burdinski G, Havens JR, Onofrey TJ, Wang D, Schweitzer M (2002) Hydrazide oligonucleotides: new chemical modification for chip array attachment and conjugation. Nucl Acid Res 30:4793-4802
41. Zhao X, Nampalli, Serino AJ, Kumar S (2001) Immobilization of oligonucleotides with multiple anchors to microchips. Nucl Acid Res 29:955-959
42. Livache T, Fouque B, Roget A, Marchand J, Bidan G, Téoule R, Mathis G (1998) Polypyrrole DNA-chip on a silicon device. Anal Biochem 255:188-194
43. Beaucage SL (2001) Strategies in the preparation of DNA oligonucleotide arrays for diagnostic applications. Curr Med Chem 8:1213-44
44. Miyachi H, Hiratsuka A, Ikebukuro K, Yano K, Muguruma H, Karube I (2000) Application of polymer-ambedded proteins to fabrication of DNA array. Biotechnol Bioeng 69:323-329
45. Zammatteo N, Jeanmart L, Hamels S, Courtois S, Louette P, Hevesi L, Remacle J (2000) Comparison between different strategies of covalent attachment of DNA to glass surfaces to build microarrays. Anal Biochem 280:143-150
46. Rogers YH, Jiang-Baucom P, Huang ZJ, Bogdanov V, Anderson S, Boyce-Jacino MT (1999) Immobilization of Oligonucleotides onto Glass Support via Disulfide Bonds: a Method for Preparation of DNA Microarrays. Anal Biochem 266:23
47. Jung A, Stemmler I, Brecht A, Gauglitz G (2001) Covalent immobilisation strategy of DNA-microspots suitable for microarrays with label-free and time-resolved optical detection of hybridisation. Fres J Anal Chem 371:128-136
48. Lindroos K, Liljedahl U, Raitio M, Syvänen A (2001) Minisequencing on oligonucleotide microarrays: comparision of immobilisation chemistries. Nucl Acid Res 29:e69
49. Dolan PL, Wu Y, Ista LK, Metzenberg RL, Nelson MA, Lopez GP (2001) Robust and efficient synthetic method for forming DNA microarrays. Nucl Acid Res 29:e107
50. Benters R, Niemeyer CM, Drutschmann D, Blohm D, Wöhrle D (2002) DNA microarrays with PAMPAM dendritic linker systems. Nucl Acid Res 30:e10
51. Sung JC, Chang KO, Joon PW (2002) Characteristics of DNA microarrays fabricated on various aminosilane layers. Langmuir 18:1764-1769
52. Wu LL, Zhou JZ, Luo J, Lin ZH (2000) Electrochim Acta 45:2923
53. Southern E, Mir K, Shchepinov M (1999) Molecular interactions on microarrays. Nature Gen Suppl 21:5
54. Steel AB, Levicky RL, Herne TM, Tarlov MJ (2000) Immobilization of nucleic acids at solid surfaces: effect of oligonucleotide length on layer assembly. Biophys J 79:975-981
55. Kodadek T (2001) Protein microarrays: prospects and problems. Chem Biol 8:105-115
56. Schaeferling M, Schiller S, Paul H, Kruschina M, Pavlickova P, Meerkamp M, Giammasi M, Kambhampati D (2002) Application of self-assembly techniques in the design of biocompatible protein microarray surfaces. Electrophor 23:3097-3105
57. Angenendt P, Glökler J, Murphy D, Lehrach H, Cahill DJ (2002) Toward optimized antibody microarrays: a comparison of current microarray support materials. Anal Biochem 309: 253-260
58. Piehler J, Brecht A, Valiokas R, Liedberg B, Gauglitz G (2000) A high-density poly(ethylene glycol) polymer brush for immobilization on glass-typer surfaces. Biosens Bioelectr 15:473-481
59. Birkert O, Haake H-M, Schuetz A, Mack J, Brecht A, Jung G, Gauglitz G (2000) A streptavidin surface on planar glass substrates for the detection of biomolecular interaction. Anal Biochem 282:200-208

60. Akkoyun A, Bilitewski U (2002) Optimisation of glass surfaces for optical immunosensors. Biosens Bioelectr 17:655-664
61. Schneider BH, Dickinson EL, Vach MD, Hoijer JV, Howard LV (2000) Highly sensitive optical chip immunoassays in human serum. Biosens Bioelectron 15:13-22
62. Eickhoff H, Schürenberg M, Nordhoff E (2001) 2D/3D-BioChips-Neue Werkzeuge für die funktionelle Genom- und Proteomanalyse. Transkript Nr.III
63. Cheung VG, Morley M, Aguilar F, Massimi A, Kucherlapati R, Childs G (1999) Making and reading microarrays. Nature Genet 21:15-19
64. Liebermann T, Knoll W (2000) Surface-plasmon field-enhanced fluorescence spectroscopy. Colloids Surfaces A 171:115-130
65. Piehler J, Brecht A, Gauglitz G, Zerlin M, Maul C, Thiercke R, Grabley S (1997) Label-free monitoring of DNA-ligand interactions. Anal Biochem 249:94-102
66. Sauer M, Brecht A, Charissé K, Maier M, Gerster M, Stemmler I, Gauglitz G, Bayer E (1999) Interaction of Chemically Modified Antisense Oligonuceotides with Sense DNA: A Label Free Interaction Study with Reflectometric Interference Spectroscopy. Anal Chem 71:2850-2857
67. Mayer C, Stich N, Palkovits R, Bauer G, Pittner F, Schalkhammer T (2001) High-througput assays on the chip based on metal nano-cluster resonance transducers. J Pharmac Biomed Analysis 24:773-783
68. Taylor JR, Fang MM, Nie S (2000) Probing specific sequences on single DNA molecules with bioconjugated fluorescent nanoparticles. Anal Chem 72:1979-1986
69. Köhler JM, Csáki A, Reichert J, Möller R, Straube W, Fritzsche W (2001) Selective labeling of oligonucleotide monolayers by metallic nanobeads for fast optical readout of DNA-chips. Sens Act B 76:166-172
70. Taton TA, Lu G, Mirkin CA (2001) Two-Color Labeling of Oligonucleotide Arrays via Size-Selective Scattering of Nanoparticle Probes. J Am Chem Soc 123:5164-65
71. Taton TA, Mirkin CA, Letsinger RL (2000) Scanometric DNA array detection with nanoparticle probes. Science 289:1757-1760
72. Han M, Gao X, Su JZ, Nie S (2001) Quantum-dot-tagged microbeads for multiplexed optical coding of biomolecules. Nature Biotechnol 19:631-635
73. Chan WC, Nie SM (1998) Quantum dot bioconjugates for ultrasensitive nonisotopic detection. Science 281:2016-2018
74. Mitchell GP, Mirkin CA, Letsinger RL (1999) Programmed assembly of DNA functionalized quantum dots. J Am Chem Soc 121:8122-8123
75. Pathak S, Choi SK, Arnheim N Thompson ME (2001) Hydroxylated quantum dots as luminescent probes for *in situ* hybridization. J Am Chem Soc 123:4103-4104
76. Preininger C, Sauer U (2003) Quality control of chip manufacture and chip analysis using epoxy-chips as a model. Sens Actuators B, 90: 98–103
77. Quackenbush J (2001) Computational Analysis of Microarray Data. Nature Reviews Genetics 2:418-427
78. Colantuoni C, Henry G, Zeger S, Pevsner S (2002) *SNOMAD* (Standardization and NOrmalization of MicroArray Data): Web-accessible Gene Expression Data Analysis. Bioinformatics 18:1540-1541
79. Sánchez-Carbayo M, Bornmann W, Cordon-Cardo C (2000) DNA Microchips: Technical and Practical Considerations. Current Organic Chemistry 4:945-971
80. Aharoni A, Vorst O (2001) DNA microarrays for functional plant genomics. Plant Mol Biol 48:99-118
81. Tran PH, Pfeiffer DA, Shin Y, Meek LM, Brody JP, Cho KWY (2002) Microarray optimizations: increasing spot accuracy and automated identification of true microarray signals. Nucl Acid Res 30:e54
82. Kane MD, Jatkoe TA, Stumpf CR, Lu J, Thomas JD, Madore SJ (2000) Assessment of the sensitivity and specificity of oligonucleotide (50mer) microarrays. Nucl Acid Res 28: 4552-4557

83. Khan J, Saal LH, Bittner ML, Chen Y, Trent JM, Meltzer PS (1999) Expression profiling in cancer using cDNA microarrays. Electrophoresis 20:223-229
84. Urakawa H, Noble PA, El Fantroussi S, Kelly JJ, Stahl DA (2002) Single-Base-Pair Discrimination of Terminal Mismatches by Using Oligonucleotide Microarrays and Neural Network Analyses. Appl Environ Microbiol 68:235-244
85. Schuchhardt J, Beule D, Malik A, Wolsji E, Eickhoff H, Lehrach H, Herzel H (2000) Normalization strategies for cDNA microarrays. Nucl Acid Res 28:e47
86. Hegde P, Qi R, Abernathy K, Gay C, Dharap S, Gaspard R, Earle Hughes J, Snesrud E, Lee N, Quackenbush J (2000) A Concise Guide to cDNA Microarray Analysis. Biotechniques 29:548-562
87. Yang YH, Dudoit S, Luu P, Lin DM, Peng V, Ngai J, Speed T (2002) Normalization for cDNA microarray data: a robust composite method addressing single and multiple slide systematic variation. Nucl Acid Res 30:e15
88. Kerr MK, Mitchell M, Churchill GA (2000) Analysis of Variance for Gene Expression Microarray Data. J Computat Biol 7:819-837
89. Raychaudhuri S, Sutphin PD, Chang JT, Altman RB (2001) Basic microarray analysis: grouping and feature reduction. Trends Biotechnol 19:189-193
90. Brazma A, Vilo J (2000) Gene expression data analysis. FEBS Letters 480:17-24
91. Lönnstedt I, Speed TP (2002) Replicated Microarray Data. Statistica Sinica 12:31-46
92. Long AD, Mangalam HJ, Chan BYP, Tolleri L, Hatfield GW, Baldi P (2001) Improved Statistical Inference from DNA Microarray Data Using Analysis of Variance and A Bayesian Statistical Framework. J Biol Chem 276:19937-19944
93. Alon U, Barkai N, Notterman DA, Gish K, Ybarra S, Mack D, Levine AJ (1999) Broad patterns of gene expression revealed by clustering analysis of tumor and normal colon tissues probed by oligonucleotide arrays. Proc Natl Acad Sci 96:6745-6750
94. Kerr MK, Churchill GA (2001) Bootstrapping cluster analysis: Assessing the reliability of conclusions from microarray experiments. Proc Natl Acad Sci 98:8961-8965
95. Eisen MB, Spellman PT, Brown PO, Botstein D (1998) Cluster analysis and display of genome-wide expression patterns. Proc Natl Acad Sci 95:14863-14868
96. Golub TR, Slonim DK, Tamayo P, Huard C, Gaasenbek M, Mesirov JP, Coller H, Loh ML, Downing JR, Caligiuri MA, Bloomfield CD, Lander ES (1999) Molecular classification of cancer: class discovery and class prediction by gene expression monitoring. Science 286:531-7
97. Tavazoie S, Hughes JD, Campell MJ, Cho RJ, Church GM (1999) Systematic determination of genetic network architecture. Nature Genet 22:281-285
98. Herrero J, Valencia A, Dopazo J (2001) A hierarchical unsupervised growing neural network for clustering gene expression patterns. Bioinformatics 17:126-136
99. Chu T, Weir B, Wolfinger R (2002) A systematic statistical linear modeling approach to oligonucleotide array experiments. Math Biosci 176:35-51
100. Jain AN, Tokuyasu TA, Snijders AM, Segraves R, Albertson DG, Pinkel D (2002) Fully Automatic Quantification of Microarray Image Data. Genome Res 12:325-332
101. Getz G, Levine E, Domany E (2000) Coupled two-way clustering analysis of gene microarray data. Proc Natl Acad Sci 97:12079-12084

CHAPTER 5

Rapid, Multiplex Optical Biodetection for Point-of-Care Applications

FRANK Y. S. CHUANG, BILL W. COLSTON, JR.

1 Need for Advanced Biodetection

Despite hopes that antibiotic and vaccine therapies might one day lead to the complete eradication of infectious disease [1], new and emergent pathogens continue to pose a global threat to public health [2]. Even in the present day, infectious disease remains among the top ten causes of death in the United States for all age groups [3]. In part due to increased virulence of recent strains, but also due to a growing cohort of individuals susceptible to severe infection, the U.S. mortality rate caused by influenza has risen to roughly 36,000 per year, exceeding the current rate of HIV/AIDS related deaths in the country [4, 5]. While the victories over bubonic plague (*Yersinia pestis*), whooping cough (*Bordetella pertussis*), polio and smallpox (variola) are clearly significant, new diseases represented by human immunodeficiency virus (HIV), as well as the tragic reality of biological agents used as weapons of terrorism and mass destruction [6] – offer sobering evidence that the battle against infectious disease is far from over [7].

More than ever, the task of effectively controlling communicable diseases ultimately rests on the ability to rapidly and accurately detect infection at the earliest stages. Rapid diagnostics are crucial for identifying outbreaks of infectious disease which could potentially escalate to a full-scale epidemic or global pandemic. Various surveillance programs and strategies have been proposed or are already in place to monitor influenza at state and national levels [8, 9]. At the lower (but no less critical) local level, hospitals are concerned with highly contagious and potentially lethal nosocomial infections which can emerge or enter the clinical setting. Antimicrobial resistance in tuberculosis and methicillin-resistant *Staphylococcus aureus* (MRSA) pose particular problems since effective vaccination and/or antibiotic treatments are either limited or nonexistent [10]. Rapid identification to contain such diseases is essential for reducing the clinical workload at metropolitan hospitals [11]. In the event of a sudden or unexpected epidemic, vaccines and antibiotics may be in short supply, and so timely assessment and triage of individuals suspected of infectious disease exposure is critical for appropriate allocation of valuable resources.

Another rationale for rapid detection of infectious disease is based on the pharmacokinetics of new antimicrobial agents. For example, Tamiflu® (oseltamivir) is shown to be effective in shortening the clinical course of influenza, but only when administered within 48 hours of the onset of symptoms [12]. Since standard laboratory culture and analysis requires up to 10 days for definitive results, a primary-care physician obliged to treat an acutely ill patient must make an "educated guess" based on the clinical setting, the patient's signs and reported symptoms. While most viral respiratory infections are generally self-limited, respiratory tract infections (RTI) which are bacterial in origin can be life-threatening without appropriate treatment [13, 14]. Nevertheless, in a recent survey conducted in the U.K., primary-care physicians lacking clinical serologic data were able to correctly rule-out bacterial RTIs only 60% of the time, and correctly distinguish bacterial from viral RTIs only 50% of the time [15].

Apart from speed and accuracy, we propose that multiplex capability in a biodetection platform is a practical necessity for today's diagnostic needs and applications. In the context of infectious disease detection, the list of causative pathogens which must be considered for any given clinical presentation grows with the emergence of new microbial species or strains and the introduction of potential biothreat agents into the civilian population [16]. Failure to include such "orphan" pathogens in the differential diagnosis only potentiates the risk and devastating health impact associated with these new and unfamiliar diseases [17]. However, the cost in both time and resources of performing serial diagnostic tests using conventional laboratory methods would be prohibitive and used only as an option of last-resort. Modern molecular in-vitro diagnostic techniques (which we will describe in greater detail) can be scaled-down to work with extremely small sample quantities. Moreover, these assay methods can be reduced and configured into "microarrays" that allow hundreds to thousands of individual tests to be performed on a single biological sample.

Finally, in the context of safeguarding the public health against the spread of infectious disease, we believe that a critical need exists for biodetection technology which can be scaled for portable or "point-of-care" applications. While rapid, multiplex in-vitro diagnostic instruments are readily available, the majority of these commercial systems emphasize performance over portability – incorporating delicate optics, electronics, and/or microfluidics, which ultimately relegates their use to clinical or research laboratories only [18]. A portable, albeit simpler, diagnostic instrument could conceivably play a vital role in environmental testing or field medicine, in which access to laboratory facilities is restricted or impractical for transport of contagious or contaminated materials [19]. A simpler, more economical diagnostic system offers a practical solution for widespread screening of individuals in large populations, enabling a more rapid and direct response to the emergence of infectious disease.

2 Fundamental Principles of Biodetection

The current "state of the art" in biodetection encompasses an impressive array of diagnostic instrumentation to identify and measure minute quantities of biological material on the molecular scale [20]. In the context of infectious disease, the target material can consist of the whole pathogen itself, or components – such as protein, lipid or genetic material derived from the microorganism. Infectious disease can also be characterized by changes which occur in the affected individual, particularly when the host immune response is triggered. Changes in the serum concentration of certain host biomolecules, including cytokines and acute phase reactants, may reflect an active infectious disease process and thus represent important markers for detection [21]. Most biodetection devices currently use one of two basic molecular detection methods: nucleic acid hybridization and immunoassay.

Nucleic acid hybridization utilizes the strong binding affinity between complementary strands of DNA or RNA to capture and detect specific genetic material in complex biological solutions. This method is frequently used in conjunction with a laboratory procedure known as polymerase chain reaction (PCR, Fig. 1), to quickly replicate select genetic sequences which may be found in solution. In principle, PCR amplification of genetic material greatly improves assay sensitivity, since even a single strand of DNA could be amplified in vitro to easily detectable levels.

It should be noted, however, that PCR does not indiscriminately amplify all DNA sequences in solution. Rather, PCR amplification occurs when "primers" containing short sequences of nucleic acid hybridize to longer strands of DNA in a test solution. These primers ultimately define the terminal endpoints of the DNA chain which is replicated by PCR. Therefore, the process of PCR amplification itself plays a role in selecting specific genetic material from complex biological mixtures.

The quality of a DNA assay using PCR and hybridization depends essentially on the specificity of the primer and probe sequences used in the assay. Genetic sequences which uniquely identify a particular pathogen become known

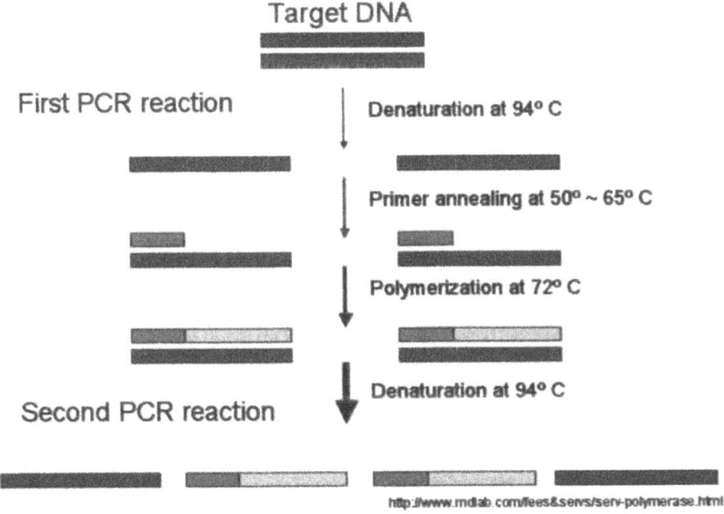

Fig. 1. Gene amplification by polymerase chain reaction (PCR, Schematic adapted from http://www.mdlab.com/fees&servs/serv-polymerase.html)

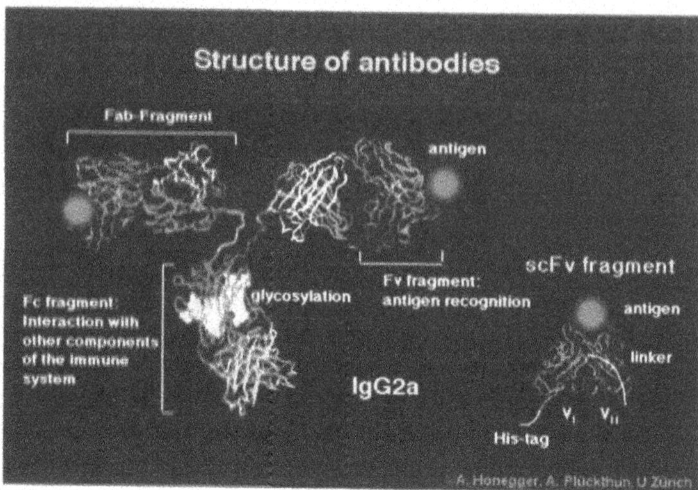

Fig. 2. Molecular/functional structure of antibodies (http://www.unizh.ch/~pluckth/slide_shows/Slides/Miniantibodies/index.htm)

as "signature" or "fingerprint" sequences for that microorganism. Signature sequences do not share homology with other genes or sequences which might also appear in the test fluid. Moreover, PCR technology has been miniaturized to the handheld scale, enabling the development of new point-of-care DNA detection instruments [18].

For the detection of non-genetic biomolecules, such as proteins, sugars and lipids, immunoassays are the method of choice. Antibodies (also known as immunoglobulins, Fig. 2) constitute a broad class of proteins whose natural function is to recognize and bind other biomolecules (antigens).

Antibodies play a critical role in the host defense, in part by monitoring the blood circulation for the presence of foreign (and possibly harmful) material that enters the body. Antibodies are sensitive and specific for their target antigen and, upon binding, can recruit other components of the immune system to respond against the foreign agent. While the pool of naturally-occurring antibodies in circulation are randomly generated and carry different specificities, certain antibody "clones" will predominate, depending on the history of recent acute exposures to selected foreign antigens. These monoclonal antibodies can thus be harvested from the blood serum and subsequently utilized for laboratory-based in vitro detection of the target antigen.

In general, DNA hybridization assays and immunoassays are complementary techniques. But since antibodies can be raised against nearly any type of biomolecule – including DNA – immunoassays have a broader range of application. (To be fair, DNA *aptamers* represent a novel application of synthetic oligonucleotides whose tertiary structure enables sensitive and specific binding to other biomolecules.) By selecting antibodies which are specific for epitopes which are exposed on the pathogen surface, immunoassays can be directly performed on biological or clinical samples, without prior sample preparation.

3
Development of Optical Methods for Biodetection

Bacteria are considered to be the smallest living cells and range in size from 0.1 to 10 μm. Traditionally these microorganisms are identified by direct visualization using light microscopy with Gram stain. The staining process reveals biochemical characteristics of the bacterial cell wall and facilitates the proper identification of bacterial species. In contrast to bacteria, viruses are much smaller (100 nm or less) and require enhanced techniques, such as electron microscopy or fluorescent labeling, for adequate visualization.

Laboratory culture plays a key role in standard detection methods. By allowing the pathogen to multiply and replicate, one effectively amplifies the signal to be detected by bioassay. Furthermore, laboratory culture itself serves as a functional assay to determine pathogen viability. Viruses, for example, are too small to be visualized by light microscopy. However, when cultured on cell monolayers, live viruses will infect and lyse cells, creating voids or "plaques" which are readily (in)visible on the the culture dish. Despite its establishment as a reference standard, laboratory culture and direct microscopic identification is a relatively costly procedure, both in time and resources. For this reason, the method is reserved for clinical cases in which more serious bacterial infections are suspected. For general or routine screening purposes a much less cumbersome diagnostic technique is necessary.

Representing a first step towards this goal, direct fluorescent assay (DFA) kits utilize specific antibodies to label various pathogens on a microscope slide. Different fluorophores covalently conjugated to different monoclonal antibodies enables several pathogens to be tested simultaneously, and the results interpreted by the color-coding determined by the fluorescent conjugation scheme. However, these kits can only be used by trained laboratory technicians and are again impractical for high-throughput applications.

3.1
Sandwich Immunoassays – ELISA

Immunoassays can be performed in various configurations; one of the most commonly used is called a *sandwich* immunoassay. The term is derived from the arrangement of stacked layers which form in the presence of the target antigen. The base layer is comprised of antibodies which are fixed onto a solid substrate. When this surface is exposed to a biological fluid specimen, the antibodies serve to capture target antigens which may be present in solution. After the treated surface is washed, the presence of bound antigen is demonstrated by incubation with detection antibodies which are labeled with either a fluorescent dye or a chromogenic enzyme. The *enzyme-linked immunosorbent assay* (ELISA, Fig. 3) has become a laboratory reference standard for measuring dilute quantities of target antigen in solution [22].

Optimized ELISA tests utilize immunoassay antibodies to the fullest extent, measuring antigen concentrations in the picomolar range. The cost of developing this level of sensitivity into an ELISA assay is a significant investment of

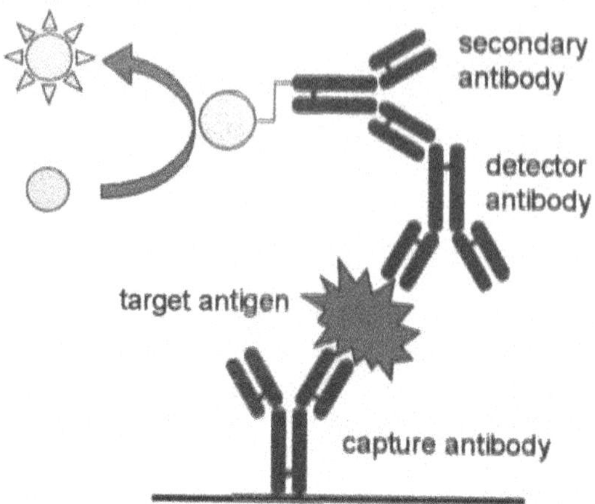

Fig. 3. Basic components of ELISA sandwich immunoassay

time and effort. ELISA assays require extensive pre-calibration with laboratory standards, and generally take half a day for results. While this is clearly an improvement over standard laboratory culture and analysis, it is still too costly to be considered for routine screening.

3.2
Lateral Flow Assays – "Strip" Tests

In order to gain speed in testing, ELISA sandwich immunoassays have been reduced to the form of lateral flow assays, commonly known as "strip" tests. Various immunoassays have been commercialized utilizing this format: for example, the i-STAT 1 handheld blood analyzer, the QuickVue influenza test by Quidel Corporation Ltd. UK, and the ZstatFlu™ test by ZymeTx, Inc. Arguably the most successful commercial implementation of the lateral flow assay concept are "home pregnancy" tests, which use monoclonal antibodies to detect the presence of human chorionic gonadotropin (hCG) in urine.

A particularly refined example of a lateral flow immunoassay is the chromatography-based handheld "smart ticket" device (Fig. 4) which is a key component of the Joint Biological Point Detection System (JBPDS) developed by the United States military (in collaboration with Canada and the United Kingdom) to detect biological warfare agents remotely.

Each JBPDS smart ticket is designed to detect one selected type of bioagent. Blue latex particles are coated with detector antibody, in order to attract and bind any targeted bioagent that may be present in a liquid sample. To determine whether any bioagent was actually found, the detection microparticles are allowed to flow across a nitrocellulose membrane strip. The strip is coated with two lines of capture antibodies: one directed also at the targeted bioagent (T) and one negative control (C) which is an antibody that directly binds to the detector antibody coated on the latex microparticle. The two lines of capture antibodies

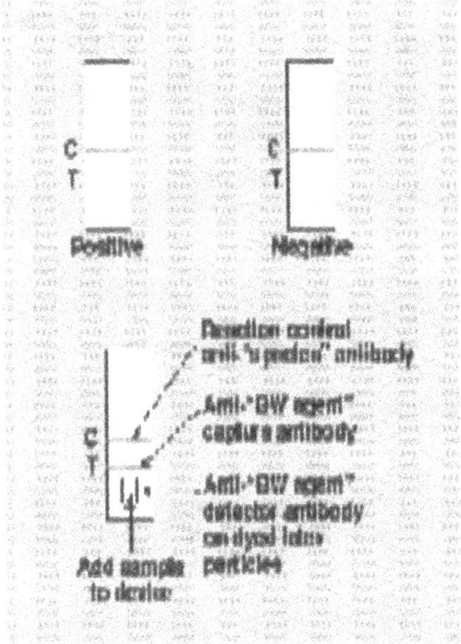

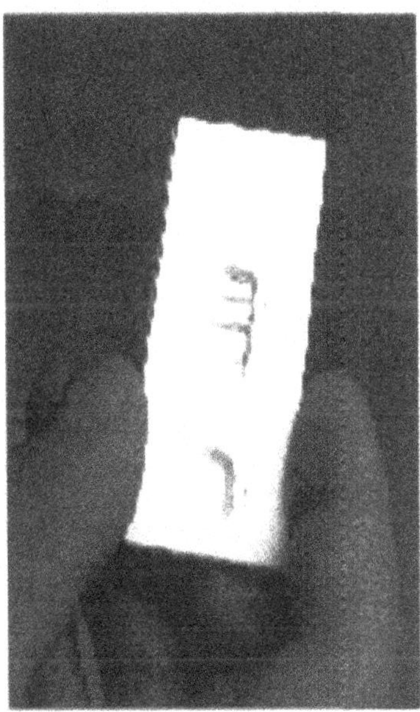

Fig. 4. JBPDS smart ticket

are initially colorless, but upon exposure to the suspended microparticles, will form either one blue line (negative result) or two (positive result), depending on the presence or absence of bioagent attached to the microparticles.

In order to expand the scope of the smart ticket device to include multiplex analysis, multiple smart tickets must be employed. The JBPDS platform uses a mechanical carousel with automated fluidics to run up to 9 smart tickets simultaneously. It should be noted that in such an arrangement, the biological sample must be divided into 9 fractions, which may adversely affect detection sensitivity.

In general, the measurements reported by lateral flow assays are less quantitative than laboratory-based ELISAs. However, for end-applications such as infectious disease screening, the device needs not so much to be precise as it should be accurate. The threshold for positive detection can be tuned to a value which is clinically relevant for proper management of the infected individual. The goal of developing a biodetector which is more rapid, more convenient and less expensive to use is to promote widespread and more frequent screening of the population to safeguard public health, and to distinguish the "worried-well" individuals from those who truly need immediate medical attention.

3.3 Fixed Microarrays – DNA Gene Chip

Advances in engineering technology have enabled new devices to operate on scales much smaller than previously thought possible [23]. As a result, both immunoassays and DNA-based tests can now be configured as microarrays, in

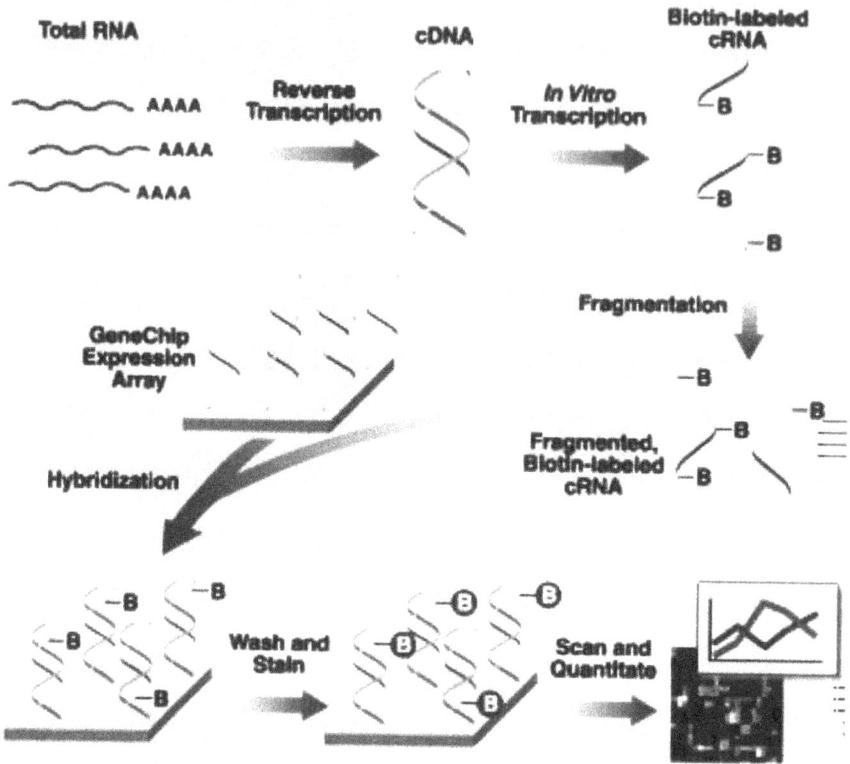

Fig. 5. "Gene chip" arrays for gene expression analysis (Affymetrix, Inc. http://www.affymetrix.com/technology/ge_analysis/index.affx)

which each array component is functionally equivalent to an ordinary strip test. The benefit of microarray technology for biomolecular detection is that detection sensitivity can be retained despite assay multiplexing.

One successful commercial application of the microarray concept is the DNA "gene chip" (Fig. 5) developed by Affymetrix Corp. [24, 25]. Fodor and colleagues first described a method of densely packing chemical probe compounds on a silicon substrate [26]. By combining solid phase chemical synthesis with photolithographic fabrication techniques, Affymetrix successfully assembled an array of 65,000 discrete nucleic acid probes into an area no larger than a few square centimeters. Each 50 µm x 50 µm microarray element was estimated to contain thousands of oligonucleotide probes. The DNA chip was designed for massively-multiplex assay applications such as gene expression profile analysis.

By incubating the gene chip with a complex biological solution such as a cell lysate, the fixed oligonucleotide probes are allowed to hybridize with segments of DNA that are present in solution. In a manner similar to the sandwich immunoassay, the surface of the gene chip is then washed, and subsequently probed for the presence of hybridized DNA. The microarray is optically analyzed by raster scanning, and finally the data is decoded by comparing the measured signals to the array pattern in which the oligonucleotide probes were deposited.

Because the individual array elements of the DNA chip are miniscule compared to the typical test strips of a lateral flow assay device, a sophisticated instrument is required to properly read the DNA microarray. The cost of purchasing such a system is more than offset by the savings in time and resources, if the same assortment of DNA tests were to be performed separately using conventional laboratory methods. But since the instrument reader is quite large and contains delicate optics, it is not rugged for portable field applications. Furthermore, a single chip design taken from concept to product can cost as much as $400,000 USD, and cannot be modified without a complete redesign. Thus in applications such as bioforensics or proteomics – in which experimental protocols are constantly evolving with the influx of new information – the cost of continually updating a DNA chip design could prove prohibitive.

3.4 Liquid Microarrays – Luminex Flow System

It is important to realize that microarrays need not be fixed in 2 dimensions [27]. The argument can be made that, for certain fixed microarray designs, individual test elements may not have equal access to the biological fluid sample [28]. In situations where the probe molecules compete for a common target ligand, the detection would be biased towards the probe element which first encounters the sample solution. One method for addressing this issue is to "un-

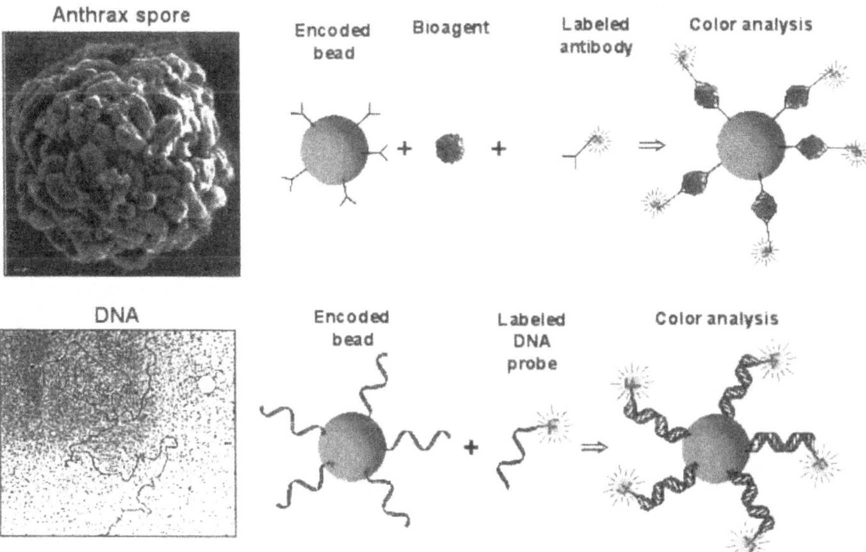

Fig. 6. Microbead-based detection of protein and DNA

fix" the microarray, allowing all probe elements to mix freely and equally with the biological fluid sample [29].

Luminex Corporation developed a multiplex bioassay system around this "liquid array" concept. Instead of designing assays to be performed on flat surfaces, as in ELISA or the Affymetrix gene chip, Luminex developed a system platform based on 5-μm diameter polystyrene microspheres that are surface-modified to facilitate chemical coupling of either capture antibodies or oligonucleotide probes (Fig. 6). The functional nature of these particles is thus similar to that of the labeled particles used by the JBPDS lateral flow assay. However, in contrast to the JBPDS method of analysis, the Luminex platform analyzes the microspheres using a customized flow cytometer (Fig. 7).

Flow cytometers are commonly found in biomedical research laboratories, and are used to analyze particles (usually living cells) suspended in solution. The core mechanism of the flow cytometer is an optical system which interrogates particles that pass through a thin glass capillary tube. Using various forms of illumination (visible or UV laser, for example), the flow analyzer is able to count and measure parameters such as particle dimension, optical absorbance and fluorescence. The microfluidic subsystem of the flow analyzer uses an inert carrier solution to draw up the biological sample at a rate and in a manner which guides the suspended particles to pass through the capillary channel in single file. In normal function, the flow cytometer can analyze thousands of particles in a matter of minutes.

In order to incorporate multiplexing capability into their diagnostic platform, Luminex introduced a method of optically encoding these latex microspheres, using discrete amounts of red and orange fluorescent dyes impregnat-

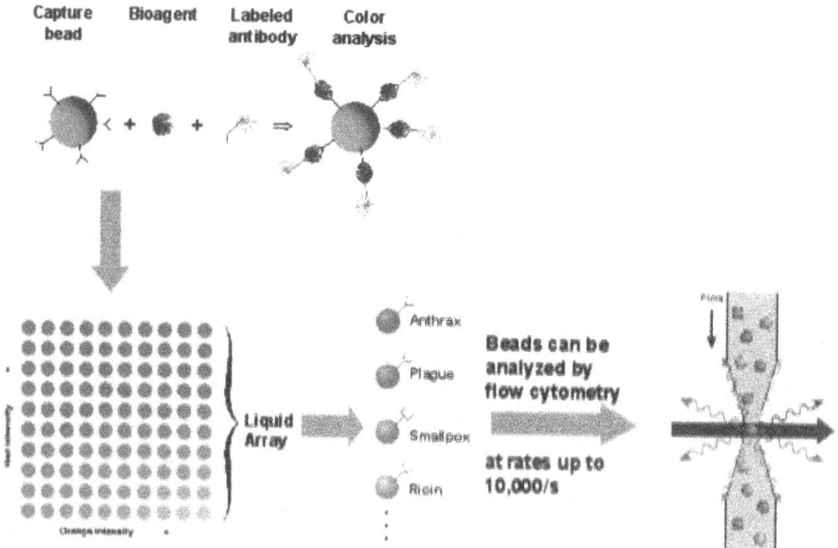

Fig. 7. Overview of microbead-based multiplex flow analysis

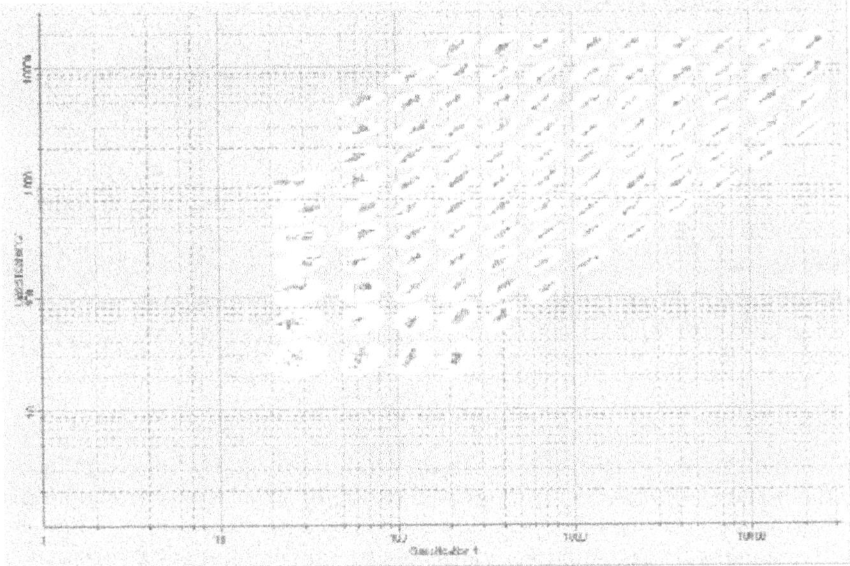

Fig. 8. LabMAP 100-plex bead map (Luminex Corporation, Austin TX)

ed into the material of each microbead. Exactly 100 distinct microbead classes were thus formed, enabling up to 100 different bioassays to be performed simultaneously on a single fluid sample (Fig. 8).

The last component of the Luminex platform is the detection element. In a typical multiplex immunoassay, an assortment of bead classes are selected for derivatization with the necessary capture antibodies. These reactive beads are then allowed to mix and diffuse freely in solution with the biological sample. After the microbeads are washed (either by centrifugation or filtration, followed by resuspension), they are subsequently treated with detector antibodies to indicate whether or not any target antigen is bound to the bead surface. The detector antibodies are linked to a common green fluorescent reporter molecule, phycoerythrin (PE).

Ultimately the assay microbeads are analyzed by the Luminex flow cytometer. Using a red laser source to measure the red and orange fluorescence intensities, the customized flow analyzer correctly determines the classification (and thus the corresponding bioassay) of each passing microsphere. Immediately following classification, a green laser source interrogates the microsphere for its corresponding reporter fluorescence – a high measured value thus indicating a positive assay result.

Because each individual microbead constitutes an independent measurement by the Luminex instrument, the data generated by a routine multiplex assay involving tens of thousands of microbeads carries much greater statistical significance, compared to other multiplex assays which may only perform measurements in duplicate or triplicate.

Table 1. Summary of Biodetection Methods

Method	Time	Multiplex	Portable	Comments
Laboratory culture	7–10 days	no	no	laboratory reference
Direct fluorescence	1 hour/ assay	serial or parallel	no	cannot automate
ELISA	<1 day	no	no	laboratory reference
Lateral flow assays	1 hour	serial or parallel	yes	least expensive
Fixed microarray	<1 day	yes	no	
Liquid array / flow analysis	1–2 hours	yes	no	
MIDS	1–2 hours	yes	yes	

Of the systems described thus far, the Luminex diagnostic platform incorporates many key features which are well suited for advanced biodetection: test results can be obtained fairly quickly, individual assays are simple to develop and modular such that multiplex panels can be built up from any assortment of derivatized microbeads, and the liquid array makes optimal use of the biological sample at hand.

Key features which the Luminex platform lacks are ruggedness and portability. Understandably, the Luminex LX100 was designed to be a sophisticated research tool of similar caliber to the Affymetrix system. However, the instrument's delicate microfluidics and optics, along with its operating power requirements, constrain its use to a laboratory benchtop environment.

4 Multiplex Immunoassay Diagnostic System (MIDS)

In response to the need for a rapid, portable, multiplex biodetection system in public health, clinical medicine and national security, we investigated the strategies developed to this point and concluded that a new system should leverage the strengths of the liquid array technology which were advanced by Luminex. Optically-encoded microbead-based immunoassays retain the advantages of speed, accuracy, cost, and multiplex-capability over other biodetection methods. However, we propose that a CCD-based optical instrument which can analyze the microbeads arranged in a 2-dimensional array is technically more rugged, scalable, and portable than a flow cytometer – and would thus be well suited for point-of-care applications. The working name for this developmental platform is Multiplex Immunoassay Diagnostic System, or MIDS (Fig. 9).

The MIDS platform consists of 3 major components which will be described in greater detail: (1) a disposable sample preprocessing unit, (2) CCD-based optical detection hardware, and (3) digital image analysis software.

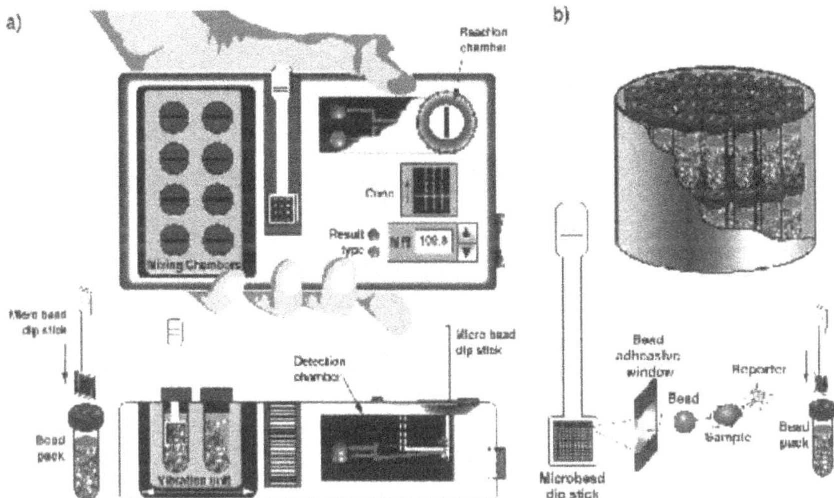

Fig. 9. Overview of MIDS concept

4.1 Disposable Sample Collection Unit

While sample collection and handling may at first seem peripheral to the discussion of optical biosensors, we put forth the argument that the ease or difficulty of detection and measurement depends entirely on the quality of sample preparation.

The disposable sample collection unit for MIDS is under active development; however, its basic roles and design have already been defined. The disposable unit performs two critical functions: (1) to receive a biological fluid specimen for immunoassay using the capture microbeads and secondary detection reagents; (2) to transfer the treated microbeads into a 2-dimensional array that will be analyzed by the optical reader.

Since the basic design of the disposable module depends primarily on the method of sample collection and on the physical characteristics of the fluid sample (i.e. blood, saliva, urine, etc.), we selected two methods most commonly used in hospitals for assessment of infectious disease: blood serum samples and nasopharyngeal (NP) swabs. Both types of fluid samples pose unique challenges for preprocessing.

Whole blood poses a unique problem for sample preparation, because the latex microspheres are similar in size to blood cells. Although filtration is a common technique to efficiently remove the microbeads from solution, whole blood contains up to 40% red blood cells by volume, and would immediately occlude the filter membrane, interfering with microsphere recovery. To address this issue, a preliminary step to prefilter or otherwise remove intact cells from the blood sample would have to be included in the sample handling procedure.

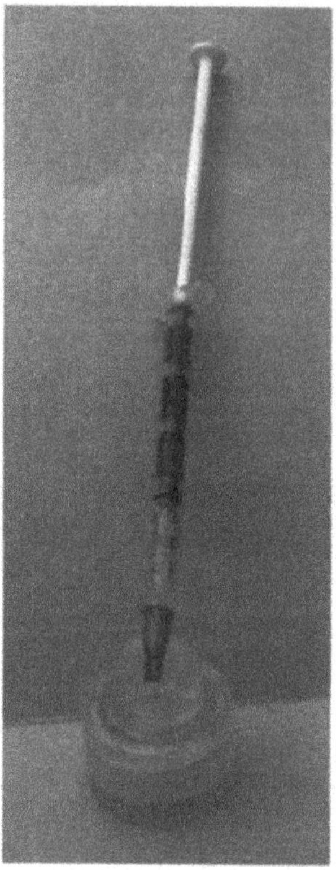

Fig. 10. Prototype disposable sample collector for MIDS

For nasopharyngeal swab samples, an extraction step is required to transfer the captured pathogens into the collection chamber. It is necessary to introduce an extraction buffer which contains a mucolytic agent to help liquefy the sample and promote release of particulates trapped by swab material. We are also considering special swab designs which will greatly improve pathogen capture and release. Because our near-term goal is to demonstrate MIDS-based detection of respiratory viruses in the hospital emergency room setting, we decided to focus development around NP swab collection, since this sampling method is more likely to detect early stages of upper respiratory tract infections, compared to blood serum samples.

The initial prototype for a NP swab-collecting disposable unit (Fig. 10) is built around a simple filter-bottom well design. The well forms a receptacle for a nasopharyngeal swab, and has a working volume of approximately 1 cm^3. Once the sample is introduced, the test microbeads are added and allowed to incubate for 20 minutes at ambient temperature. After incubation, the unbound antigen and remaining fluid sample are removed by filtration, either by applying negative pressure with a small syringe pump, or by introducing an absorbent material on the opposite surface of the filter. The microbeads are drawn down onto the surface of the filter bottom by bulk flow across the filter membrane. The detection antibodies and reporter label are subsequently added to the resuspension medium, with intermediate wash steps, using the same filtration method.

Reagent packaging, storage and delivery is another design component which we must consider for long term development of the disposable sample collector. Eventually we anticipate reagents to be supplied in dry, lyophilized form requiring rehydration prior to dispensing. In the near-term, we have developed multiple concepts for delivery of fluid reagents which require further evaluation and refinement. In general, the reagent packs must allow for isolated storage of multiple solutions and subsequent dispensing at various time intervals and rates. One method employs a syringe pre-loaded with multiple reagents in series separated by intermediate plugs. Computer control of the syringe pump would enable the various solutions to be dispensed at preset intervals between incubations. A similar strategy has been developed using a carousel mechanism which holds individually-packaged reagent "blister packs" that are ruptured and dispensed as needed. Preliminary results with the carousel design have been encouraging, but we will continue to examine details of inter-compartmental contamination, complete expulsion of the bead reagents, consistency of flow versus pressure, and accuracy of delivering small microtiter volumes. We will also consider other methods of reagent storage and dispensing, with appropriate screening of commercially available solutions – with the eventual goal of completely automating reagent dispensing, by using a simple and inexpensive disposable reagent pack that requires minimal user interface

The other primary role of the disposable sample handling unit is to transfer to assay microbeads from suspension into a planar array for imaging and analysis by the optical detector. We had previously evaluated several methods to

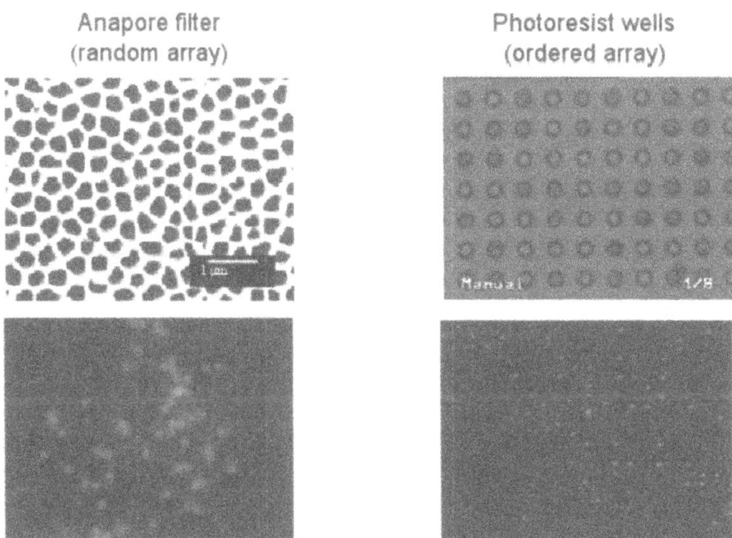

Fig. 11. Random vs. ordered 2-dimensional arrays

deposit suspended microbeads onto flat surfaces (Fig. 11). Initial investigations demonstrated partial success in immobilizing beads onto glass surfaces; however, neither surface tension nor chemical adhesion (for example, long-chain biotinylation of silanated glass slides with streptavidin-coated beads) permitted the microbeads to withstand vigorous washing. We also investigated photoresist surface technology, spinning a 5–10 μm layer of photoresist material onto glass slides and etching microscopic (10 μm diameter) wells in an ordered array across the photoresist. We found that the microbeads placed in suspension over this surface precipitated out of solution and into the wells. The resulting array was well-ordered and stable throughout moderate washing. Several disadvantages were that bead sedimentation was essentially a passive process and not well characterized. Also, the production of the microwells by photolithography was labor-intensive and would ultimately raise the cost of the disposable component of MIDS

We found a promising solution with a new filter technology by Whatman Inc., called Anopore™. Unlike conventional paper filters, Anopore is made of a semicrystalline aluminum oxide, which under electron microscopy appears almost

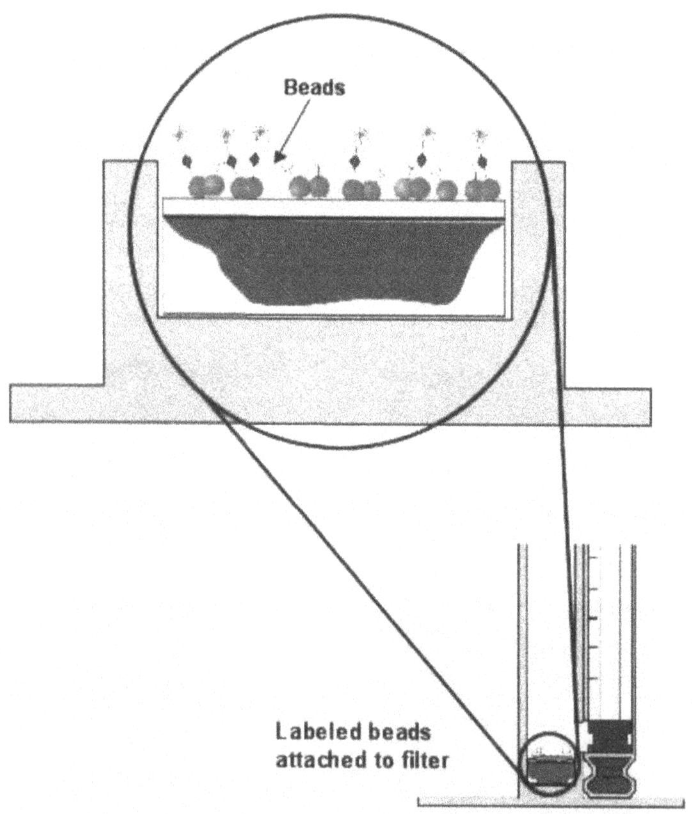

Fig. 12. Generating a 2-D microbead array by direct filtration on a flat surface

as a honeycomb structure with densely packed pores averaging 0.2 μm in diameter (compared to the 5.5 μm diameter microspheres). Furthermore, the Anopore™ material is milled flat to exacting tolerance and is optically transparent when wet. This made for an intriguing combination of features that seemed perfectly suited for the MIDS disposable device. Not only could the Anopore™ medium provide high flow rate filtration to retain the encoded microbeads during the immunoassay procedure, but it could form the substrate that would allow the derivatized beads to be imaged directly on the flat surface of the filter.

In practice, the new filter works well in the disposable sample chamber for mixing and exchanging suspension buffers during the immunoassay procedure. However, the deposition of the treated microbeads into a 2-dimensional array proved to be a more complex task. While our goal was to deposit the beads in a random but uniformly distributed monolayer on the Anopore™ surface, we found that the assay microbeads (which in the final state are coated with antibody protein, captured pathogens and reporter label) tended to form aggregates in aqueous solution. While beads which form the first layer on top of the filter surface are acceptable for image analysis, extra beads which stack on top of the first monolayer can disrupt bead classification and analysis by contributing additional fluorescence signals from their vicinity. This issue has been addressed first by adjusting the bead concentration per immunoassay, such that the total number in a given sample well is less than that required to form a complete monolayer covering the filter surface. Since the suspended latex microbeads will follow the bulk flow of solution in the reaction chamber, we have taken into consideration possible methods of controlling the quality of liquid flow across the filter membrane to promote more even distribution of the microbeads on the Anopore™ material (Fig. 12).

4.2
CCD-based Optical Hardware

We selected image-based cytometry as a basis for the MIDS detection instrument, because the hardware supporting charge coupled device camera (CCD) technology is more amenable to building small scale or portable instruments (Fig. 13). It is worth noting that the tremendous sensitivity of the photomultiplier tube used for reporter detection in flow cytometers can be matched by the relatively insensitive CCD due to the fact that CCD-based measurements can integrate fluorescence signals for many seconds, while beads in a flow system are illuminated for less than a millisecond.

For a given number of excitation sources, the degree of multiplexing for MIDS is inherently less than that of the Luminex system. There are two primary reasons for this. First, the dynamic range of the CCD is less than that of the avalanche photodiodes or photomultiplier tubes that are used in flow systems. Secondly, the distribution of fluorescence intensity for any particular (Luminex) bead class is more disperse with the imaging system because of inherent spatial variations in the excitation level over the much larger field of illumination. (A flow system concentrates the excitation in a 50-μm spot, whereas the imaging system illuminates a field measuring a millimeter or two.) In its current form,

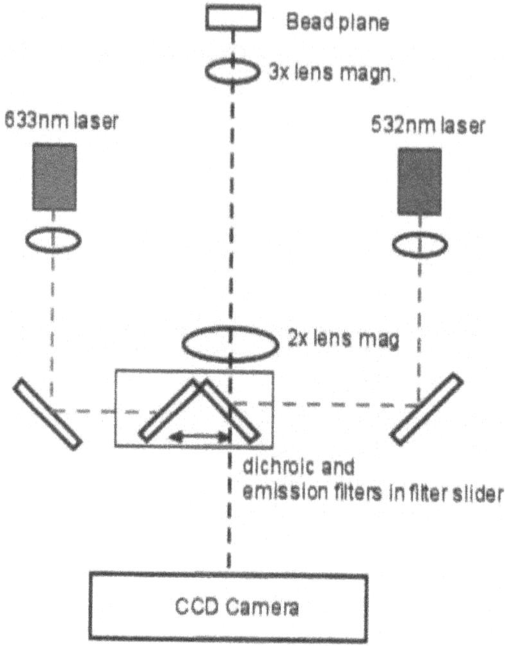

Fig. 13. Schematic of MIDS optical detection system

the benchtop prototype instrument is capable of performing the equivalent of a Luminex 6-plex assay. We are confident that the MIDS system will be able to accommodate an 8-plex assay in a basic package. By upgrading the lasers, beam conditioning, and the CCD, we estimate that up to a 32-plex will be possible. In either case, the MIDS is well suited for point of care applications which we anticipate to require 10–20 assays/panel.

The optical system consists of the following components: an illumination source, detection electronics, analysis package, and user interface. The simplest types of light sources include light emitting diodes (LEDs), lasers, laser diodes, and filament lamps. These sources can be used in conjunction with optical filters, diffraction gratings, prisms, and other optical components to provide a specified spectral component of light. Alternative forms of radiation such as bioluminescence, phosphorescence, and others could also potentially be employed. Although typical fluorophores require excitation wavelengths in the visible portion of the spectrum (300–700 nm wavelength), other wavelengths in the infrared and ultraviolet portion of the spectrum could also prove useful for illuminating the dipstick microbead array. The transmitted, reflected, or re-emitted light from the trapped microbeads must then be propagated to an optical apparatus for detection, using photosensitive detectors such as photodiodes or photomultiplier tubes, in combination with some type of spectral and/or spatial filtering. Spatial filtering of the light is possible either by transverse scanning of the dipstick microbead array or with two-dimensional detectors such as CCD and video cameras.

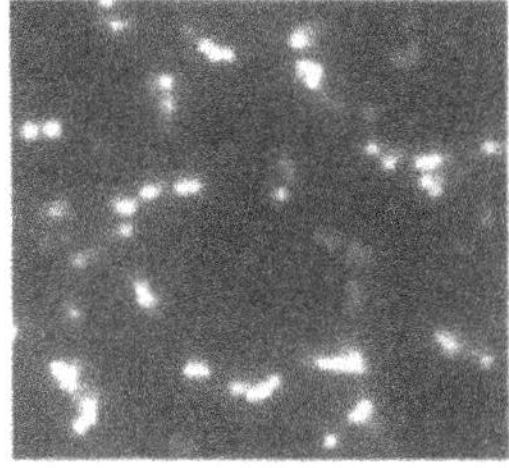
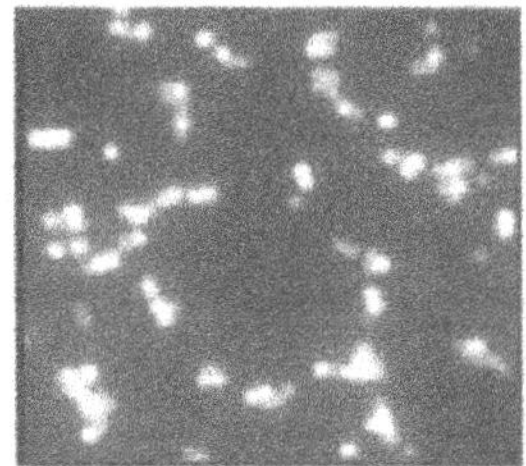
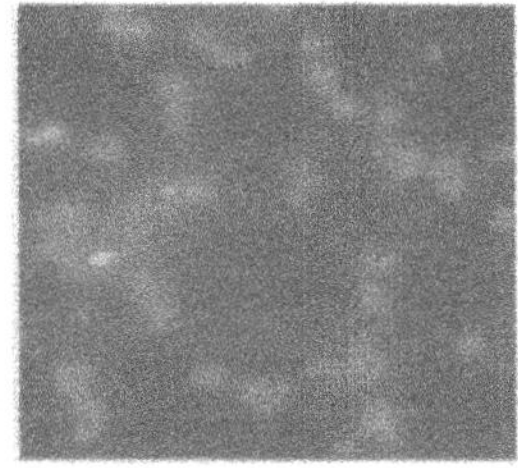

Fig. 14. Red, orange, and green component fluorescent images of random bead array (pseudocolor representation)

The criteria for evaluating system performance can be subdivided under the orthogonal functions of classification and reporting. The key variables determining system performance relate to image quality, image processing and assay analysis. System specifications are addressed concurrently by considering issues such as size, weight, cost and complexity. Lasers require power supplies and controllers, as do shutters and motorized stages. Furthermore, CCDs need need supporting mechanisms for cooling and control as well.

The optical system uses a form of Köhler illumination, by which a fluorescent sample is simultaneously illuminated from the front by a collimated laser source and imaged from the front to a CCD camera. This is done for a series of three timed exposures using a combination of dichroic beam splitters and optical bandpass filters to generate a "red-orange" image pair for bead classification and a "green" image for reporting the level of attached reporter label (Fig. 14). These images are subsequently processed and analyzed to identify bead classes and the presence of phycoerythrin reporter dye in the original sample.

A three-color component fluorescence image set must be taken for each microbead array to be analyzed. Ideally, in addition to being in focus, they would contain thousands of beads, with hundreds of CCD pixels devoted to each bead. While "megapixel" CCDs exist with necessary pixel resolution to accomplish this, such devices are both relatively large and expensive. Furthermore, the system would require a relatively high (40×) magnification, which would place greater demands on the quality of optics, beam uniformity across the image field and would reduce the depth of focus. A larger field of view also means that the available laser power is divided over a larger area, which requires longer integration times or more expensive lasers.

One of the most significant differences between Luminex's flow-based system and the MIDS' image-based method is the number of microspheres which can be analyzed per assay, which plays a large role in determining the instrument's precision. On the MIDS system the factor limiting the number of beads per image is the density of isolated beads that can be produced on a sample before it is even inserted into the optical system. The random arrays generated by bead filtration are likely to have relatively large unoccupied regions as well as those in which beads are stacked and difficult to reliably analyze. We previously described our approach to these challenges. Clearly, an ordered array is far superior for image analysis but is difficult to achieve and to fabricate when the design calls for a simple, disposable platform.

Given the constraints placed on hardware design, we must investigate other methods to compensate for the number of data points which may be collected in one assay. One option is to take several image triplets (red, orange and green) of the same sample, measuring multiple regions to collect more data points for analysis. The key drawbacks to implementing this solution are that an extra motor drive would be required as well as extra time for re-iterative image acquisition and analysis. We anticipate that the time required to collect one triplet set of images will be approximately one minute.

4.3 Digital Image Analysis Software

Currently the software for MIDS is divided into three components. LabView™ (National Instruments) controls the operation of the benchtop instrumentation, including CCD camera, shutter controls and filter wheel. As previously described, assay microbeads were deposited on Anopore™ filters and the resultant arrays viewed under the fluorescent imaging setup. Using the appropriate filter sets to extract the necessary red, orange and green fluorescence from the bead array, digital images were captured to computer with a Pixel CCD camera. Intensity measurements were made on the acquired images using subroutines written in IP-Lab™ (Scanalytics, Inc.) software. Fluorescent microbeads are identified by means of a "smart search" algorithm which sequentially raster scans each color image component. Microbead targets are identified by virtue of size, shape, and expected fluorescence intensity profile when imaged by the MIDS system. The red, orange and green images are placed into registration so that fluorescent microbeads may be sorted and their corresponding classification and reporter values saved to a data file. When displayed on a scatter plot of red vs. orange fluorescence intensity, the bead mixture can be sorted into distinct clusters of data points – resembling those of the pre-calibrated bead maps displayed using Luminex Data Collector software. The task of bead classification is performed by IgorPro (WaveMetrics, Inc.), which compares incoming data to a pre-calibrated bead map and decodes the classification of each identified bead. IgorPro also performs statistical analysis of the reporter fluorescence intensity for each bead class, ultimately reporting which classes (and hence assays) were "positive" by multiplex immunoassay.

We plan eventually to port the entire image acquisition and analysis to the LabView™ software platform, taking advantage of new functional capabilities added with the IMAQ Vision™ imaging toolset for LabView™.

4.4 Preliminary Results

We have successfully characterized the MIDS prototype using six fluorescent-encoded bead classes from Luminex. The six classes were derivatized with various capture antibodies and control reagents to simulate a general biological sensor which could detect 3 types of pathogenic material: bacterial spores (of *Bacillus globigii*), virus (MS2 bacteriophage), and protein (ovalbumin). The bead classes were derivatized according to the following scheme:

Bead class	surface conjugate	function
C200	Bovine serum albumin (BSA)	negative control
C198	r-phycoerythrin	positive detector control
C194	anti-MS2	viral probe
C170	anti-*Bacillus globigii*	bacterial probe
C168	anti-ovalbumin	protein probe
C164	anti-human IgA	positive sample control

In addition to the three experimental bead sets which probe for target pathogen, one bead class was coupled to BSA as a negative control to demonstrate nonspecific binding which might occur. C198-phycoerythrin would serve as a positive control and calibration standard for the MIDS detector. Finally, to confirm whether a biological sample was indeed placed into MIDS, a bead set was added to test for human IgA – a ubiquitous protein which is secreted by the nasal mucosa.

The simulant panel was tested in the laboratory, using calibrated solutions containing various concentrations of antigen. Briefly, aliquots of the 6-plex bead panel were incubated with variable dilutions of stock antigen (Bg or MS2, for example) in 96-well filter bottom plates for 30 minutes at room temperature on an orbital microplate shaker. After antigen was allowed to bind the capture microbeads, the wells were washed with phosphate-buffered saline, filtered, and then resuspended for 30 minutes in a solution containing the detector antibody cocktail. These detector antibodies are identical to the ones covalently bonded to the surface of the capture microspheres, but are instead biotinylated – so that they will attach to the final reporter label, which is phycoerythrin-coupled to streptavidin (Sa-PE).

The assay beads were washed and transferred to round-bottom 96-well plates for measurement on a Luminex LX100 flow cytometer. Analysis of the various bead samples showed that the reporter fluorescence intensity associated with bead classes C194 and C170 increased in proportion to the concentrations of MS2 and Bg, respectively. Positive and negative control beads showed consistent behavior – and because the samples were prepared from laboratory standards, no human IgA was present in solution, so that C164 showed little reporter fluorescence. The graph in Fig. 15 depicts the 6-plex microbead response to high concentrations ($\approx 10^7$ pfu/ml) of MS2 bacteriophage in solution.

After the Luminex flow analysis, the same bead samples were individually recovered, and then processed into 2-dimensional arrays by Anopore™ filtration, using the prototype disposable sample collectors. While remaining attached to the filter bottom, the random bead arrays were placed in the MIDS optical detector, and fluorescent red, orange, and green image triplets were acquired. The sample chamber was rotated manually several times in the holder, so that several fields could be analyzed for each microbead sample.

The composite data in Fig. 16 shows the equivalent MIDS results for the same Luminex bead assay depicted in Fig. 15. The distribution of red and orange classification intensities for the microbeads detected on the MIDS platform is similar to that determined on the Luminex flow cytometer. Also similar in fashion to the

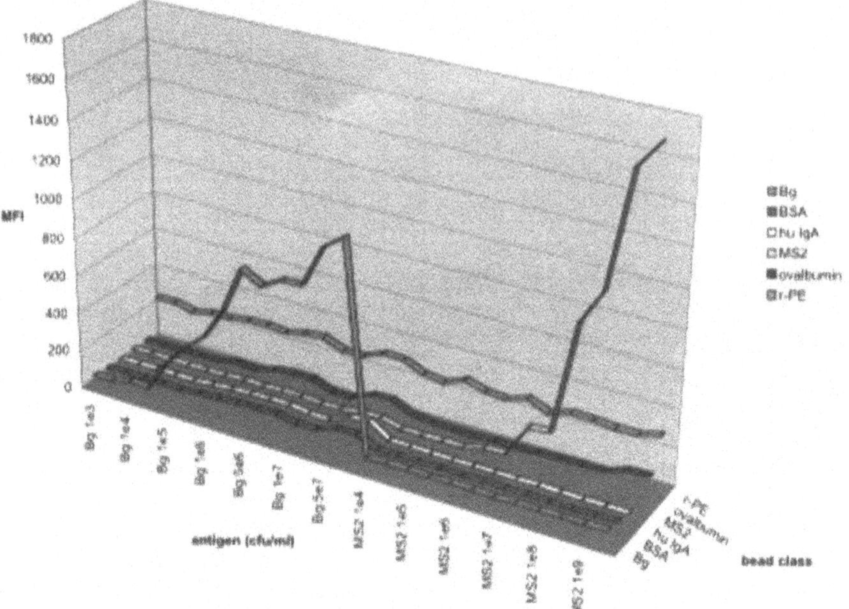

Fig. 15. Specific, concentration-dependent response of 6-plex bead assay to simulants. Luminex flow analysis

Luminex bead maps, polygonal regions are drawn to identify and distinguish the 6 bead classes. On the chart showing the distribution of reporter intensities versus bead class, C194 (anti-MS2) shows increased median reporter fluorescence as expected – while C198 (rPE) shows an intermediate level of reporter fluorescence. It should be noted that the multiplex bead set reveals inherently large variability in the reporter fluorescence intensity for any given measurement. However, the median reporter fluorescence intensity (MFI) has been shown to be a suitable population statistic that is significant for measuring antigen concentration. An arbitrary threshold level of 100 counts was set in this particular instance to demonstrate a possible method of automating MIDS detection. We expect that separate thresholds will have to be determined, based on the binding characteristics of each capture antibody used in the multiplex assay.

Since we designed MIDS to be used in point-of-care diagnostic applications, we have developed a "real" multiplex microbead panel to assess respiratory viruses that may be contained in nasopharygeal swab samples. The panel is based on the commercially available Bartels direct fluorescence assay (DFA) kit, and includes specificity against 7 common viruses: Influenza types A and B, Parainfluenza types 1, 2, and 3, adenovirus, and respiratory syncytial virus (RSV) [30]. Since the DFA kit is fully manual and requires sequential testing for all 7 targets, the development of a multiplex-capable bead panel to be used in conjunction with either MIDS or Luminex would itself be considered an advancement in diagnostic tech-

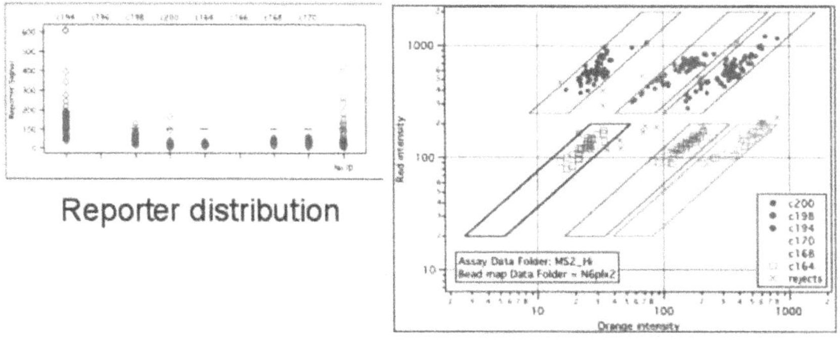

Bead_	Assay	Conc	found	Positives	Negatives	Rep_mean	Rep_medn	Rep_sdev	Level
c194			44	30	14	157.2	137.0	124.8	100
c196									
c198			47	2	45	47.5	45.0	20.7	100
c200			56	1	55	22.5	18.5	22.6	100
c164			26	0	26	19.7	20.0	6.1	100
c166									
c168			47	0	47	26.9	25.0	10.4	100
c170			37	0	37	30.3	29.0	12.2	100
	No ID		54	17	37				100

Fig. 16. Simulant 6-plex results by MIDS assay

Table 2. Respiratory virus panel.

Target Antigen	Monoclonal Antibody	Bead Class
Influenza A	A1 (Chemicon)*	151
	A3 (Chemicon)*	158
	c102 (Adv ImmunoChem)	133
Influenza B	B2 (Chemicon)*	153
	22D5 (Chemicon)*	156
	4H7 (Adv ImmunoChem	129
RSV	133/1H (Chemicon)*	150
	131/2G (Chemicon)*	155
	130/12H (Chemicon)*	147
	8C5 (Adv ImmunoChem)	142
	9C5 (Adv ImmunoChem)	139
	8B10 (Adv ImmunoChem)	140

nology. Together with positive and negative control bead sets, the respiratory panel would constitute a minimum 10-plex assay. However, in practice, respiratory viruses such as influenza have multiple strains which may have different specificities and binding affinities with different monoclonal antibodies. As a result, it may be difficult (if not altogether unreasonable) to expect just one monoclonal antibody to efficiently bind *all* strains of a chosen viral species. For this reason, 12 different monoclonal antibodies already have been selected for use in detecting only three (Flu A, Flu B, and RSV) of the seven targeted species.

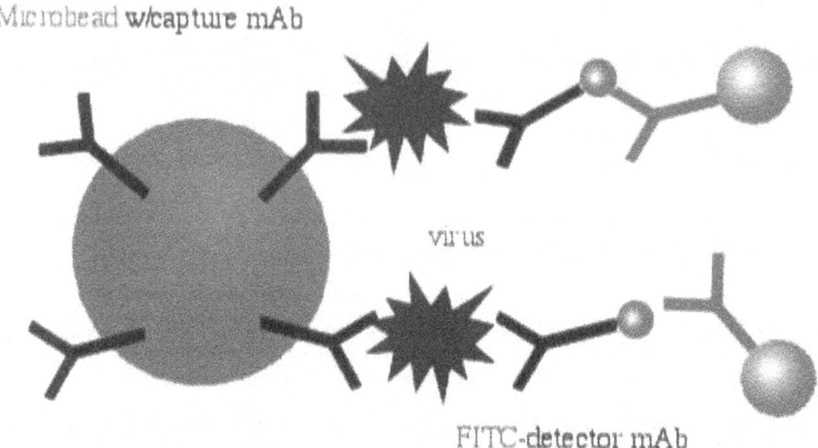

Fig. 17. Binding schematic for microbead-based respiratory viral assay

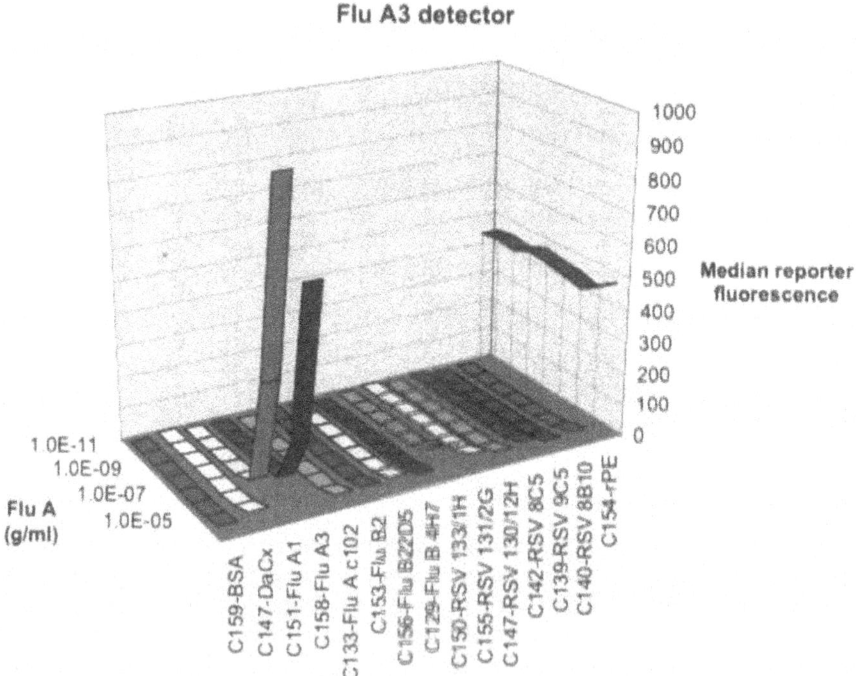

Fig. 18. Multiplex assay results by flow analysis. Representative titration curves show specific response by microbeads coated with anti-Flu A antibodies

Bead class 154 was directly coupled with R-phycoerythrin for use as a positive fluorescent control for the flow cytometer, while bead class 159 was coated with serum albumin (BSA) as a negative control bead to test non-specific reporter binding.

The charts shown below (Fig. 19) demonstrate bead capture efficiency as a function of antibody density on the microbead surface. The density is inferred by the concentration of capture antibody used when first derivatizing carboxylated Luminex microbeads. For each monoclonal antibody, 3 bead sets were derivatized with one of 3 antibody concentrations: high (1 mg/mL), medium (0.5 mg/mL) and low (0.1 mg/mL). Although the data suggests that antigen capture and binding efficiency are nearly saturated at even the lower antibody concentrations, we take advantage of the opportunity to demonstrate specificity and uniformity of response of each bead class for its respective target antigen.

4.5 Discussion

In summary, we have developed a benchtop optical system using conventional hardware currently available in our lab to demonstrate fluorescent imaging of two-dimensional microbead arrays. This setup is adequate to demonstrate fea-

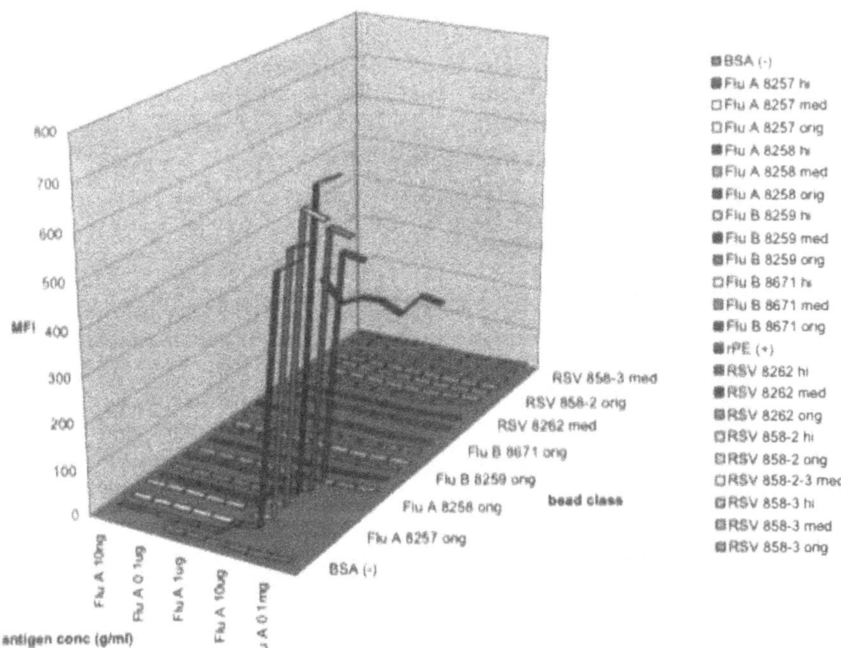

Fig. 19. Comparison of viral detection efficiency vs. capture antibody density

sibility, but components chosen specifically for this application will yield significantly better results and enable miniaturization of the entire system.

We have focused development efforts in three primary areas: sample preparation and bead arrays, optical imaging, and application development. We have successfully demonstrated bead deposition in an ordered array onto photoresist-coated glass substrates with binding efficiencies of >80%. We have also successfully captured beads in random planar arrays onto specialized filter substrates. Imaging bead arrays with a CCD camera, we have demonstrated fluorescence intensity based discrimination of up to 6 different bead classes. We are currently extending this bead decoding algorithm to allow simultaneous identification of up to 20 pathogens within a single test. Finally, we have been exploring suitable applications for MIDS such as infectious disease detection, cancer screening, and other biomarkers, with the intention of fully leveraging this unique technology.

Our preliminary studies strongly suggest that a rapid portable MIDS platform is a viable approach for point-of-care diagnostic applications. For a given microbead assay, the results achieved thus far with MIDS successfully approximate, if not match, the results obtained by conventional benchtop analysis with the Luminex flow cytometer. As the Luminex diagnostic system gains broader acceptance in clinical laboratories and is validated against other reference standards, we believe that the MIDS device concept will also gain acceptance. It should be emphasized that the objective of developing the MIDS platform is *not* to supplant the Luminex flow cytometric system – rather, MIDS will perform a critical role in point-of-care diagnostics and field medicine, where the Luminex flow analyzer cannot readily function. As an initial screening tool, MIDS must be able to detect the presence or absence of microbial pathogens in a clinical sample. More importantly, it must be able to correctly rule-out cases that do not require further work-up. For such an application, we are confident that the technology in MIDS will provide satisfactory results.

5
Conclusions and Future Directions

Not only for infectious disease applications, but for broader clinical diagnostic and biomedical research applications – multiplex biodetection using microarray technology will continue to evolve and advance. Currently the Luminex microsphere-based diagnostic system is the most mature technology applicable to the liquid array format. However, the method of optically-encoding beads by discretizing fluorescence intensity is better suited for analysis by flow-based (rather than image-based) cytometry, because of the greater inherent variability in accurately measuring fluorescence intensity across a 2-dimensional field. Flow cytometers bypass this issue by analyzing arrays which may be considered zero-dimensional, since the relative orientation of microbead and detector is fixed in space.

We will investigate alternative methods of designing multiplex-capable optical probes. Some alternative bead-based technologies include varying bead size or bead material (i.e., polystyrene, silica, etc.), and incorporating ferromagnetic particles into the bead material to enable new methods of manipulating beads in a microarray. We will also follow developments of Quantum Dot technology (spectral coding) and NanoBarCodes (pattern coding) and other multiplex methods for appropriate insertion into the MIDS platform should they prove to have significant advantages. We are, however, encouraged that the rationale for the basic components of the MIDS concept remains essentially valid for all types of multiplex optical probes. That is, regardless of the particle-labeling technology, samples need to be introduced to the instrument, mixed with particles and other reagents, and optically measured. We believe that this new paradigm will play a critical role in guiding the future development of optical biosensors .

References

1. WHO (2002) Communicable Disease Surveillance and Response. World Health Organization Regional Office for Europe, Copenhagen
2. Noah D, Fidas G (2000) The Global Infectious Disease Threat and Its Implications for the United States. Gordon DF, Ed. National Intelligence Council, Washington, DC
3. Anderson RA (2001) Deaths: Leading Causes for 1999. National Vital Statistics Reports, Vol. 49, No. 11. CDC
4. Simonsen L, Fukuda K, Schonberger LB, Cox NJ (2000) J Infect Dis 181: 831
5. Ghendon Y (1992) World Health Stat Q 45: 306
6. Eldad A (2002) Harefuah 141: 21
7. Binder S, Levitt AM (2002) Emerging Infectious Diseases: A Strategy for the 21st Century. Centers for Disease Control
8. Meltzer MI, Cox NJ, Fukuda K (1999) Emerg Infect Dis 5: 659
9. Snacken R, Kendal AP, Haaheim LR, Wood JM (1999) Emerg Infect Dis 5: 195
10. Fong WK, Modrusan Z, McNevin JP, Marostenmaki J, Zin B, Bekkaoui F (2000) J Clin Microbiol 38: 2525
11. Brundtland GH (2000) World Health Organization Report on Infectious Diseases 2000
12. Uphoff H, Metzger C (2002) Dtsch Med Wochenschr 127: 1096
13. Goldmann DA (2001) Emerg Infect Dis 7: 249
14. Greenberg SB (2002) Curr Opin Pulm Med 8: 201
15. Lieberman D, Shvartzman P, Korsonsky I, Lieberman D (2001) Br J Gen Pract 51: 998
16. Franz DR, Zajtchuk R (2002) Dis Mon 48: 493
17. Gensheimer KF, Fukuda K, Brammer L, Cox N, Patriarca PA, Strikes RA (2002) Vaccine 20: S63
18. Nadder TS, Langley MR (2001) Clin Lab Sci 14: 252
19. Petersen K, McMillan W (2002) IVD Technology
20. Tang YW, Procop GW, Persing DH (1997) Clin Chem 43: 2021
21. McElhaney JE, Gravenstein S, Krause P, Hooton JW, Upshaw CM, Drinka P (1998) Clin Diagn Lab Immunol 5: 840
22. al-Nakib W, Dearden CJ, Tyrrell DA (1989) J Med Virol 29: 268
23. Mazzola LT, Fodor SP (1995) Biophys J 68: 1653
24. Lipshutz RJ, Fodor SP, Gingeras TR, Lockhart DJ (1999) Nat Genet 21: 20
25. Gabig M, Wegrzyn G (2001) Acta Biochim Pol 48: 615
26. Fodor SP, Read JL, Pirrung MC, Stryer L, Lu AT, Solas D (1991) Science 251: 767
27. Vignali DA (2000) J Immunol Methods 243: 243
28. Schaertl S, Meyer-Almes FJ (2001) Expert Rev Mol Diagn 1: 456
29. Pickering JW, Martins TB, Schroder MC, Hill HR (2002) Clin Diagn Lab Immunol 9: 872
30. Irmen KE, Kelleher JJ (2000) Clin Diagn Lab Immunol 7: 396

CHAPTER 6

Multi-functional Biochip for Medical Diagnostics and Pathogen Detection

Tuan Vo-Dinh, Guy Griffin, David L. Stokes, Dimitra N. Stratis-Cullum, Minoo, Askari, Alan Wintenberg

1 Introduction

There is an urgent need to develop monitoring devices capable of screening multiple medical diseases and detecting multiple infectious pathogens simultaneously. A critical factor in biomedical diagnostics involves selectivity and sensitivity for detection of a wide variety of biochemical substances (e.g., proteins, metabolites, nucleic acids), biological species or living systems (bacteria, virus or related components) at trace levels in complex biological samples (e.g., tissues, blood, other body fluids, and environmental biosamples). Biosensors are diagnostic devices that employ the powerful molecular recognition capability of bioreceptors such as antibodies, DNA, enzymes and cellular components of living systems. The operating principle of a biosensor involves detection of this molecular recognition and transforming it into another type of signal using a transducer that may produce either an optical signal (i.e., optical biosensors) or an electrochemical signal (i.e., electrochemical biosensors). A biosensor that involves the use of a microchip system for detection is often referred to as a biochip.

In general biosensors and biochips employ only one type of bioreceptor as probes, i.e., either nucleic acid or antibody probes [1–10]. Biochips with DNA probes are often called gene chips, and biochips with antibody probes are often called protein chips. We have previously developed an integrated DNA biochip that combines nucleic acid probes and a detection system into a self-contained microdevice [11–16]. This manuscript describes a unique biochip system that uses multiple bioreceptors with different functionalities on the same biochip, allowing simultaneous detection of several types of biotargets on a single platform. This device is referred to as the multi-functional biochip (MFB).

2 The Multi-functional Biochip

Two fundamental operating principles of a biosensor are: (1) "biological recognition" (bioreceptor) and (2) "sensing" (transducer). The basic principle of a biosensor is to detect this molecular recognition and to transform it into another type of signal using a transducer. There are different types of transducers, which may produce either an optical signal (i.e., optical biosensors) or an electrochemical signal (i.e., electrochemical biosensors), or a mass-based signal (e.g., microbalances, surface-acoustic wave devices, microcantilevers). There are also different types of bioreceptors, which may consist of an enzyme, an antibody, a nucleic acid fragment, a chemoreceptor, a tissue, an organelle or a microorganism. Sometimes a synthetic molecule, often called biomimetic receptor (e.g., synthetic antibody, molecular imprint) can be used to mimic the properties of biological receptors.

The MFB is an integrated multi-array biochip, which is designed by combining integrated circuit elements, an electro-optics excitation/detection system, and bioreceptor probes into a self-contained and integrated microdevice. Fig. 1 depicts a schematic diagram of the MFB device, which includes the following elements: 1) an excitation light source, 2) multiple bioprobes having different types of bioreceptors, 3) a sampling platform, 4) sensing elements, and 5) a signal amplification and treatment system. The development of the multichannel sampling platform involves immobilization of bioprobes on multi-array (4 × 4 or 10 × 10 channels for current biochips) substrates, which can

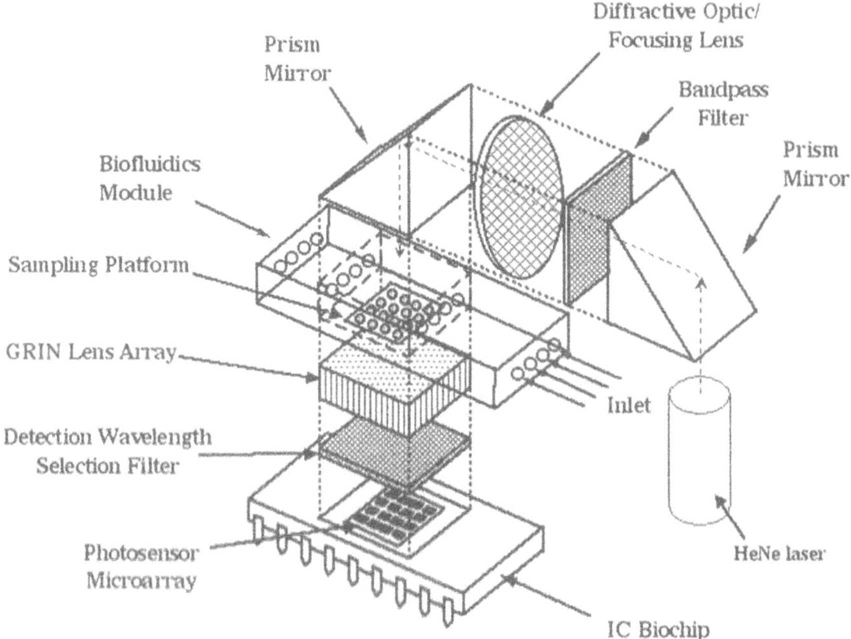

Fig. 1. Schematic diagram of the Multi-Functional Biochip (MFB) optical system.
The MFB includes the following elements: 1) an excitation light source, 2) multiple bioprobes having different types of bioreceptors, 3) a sampling platform, 4) sensing elements, and 5) a signal amplification and treatment system. The biochip has a heterogenous functional diagnostic capability due to the different types of bioprobes: antibody, DNA, enzymes, tissues, organelles, and other receptor probes. The biochip in this work uses DNA and antibody probes.
(Source: Stokes DL, Griffin GD, Vo-Dinh T (2001) Fresenius' J Anal Chem 369:295)

be performed on a transducer detection surface to ensure optimal contact and maximum detection. When immobilized onto a substrate, the bioprobes are stabilized and, therefore, can be reused repetitively. Labeled and unlabeled DNA probes were prepared in our laboratory when needed, or were purchased from a commercial source (Oligos Etc., Wilsonville, Oregon). The desired strands of oligonucleotides were synthesized and labeled with fluorescent dyes (e.g., fluorescein and Cy5 dyes) (further details are described later in the Methods Section). Several methods have been investigated to bind bioprobes to different supports. The method commonly used for binding bioprobes to glass involves silanization of the glass surface followed by activation with carbodiimide or glutaraldehyde. Immobilization of the bioreceptor probes onto a substrate or membrane and subsequently attaching the membrane to the transducer detection surface is another approach that can be used. Experimental details of probe preparation procedures have been described previously [6, 12].

2.1 Integrated Circuit Development of the Biochip

The integrated electro-optic microchip system developed for this work involved integrated electrooptic sensing photodetectors for the biosensor microchips. Such an integrated microchip system with on-board integrated circuit (IC) electronics is not currently available commercially. Therefore, IC electrooptic systems for the microchip detection elements were designed at Oak Ridge National Laboratory. Highly integrated biosensors are made possible partly through the capability of fabricating multiple optical sensing elements and microelectronics on a single IC [11, 12]. Such an integrated microchip system is not currently available commercially.

To develop a biochip system with optimized performance, several biochip IC systems based on photodiode circuitry were designed, fabricated and evaluated, one system having 16 channels (4 × 4 array), and another having 64 channels (8 × 8 array) having four types of electronic circuits on a single platform. The biochips include a large-area, n-well integrated amplifier-photodiode array that has been designed as a single, custom IC, fabricated for the biochip. This IC device is coupled to the multiarray sampling platform and is designed for monitoring very low light levels. The individual photodiodes have 900-μm square size and are arrayed on a 1-mm spacing grid. The photodiodes and the accompanying electronic circuitry were fabricated using a standard 1.2-micron n-well complementary metal-oxide semiconductor (CMOS) process. The use of this type of standard process allows the production of photodiodes and phototransistors as well as other types of analog and digital circuitry in a single IC chip. This feature is the main advantage of the CMOS technology in comparison to other detector technologies such as charge-coupled devices or charge-injection devices. The photodiodes are produced using the n-well structure that is generally used to make resistors or as the body material for transistors. Since the anode of the diode is the p-type substrate material, which is common to every circuit on the IC chip, only the cathode is available for monitoring the photocurrent and the photodiode is constrained to operate with a reverse bias.

An analog multiplexer was designed that allows any of the elements in the array to be connected to an amplifier. In the final device, each photodiode could be supplied with its own amplifier. The multiplexer is made from 16 cells for the 4 × 4 array device. Each cell has a CMOS switch controlled by the output of the address decoder cell. The switch is open when connecting the addressed diode to an amplifier. This arrangement allows connecting a 4 × 4 (or 10 × 10) array of light sources (different fluorescent probes, for example) to the photodiode array and reading out the signal levels sequentially. With some modification, a parallel reading system can be designed. Using a single photodiode detector would require mechanical motion to scan the source array. The additional switches

Fig. 2. Photograph of the 8 × 8 integrated circuit microchip

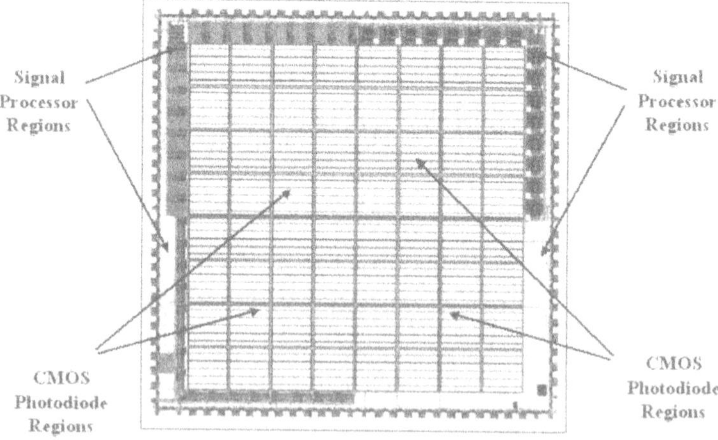

Fig. 3. Schematic of the electronic design of the 8 × 8 microchip with CMOS photodiode regions and signal processor regions

and amplifier serve to correctly bias and capture the charge generated by the other photodiodes. The additional amplifier and switches allow the IC to be used as a single, large area (nearly 4 mm square) photodetector.

To evaluate and select an improved IC system for the biochip, a chip with an 8 × 8 CMOS sensor array was designed and fabricated. This microchip contains 4 quadrants, each having a different electronic design, which was evaluated for optimal performance. Figure 2 shows a photograph of the 8 × 8 IC microchip. Figure 3 shows the design of the different CMOS photodiode regions and signal processor regions of the 8 × 8 microchip. With the CMOS technology, highly integrated biosensors are made possible partly through the capability of fabricating multiple optical sensing elements and microelectronics on a single IC. A two-dimensional array of optical detector-amplifiers was integrated on a single IC chip.

3 Experimental Systems and Procedures

3.1 Instrumentation

3.1.1 Optical Setup

Figure 1 illustrates a schematic diagram of the IC biochip detection system with associated excitation and signal collection optics. A He-Ne laser (Spectra-Physics, Inc., Model 106-1, Eugene, OR) was used for excitation of the Cy5 label at 632.8 nm. The laser beam was filtered with a 632.8-nm bandpass filter (Cat. No. P3-633-A-X516, Corion, Franklin, MA) and directed through a diffractive optic device. The diffractive optic formed a 4 × 4 array of equally intense laser beams, which was focused on the sampling platform (e.g. membrane). The intensity of each laser spot was estimated to be 0.2 mW, and the spacing between the spots was approx. 1 mm. An image (1:1) of the laser spot array was projected onto the corresponding 4 × 4 array of photosensors of the IC biochip via a gradient index microlens array (Cat. No. 024-5680, OptoSigma®, Santa Ana, CA). A combination of a 633-nm holographic notch filter (Cat. No. HNPF-633-1.0, Kaiser Optical Systems, Inc., Ann Arbor, MI) and a thin-film dielectric filter with a high-pass at 645 nm (Visionex, Atlanta, GA) was used to isolate the Cy5 emission signal from the laser line. Voltage output from the IC biochip was recorded with a strip chart recorder (Model BD40, Kipp and Zonen, Delft, The Netherlands) or from a digital multimeter (Model 506, Protek). For some studies, the system was modified with a biofluidics system for on-chip monitoring of bioassays. The sampling chamber of this biofluidics system was placed in the optical system as shown in Fig. 1, with the focus of the excitation beam array being maintained at the sampling platform (placed within the sampling chamber).

3.1.2
The Biofluidics System

A schematic diagram of the biofluidics system for on-chip immunoassays is illustrated conceptually in Fig. 4. A pump and valve system enables the sequential delivery of the fluid-based sample, the reagents and probes required for on-chip DNA hybridizations or antibody conjugations, and the rinsing solutions. Central to the system is a sampling chamber block. The sampling chamber block was fabricated from a 1" × 1" × 1/8" Plexiglas slab (OPTIX®, Plaskolite, Inc., Columbus, OH). A 1/2" diameter hole was drilled through the center of the block to form the sampling chamber, resulting in a 0.4 mL total sampling volume. Several 1/16" diameter holes were also drilled from the edges of the Plexiglas slab to the central sampling chamber. Some of these channels served as conduits for reagent or sample delivery or purging. For these studies, the plumbing was 1/16" o.d. × 1/50" i.d. polyetheretherketone (PEEK) tubing (Cat. No. 1532, Upchurch Scientific, Oak Harbor, WA). Inlet plumbing was also attached to a peristaltic pump capable of a flow rate of 1 mL min^{-1}. Other channels served as ports for various accessories, such as a nichrome heating filament and a thermocouple temperature probe (Aeropak). Once all plumbing and other accessories were seated in the Plexiglas block, the channels were all sealed with 5-minute epoxy. The sampling chamber was sealed with a pair of optical windows cut from microscope slides. While the bottom window was held permanently in place by 5-minute epoxy, the top window was intended to be removable for easy

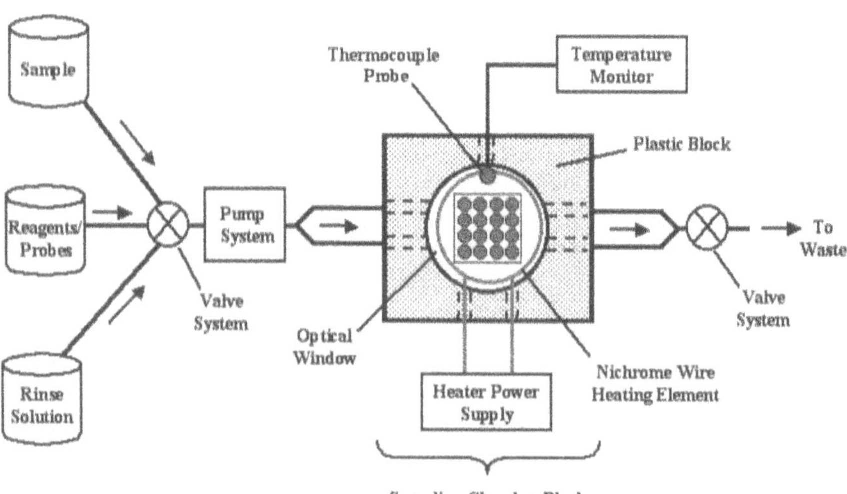

Fig. 4. Schematic diagram of the biofluidic system featuring a single-chamber sampling module with internal heating and temperature minotoring

access to the sampling chamber. High vacuum grease (Dow Corning Corp., Midland, MI) was used to temporarily hold the top window in place during on-chip assays. Consistent with the optical system depicted in Fig. 1, the laser excitation was introduced through the top window and the fluorescence signal was collected through the bottom window of the sampling chamber.

3.2 Preparation of DNA Probes

Nucleic acid probes were purchased from commercial sources (Oligo,Etc., Inc.) whenever available. Laboratory-prepared oligonucleotides were synthesized using an Expedite 8909 DNA synthesizer (Millipore). Oligonucleotides with amino linkers were synthesized using either C_3 aminolink CPG for 3' labeling or 5' amino modifier C_6 (Glenn Research, Sterling, Virginia) for 5' labeling. All oligonucleotides were synthesized using Expedite reagents (Millipore) and were de-protected and cleaved from the glass supports using ammonium hydroxide. The de-protected oligonucleotides were concentrated by evaporating the ammonium hydroxide in a Speedvac evaporator (Savant) and resuspended in 100 μL distilled H_2O. Further purification was performed by isopropanol precipitation of the DNA as follows: 10 μL of 3-M sodium acetate pH 7.0 and 110 μL isopropanol was added to 100 μL of a solution of DNA. The solution was then frozen at –70 °C. The precipitate was collected by centrifugation at room temperature for 15 min and was washed 3 times with 50% isopropanol. Residual isopropanol was removed by vacuum drying in the Speedvac and the DNA resuspended in sterile distilled water at a final concentration of 10 μg μL^{-1}. These stock solutions were diluted in the appropriate buffer at a 1:10 dilution to give a DNA concentration of 1 μg μL^{-1}.

To label DNA with the Cy5 dye (Amersham Life Sciences, Arlington Heights, Illinois), modified oligonucleotides containing alkyl amino groups were derivatized as follows: 30 pmoles of the DNA was dissolved in 250 μL 0.5-M sodium chloride and passed through a Sephadex G10 (1 cm diameter, 10 cm long) (Pharmacia, San Diego, CA) column equilibrated with 5 mM borate buffer (pH = 8.0). The void volume containing the oligonucleotides was collected and concentrated by evaporation. The resulting solution was dissolved in 100 μL 0.1-M carbonate buffer (pH = 9.0). Cy5 (1 mg in carbonate buffer) was added to the oligonucleotides and the conjugation reaction was performed at room temperature for 60 min with occasional mixing. The conjugated oligonucleotide was separated from the free dye using a Sephadex G10 column as described above. The fractions containing the labeled DNA were collected and concentrated using a Speedvac evaporator.

3.3 Protocol for DNA Studies

We have investigated several methods to bind DNA to different supports that can be used as materials for the biochip sampling platform. One method for binding DNA to glass has involved silanization of the glass surface followed by activation with carbodiimide or glutaraldehyde. In early studies, we have used the silanization methods for binding to glass surfaces using 3-glycidoxypropyltrimethoxysilane (GOPS) or aminopropyltrimethoxysilane (APTS) and attempted to covalently link DNA via amino linkers incorporated either at the 3' or 5' end of the molecule during DNA synthesis. More recently, we have investigated binding the DNA probe onto a membrane and subsequently attaching the membrane to the transducer detection surface. This approach avoids the need to bind the bioreceptor onto the transducer and could possibly allow easier large-scale production. Several types of membranes are commercially available for DNA binding: nitrocellulose, charge-modified nylon, etc. The DNA probe is then bound to the membrane using ultraviolet activation.

In these studies, we have used Zeta-Probe membranes for the immobilization of DNA capture probes. Small pieces (0.4 cm square) of Zeta-Probe membrane were prepared from Zeta-Probe sheets stored in a dessicator. Two µL of various dilutions of a 152-mer single-stranded oligonucleotide fragment of the Bac 813 gene of *B. anthracis* DNA (which was used as a model for the detection of *B. anthracis* genomic DNA [17])was spotted onto individual membrane squares. The concentration of the gene fragment spotted onto the individual squares ranged from 50 pmoles to 2 attomoles. The DNA fragment was subsequently immobilized by UV crosslinking for 4 min in a UV irradiation chamber. The membrane squares were then blocked by incubation in a solution of 6 × SSC (1 × SSC = 15 mM sodium citrate, 150 mM sodium chloride, pH 7.0), 1% bovine serum albumin, and 0.2% sodium dodecyl sulfate (SDS) for 1 h at 37 °C. After blocking, hybridization was carried out in the following manner. A hybridization probe of the following sequence (GAT GGG ATT TCT TTC TGA CTT GG) labeled with Cy5 on the 5' end (probe selected using the Probe Selection Tool found at http://lces.med.umn.edu/rawprimer.html) was diluted in the 6 × SSC, 1% bovine serum albumin, 0.2% SDS solution used for blocking. The final concentration of the probe in solution was 0.27 µmolar. Individual membrane squares were placed in microcentrifuge tubes (1.5 mL) and 0.85 mL of probe-containing solution was added to each tube. Hybridization was carried out for 18 h at 50 °C. Following hybridization, the individual squares, still in the microcentrifuge tubes were washed for 15 min at 24 °C with 5 × SSC + 0.1%SDS. The membrane squares were then washed in distilled water 2×, for 1 min each, and then the fluorescence was determined as explained elsewhere. Blanks were made up of membrane squares which either were not spotted with any DNA, or were spotted with herring sperm DNA, and treated subsequently as described above.

3.4 Protocol for Antibody Studies

3.4.1 *Assay for E. coli*

A two-component Cy5-labeled antibody sandwich was used in this immunoassay, as demonstrated in Fig. 5. Heat-killed *E. coli* 157:H7 whole cells were immobilized and, following blocking, were coated with goat anti-*E. coli* antibody, which, in turn, was reacted with Cy5-labeled rabbit anti-goat IgG antibody. A 96-well assay plate with cellulose ester membrane bottoms housed the reactions. The membranes in the wells were pre-wetted for 1 min with 200 μL of phosphate buffered saline (PBS), followed by removal using the multi screen vacuum manifold (Cat. No. MAVM 09601, Millipore Corp., Bedford, MA 01730). Various dilutions of the antigen were then pipetted into the wells. The antigen used was a preparation of heat-killed *Escherichia coli 0157:H7* organisms (Cat. No. 50-95-90, Kirkeguard and Perry Laboratories, Inc., Gaithersburg, MD 20879), containing 7.0×10^9 organisms mL^{-1}. The cell suspension was diluted in

Method for On-chip Antibody Detection Via Sandwich Immunoassay

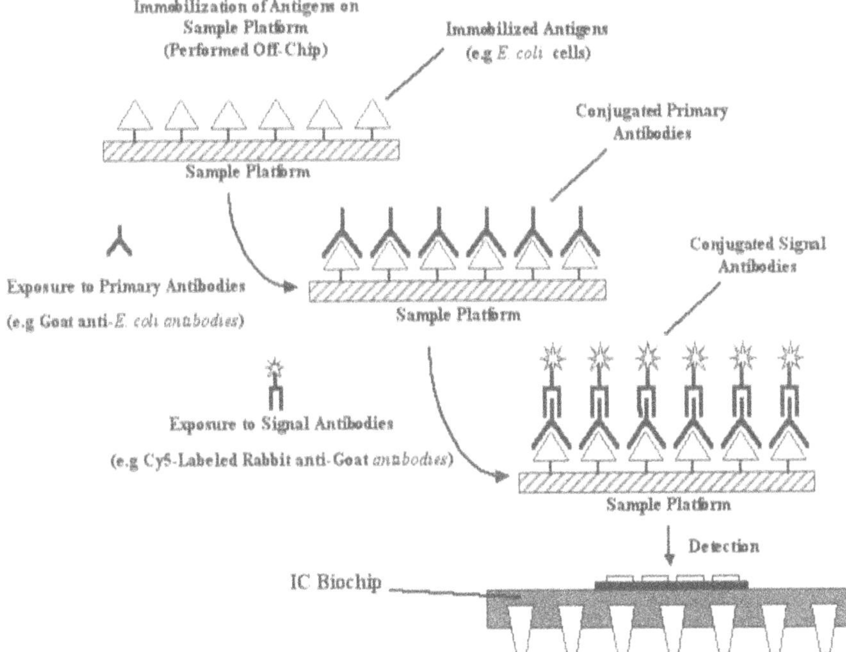

Fig. 5. Schematic diagram of a sandwich immunoassay for *E. coli* detection using Cy5-labeled antibody probes

0.1 M sodium carbonate buffer, pH 9.6. Aliquots of 80 μL of varying dilutions of *E. coli* cell suspension were pipetted into separate wells of the Multi Screen plates. Following 1 h of incubation at room temperature (24 °C), the antigen suspension was removed using the vacuum manifold. 200-μL aliquots of a solution of 0.25% BSA, 0.05% Tween 20 in PBS were added to each well to block any unoccupied binding sites, and the plate was incubated for 1 h at room temperature, followed by removal of the blocking solution. Dilutions of a goat affinity-purified antibody to *E. coli 0157:H7* (Kirkeguard and Perry Laboratories, Cat. No. 01-95-90, Gaithersburg, MD 20879) were prepared in the same buffer used for the blocking step. Usual dilutions were 1:25 or 1:50. All wells received 80-μL aliquots of this primary antibody. After incubation for 1–1.5 h at room temperature, the antibody solution was removed using the vacuum manifold. Each well was then washed 4× using 0.5% Tween 20 in PBS, by filling the well by means of a Pasteur pipette and removing the wash solution by vacuum filtration. A secondary anti-species antibody with covalently attached dye molecules was used to complete the immunoassay sandwich. This antibody was a purified rabbit anti-goat IgG(H+L) antibody (ZyMax Grade) conjugated with Cy5 (Cat. No. 81-1616, Zymed Laboratories, San Francisco, CA 94080). This antibody was diluted (usually 1:25 or 1:50) in the same buffer as was used for the primary antibody, and 80-μL aliquots were dispensed into appropriate wells. After incubation for 1–1.5 h at room temperature in the dark (i.e. plate wrapped in aluminum foil), the wells were emptied by vacuum filtration, and all wells were washed 5–6× with 0.5% Tween in PBS. Negative controls were prepared by treating other wells with blocking solution, primary antibody and secondary antibody, but no *E. coli* antigen.

3.4.2
Assay for FHIT Protein

Two cell lines, MKN74-PRC-FHIT A66 (MKN/FHIT) and MKN74-PRC-E4 (MKN/E4) were kindly provided by Dr Kay Hubner (Kimmel Cancer Center, Philadelphia, PA). Both cell lines were propagated from gastric carcinoma-derived cells. The MKN/FHIT cell line was transfected with a vector for the expression of the fragile histidine triad (FHIT) gene. The MKN/E4 cell line was transfected with a null vector. Except for the missing FHIT gene fragment in the MKN/E4 line, the two cell lines were identical. They were maintained in DMEM media supplemented with 10% fetal bovine serum (FBS) in the presence of 200 μg mL^{-1} Geneticin (GIBCO/BRL) and in a 5% CO_2 atmosphere at 37 °C. The cells were lysed via sonication in a lysis buffer. Cytoplasmic proteins were isolated from the mixture by centrifuging the suspension. The supernatant was analyzed for total protein content using the standard, chemiluminescence-based Pierce-BCA Assay. Extracts of both cell lines were also analyzed via Western Blot assay after gel electrophoresis to confirm the presence and absence of FHIT protein in the MKN/FHT and MKN/E4 lines, respectively. After these preliminary analyses, cell line supernatants were diluted and spotted on Immobilon-P membranes. The membranes were then treated with a blocking solution of 5% nonfat dry milk, 10% FBS in PBS (pH 7.2). The membranes were finally placed in the sampling chamber of the biofluidics system for on-chip bioassays. For on-chip analysis, the membrane was simply exposed to a Cy5-labeled anti-FHIT antibody and rinsed.

3.5 Protocol for DNA/Antibody Combined Assay

DNA (fragment of Bac 813 gene of *B. anthracis* as above) and antigen (*E. coli* 0157 heat-killed cells) were spotted in 2 μL spots onto adjoining segments of a rectangular piece of Zeta-Probe membrane (0.5 by 1.5 cm). Specifically, the amount of DNA spotted was 206 pmoles, and the amount of *E. coli* cells was 1.4×10^6. Following UV fixation and blocking as described above, the membrane was processed through the hybridization protocol also detailed above, except the temperature at which hybridization occurred was 37 °C. Following the initial washing step in 5 × SSC, 0.1% SDS, the membrane strip was carried through an immunoassay protocol as follows. The membrane strip was placed in a microcentrifuge tube, and 0.8 mL of a goat affinity purified antibody to *E. coli* 0157(KPL Inc) diluted to 40 μg mL^{-1} in BSA Diluent/Blocking solution (diluted 1:15 in distilled water) (KPL Inc) was added. The antibody binding step was allowed to occur for 1 h at 37 °C. Following washing in PBS + 0.5% Tween 20, the membrane strip was transferred to another tube, and was incubated for 1h at 37 °C with a Cy5 -labeled rabbit antibody to goat IgG (H + L) (Zymed Laboratories), used at a final concentration of 50 μg mL^{-1} in the BSA Diluent described above. Following extensive washing, the fluorescence was determined. Part of the membrane which was not spotted either with DNA or antigen served as the blank for assay purposes.

3.6 Protocol for ELISA-based Detection of *B. globigii*

In order to provide both a sensitive and selective analysis, an enzyme-linked immunosorbent assay (ELISA) for antibody-based capture and identification of *Bacillus globigii (B.g.)* spores (a surrogate species for *Bacillus anthracis*), was used in conjunction with the biochip detection system. As illustrated in Fig. 6, antibodies specific to a surface antigen on the *B.g.* spore (goat anti-*B. globigii*, lot T 190999-03 purchased from Tetracore, MD) were immobilized onto a Nunc maxisorp protein binding platform (Nunc Maxisorp 96 well plates) overnight at 4 °C. The goat antibody was diluted to 10 μg mL^{-1} in 0.1M carbonate buffer, pH 9.6. The remaining binding sites were blocked for 1 h at room temperature using a BSA diluent/blocking solution concentrate, diluted 1:10 in distilled water (KPL, Gaithersburg, MD). Following blocking, the immobilized antibodies were then incubated with the *B.g.* spores (diluted to various concentrations in PBS) for 1 h at 37 °C, and then washed thoroughly in PBS to remove any unbound target species. Subsequently, a detector antibody (rabbit anti-*B. globigii*, lot # T181099-01, Tetracore, MD) recognizing another epitope on the *B.g.* spore surface, was diluted in BSA diluent/blocking solution concentrate (1:15 dilution in distilled water) to a final concentration of 5 μg mL^{-1}, and similar to before incu-

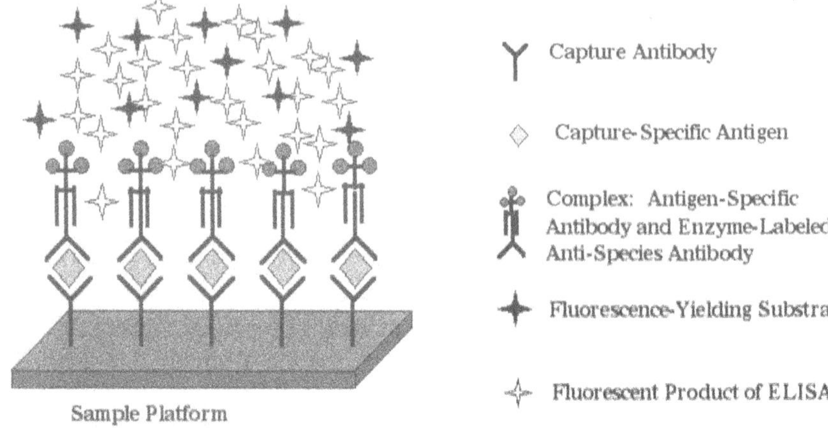

Fig. 6. Schematic diagram of an ELISA assay for *B. globigii* detection using an alkaline phosphatase-conjugated detector antibody and DDAO-P substrate

bated at 37 °C for 45 min followed by a series of washes (PBS + 0.5%Tween 20). The final antibody, goat anti-rabbit IgG (H+L) conjugated with alkaline phosphatase (Jackson Immunoresearch Laboratories, Avondale, PA), was diluted 1: 3000 in AP stabilizer (KPL), and was incubated with the sample complex for 45 min and the unbound enzyme-antibody conjugate was removed through several washes as described above. Finally, the fluorogenic substrate dimethylacridinone phosphate (DDAO-P) 20 μM in 0.1 M Tris, pH 9.9 + 1 mM $MgCl_2$ was incubated at 37 °C with the immunocomplex to yield a detectable fluorescence product. Upon enzymatic cleavage, DDAO-P produces a product (DDAO) with a shift in absorption maximum of over 200 nm relative to the unreacted substrate, allowing the two species to be easily distinguished.

4 Results and Discussion

4.1 Fundamental Evaluations of the IC Biochip via Off-chip Bioassays

We have performed measurements to evaluate the analytical figures of merit of the biochip detection array system. Since the chip design includes on-board data filtering/amplification circuitry, the signals from the biochip were directly recorded without the need of any electronic interface system or signal amplification device. This unique feature of the integrated MFB biochip developed in our laboratory differentiates it from many commercially available chip systems. Figures 7 and 8 demonstrate the quantitative capability of the biochip in monitoring the immobilization of various biomolecules of medical and environmental interest. Figure 7 shows a calibration curve for the detection of a segment of the Bac 816 gene of *Bacillus anthracis* through hybridization of a Cy5-labeled

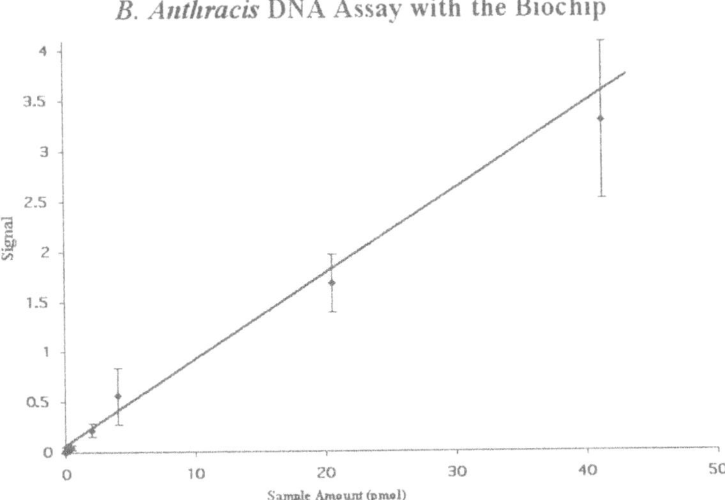

Fig. 7. Calibration curve of for the detection of *Bacillus anthracis* using DNA probes

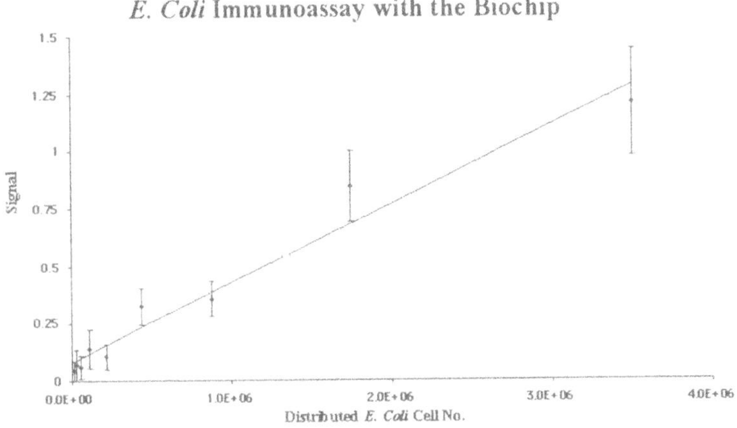

Fig. 8. Calibration curve for the detection of *Escherichia coli* using antibody probes

DNA probe. Each point in the calibration curve represents an average of 16 blank-subtracted sample platform signals acquired with the 4 × 4 photosensor array. The blank was a membrane which had been treated with herring sperm DNA and blocking agent before being exposed to the Cy5-labeled DNA probe, thus enabling correction for nonspecific binding. As demonstrated in our calibration measurements, the biochip exhibited a linear dynamic range of more than 4 orders of magnitude for immobilized DNA. In addition, an average relative standard deviation of 15% was observed for element-to-element response within the photosensor array. This reproducibility is quite good considering the variability due to the inhomogeneity of the membrane substrate.

Figure 8 shows a calibration curve for the detection of immobilized *Escherichia coli* 157:H7. In contrast to the *Bacillus anthracis* study described above, this screening method involved antibody/antigen interaction, analogous to the Western Blot assay. A two-part Cy5-labeled antibody probe was used in this three-part sandwich assay. Heat-killed *E. coli* 157:H7 whole cells were immobilized and, following blocking, were coated with goat anti-*E. coli* antibody, which, in turn, was reacted with Cy5-labeled rabbit anti-goat IgG antibody. Each point in the calibration curve represents an average of 16 blank-subtracted sample platform signals. The blank was a membrane, which had been treated with blocking agent and goat anti-*E. coli*. antibody before being exposed to the Cy5-labeled rabbit anti-goat IgG antibody probe. In this study, the biochip yielded a linear dynamic range of approximately 2.5 orders of magnitude (1.4×10^4 cells – 3.5×10^6 cells) for immobilized *E. coli*. The upper limit of the linear dynamic range can be attributed not to the linear range of the photosensors of the biochip, but rather to fluorescence quenching or self-absorption at higher concentrations. As before, we have observed an average relative standard deviation of 15% for element-to-element response exhibited by the photosensor array in this three-part assay. This result is particularly interesting due to the fact that not only a different type of sampling platform (membrane) was used, but also that the extra „layer" of this „sandwich" assay likely introduced a higher potential for variability. The results of this study demonstrate the feasibility of quantitative determination of pathogenic agents using the biochip.

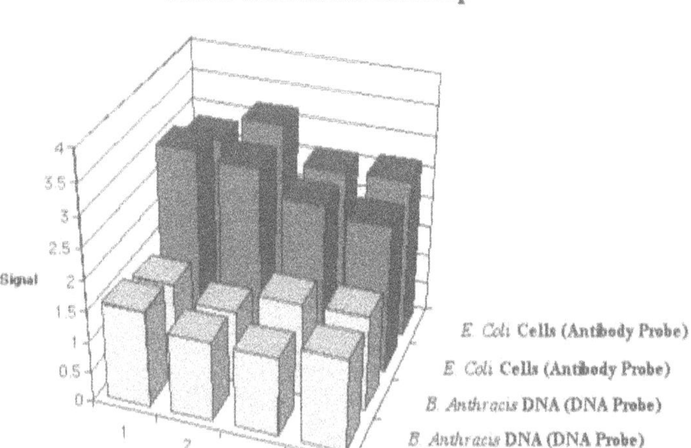

Fig. 9. Detection of *E. coli and B. anthracis* using the multifunctional biochip system. Two rows of the biochip were used for the detection of 1 x 10^6 *E. coli* cells via Cy5-labeled antibody probes, while the other two rows served for the detection of the 41.2 pmol *B. Anthracis* using Cy5-labeled DNA probes

Figure 9 illustrates the detection of *E. coli* and *B. anthracis* with the multi-functional biochip system using antibody and DNA probes, respectively. Each signal bar of the graph represents the signal acquired with an individual detection element of the 4×4 photosensor array. Sensing arrays of half of the biochip show the detection of 1×10^6 *E. coli* cells via Cy5-labeled antibody probes, while the other half was used for the detection of the 41.2 pmol *B. anthracis* gene fragment via Cy5-labeled DNA probes. Samples were immobilized on sample platforms as described for the establishment of the individual calibrations shown in Figs. 7 and 8.

The unique feature of the biochip is the capability to perform different types of bioassay on a single platform using DNA and antibody probes simultaneously. Hybridization of a nucleic acid probe to DNA biotargets (e.g., gene sequences, bacteria, viral DNA) offers a very high degree of accuracy for identifying DNA sequences complementary to that of the probe. In addition to DNA probes, the MFB uses another type of bioreceptor, i.e., antibody probes that take advantage of the specificity of the immunological recognition. Antibodies have been used previously in immunosensors for chemical and biological analysis [18–20]. Just as specific configurations of a unique key enable it to enter a specific lock, so in an analogous manner, an antigen-specific antibody "fits" its unique antigen. Thus an antigen-specific antibody interacts with its unique antigen in a highly specific manner, so that the total three-dimensional biochemical conformation of antigen and antibody are complementary. This molecular recognition capability of antibodies is the key to their usefulness in immunosensing channels of the MFB; molecular structural recognition allows one to develop antibodies that can bind specifically to chemicals, biomolecules, and microorganism components. The results of this study using antibody against *E. coli* and DNA probes for *B. anthracis* demonstrate the feasibility of the multi-functional biochip for the detection of multiple biotargets of different functionality (DNA, proteins, etc.) using a single biochip platform.

4.2 Application of the ELISA Technique to Biochip-based Detection

Enzyme-linked immunosorbent assay (ELISA), is a method capable of a selective, yet sensitive analysis of biological samples. Typically, the two most common enzymes that are utilized in an ELISA protocol are alkaline phosphatase and horseradish peroxidase. Both of these enzymes are ideally suited for many applications since they possess very high turnover rates, allowing for a rapid accumulation of a measurable product that can be correlated to the concentration of the target present in the sample. Recently, a novel fluorogenic alkaline phosphatase substrate has become available as an alternative to fluorescein diphosphate (FDP) and 4-methylumbelliferyl phosphate (MUP), which have been in widespread use for many years. DDAO-P has a number of advantages over other fluorogenic substrates including avoidance of the biphasic kinetics of fluorescein- and rhodamine-based substrates due to a single hydrolysis-sensitive moiety. Upon enzymatic cleavage, DDAO-P produces a product with a shift in absorption maximum of over 200 nm relative to the unreacted sub-

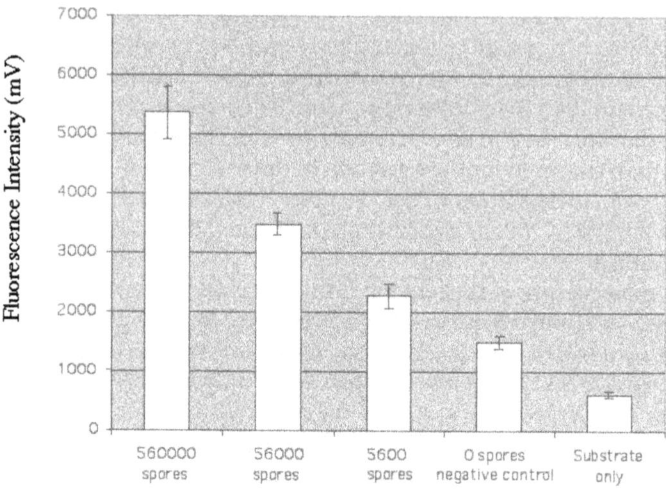

Fig. 10. Detection of *B. globigii* spores using ELISA assay and biochip detection

strate, allowing the two species to be easily distinguished. Unlike other fluorogenic substrates, DDAO-P yields a product that fluoresces in the red (emission maxima 659 nm), and the resulting product is conveniently excited by the 633 nm spectral line of a helium neon laser or the 635 nm line of a small diode laser (absorption maxima 646 nm).

Figure 10 shows the fluorescence intensity of the DDAO product for different concentrations of target *B.g.* spores obtained using an ELISA and the biochip detection. The last bar depicts the fluorescence intensity of the substrate, which has not been hydrolyzed to the DDAO product. The negative control exhibits some non-specific binding, as evidenced by the slightly higher fluorescence intensity relative to that observed for the substrate. However, the fluorescence intensity for the 5600 spore sample is clearly distinguishable from the negative control. In addition, the fluoresence intensity was found to increase with increasing concentrations of spores. Enzyme-based amplification offers an attractive alternative to nucleic acid- based amplification methods, such as polymerase chain reaction, since it eliminates the need for a cell lysing step that is particularly difficult to achieve with spores, such as *Bacillus anthracis.* In addition, the ability to perform this analysis using a small diode laser for excitation of the product, along with the self-contained ORNL biochip design, allows for a small, compact system for biological warfare agent detection.

4.3 Evaluation of the Biofluidics-based Biochip System for On-chip Bioanalysis

We have developed a biofluidics system for sample and regent delivery to the sensing area of the biochip for on-chip, real-time monitoring of bioassays. With such a system, potential applications of the IC biochip may be extended to on-line monitoring of liquid or gaseous flow streams which ultimately end in liquid-based assays. Furthermore, such a system can also aid in fundamental studies which can facilitate optimization of sample platform design and various bioassay reaction parameters (e.g. temperature, reagent concentrations, incubation periods, reagent flow rates, etc.).

4.3.1
Assay for E. coli

The sandwich immunoassay method for detection of *E. coli 0157:H7* has been performed with the biofluidics system to demonstrate the feasibility of on-chip monitoring of bioassays. A practical approach to biofluidics-based detection is to pretreat sampling platforms with bioreceptors prior to installation in the reaction chamber of the biofluidics system. Sample and sensing probe introduction steps would subsequently be performed with the biofluidics system. In this study, we closely mimicked this approach by pretreating a membrane with *E. coli* organisms and *E. coli*-specific antibodies prior to placement in the reaction chamber of the biofluidics module (the first two steps of the sandwich immunoassay). The pretreated membrane was subsequently incubated with the Cy5-labeled anti-species antibody and rinsed on-chip via the biofluidics system.

Figure 11 illustrates signal traces for an Cy5-labeled antibody-based assay for *E. coli* [15]. These profiles are real-time stripchart recordings corresponding to the signal output from individual detection elements of the biochip. The initial baseline region of the profiles corresponds to the signal yielded by the pretreated membrane immersed in PBS buffer, which included approximately 20,000 organisms per probed area (as defined by the laser spot size). The sharp increase in signal corresponds to the inflow of Cy5-labeled antibody solution (0.04 μg mL^{-1}) in the reaction chamber, while the ensuing high plateau region corresponds to a stop-flow 30 min incubation period. The sharp decrease in signal at the end of the incubation period marks the onset of rinsing with the Tween/PBS buffer solution. The intensity of Cy5 fluorescence signal is thereafter recorded for 15 min of rinsing with a flow rate of 1 mL min^{-1} (note: the time scale is different for the incubation and rinsing regions of the profile). As demonstrated by the figure, the signal profiles for the *E. coli* treated membrane and the blank take on significantly different characteristics during the rinsing step. The blank membrane was pretreated with blocking reagent and the *E. coli*-specific antibody, but no *E. coli* organisms. The fluorescence profile for the blank membrane illustrates some significant signal from residual nonspecific interaction. Nevertheless, the low noise of the traces permits the confirmation of *E. coli* detection, particularly after 15 min of rinsing.

Using the procedure described above, the biofluidics system was used to measure signals for various numbers of probed *E. coli* organisms. The results are illustrated in Fig. 12 [15]. The signal levels depicted by the figure correspond to blank-subtracted signals obtained after 15 min of rinsing. Repeated measurements performed with the biochip have produced a relative standard deviation of approximately 15%. While a proportional response of signal level to the

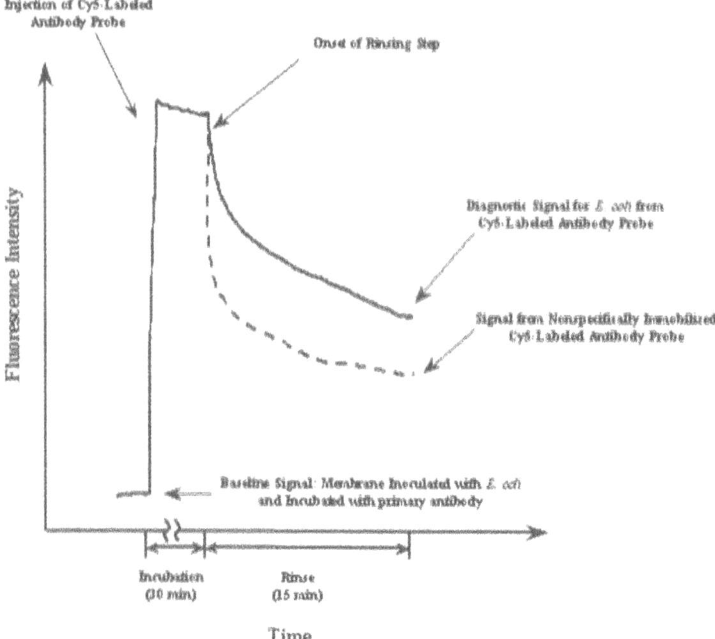

Fig. 11. Real time monitoring of an on-chip bioassay for *E. coli* using Cy5-labeled antibody probes. The fluorescence profiles were generated from two membranes. One membrane was innoculated with approximately 20,000 organisms (solid line) and the other membrane was the control (dashed line). (Source: Stokes DL, Griffin GD, Vo-Dinh T (2001) Fresenius' J Anal Chem 369:295)

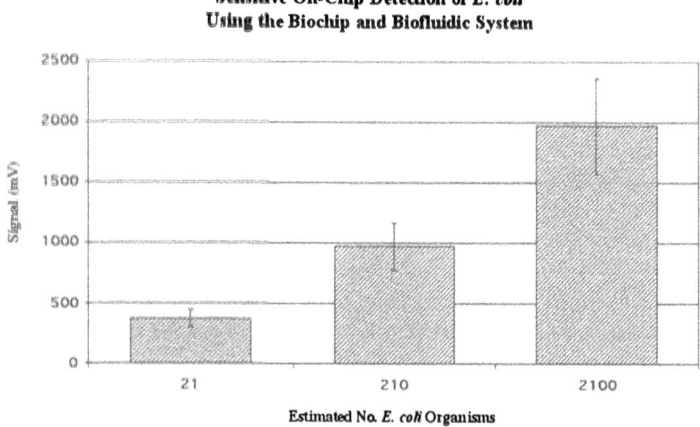

Fig. 12. Fluorescence signals (blank-subtracted) observed for various numbers of *E. coli* organisms following a sandwich immunoassay with Cy5-labeled antibody probes. Signals were acquired after 15 min of rinsing. (Source: Stokes DL, Griffin GD, Vo-Dinh T (2001) Fresenius' J Anal Chem 369:295)

number of probed organisms is observed, the true highlight of this experiment was the detection of as little as 21 *E. coli* organisms using the sandwhich immunoassay technique.

4.3.2
Assay for FHIT Protein

The biofluidics system has also been used for on-chip monitoring of an assay for the FHIT protein. Human FHIT protein, a tumor suppressor, has recently become a significant factor in cancer research. This protein is composed of 147 amino acids (16.8 kDa mass) assembled from a 1.1 kilobase message, transcribed from a 1 megabase strand of genomic DNA. The genomic strand is located in chromosome region 3p14.2. Alterations of the FHIT gene have been associated with both early and late events in the development of a variety of cancers [21–28]. As a result, loss of FHIT gene expression has been correlated with various carcinomas (e.g. lung, esophageal). Indeed, the overall frequency of FHIT expression or inactivation in pre-malignant and malignant tissues might provide important diagnostic and prognostic information [29], ultimately aiding in early-detection of FHIT-associated cancers. Clearly, FHIT protein detection could be a valuable medical diagnostic application of the IC biochip, particularly when coupled to the biofluidics system for assays performed in a clinical setting.

Figure 13 illustrates the detection of FHIT protein achieved with an antibody-based assay, performed on-chip via the biofluidics system. Each bar represents the total signal observed from the total protein sample deposited on

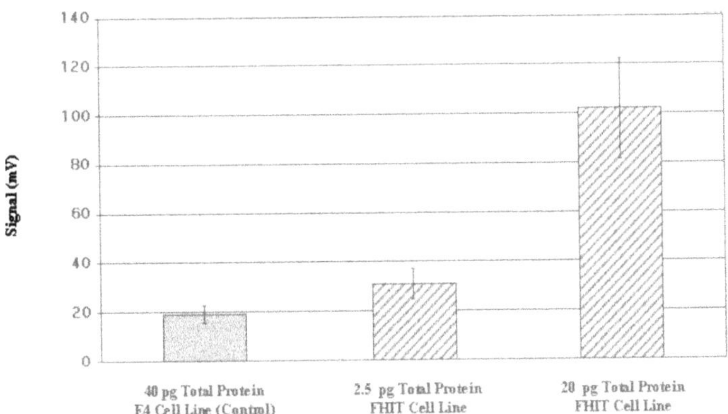

Fig. 13. Fluorescence signals observed for Cy5-labeled anti-FHIT antibody probes immobilized by various amounts of cytoplasmic protein samples obtained from MKN/FHIT and MKF/E4 cells. The MKF/E4 sample serves as the control for FHIT protein detection, lacking only the FHIT protein component of the cytoplasm. Signals were acquired after 15 min of rinsing

the sampling platform. Total protein estimates for each assay were based on the original protein concentrations for each cell line (as determined with the Pierce BCA assay), the dilution factor, the spotted sample volume, and the surface area probed by the excitation laser spot. It is important to note that each spotted protein sample contained all proteins present in the supernatant following centrifugation of lysed cells. This isolated portion should have included all cytoplasmic proteins. Therefore, the results depicted in Fig. 13 demonstrate selective detection of FHIT protein as immobilized from a complex sample of a cellular proteins. Signals from protein samples of both the MKF/E4 and MKF/FHIT cell lines are shown in the figure. The samples derived from the two cell lines should have been equivalent except for the presence of the FHIT protein in the MKF/FHIT sample; hence, the MKF/E4 sample could have been considered a nearly ideal control. Fluorescence signal observed from this control could therefore be attributed to nonspecific binding of the Cy5-labeled probe on the sampling platform. Nevertheless, the signal observed for a 40 pg sample of the MKF/E4 cell line protein was significantly less than the signal observed for a much smaller 2.5 pg sample of the MKF/FHIT cell line. Furthermore, the FHIT protein portion of the 2.5 pg total protein sample could have been very small. In other words, the actual amounts of FHIT protein detected was likely to have been much smaller than the total protein amounts reported in Fig. 13, hence demonstrating the potential for highly sensitive detection of FHIT protein with the biofluidics-based IC biochip system. Finally, a proportional signal response with respect to the amount of sample protein illustrates the potential for quantitative detection.

4.4
Portable IC Biochip Prototype with Biofluidic System

The IC Biochip optical detection system and biofludics system has been packaged in a portable device for field studies. The size of this new system, illustrated in Fig. 14a, is 20 cm × 20 cm × 30 cm and weighs approximately 4.5 kg. It represents a significant advance in functionality, miniaturization and compactness in comparison to typical laboratory-based systems used for the many studies presented in this work. One factor in this miniaturization has been the development of miniature diode lasers operating at 635-nm and powered by low voltage (e.g. 5 V) power supplies. Using this type of laser, we have achieved sensitivities comparable to those observed with the larger HeNe laser-based systems. Furthermore, the portable system has an electronic control board, which allows for computer control of all fluidic operations and heating processes during the assay steps. Finally, the biofluidics system in this new device features a free-standing, disposable assay chamber in the form of a unique self-sealing "snap-in" cartridge (Fig. 14b). This cartridge integrates heating elements and fluidic connections through a septum arrangement as the cartridge is snapped into place through the side of the box. All of these features, along with relatively low power consumption and cost-effectiveness, make the multifunctional biochip technology an attractive tool for medical diagnostics and pathogen detection in clinical and field settings.

A Handheld Biochip Device with Biofluidics Cartridge

(a)

(b)

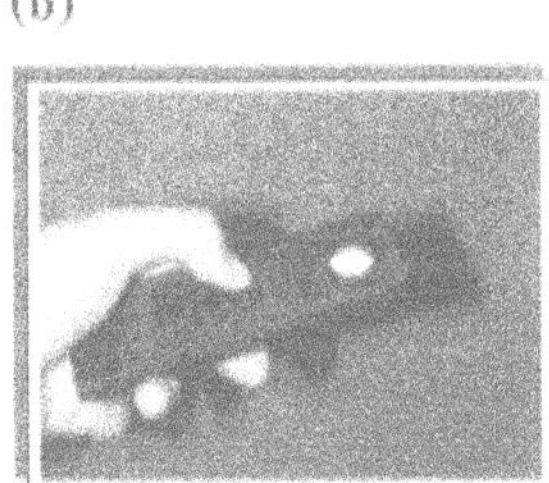

Fig. 14. Photographs of an IC Biochip prototype device packaged in a compact box (a) and featuring a free-standing, self-sealing cartridge (b) which houses the sampling platform and snaps into place through the side of the unit for on-chip bioassays

5 Conclusion

There is an increasing need to develop rapid, simple, cost-effective medical devices for screening multiple medical diseases simultaneously and to monitor infectious pathogens for early medical diagnosis. Such a system will be useful in physician offices or for use by relatively unskilled personnel in the field for human health protection. The MFB, which is a truly integrated biochip system that comprises probes, samplers, detector as well as amplifier, and logic circuitry on board, could enable a rapid and inexpensive test for multiple diseases and for a wide variety of applications. With its multi-functional capability, the MFB technology is the only current system that allows simultaneous detection of multiple biotargets simultaneously. Such a device could provide information on both gene mutation (with DNA probes) and gene expression (with antibody probes against proteins) simultaneously.

Acknowledgements. This work was sponsored by by the U.S. Department of Energy, Chemical and Biological National Security Program, under contract DE-AC05-00OR22725 with UT-Battelle, LLC.

List of Abbreviations

AP	alkaline phosphatase
APTS	aminopropyltrimethoxysilane
B. anthracis	*Bacillus anthracis*
BCA	bicinchoninic acid
B.g.	*Bacillus globigii*
BSA	bovine serum albumin
CMOS	complementary metal-oxide semiconductor
CPG	controlled-pore-glass
Da	Dalton
DDAO	dimethylacridinone
DDAO-P	dimethylacridinone-phosphate
DMEM	Dulbecco's Modification of Eagle's Medium
E. coli	*Escherichia. coli*
FBS	fetal bovine serum
FDP	fluorescein diphosphate
FHIT	fragile histidine triad
GOPS	3-glycidoxypropyltrimethoxysilane
IC	integrated circuit
IgG	immunoglobulin G
MFB	multi-functional biochip
MUP	4-methylumbelliferyl phosphate
PBS	phosphate buffered saline
PEEK	polyetheretherketone
SDS	sodium dodecyl sulfate
SSC	saline- sodium citrate
Tris	tris (hydroxymethyl) aminomethane
UV	ultraviolet

References

1. Fodor SPA, Read JL , Pirrung MC, Stryer LT, Lu A, Solas D (1991) Science 251:767
2. Anderson RC, McGall G, Lipshutz RJ (1997) Polynucleotide arrays for genetic sequence analysis. In: Manz A, Becker H (eds) Microsystem technology in chemistry and life science. Springer Verlag, Berlin, p 117
3. Schena M, Shalon D, Davis RW, Brown PO (1995) Science 270:467
4. Piunno PAE, Krull UJ, Hudson RHE, Damha MJ, Cohen H (1995) Anal Chim Acta 228: 205
5. Vo-Dinh T, Tromberg BJ, Griffin GD, Ambrose KR, Sepaniak MJ, Gardenhire EM (1987) Appl Spectrosc 41:735
6. Vo-Dinh T, Griffin GD, Sepaniak MJ (1991) Fiberoptic immunosensors. In: Wolfbeis OS (ed) Fiber optic chemical sensors and biosensors, vol 2. CRC Press, Boca Raton, Florida, p 217
7. Kumar P, Wilson RC, Valdes JJ, Chambers JP (1994) Mat Sci Eng C1:187
8. Eggers M, Hogan M, Reich RK, Lamture J, Ehrlich D, Hollis M, Kosicki B, Powdrill T, Beattie K, Smith S, Varma R, Gangadharam R, Mallik A, Burke R, Wallace D (1994) Biotechniques 17:516

9. Wolfbeis OS (ed) (1991) Fiber optic chemical sensors and biosensors. CRC Press, Boca Raton, Florida
10. Vo-Dinh T, Isola NR, Alarie JP, Landis D, Griffin GD, Allison S (1998) Instrum Sci Technol 26:503
11. Vo-Dinh T (1998) Sens Actuators B 51:52
12. Vo-Dinh T, Alarie JP, Isola NR, Landis D, Wintenberg AL, Ericson MN (1999) Anal Chem 71:358
13. Vo-Dinh T (2003) Biodrugs (in press)
14. Allain LR, Askari M, Stokes DL, Vo-Dinh T (2001) Fresenius' J Anal Chem 371:146
15. Stokes DL, Griffin GD, Vo-Dinh T (2001) Fresenius' J Anal Chem 369:295
16. Vo-Dinh T, Askari M (2001) Cur Genomics 2:399
17. Patra G, Sylvestre P, Ramisse V, Therasse J, Guesdon JL (1996) FEMS Immunol Med Mic 15:223
18. Gopel G, Hesse J, Zemel JN (eds) (1992) Sensors, vol.3. VCH, New York, NY, 1991
19. Tromberg BJ, Sepaniak MJ, Vo-Dinh T, Griffin GD (1987) Anal Chem 59:1226
20. Rowe CR, Scruggs SB, Feldstein MF, Golden JP, Ligler FS (1999) Anal Chem 71:433
21. Huebner K, Garrison PN, Barnes LD, Croce CM (1998) Ann Rev Genet 32:7
22. Ingvarsson S, Agnarsson BA, Sigbjornsdottir BI, Kononen J, Kallioniemi OP, Barkardottir RB, Kovatich AJ, Schwarting R, Hauk WW, Huebner K, McCue PA (1999) Cancer Res 59:2682
23. Huiping C, Jonasson JG, Angarsson BA, Sigbjornsdottir BI, Huebner K, Ingvarsson S (2000) Eur J Cancer 36:1552
24. Mangray S, King TC (1998) Front Biosci 3:D1148
25. Mueller J, Werner M, Siewert JR (2000) Recent Results Cancer Res 155:29
26. Gartenhaus RB (1997) Oncogene 14:375
27. Sozzi G, Pastorino U, Moiraghi L, Tagliabue E, Pezzella F, Ghirelli C, Tornielli S, Sard L, Huebner K, Pierotti MA, Croce CM, Pilotti S (1998) Cancer Res 58:5032
28. Mori M, Mimori K, Shiraishi T, Alder H, Inoue H, Tanaka Y, Sugimachi K, Huebner K, Croce CM (2000) Cancer Res 60:1177
29. Sozzi G, Musso K, Ratcliffe C, Goldstraw P, Pierotti MA, Pastorino U (1999) Clin Cancer Res 5:2689

CHAPTER 7

Surface Plasmon Resonance Biosensors for Food Safety

Jiří Homola

1 Introduction

Technology for early detection and identification of biological substances is urgently needed in fields such as environmental protection, biotechnology, medicine, and food and drug screening. In the United States, foodborne illnesses caused by chemical contaminants, toxins and bacterial pathogens result in medical and lost productivity costs of up to $ 22 billion annually [1]. Therefore, detection of food safety-related substances is of paramount importance to food producers, processors, distributors and regulatory agencies. Although numerous detection methods have been developed and implemented in centralized testing sites (high performance liquid chromatography, gas chromatography mass spectroscopy, culturing including Gram-staining and microscopic examination, *etc.*), the intensive search for cost-effective and practical methods capable of detecting very low concentrations of chemical contaminants, toxins and bacterial pathogens in food samples in the field continues. In recent years, various sensor technologies have been developed (electrochemical sensors [2], piezoelectric sensors [3], electrical impedance sensors [4], optical sensors [5]) and tested for detection of analytes implicated in food safety [6, 7]. Optical sensors offer several important advantages. The performance of optical sensors is insensitive to electromagnetic interference. In addition, optical sensors do not require electrical signals in the sensing area and therefore can be operated in hazardous environments of industrial plants. Several types of optical biosensors have been demonstrated for detection of chemical contaminants, bacterial pathogens and toxins in food. These include fluorescence-based sensors [8], grating coupler sensors [9], resonant mirror sensors [10], and surface plasmon resonance (SPR) sensors [11]. The fluorescence-based sensors offer high sensitivity, but due to the use of labels, they require either multi-step detection protocols resulting in longer detection times or delicately balanced affinities of interacting biomolecules for displacement assays [12] causing sensor cross-sensitivity to non-target analytes. Grating coupler, resonant mirror and SPR sensors are label-free sensor technologies and thus, in principle, allow for direct and continuous detection.

This paper focuses on the surface plasmon resonance biosensor technology and evaluation of its potential for food safety applications. The report reviews fundamentals of SPR sensing (Section 2), presents main implementations of SPR biosensors (Section 3), and discusses applications of SPR biosensor technology for detection of analytes implicated in food safety (Section 4).

2 Fundamentals of Surface Plasmon Resonance (SPR) Biosensors

SPR biosensors exploit optical spectroscopy of special electromagnetic waves, surface plasmon-polaritons, to measure refractive index changes produced by binding events involving a target analyte.

2.1 Surface Plasmon-Polaritons and their Excitation by Light Waves

Surface plasmon-polaritons (SPP) or surface plasma waves (SPW) are special modes of electromagnetic field which can exist at the interface between a dielectric and a metal that behaves like nearly-free electron plasma. An SPP is a transverse-magnetic (TM) mode (the magnetic vector is perpendicular to the direction of propagation of the wave and parallel to the plane of the interface) and is characterized by its propagation constant and field distribution. The propagation constant of an SPP, β can be expressed as follows:

$$\beta = \frac{\omega}{c}\sqrt{\frac{\varepsilon_M \varepsilon_D}{\varepsilon_M + \varepsilon_D}}, \tag{1}$$

where ω is the angular frequency, c is the speed of light in vacuum, and ε_D and ε_M are dielectric functions of the involved dielectric and metal [13, 14]. This equation describes an SPP that propagates along the interface if the real part of ε_M is negative and its absolute value is smaller than ε_D. At optical wavelengths, this condition is fulfilled for several metals, of which gold is most commonly used in SPR biosensors. The real part of the propagation constant is related to the effective refractive index, N:

$$N = \frac{c}{\omega}\mathrm{Re}\{\beta\} = \mathrm{Re}\left\{\sqrt{\frac{\varepsilon_M \varepsilon_D}{\varepsilon_M + \varepsilon_D}}\right\}, \tag{2}$$

where Re{} denotes the real part of a complex number. The imaginary part of the propagation constant is related to the modal attenuation b (in dB/cm if β is in 1/m):

$$b = \mathrm{Im}\{\beta\}\frac{0.2}{\ln 10} = \mathrm{Im}\left\{\sqrt{\frac{\varepsilon_M \varepsilon_D}{\varepsilon_M + \varepsilon_D}}\right\}\frac{0.2\omega}{c\ln 10}, \tag{3}$$

where Im{} denotes the imaginary part of a complex number.

Spectral dependencies of the effective refractive index and mode attenuation for an SPP propagating along a gold surface are shown in Fig. 1. As follows from Fig. 1b, the SPP undergoes substantial attenuation in the direction of propagation. The propagation length of an SPP – the distance in the direction of prop-

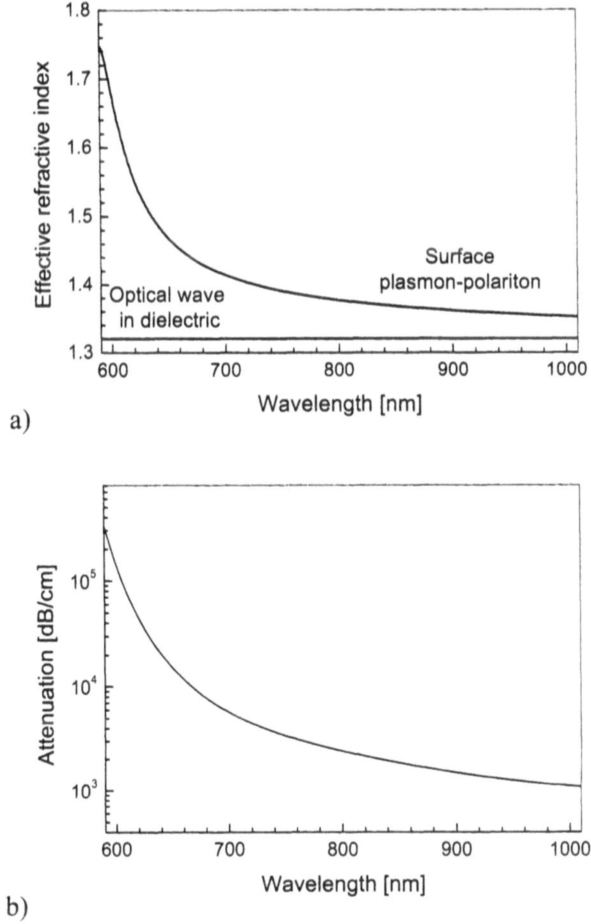

Fig. 1. Effective refractive index (a) and attenuation (b) of an SPP as a function of wavelength for an SPP propagating along the interface between gold and a non-dispersive dielectric (refractive index - 1.32). The dispersion curve for an optical wave in the dielectric medium is shown for comparison

agation at which the energy of the SPP decreases by a factor of 1/e – is less than 40 μm for wavelengths in visible and near infrared regions.

Figure 2, showing the field profile of an SPP, suggests that the SPP field is concentrated at the metal-dielectric interface and decreases exponentially into both media with an increasing distance from the interface. The electromagnetic field of an SPP is distributed in a highly asymmetric fashion and most of the field is concentrated in the dielectric. For SPP's at the gold-aqueous environment interface, the penetration depth (the distance from the interface at which the amplitude of the field falls to 1/e of its value at the surface) is typically 20–30 nm and 100–500 nm in metal and dielectric, respectively, in visible and near infrared

a)

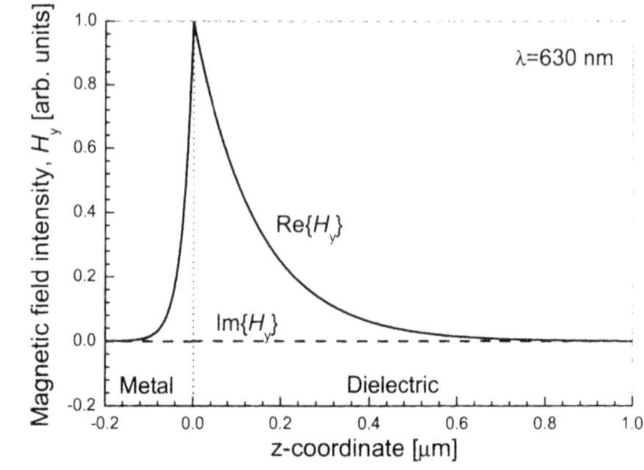

b)

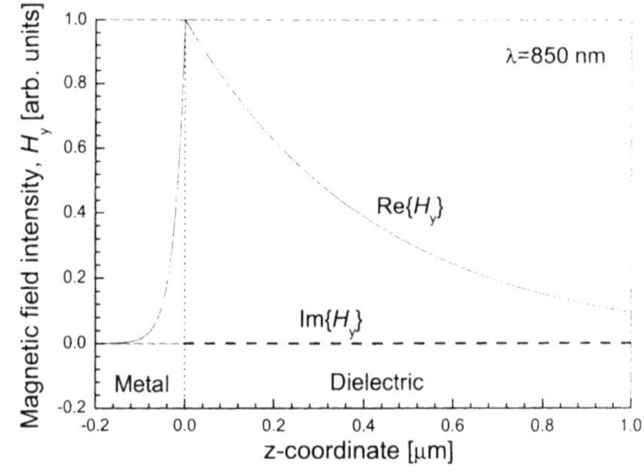

Fig. 2. Distribution of the magnetic field intensity for an SPP at the interface between gold and dielectric (refractive index of the dielectric – 1.32) in the direction perpendicular to the interface (x-y plane), calculated for the wavelengths of 630 nm (a) and 850 nm (b)

regions. The later quantity is particularly significant for SPR sensing, as it represents the depth probed by optical sensors using SPPs.

An optical wave can excite an SPP if the component of the light's wave vector that is parallel to the interface, matches that of the SPP. As the propagation constant of an SPP at the metal-dielectric interface is larger than that provided by the component of the wave vector of light in the dielectric (see Fig. 1a), the SPP cannot be excited directly by light incident onto a smooth metal surface. In order to allow for excitation of an SPP by a light wave, the light's wave vector needs to be enhanced to match that of the SPP. An enhancement of a light wave's wave vector can be produced, for instance, by passing light through an optically denser medium (Fig. 3a). A light wave passes through a high refraction index prism and is totally reflected at the prism base, generating an evanescent wave which penetrates a metal film. This evanescent wave propagates along the inter-

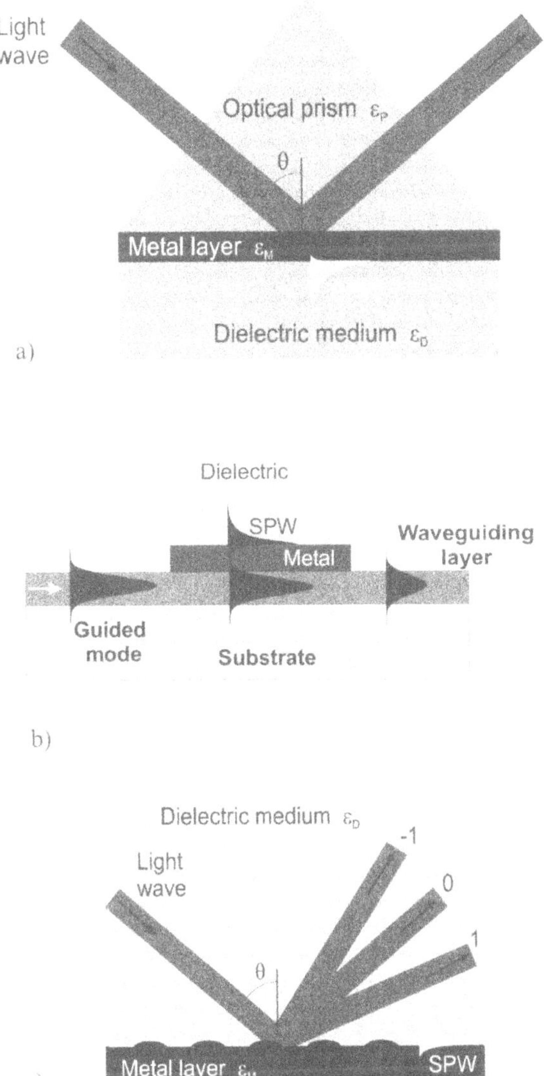

Fig. 3. Excitation of surface plasmon-polaritons (SPP). a) Excitation by a light beam in the Kretschmann geometry of attenuated total reflection. b) Excitation by a guided mode of optical waveguide. c) Excitation by light diffraction on the surface of a diffraction grating

face with a propagation constant, which can be adjusted to match that of the SPP by controlling the angle of incidence. Thus, the matching condition can be fulfilled, allowing light to be coupled to the SPP. This method is referred to as the attenuated total reflection (ATR) method [14]. The process of exciting an SPP in an optical waveguide-based SPR structure (Fig. 3b) is similar to that in the ATR coupler. A light wave is guided by the waveguide and, when entering the region

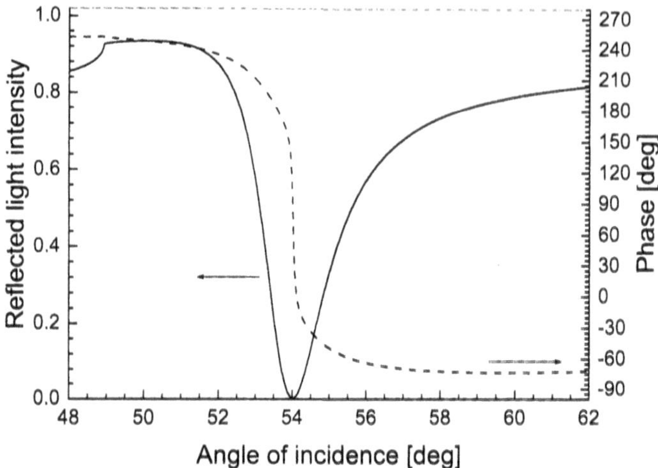

Fig. 4. Intensity and phase of light wave reflected in the Kretschmann geometry consisting of an SF14 glass prism (refractive index – 1.65), a gold layer (thickness – 50 nm), and a low refractive index dielectric medium (refractive index – 1.32), wavelength – 682 nm

with a thin metal overlayer, it evanescently penetrates through the metal layer. If the SPP at the outer boundary of the metal layer and the guided mode are phase-matched, the light wave excites the SPP. Enhancement of a light wave's wave vector can be also accomplished by diffraction on a diffraction grating, Fig. 3c. If a light wave is made incident on a diffraction grating, a series of diffracted waves is produced [15]. The component of the wave vector of these waves parallel to the interface is diffraction-increased by an amount which is inversely proportional to the period of the grating. Thus the wave vector of diffracted light can be matched to that of an SPP, allowing light coupling to the SPP.

The excitation of an SPP is accompanied by the transfer of the light wave energy into the energy of the SPP and its dissipation in the metal film, producing a characteristic absorption dip in the spectrum of the light wave interacting with the SPP. A typical SPR dip in the angular spectrum of light reflected in the three-medium Kretschmann configuration is illustrated in Fig. 4. It should be noted that besides the change in amplitude, the light wave exciting an SPP undergoes also a change in phase.

2.2 Surface Plasmon Resonance Sensors

When an SPP propagates along a metal-dielectric interface, its field probes the dielectric medium. Any change in the refractive index of the dielectric results in a change in the propagation constant of the SPP. Optical sensors based on resonant excitation of surface plasmons, often referred to as surface plasmon resonance (SPR) sensors, take advantage of this phenomenon and measure changes in the propagation constant of an SPP to determine changes in the refractive index. Changes in the propagation constant of the SPP are observed as changes in characteristics of the light wave interacting with the SPP. Based on which characteristic of a light wave interacting with the SPP is measured, SPR sensors can be classified as follows:

1. *SPR sensors with angular modulation.* The component of the light wave's wave vector parallel to the metal surface matching that of the SPP is determined by measuring the coupling strength at a fixed wavelength and multiple angles of incidence and determining the angle of incidence yielding the strongest coupling, [16, 17]. Figure 5 illustrates implementation of this approach in the ATR method.

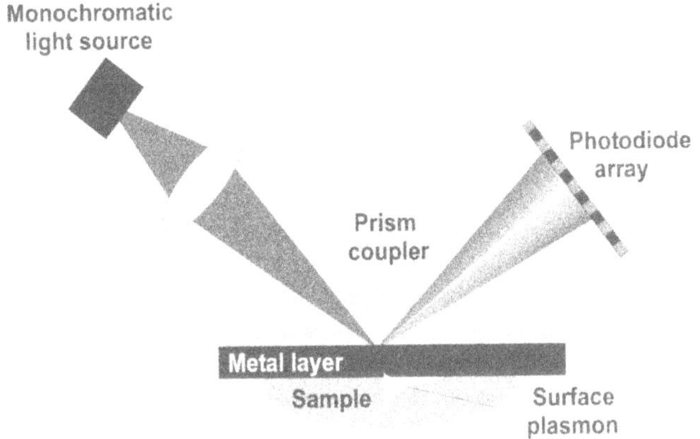

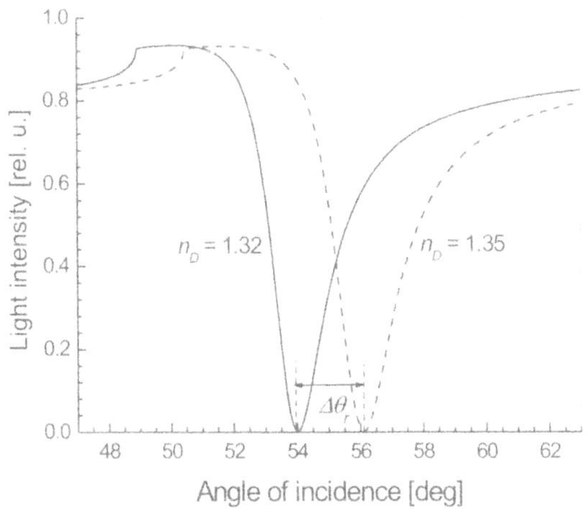

Fig. 5. SPR sensor based on angular modulation and the Kretschmann geometry of the ATR method (structure: SF14 glass prism – 50 nm thick gold layer – sample) and intensity of a transverse magnetic (TM) light wave exciting an SPP as a function of the angle of incidence for two different refractive indices of the sample; wavelength – 682 nm

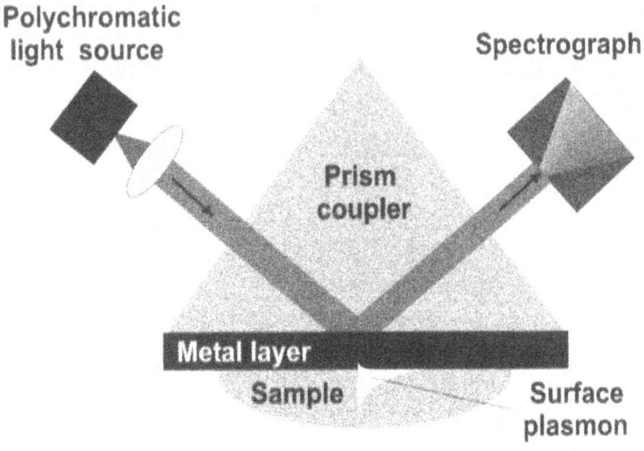

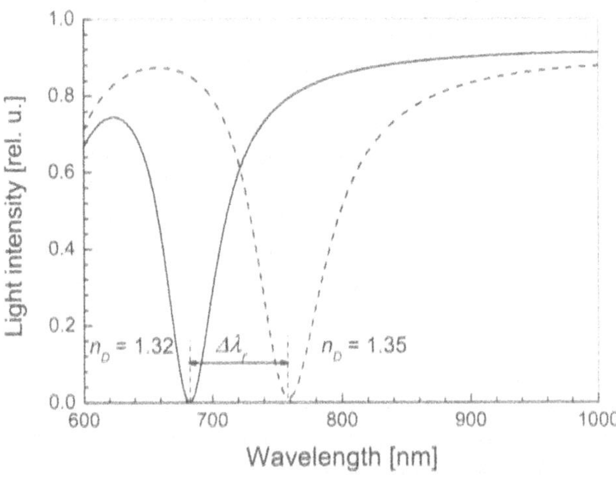

Fig. 6. SPR sensor based on wavelength modulation and the Kretschmann geometry of the ATR method (structure: SF14 glass prism – 50 nm thick gold layer – sample) and intensity of a transverse magnetic (TM) light wave exciting an SPP as a function of the wavelength for two different refractive indices of the sample; angle of incidence – 54 deg

2. *SPR sensors with wavelength modulation.* The component of the light wave's wave vector parallel to the metal surface matching that of the SPP is determined by measuring the coupling strength at a fixed angle of incidence and multiple wavelengths and determining the wavelength yielding the strongest coupling [18], Fig. 6.
3. *SPR sensors with intensity modulation.* A change in the intensity of the light wave interacting with the SPP is measured, [19, 20]. Both the angle at which the light wave is incident onto the metal film and the wavelength are kept constant, Fig. 7.

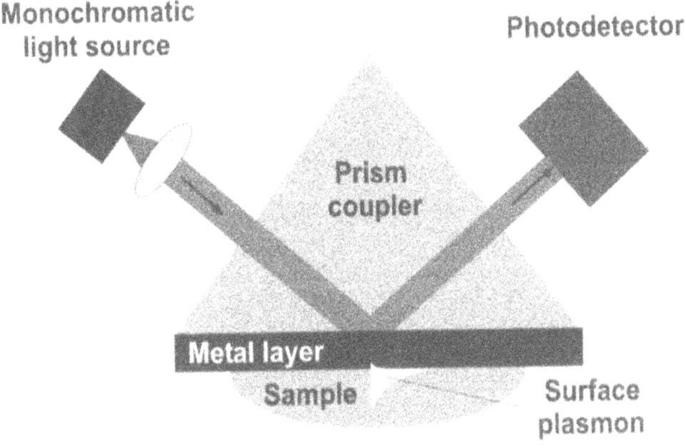

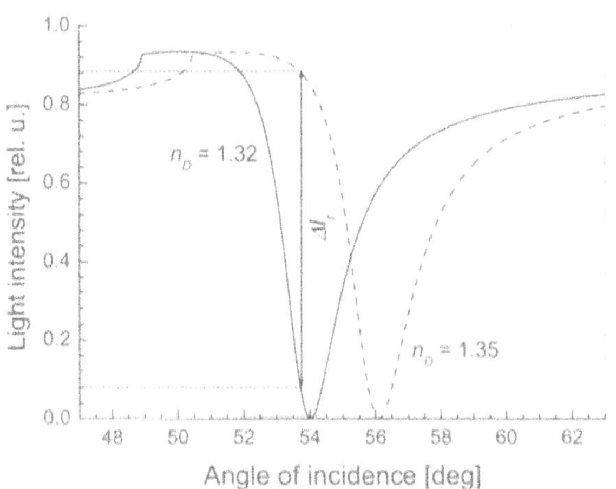

Fig. 7. SPR sensor based on intensity modulation and the Kretschmann geometry of the ATR method (structure: SF14 glass prism – 50 nm thick gold layer – sample) and intensity of a transverse magnetic (TM) light wave exciting an SPP as a function of the angle of incidence for two different refractive indices of the sample; wavelength – 682 nm

4. *SPR sensors with phase modulation.* A shift in phase of the light wave interacting with the SPP is measured [21, 22] (Fig. 8). Both the angle at which the light wave is made incident onto the interface and its wavelength are kept constant.
5. *SPR sensors with polarization modulation.* The amplitude and phase of the TM polarized wave interacting with the SPP change if the propagation constant of the SPP changes. TE-polarized light wave does not interact with the SPP and thus exhibits no resonant amplitude and phase variations. Therefore, the polarization state of the incident light wave consisting of both the polar-

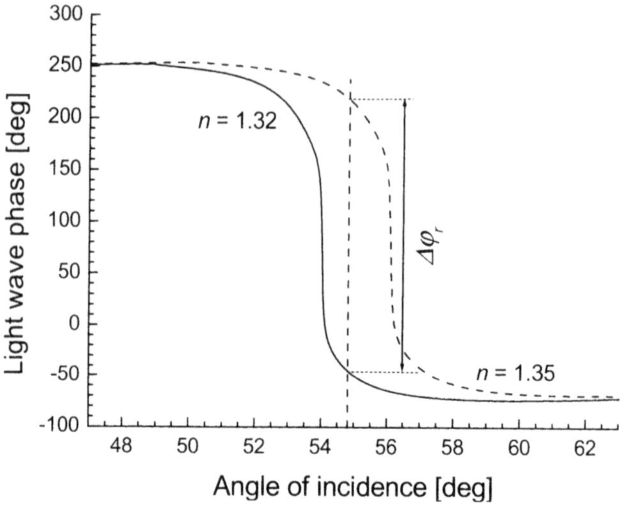

Fig. 8. Phase of a transverse magnetic (TM) light wave exciting an SPP in the Kretschmann geometry (SF14 glass prism – 50 nm thick gold layer – dielectric) as a function of the angle of incidence for two different refractive indices of the dielectric; wavelength – 682 nm

izations would also be sensitive to variations in the propagation constant of the SPP [23].

Today, most SPR sensors use angular and wavelength modulations, as these approaches incorporate multiple data measurement which offers better signal to noise figures than simple intensity modulation. Recent publications have demonstrated the potential of phase modulation-based SPR sensors, [21, 22]; however, their utilization is still rather limited.

2.3 Surface Plasmon Resonance Biosensors

SPR biosensors belong to the affinity biosensors. They are composed of an SPR transducer and a biological recognition element (*e.g.* antibody, receptor protein, biomimetic material, DNA) which recognizes and is able to interact with a selected analyte. One of the interacting molecules (usually the biomolecular recognition element except in the inhibition assay detection format, Section 3.4) is immobilized on the sensor surface while the other is contained in a solution. When the solution is brought in contact with the sensor surface, interaction between the biomolecular recognition element and analyte occurs. The binding produces a change in the refractive index at the sensor surface. This change produces a change in the propagation constant of an SPP excited at the sensor surface and is eventually measured by measuring a change in one of the characteristics of the light wave interacting with the SPP – resonant wavelength, resonant angle, intensity, phase, and polarization, Fig. 9.

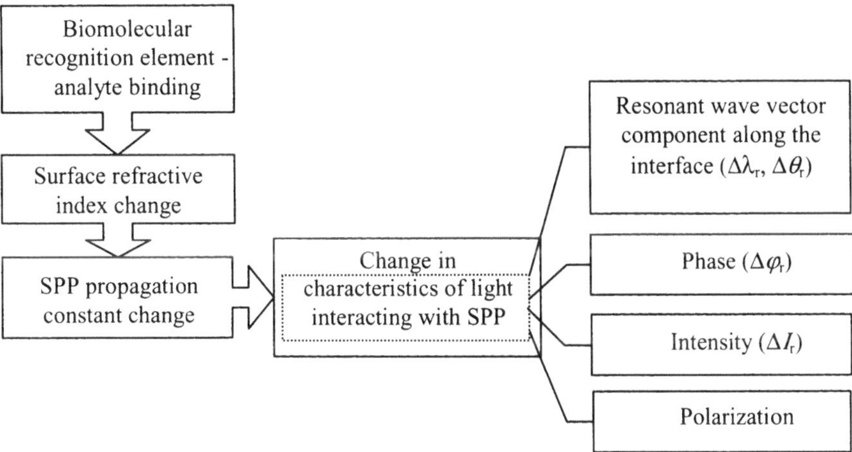

Fig. 9. Surface plasmon resonance (SPR) biosensor – principle of operation

2.4 Advantages and Drawbacks of SPR Biosensors

In principle, SPR biosensor technology can detect any analyte, providing a biomolecular recognition element recognizing the analyte is available. A target analyte does not have to exhibit any special properties such as fluorescence or characteristic absorption and scattering bands. In addition, SPR biosensors are label-free detection devices – binding between the biomolecular recognition element and analyte can be observed directly without the use of radioactive or fluorescent labels. The binding event can be observed in real-time. Moreover, SPR sensor can perform continuous monitoring as well as one-shot analyses.

The main limitation of SPR biosensors is in the specificity of detection, which is solely based on the ability of biomolecular recognition elements to recognize and capture target analytes. Therefore, the accuracy of SPR measurements can be compromised by interfering effects which produce a change in the refractive index but are not associated with the capture of target analyte. One source of interference is non-specific interaction between the sensor surface and sample, which includes adsorption of non-target molecules to the sensor surface and binding of structurally similar but not target molecules to the biomolecular recognition elements. Interference may also be produced by effects generating background refractive index variations such as sample temperature and composition fluctuations. It is a challenge for SPR biosensor research to eliminate or compensate for these effects to advance the utility of SPR biosensors for analysis of crude samples in out-of-laboratory conditions.

3 Implementations of SPR Biosensors

In general, an SPR biosensor consists of an optical system for excitation and interrogation of SPPs, a biospecific coating which interacts with biomolecules in a liquid sample, and a fluidic system comprising flow-cell or cuvette for sample confinement at the sensing surface and a fluidic system for sample collection, processing and delivery, Fig. 10.

In this chapter, we shall focus on SPR sensor instrumentation, biomolecular recognition elements and SPR measurement formats as these aspects are common to all SPR biosensors. Sample collection and processing are, on the contrary, application-specific and depend substantially on the sample matrix. The sample processing may range from no preparation or a simple filtration for liquid samples (water, milk, *etc.*) to complex multi-step procedures for extraction of liquid samples from solid food matrices (meat, corn, *etc.*). These issues are beyond the scope of this work and will not be discussed here.

3.1 Surface Plasmon Resonance Sensor Platforms

SPR sensor platforms are based on excitation of SPPs by means of prism and grating couplers and optical waveguides, Fig. 3.

3.1.1 *SPR Sensors Using Prism Couplers*

The Kretschmann geometry of the attenuated total reflection (ATR) method has been found particularly suitable for sensing and has become the most widely used geometry in SPR sensors. All the main modulation approaches have been demonstrated in SPR prism-based sensors: angular modulation [16, 17], wave-

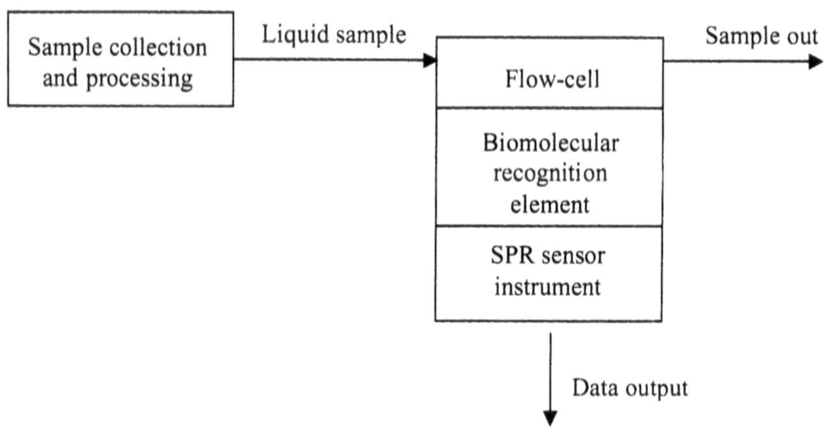

Fig. 10. Surface plasmon resonance (SPR) biosensor system

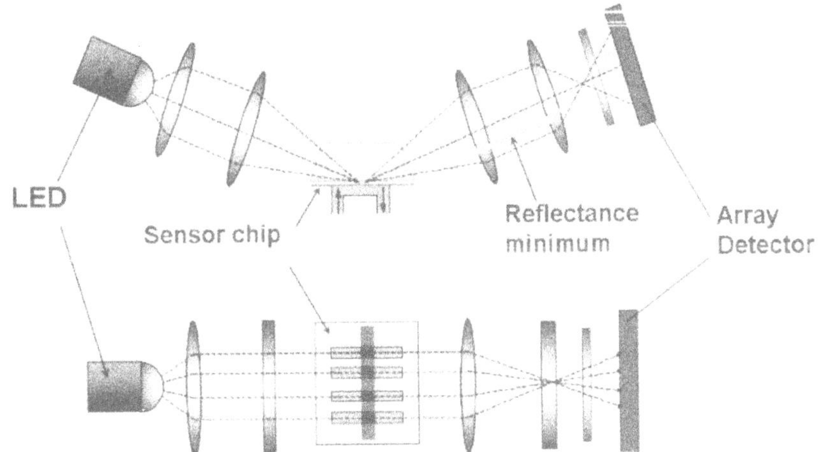

Fig. 11. SPR sensor with four parallel sensing channels. (Figure provided by S. Löfås, Biacore AB)

length modulation [18], intensity modulation [19, 20], phase modulation [21, 22], and polarization modulation [23]. The angular modulation has been most frequently reported to be used in SPR sensors and has been exploited in several commercial SPR sensor instruments. These include systems developed by Biacore International AB (Sweden), the IBIS system developed by British Windsor Scientific Ltd. (UK), SPR-670 and SPR-CELLIA systems by Nippon Laser and Electronics Laboratory (Japan), the SPR system developed by Johnson & Johnson Clinical Diagnostics Ltd. (UK), and the Spreeta SPR sensor developed by Texas Instruments Inc. (USA). The best laboratory SPR sensors based on prism coupling and angular modulation provide refractive index resolution better than 3×10^{-7} RIU (RIU – refractive index unit) [24]. The use of wavelength modulation in SPR sensors has been increasingly popular as in the last decade inexpensive high-resolution spectrometers have become a commercial reality. Laboratory SPR systems with wavelength modulation have been developed in several laboratories [25, 26]. The best of these systems offer performance comparable with the best angular modulation-based SPR systems.

Prism couplers have been used in the first multichannel SPR sensing device, Fig. 11. In this approach, SPPs are excited simultaneously in multiple areas which are arranged perpendicularly to the direction of propagation of SPPs [17]. Light reflected from each area is separately analyzed to yield information about SPR in every location, Fig. 11. A similar parallel channel architecture multichannel SPR sensor with wavelength modulation has also been reported [27].

Recently, an alternative approach to development of multichannel SPR sensors with wavelength modulation has been developed. This approach uses the wavelength division multiplexing technique (WDM) in which each particular sensing channel is assigned a distinct spectral region. The spectral encoding can be accomplished by sequentially exciting SPPs in different areas of a sensing element by light wave at different angles of incidence, Fig. 12a [28], or by employ-

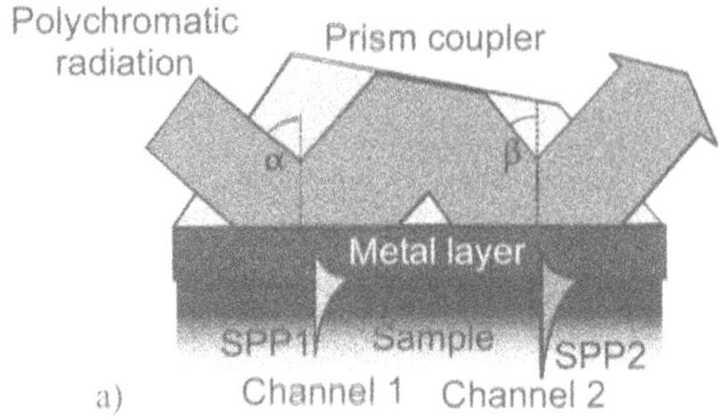

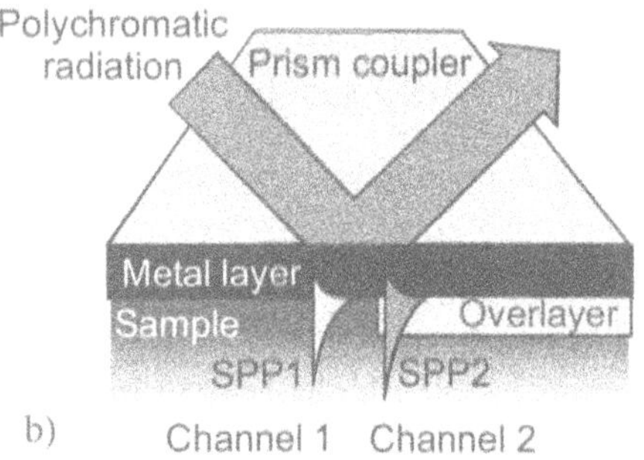

Fig. 12. SPR sensors with wavelength division multiplexing. a) Spectral encoding by means of altered angles of incidence [28]. b) Spectral encoding by means of a high refractive index overlayer [29]

ing an overlayer which shifts the resonant wavelength for a part of the sensing surface to longer wavelengths, Fig. 12b [29].

3.1.2
SPR Sensors Using Grating Couplers

Light diffraction at the surface of a diffraction grating has been used in SPR sensors to a lesser extent than the attenuated total reflection in prism couplers. This is mainly because the grating-based SPR sensors require advanced modeling and optimization tools [15, 30], and complex fabrication procedures. Examples of grating-based SPR sensors include intensity-modulated [31, 32], and wavelength-modulated [33, 34], sensing devices. Recently an interesting modification of the wavelength modulation approach using an acousto-optic modulator has been shown to provide a refractive index resolution better than 10^{-6} RIU [35].

3.1.3
SPR Sensors Using Optical Waveguides

The use of optical waveguides in SPR sensors provides numerous attractive features such as a simple way to control the optical path in the sensor system (efficient control of properties of light, suppression of the effect of stray light, *etc*), small size and ruggedness. To date various SPR sensing devices using slab [36] and channel [37] single-mode integrated optical waveguides have been reported. These sensing devices exhibit a limited operating range which, however, can

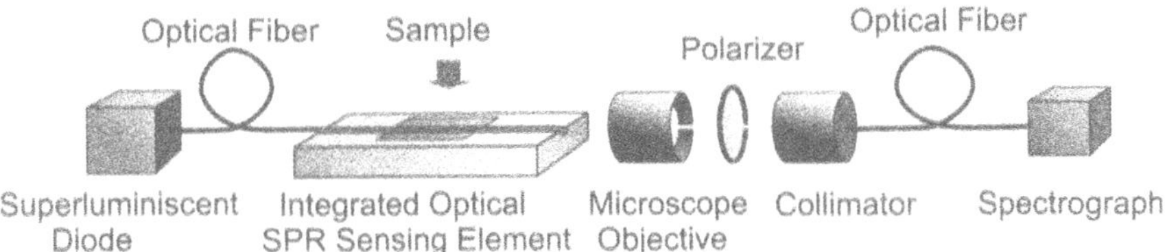

Fig. 13. Integrated optical SPR sensor with wavelength modulation [40]

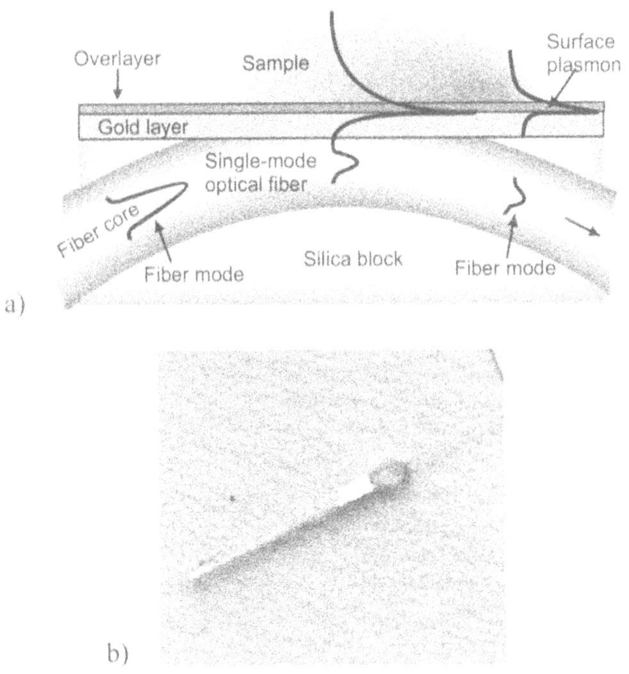

Fig. 14. SPR sensor based on a single-mode optical fiber. a) Fiber optic sensing structure. b) Detail of miniature fiber optic SPR probe [45]

be increased by using waveguides fabricated in low refractive index glass [37], waveguides with a buffer layer [36], or a high refractive index overlayer [38]. To improve robustness of the integrated optical SPR sensors, an intensity-based integrated optical SPR sensor with a reference arm compensating for variations of light levels in the waveguide input [39], and a sensor based on wavelength modulation [40], have been developed. A refractive index resolution as low as 1×10^{-6} RIU has been demonstrated with the wavelength modulation-based SPR integrated-optical sensor [40].

The first fiber optic SPR sensor was based on a multimode optical fiber and wavelength modulation. The sensor used a conventional polymer clad silica (PCS) fiber with partly removed cladding and a metal film deposited symmetrically around the exposed section of fiber core [41]. This sensor has been shown to be able to detect refractive index variations with a resolution up to 5×10^{-5} RIU. A similar geometry, in which the sensing area is formed not at the tip but in the middle of an optical fiber, has been used for the development of an intensity modulation-based SPR sensor [42]. SPR sensors based on single-mode optical fibers have also been reported [43, 44]. These sensors employ an optical fiber with a locally removed cladding and an SPP-active metal film. A guided mode propagates in the fiber and excites an SPP at the outer boundary of the metal film. This sensor can be designed to operate as a transmissive sensor [44] or a fiber optic probe [45], Fig. 14. A wavelength modulated-version of this SPR sensor has been demonstrated to be able to attain a refractive index resolution as low as 2×10^{-6} RIU.

3.2 Biomolecular Recognition Elements and their Immobilization

Several types of biomolecular recognition elements have been used in SPR biosensors. These include antibodies, biomimetic materials, and nucleic acids. Antibodies are the most widely reported biological recognition elements used in SPR biosensors because of their high affinity, versatility, and commercial availability. Antibodies are immobilized on the gold SPR-active layer on the sensor surface in the form of a monolayer or a three-dimensional biomolecular assembly. Various immobilization chemistries that provide stable and defined attachment of antibodies have been developed. One approach is to use a streptavidine monolayer immobilized on a gold film with biotin [46], which may be further functionalized with biotinylated biomolecules. Another approach is to form a self-assembled monolayer of thiol molecules (e.g. alkanethiols) with suitable reactive groups on one end of the molecule and a gold-complexing thiol on the other [47]. Then the antibody can be attached to the thiols. Another approach uses a hydrogel matrix composed of carboxyl-methylated dextran chains to yield a three-dimensional matrix for antibody attachment [48]. Carboxyl groups on the dextran are easily modified using standard coupling chemistries allowing antibodies to be attached via surface-exposed amine, carboxyl, sulfhydryl, and aldehyde groups. SPR sensing surfaces may also be functionalized by thin polymer films to which antibodies can be coupled via amino groups [49].

3.3
Biomolecular Interactions

As typical concentrations of free (target) molecules in the solution exceed concentrations of the molecular recognition elements (e.g. antibodies) immobilized on the sensor surface, the biomolecular recognition element-analyte interaction can be described by the pseudo-first-order kinetic equation:

$$\frac{\mathrm{d}R}{\mathrm{d}t} = k_a c(1-R) - k_d R, \tag{4}$$

where R is the relative amount of bound analyte, c is analyte concentration, and k_a and k_d are the association and dissociation kinetic rate constants, respectively [50]. After integration this equation yields:

$$R(t) = \left[\frac{k_a c}{k_a c + k_d} - R_0\right]\left(1 - e^{-(k_a c + k_d)t}\right) + R_0, \tag{5}$$

where R_0 denotes the initial amount of analyte bound at the time $t = 0$; [50]. This model assumes 1:1 binding and rapid mixing of the analyte from the bulk phase to the sensor surface layer and single-step binding. Observed binding, however, may deviate from this simple model due to more complex mechanisms of interaction and mass transport limitations [51].

3.4
Detection Formats used in SPR Biosensors

SPR affinity biosensors have been constructed to detect an analyte in a variety of formats. The main detection formats include direct detection, sandwich assay and inhibition assay. In general, for macromolecular analytes (molecular weight > 10,000), direct detection and sandwich assay formats are used. For smaller analytes (molecular weight < 10,000) the inhibition assay format is used.

Direct detection presents the most straightforward detection format. In this format, analyte in a sample interacts with the biomolecular recognition element (antibody) immobilized on the sensor surface, Fig. 15. The binding generates refractive index change, which is directly proportional to the concentration of analyte.

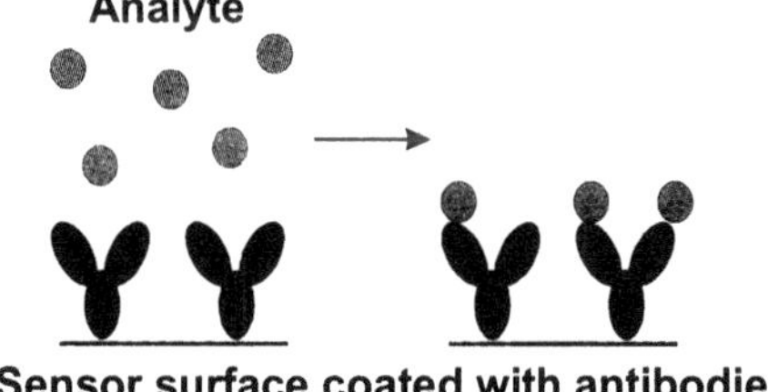

Fig. 15. Direct detection

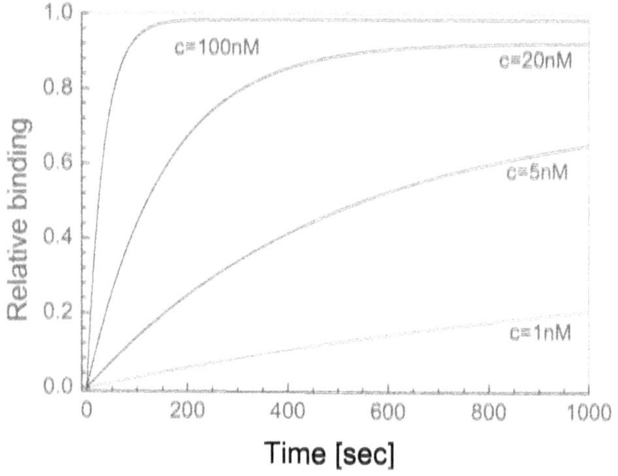

Fig. 16. Direct detection. Binding between antibody and analyte calculated for four different concentrations of analyte. Assumed rate constants: $k_a=3\times10^5$ M^{-1} s^{-1}; $k_d=5\times10^{-4}$ s^{-1}

Figure 16 shows a kinetic model of the interaction between antibody and analyte simulated using Eq. 5. As follows from Fig. 16, the binding between the target analyte and antibody is fast initially. As the interaction progresses, the binding rate gradually decreases and eventually the system reaches a state in which the association and dissociation processes are in equilibrium. The time required for the interaction to reach equilibrium depends on the concentration of analyte and is longer for lower concentrations of analyte.

Figure 17a shows the dependence of the relative binding at equilibrium on the concentration of analyte. At low analyte concentrations, the equilibrium binding increases with the analyte concentration in a linear fashion. At higher analyte concentrations, the binding sites provided by the biomolecular recognition elements are saturated and a further increase in the analyte concentration produces a smaller increase in the amount of bound analyte. The initial binding rate $dR/dt(t=0)$ is directly proportional to the association rate constant and analyte concentration (Fig. 17b). Both the amount of analyte bound at equilibrium and the initial binding rate can be used to determine analyte concentration. In principle, the measurement of the binding rate is faster and offers a larger dynamic range than the measurement of the equilibrium binding.

In the *sandwich assay format* the SPR sensor surface is functionalized with antibodies for a specific analyte. In the first detection step, the sample containing analyte is brought into contact with the sensor surface and the analyte molecules bind to the antibodies on the sensor surface.

Then the sensor surface is incubated with a solution containing 'secondary' antibodies. The secondary antibodies bind to the previously captured analyte, producing a further increase in the number of bound biomolecules (Fig. 18).

Inhibition assay is an example of a competitive assay. In this detection format, a sample is initially mixed with respective antibodies, then the mixture is

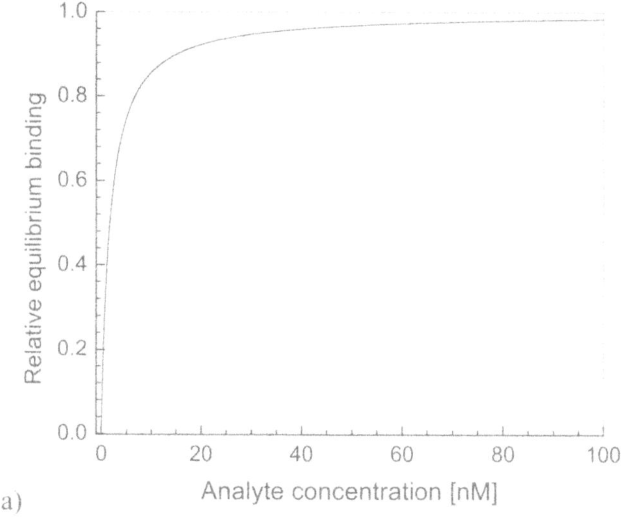

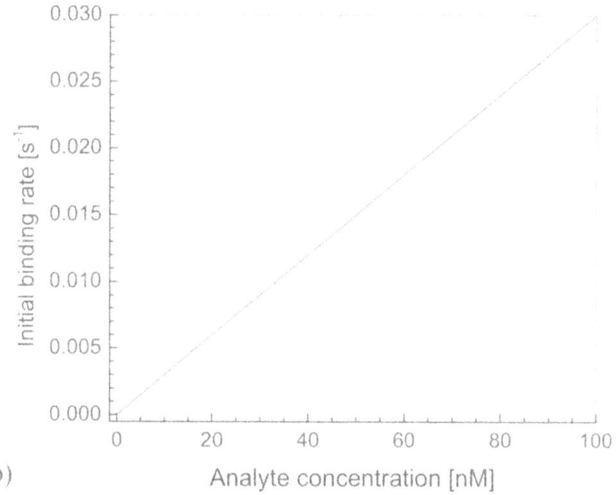

Fig. 17. Direct detection. a) Relative equilibrium binding as a function of analyte concentration. b) Initial binding rate as a function of analyte concentration. Assumed rate constants: k_a=3×10^5 M^{-1} s^{-1}; k_d=5×10^{-4} s^{-1}

brought into contact with the sensor surface derivatized with analyte molecules, so that the unoccupied antibodies could bind to the analyte molecules at the sensor surface (Fig. 19).

The amount of bound analyte versus time may be estimated by calculating the equilibrium concentration of antibody which did not bind to the analyte in the sample and then simulating the interaction between the unbound anti-

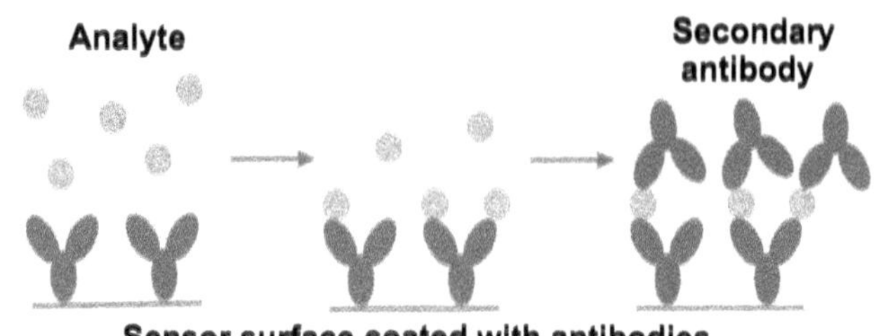

Fig. 18. Sandwich assay

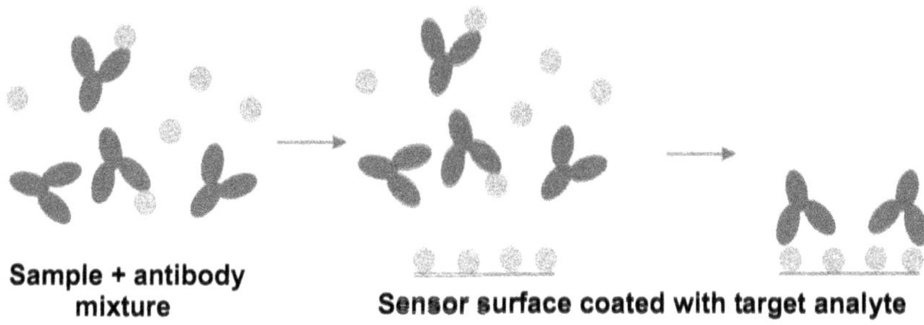

Fig. 19. Inhibition assay

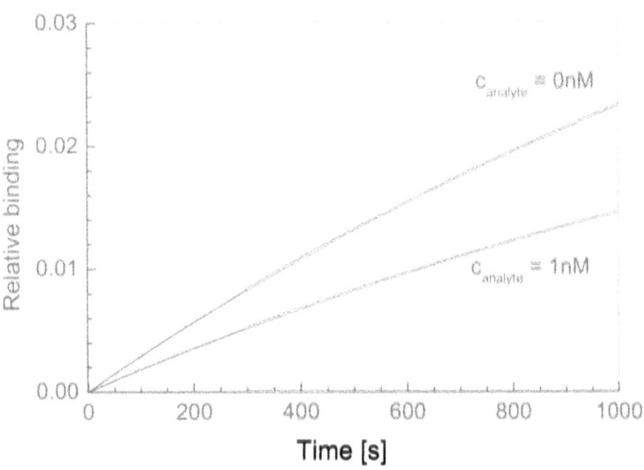

Fig. 20. Inhibition assay. Binding between antibodies previously incubated with the analyte of the concentrations of 0 nM and 1 nM and analyte immobilized to the sensor surface. Rate constants: k_a=3×10^5 M^{-1} s^{-1}; k_d=5×10^{-4} s^{-1}, initial antibody concentration: 0.1 nM

body and the analyte-derivatized surface. Figure 20 shows the kinetics of binding between the antibody and the analyte immobilized on the surface, assuming that antibody in a concentration of 0.1 nM was incubated with sample containing analyte in a concentration of 1nM and the system was provided enough time to reach equilibrium. Binding kinetics for analyte-free sample is provided for comparison. The amount of bound antibody is inversely proportional to the analyte concentration in the sample.

Figure 21 shows the equilibrium binding and initial binding rate as a function of analyte concentration, indicating that both these binding features can be unambiguously correlated with the analyte concentration.

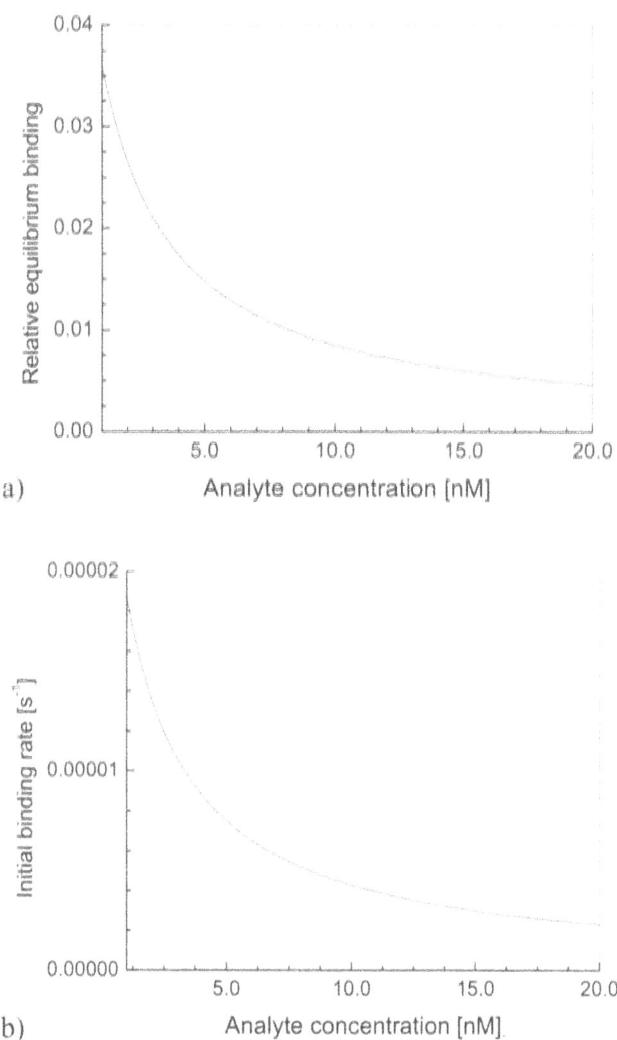

Fig. 21. Inhibition assay. a) Relative equilibrium binding of the antibody as a function of analyte concentration. b) Initial binding rate as a function of analyte concentration. Rate constants: $k_a = 3 \times 10^5$ M^{-1} s^{-1}; $k_d = 5 \times 10^{-4}$ s^{-1}, initial antibody concentration: 0.1 nM

4 SPR Biosensors for Detection of Food Safety-related Analytes

This Section presents examples of applications of SPR biosensors for detection of food safety-related analytes including chemical contaminants, foodborne pathogens and toxins.

4.1 SPR Biosensor-based Detection of Chemical Contaminants

Major chemical contaminants implicated in food safety include pesticides, herbicides, mycotoxins and antibiotics [52]. Analytical methods for detection of chemical contaminants are required to measure concentrations of these analyte down to sub-ng mL^{-1} or ng mL^{-1}. As these analytes are rather small (typical molecular weight < 1,000), direct binding to a biomolecular recognition element-coated sensor surface produces only a limited sensor response. Therefore, for detection of low concentrations of these analytes, inhibition assay is a preferred detection format. A characteristic sensor response for inhibition assay is illustrated in Fig. 22, which shows a typical sensorgram for detection of atrazine using a wavelength-modulated SPR sensor.

Examples of chemical contaminants detected by SPR biosensors are presented in Table 1. Detailed information can be found in the references. Detection limits achieved by the SPR biosensors for selected chemical contaminants (Table 1) suggest that SPR biosensors utilizing inhibition assays are capable of detecting chemical contaminants at practically relevant concentrations.

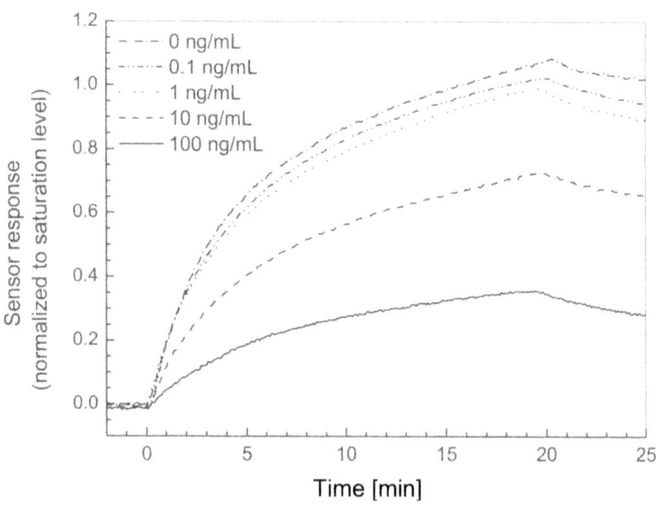

Fig. 22. Detection of atrazine using SPR biosensor and inhibition assay. Kinetic sensor response to various concentrations of atrazine, antibody concentration – 4 μgmL^{-1}

Table 1. Food chemical contaminants detected by SPR biosensors

Analyte	Sensor configuration and detection format	Matrix	Lowest detection limit	Reference
Pesticides, herbicides:				
Atrazine	BIACORE (prism optic) Inhibition assay	Water	0.05 ng mL^{-1}	[53]
Simazine	Integrated optic Inhibition assay	Water	0.1 ng mL^{-1}	[54]
Mycotoxins:				
Fumonisin B1	Prism optic Direct detection	Aqueous solution	50 ng mL^{-1}	[55]
Antibiotics:				
Sulphamethazine	BIACORE (prism optic) Inhibition assay	Milk	1 ng mL^{-1}	[56]
Sulphamethazine	BIACORE X (prism optic) Inhibition assay	Milk	2 ng mL^{-1}	[57]
Sulphadiazine	BIACORE (prism optic) Inhibition assay	Pig bile	20 ng mL^{-1}	[58]

4.2 SPR Biosensor-based Detection of Toxins

Food safety-related toxins include natural toxins (scombrotoxin, saxitoxin, *etc.*) and toxins produced by bacteria (*Clostridium botulinum, Staphylococcus aureus, Vibrio cholerae, etc.).* The most potent bacterial toxins (botulinum toxin) are lethal at levels as low as 100 ng [59], which, assuming consumption of 100 grams of food yields a lethal concentration of about 1 ng mL^{-1}. Therefore, analytical methods are needed to detect foodborne toxin in sub-ng mL^{-1} concentrations. Molecular weight of the main bacterial toxins ranges from 28,000 to 150,000, which makes it possible for most sensitive SPR biosensors to measure their concentrations directly down to ng mL^{-1} levels. Lower detection limits may be accomplished using a sandwich assay.

A typical sensorgram for Staphylococcal enterotoxins B (SEB) using sandwich assay is shown in Fig. 23. Initial binding of SEB to the sensor surface coated with respective antibodies is steady and the sensor response reaches about 0.4 nm in 30 minutes. Upon injection of secondary antibodies, the antibodies bind to the captured SEB, generating a 10-fold increase in the sensor response [60]. Figure 24 shows the equilibrium sensor response for both the direct capture of SEB and amplification by secondary antibodies as a function of SEB concentration. Figure 25 shows initial binding rates for both the direct capture of SEB and amplification by secondary antibodies as a function of SEB concentration.

Table 2 lists food toxins which have been detected by SPR biosensors and the achieved lowest detection limits.

As shown in Table 2, most recent publications indicate that SPR-based detection of bacterial toxins at practically relevant concentrations is feasible.

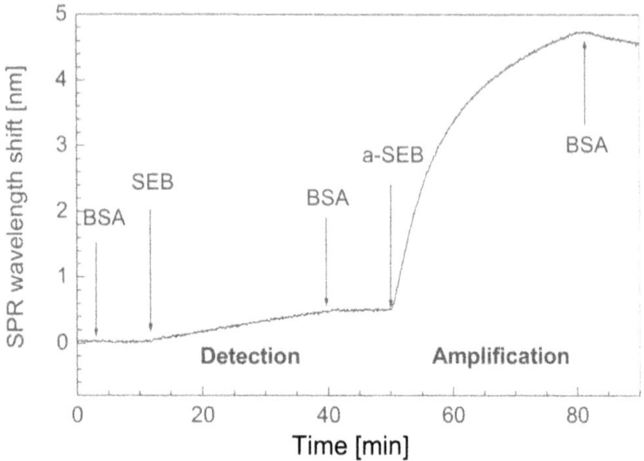

Fig. 23. Detection of Staphylococcal enterotoxin B using an SPR biosensor and sandwich assay. Resonant wavelength as a function of time illustrating direct detection phase (10–40 min) and amplification phase (50–80 min). SEB concentration – 25 ng mL^{-1}, secondary antibody concentration 3 μg mL^{-1}; [60]

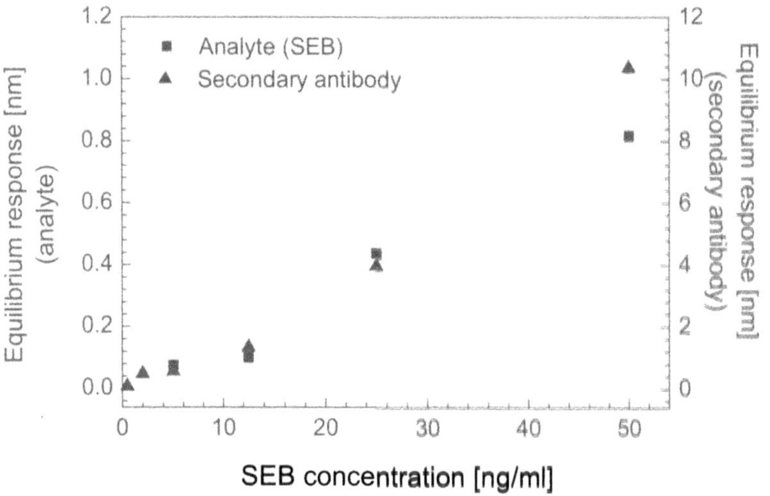

Fig. 24. Detection of Staphylococcal enterotoxin B using an SPR biosensor. Equilibrium sensor response for different concentrations of SEB

4.3 SPR Biosensor-based Detection of Microbial Pathogens

Foodborne illnesses can be caused by bacterial pathogens such as *Escherichia coli*, *Yersinia enterocolitica*, *Campylobacter jejuni*, *Salmonella*, *Shigella* and *Listeria monocytogenes*. Infectious doses for these bacterial pathogens range from 10

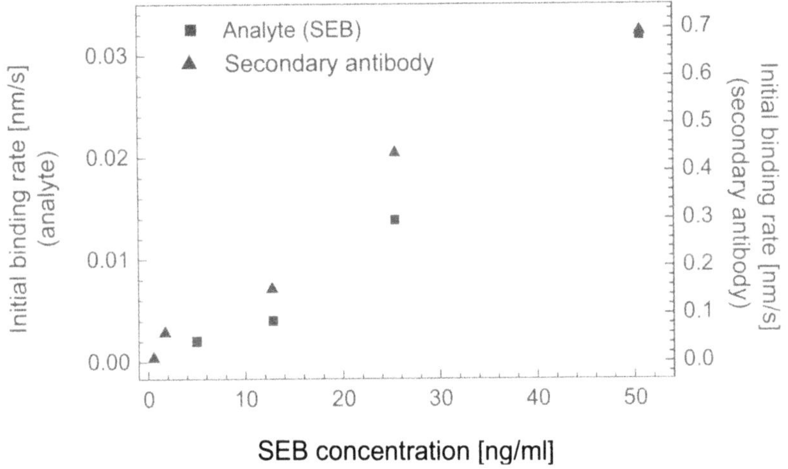

Fig. 25. Detection of Staphylococcal enterotoxin B using an SPR biosensor. Initial binding rate for different concentrations of SEB

Table 2. Food toxins detected by SPR biosensors

Analyte	Sensor configuration and detection format	Matrix	Lowest detection limit	Reference
Botulinum toxin	BIACORE X (prism optic) Direct detection	Buffer	2.5 μg mL^{-1}	[61]
E. coli enterotoxin	TI Spreeta (prism optic) Direct detection	Aqueous solution	6 μg mL^{-1}	[62]
Staphylococcal enterotoxin B	BIACORE 3000 (prism optic) Sandwich assay	Food	10 ng mL^{-1}	[63]
Staphylococcal enterotoxin B	Prism optic Direct detection Sandwich assay	Aqueous s. /milk	5 ng mL^{-1} 0.5 $ngmL^{-1}$	[60]
Staphylococcal enterotoxin B	Fiber optic Direct detection	Aqueous solution	4 ng mL^{-1}	[64]

Table 3. Foodborne pathogens detected by SPR biosensors

Analyte	Sensor configuration and detection format	Matrix	Lowest detection limit	Reference
Escherichia coli	BIACORE (prism optic) Sandwich assay	Buffer	5×10^{7} cfu mL^{-1}	[66]
Listeria monocytogenes	Prism optic Direct detection	Buffer	10^{6} cfu mL^{-1}	[67]
Salmonella enteritidis	Prism optic Direct detection	Buffer	10^{6} cfu mL^{-1}	[67]

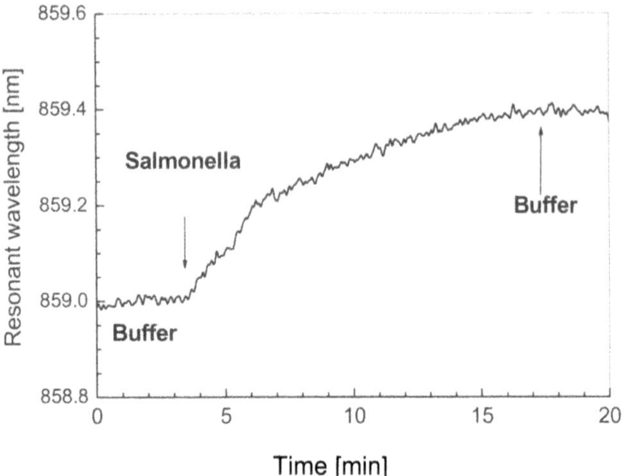

Fig. 26. Direct detection of *Salmonella enteritidis* using an SPR biosensor. Kinetic response to injection of Salmonella at the concentration of 10^6 cfu mL^{-1}

to 10^4 cells [65]. Bacterial pathogens are large organisms (≈ 1 μm). Therefore, their presence is usually measured directly with an optional amplification by secondary antibodies (sandwich assay). A typical sensorgram for direct detection of *Salmonella enteritidis* is shown in Fig. 26.

Examples of foodborne bacterial pathogens detected by SPR biosensors are presented in Table 3. Detailed information can be found in the references.

The demonstrated lowest detection limits (Table 3) are not yet satisfactory for detection of bacterial pathogens at lowest infectious concentrations. This is mainly due to: a) size of a bacterium (the bulk of the bound cell is situated outside the evanescent field of a surface plasmon), b) low concentration of the particular antigen (relative to total cellular material), and c) slow diffusion of bacterial cells to the sensor surface [68].

5 Summary

In the last two decades, surface plasmon resonance (SPR) biosensors have made great strides both in terms of instrumentation and applications. A variety of SPR sensor platforms ranging from miniature SPR fiber optic probes to high-performance laboratory systems have been developed. Numerous biomolecular recognition elements have become available, enabling the use of SPR affinity biosensors for detection of a large number of chemical and biological analytes. Usefulness of SPR biosensor technology has been established in many important fields including food and drug screening, environmental monitoring, medicine, and antiterrorism.

Detection of analytes implicated in food safety concerns numerous sectors, including agriculture, food industry, regulatory authorities, and security organizations. SPR biosensors are emerging as an important real-time analytical tool holding potential for applications in this area. SPR biosensors have been demonstrated to be able to detect small and medium size analytes such as food chemical contaminant and bacterial toxins at practically relevant concentrations. Lowest detection limits achieved for large bacterial analytes still need to be improved to meet today's needs for detection of bacterial pathogens. Therefore, research leading to improvements in the sensitivity and detection limits of SPR biosensors will be important in future. Advances in this area will enable detection of lower concentrations of target analytes and also decrease the analysis time. Other important research areas will include development of biomolecular recognition elements and referencing approaches for improving specificity and robustness of SPR measurements in crude samples.

References

1. Food Safety: Information on foodborne illnesses, Report of General Accounting Office of the United States, Report GAO/RCED-96-96 (1996)
2. Ghindilis AL, Atanasov P, Wilkins M, Wilkins E (1998) Biosens Bioelectr 13: 113
3. Chu X, Lin ZH, Shen GL, Yu RQ (1995) Analyst 120: 2829
4. Pless P, Futschik K, Schopf E (1994), J Food Protect 57: 369
5. Gauglitz G (1996): Opto-Chemical and Opto-Immuno Sensors, Sensor Update Vol. 1, VCH Verlagsgesellschaft, Weinheim
6. Ivnitski D, Abdel-Hamid I, Atanasov P, Wilkins E (1999) Biosens Bioelectr 14: 599
7. Collings AF, Caruso F (1997) Rep Prog Phys 60: 1397
8. Rowe-Taitt CA, Hazzard JW, Hoffman KE, Cras JJ, Golden JP, Ligler FS (2000) Biosens Bioelectr 15: 579
9. Clerc D, Lukosz W (1994) Sens Actuators B 19: 581
10. Cush R, Cronin JM, Stewart WJ, Maule CH, Molloy J, Goddard NJ (1993) Biosens Bioelectr 8: 347
11. Homola J, Yee S, Gauglitz G (1999) Sens Actuators B 54: 3
12. Rabbany SY, Lane WJ, Marganski WA, Kusterbeck AW, Ligler FS (2000) J Immunol Methods 246: 69-77
13. Boardman AD (1982) (ed) Electromagnetic Surface Modes, John Wiley & Sons, Chichester
14. Reather H (1983) Surface Plasmons on Smooth and Rough Surfaces and on Gratings, Springer tracks in modern physics, Vol 111, Springer Verlag, Berlin
15. Hutley MC (1982) Diffration Gratings, Academic Press, London
16. Matsubara K, Kawata S, Minami S (1988) Appl Opt 27: 1160
17. Liedberg B, Lundstrom I, Stenberg E (1993) Sens Actuator B11: 63
18. Zhang LM, Uttamchandani D (1988) Electr Lett 23: 1469
19. Nylander C, Liedberg B, Lind T (1982) Sens Actuator 3: 79
20. Manuel M, Vidal B, Lopez R, Aleggret S, Alonso-Chamarro J, Garces I, Mateo J (1993) Sens Actuator B11: 455
21. Nelson SG, Johnston KS, Yee S (1996) Sens Actuator B35-36:187
22. Nikitin PI, Beloglazov AA, Kabashin AV, Valeiko MV, Kochergin VE (1999) Sens Actuator B54, 43
23. Kruchinin AA, Vlasov YG (1996) Sens Actuator B30: 77
24. Karlsson R, Ståhleberg R (1995) Anal Biochem 228: 274
25. Homola J, Lu HB, Nenninger GG, Dostálek J, Yee S (2001) Sens Actuator B76: 403
26. Pfeifer P, Aldinger U, Schwotzer G, Diekmann S, Steinrücke P (1999) Sens Actuator B54: 166

27. Nenninger GG, Clendenning JB, Furlong CE, Yee S (1998) Sens Actuator B51: 38
28. Homola J, Dostálek J, Čtyroký J (2001) Proc SPIE 4416: 86
29. Homola J, Lu HB, Yee S (1999) Electr Lett 35: 1105
30. Moharam MG, Gaylord TK (1986) J Opt Soc Am A3: 1780
31. Cullen DC, Brown RG, Lowe CR (1987/88) Biosens Bioel 3: 211
32. Lawrence CR, Geddes NJ, Furlong DN, Sambles JR (1996) Biosen Bioelectr 11: 389
33. Vukusic PS, Bryan-Brown GP, Sambles JR (1992) Sens Actuator B8: 155
34. Jory MJ, Vukusic PS, Sambles JR (1994) Sens Actuator B17: 1203
35. Jory MJ, Bradberry GW, Cann PS, Sambles JR (1995) Meas Sci Tech 6: 1193
36. Lavers CR, Wilkinson JS (1994) Sens Actuator B22: 75
37. Harris RD, Wilkinson JS (1995) Sens Actuator B29: 261
38. Čtyroký J, Homola J, Skalský M (1997) Electr Lett 33: 1246
39. Mouvet C, Harris RD, Maciag C, Luff BJ, Wilkinson JS, Piehler J, Brecht A, Gauglitz G, Abuknesha R, Ismail G (1997) Anal Chim Acta 338: 109
40. Dostálek J, Čtyroký J, Homola J, Brynda E, Skalský M, Nekvindová P, Špirková J, Škvor J, Schröfel J (2001) Sens Actuator B76: 8
41. Jorgenson RC, Yee S (1993) Sens Actuator B12, 213
42. Trouillet A, Ronot-Trioli C, Veillas C, Gagnaire H (1996) Pure App Opt 5: 227
43. Dessy RE, Bender WJ (1994) Anal Chem 66: 963
44. Homola J (1995) Sens Actuator B 29: 401
45. Slavík R, Homola J, Čtyroký J (1998) Sens Actuator B51: 311
46. Morgan H, Taylor DM (1992) Biosens Bioelectr 7: 405
47. Duschl C, Sevin-Landais A, Vogel H (1995) Biophys J 70: 1985
48. Löfås S, Malmqvist M, Rönnberg I, Stenberg E, Liedberg B, Lundström I (1991) Sens Actuator B5: 79
49. Nakamura R, Muguruma H, Ikebukuro K, Sasaki S, Nagata R, Karube I, Pedersen H (1997) Anal Chem 69: 4649
50. Edwards PR, Leatherbarrow RJ (1997) Anal Biochem 246: 1
51. Vijayendran RA, Ligler FS, Leckband DE (1999) Anal Chem 71: 5405
52. Patel PD (2002) Trends Anal Chem 21, 96
53. Minunni M, Mascini M (1993) Anal Lett 26: 1441
54. Harris RD, Luff BJ, Wilkinson JS, Piehler J, Brecht A, Gauglitz G, Abuknesha RA (1999) Biosens Bioelectr 14: 377
55. Mullett W, Edward PC, Yeung MJ (1998) Anal Biochem 258: 161
56. Sternesjo A, Mellgren C, Bjorck L (1996) ACS Symposium Series 621: 463
57. Gaudin V, Pavy ML (1999) J AOAC Int 82: 1316
58. Elliot CT, Baxter GA, Crooks SRH, McCaughey WJ (1999) Food Agric Immunol 11: 19
59. Gill DM (1982) Microbiol Rev 46: 86
60. Homola J, Dostálek J, Chen S, Rasooly A, Jiang S, Yee S (2002) I J Food Microbiol 75: 61
61. Choi K, Seo W, Cha S, Choi J (1998) J Biochem Mol Biol 31: 101
62. Spangler BD, Wilkinson EA, Murphyb JT, Tyler BJ (2001) Anal Chim Acta 444: 149
63. Rasooly A (2001) J Food Protect 64: 37
64. Slavík R, Homola J, Brynda E (2002) Biosens Bioelectr 17: 591
65. Bad Bug Book, US. Food and Drug Administration, Center for Food Safety and Applied Nutrition, http://vm.cfsan.fda.gov.
66. Fratamico PM, Strobaugh TP, Medina MB, Gehring AG (1998) Biotechnol Techni 12: 571
67. Koubová V, Brynda E, Karasová L, Škvor J, Homola J, Dostálek J, Tobiška P, Rošický J (2001) Sen Actuator B 74: 100
68. Perkins EA, Squirrell DJ (2000) Biosens Bioelectr 14: 853

CHAPTER 8

NIR Dyes for Ammonia and HCl Sensors

Peter Šimon, Frank Kvasnik

1 Introduction

Dyes with absorption bands in the near infrared region (NIR dyes) were for a long time considered to have only a few practical applications other than as sensitisers in photographic emulsions. The situation has changed dramatically in recent years, with the rapid developments of many fields such as semiconductor laser technology, fibre optic communications, printing, data recording, photography, analytical tools for environmental and process monitoring, sensors techniques and medical diagnostics. This has resulted in research papers, books, review articles and chapters in books, all dedicated to the synthesis and applications of NIR dyes [1–5].

Photometric devices using chemical transducers in the form of solid matrix-supported dyes represent a promising class of instruments for gas monitoring over a wide range of concentrations. When combined with fibre optic technology, this approach offers the possibility of remote sensing of chemicals.

Ammonia and hydrogen chloride belong to the most important hazardous agents encountered in many areas. Ammonia is widely used in the production of explosives, fertilisers and as an industrial coolant. The toxicity of this gas is well documented, and acute poisoning can result from inhalation of only small doses of ammonia vapour. Exposure limits of 25 ppm over an 8 hour period and 35 ppm over a ten minutes period have been recommended and have recently been legislated for [6]. A number of leaks and escapes leading to unwanted exposure incidents are reported every year in locations employing industrially sized cooling systems, such as food production and storage plants. As industry is becoming increasingly safety conscious, efficient sensor devices for rapid detection of ammonia leaks and to monitor personal exposure of workers who might be at risk of contact with ammonia are desirable, as rapid evacuation of personnel from contaminated areas may be all that is required to prevent serious illness.

Hydrogen chloride (HCl) is widely used in many branches of chemical technology, such as the manufacture of vinyl chloride, fertilizers, artificial silk, dyes; it is also used in the electroplating, leather tanning, textile and rubber industries. HCl is produced in large quantities during combustion of many materials, especially materials with high chlorine content. It has sharp irritating odour detectable at 0.25 to 10 ppm. Inhalation of hydrogen chloride gas can cause severe irritation and injury to the upper respiratory tract and lungs, and exposure to high concentrations may cause death. The current exposure limit for HCl is 5 ppm over an 8 hour period [7].

Attempts have been made to develop photometric sensors in the visible region for detecting chemicals in both the gas and the liquid phase [8–21]. However, these sensors are not suitable for remote fibre optic sensing since the high attenuation of optical fibres in the visible spectral region limits the practical length of the fibre optic link. In the NIR spectral region the attenuation of optical fibres falls significantly and hence sensing over long fibre optic links is possible. Development of sensors in this spectral region can also exploit advances in communication and CD technologies. Furthermore, in the 700 nm to 900 nm spectral region the absorbance of most analytes is minimal and therefore it would not interfere with the absorbance of a NIR chromophore in NIR dyes. The development of chemical sensors operating in the NIR region was initially hampered by the lack of suitable dyes. Original studies to demonstrate the feasibility of this approach were therefore carried out on dyes that were developed either for laser or printing applications [22–24].

This chapter deals with NIR dyes that were developed for chemical sensing in this spectral region for potential use in remote fibre optic sensors.

2
NIR Transducers

Numerous substances (molecules and ions) do not have suitable optical characteristics that are easily measurable. Indicators are chemical systems whose electronic spectrum changes when they come in a contact with a particular substance. Hence, the indicator can act as a chemical transducer for the substances that cannot be detected directly by optical techniques. Ideally, the indicator should be highly sensitive and a large change in the spectrum should occur for small concentration changes of the target substance. Many indicators active in the visible spectral region are known and the fundamentals of their structures and sensing mechanisms are well established [25], but only a few of them are useful for sensing. These indicators can be categorized in many ways, for example, by their optical properties, chemical composition, sensing mechanism or applications to optical sensors.

In contrast, the number of studies dealing with the indicators in the NIR region is very limited. It is a matter of course to discover that many commercial NIR dyes could be sensitive to chemical agents. Systematic studies of the sensitivity of various commercial dyes to common chemical agents were carried out in the mid-1990s, both in solution and in the silicone rubber matrix [22, 26, 27]. Many dyes appeared to be sensitive both to ammonia and acids. However, the spectral changes of the dyes were mostly irreversible. Also, it was quite difficult to draw any conclusion about the mechanism of the dye-agent interaction since the structure of the dyes was mostly unknown.

The dyes tailor-made as NIR transducers should have an absorption band between 750 and 850 nm, they should have a reversible and highly specific spectral response to the chemical agent, be long-term stable to light, oxygen, and temperatures between 0 °C and 100 °C, and they should be compatible with common matrices for immobilisation of indicators [28]. The dyes dealt with in this chapter can be roughly divided into two groups according to their response to the target chemical:

- NIR dyes sensitive to ammonia – very often also sensitive to amines.
- NIR dyes sensitive to hydrogen chloride – very often used as NIR pH indicators.

2.1 Structure and Tests of NIR Ammonia Transducers

2.1.1 Metal Complexes

Metal complexes, mainly copper quinoline complexes, appeared to be very suitable for ammonia sensing. They possess perfect reversibility, long-term stability and they can be very simply incorporated into polymers. The first dye of this group, the Cu(II) complex of 5-(*N'*-diethylamino-phenylimino)quinolin-8-one (dye D1), was originally synthesised by Kubo et al. [29, 30]. The reaction yielding D1 is depicted in Scheme 1.

VIS/NIR solution absorption spectra were recorded using an apparatus based on the S1000 Ocean Optics CCD array spectrometer with spectral response from 400 nm to 850 nm. Sensitivity of the dyes to ammonia was tested by addition of 0.2 mL of 2.5 wt-% aqueous solution of ammonia to 10 mL of an ethanol solution of the dye. The reversibility was tested by addition of 0.02 mL of 35 wt-% HCl to the ammonia-containing sample.

The spectrum of D1 in ethanol solution is shown in Fig. 1. As it can be seen, after addition of ammonia to the ethanol solution of the complex, the absorption peak of the complex diminishes and the peak of the free ligand appears. The absorption maximum of the free ligand is situated at 618 nm, where

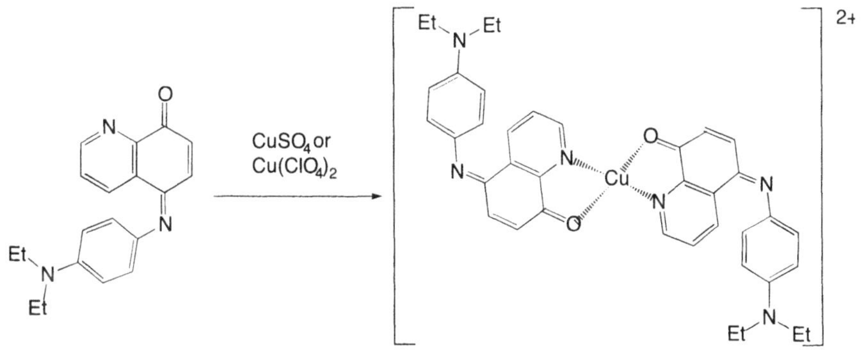

$CuSO_4$ or $Cu(ClO_4)_2$

D1

Scheme 1

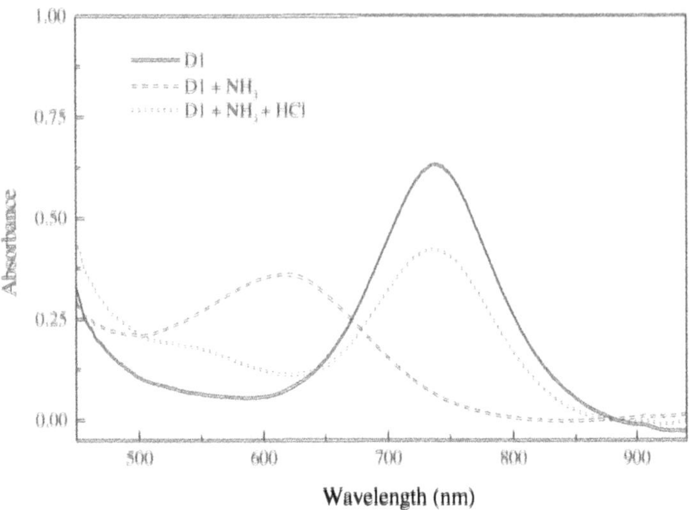

Fig. 1. Test of reversibility of the reaction of the dye D1 with ammonia in ethanol solution

as that of the complex is found at 738 nm. The molar absorptivity of D1 is $\varepsilon_{max} = 144\,000$ L mol^{-1} cm^{-1}.

Hence, the reaction of the complex with ammonia results in evolution of the free ligand. Ammonia is a stronger free electron pair donor than the ligand, and, consequently, ammonia substitutes the ligand in the ligand sphere of the complex. Addition of HCl removes free ammonia from the solution and the peak of the copper(II) complex recovers, which shows that the reaction of the complex with ammonia is reversible in solution.

Attempts to incorporate D1 into a sol-gel derived silica matrix (based on tetraethoxysilane and prepared at pH ≡ 1) were unsuccessful due to the decay of the dye in this material. Fig. 2 shows that the absorption spectra of the dye incorporated in a fluoropolymer had a broad peak between 550–750 nm for the Cu(II) complex and that the ligand liberated on interaction with ammonia had an absorption maximum at 605 nm. Interaction of the dye with ammonia vapours again results in the evolution of the free ligand. After exposure of the dye to ammonia, the peak of the free ligand slowly decreases and the peak of the Cu(II) complex recovers, which shows that the response of D1 is reversible to ammonia [30]. This reversibility is apparently due to the fact that the Cu(II) complex with ammonia is partly unstable and the equilibrium between the complex and ammonia is connected with the existence of a low partial pressure of ammonia. Diffusion of ammonia out of the polymer is accompanied by the recovery of D1.

We also synthesised a methyl analogue of D1, i.e. copper(II) complex of 5-(*N'*-dimethylaminophenylimino)quinolin-8-one. The procedure for its synthesis was practically the same as that for the synthesis of D1. The spectral characteristics and the reversibility to ammonia in solution were also the same as for D1. However, a remarkable difference in the response times has been observed

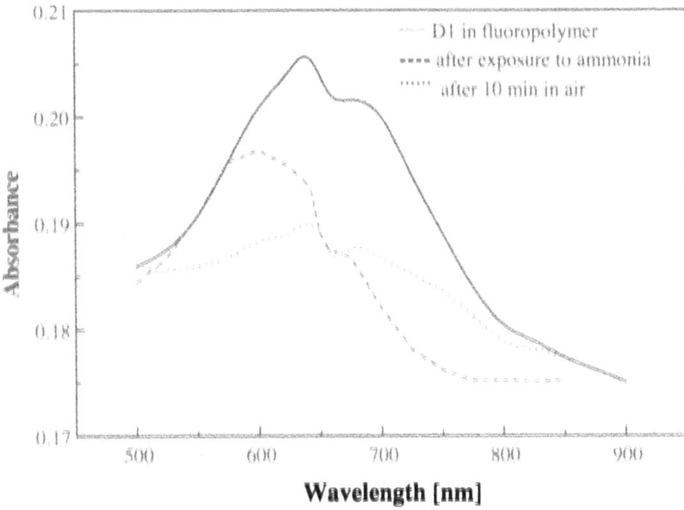

Fig. 2. Absorption spectra of D1 incorporated in the fluoropolymer before and after exposure to ammonia

Table 1. Responses of D1-type dyes to ammonia at the maximum absorption wavelength λ_{max}

Dye	D1	Methyl analogue of D1
λ_{max}	770 nm	740 nm
$\downarrow T_{90}$(100 ppm)	40 s	30 s
$\uparrow T_{90}$(100 ppm)	150 s	45 s
ΔA(100 ppm)	0.08	0.05
Range	1–100 ppm	1–100 ppm

in solid matrix. D1 and its methyl analogue were immobilized in a thin-film PVC matrix coated on a 600-μm optical fibre. Evanescent wave absorption was used to interrogate the dye under exposure to 10–100 ppm ammonia in nitrogen carrier gas. The response and recovery times were found to be in the order of seconds and the reversibility was excellent [31]. The characteristics of the two dyes when immobilised in PVC are summarised in Table 1.

The responses of the sensing coatings depended largely on the humidity of the analyte gas. The results presented in Table 1 were taken at 100 % relative humidity of the nitrogen carrier gas and a large decrease in the response of the coated fibres was seen when dry carrier was employed [31]. From the data of Table 1 it can be seen that the response times of the methyl analogue of D1 are lower that those of D1. The dye D1 has an ethyl substituent attached to nitrogen so that D1 is obviously more bulky than its methyl analogue. As the dyes interact with ammonia, they undergo a series of changes of their geometry. The more bulky the dye, the more space is needed for the occurrence of geometry changes.

Hence, the longer response time of D1 can be accounted for by a stronger interaction of bulkier D1 with the solid matrix.

Another Cu(II)-quinoline type complex tested as a transducer for ammonia was the one with 7-chloro-5-(4'-diethylamino-2'-methylphenylimino)quinolin-8-one (dye D2). The synthesis was again carried out as described by Kubo et al. [29] accordig to Scheme 2. In ethanol solution, the free ligand shown at the left side of Scheme 2 has $\lambda_{max} = 650$ nm and $\varepsilon_{max} = 23\,000$ L mol^{-1} cm^{-1}, whereas the spectral characteristics of D2 are $\lambda_{max} = 740$ nm and $\varepsilon_{max} = 120\,000$ L mol^{-1} cm^{-1}. The solution spectra of the original dye D2 and after the addition of ammonia are shown in Fig. 3.

Just like D1, the dye D2 is also reversibly sensitive to ammonia when incorporated in solid matrix, such as silicone rubber [26]. The sensing mechanism is

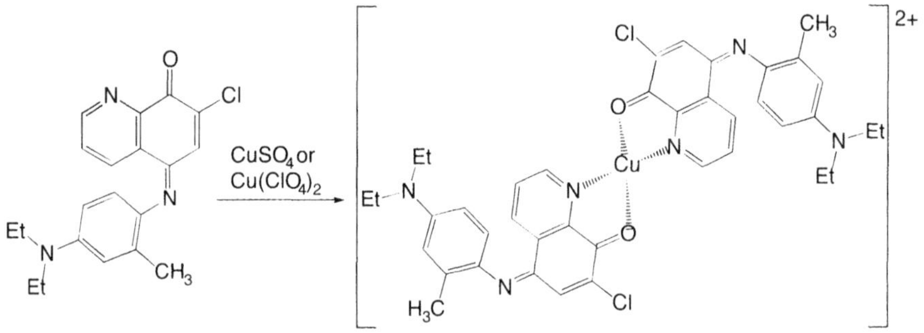

$CuSO_4$ or $Cu(ClO_4)_2$

D2

Scheme 2

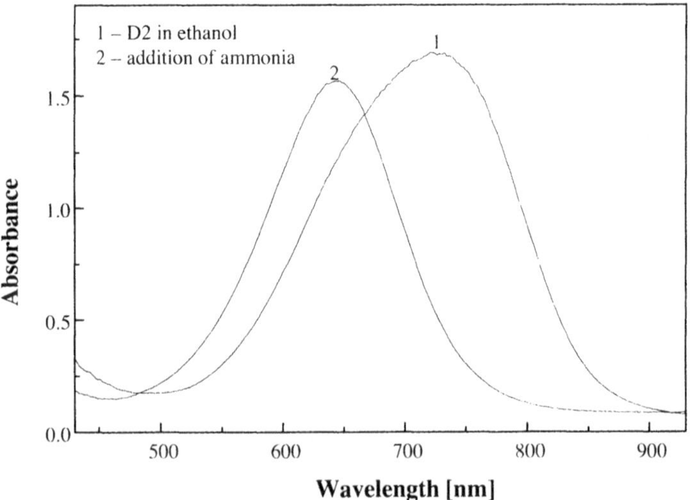

Fig. 3. Absorption spectra of D2 and the ligand liberated by an addition of ammonia in ethanol

obviously the same as for D1 and resides in liberation of free ligand when D2 interacts with ammonia.

5-(4-Diethylamino-2-methylphenylimino)-5*H*-pyrido[2,3-*a*]-phenothiazine is the ligand for the dye D3. The ligand itself has been prepared by Kubo et al. [32] from 7-chloro-(4'-diethylamino-2'-methylphenylimino)quinolin-8-one and 2-aminothiophenol in ethanol. The spectral characteristics of the ligand are λ_{max} = 640 nm and ε_{max} = 11 000 L mol^{-1} cm^{-1}. The detailed procedure for the synthesis of D3 has been described previously [33] and is summarised in Scheme 3. The spectral characteristics of D3 are λ_{max} = 840 nm and ε_{max} = 50 000 L mol^{-1} cm^{-1}. The solution spectra of D3 in ethanol are shown in Fig. 4.

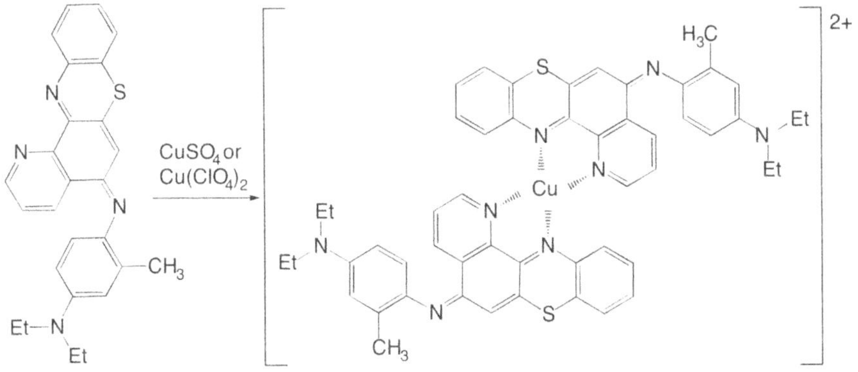

CuSO$_4$ or Cu(ClO$_4$)$_2$

D3

Scheme 3

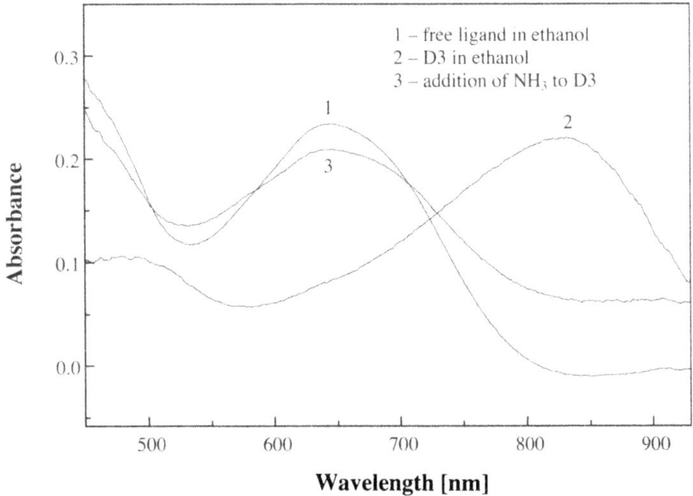

Fig. 4. Absorption spectra of D3 and the free ligand in ethanol solution

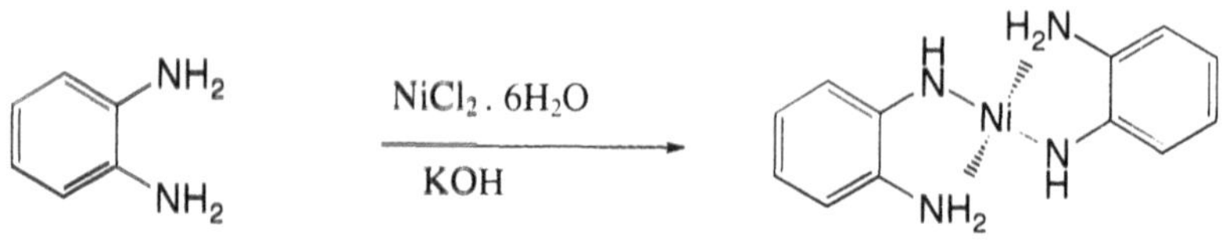

$$\text{1,2-}C_6H_4(NH_2)_2 \xrightarrow[\text{KOH}]{NiCl_2 \cdot 6H_2O} \mathbf{D4}$$

D4

Scheme 4

The complexes of Ni(II) with the ligands involved in the formation of D1 to D3 have absorption spectra similar to those described above. Since other metal ions such as Co(II) and Fe(II) also form complexes absorbing in the NIR region, attempts have been made to use these ligands for the determination of metal ions in water solutions [34].

The sensing properties of Ni(II) complexes with *o*-diamino-substituted aromatic rings have been also investigated. The complex of Ni(II) with 1,2-phenylenediamine (dye D4) was synthesised according to Scheme 4 [35].

The absorption maximum of D4 in ethanol solution is at 775 nm. We also tested the properties of the Ni(II) complex with 9,10-diaminophenanthrenene (dye D5), first synthesised by Balch and Holm [36]; its absorption maximum was observed at 1115 nm. Both D4 and D5 dyes were found to be reversible to ammonia, both in solution and in solid matrices.

The nickel-complex based NIR dyes are suitable for ammonia sensing; however, their properties are not as good as those of copper complexes. Copper(II) complexes have higher molar absorptivities and better solubilities than their nickel(II) analogues. The molar absorptivities of the dyes D1-D3 are of the order 10^5–10^6 L mol^{-1} cm^{-1} and they are well soluble in practically all solvents including water. The solubility of D4 and D5 is low in common solvents. Furthermore, the thermooxidation stability of nickel(II) complexes is lower than that of copper(II) complexes. Consequently, it is easier to incorporate the copper complexes into a polymer matrix than the nickel complexes and also the copper complexes have longer lifetimes.

The mechanism of ammonia sensing resides in replacing the ligand in the coordination sphere of the metal ions by ammonia since ammonia is a stronger free electron pair donor than the ligands are. This implies that the Cu(II) and Ni(II) complexes are reversibly sensitive not only to ammonia, but also to amines as has been verified by reactions with dimethylcyclohexyamine and triethylamine [37]. In contrast, acids destroy the complexes irreversibly by forming onium salts with the ligands.

Commercially available dye CI Acid Green 1 was also tested for its sensitivity to ammonia. This dye has a broad peak at 750 nm in ethanol solution. The dye can be assigned to the group of metal complexes, since it is a nitroso complex of Fe(II) [38]. The dye entrapped in fluoropolymer and silicone films has been found to have sensing potential for hazardous substances such as ammonia, hydrogen chloride and acetic acid [39].

2.1.2
Polymethine Dyes

1,5-Bis-(dimethylaminophenyl)divinylene carbenium perchlorate (dye D6) was first synthesised by Schmidt and Wizinger [40] according to Scheme 5.

The absorption spectra of the dye D6 in ethanol solution before and after exposure to ammonia are shown in Fig. 5. The maximum absorption is at 784 nm. D6 acts as an acid-base indicator, where the interaction of the dye D6 with ammonia results in a decrease of the absorption peak. A similar behaviour has been encountered for D6 incorporated in the TEOS silica matrix. However, D6 in the silica matrix exhibits very long response times on exposure to ammonia, typically greater than 10 minutes [30].

1,1,5,5-Tetrakis(dimethylaminophenyl)-2,4-pentadien-1-ol perchlorate (dye D7) is an analogue of D6. It has been synthesised according to Scheme 6.

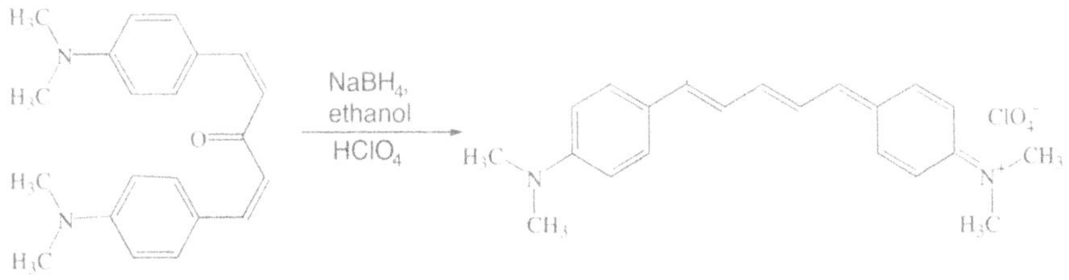

$NaBH_4$, ethanol, $HClO_4$

D6

Scheme 5

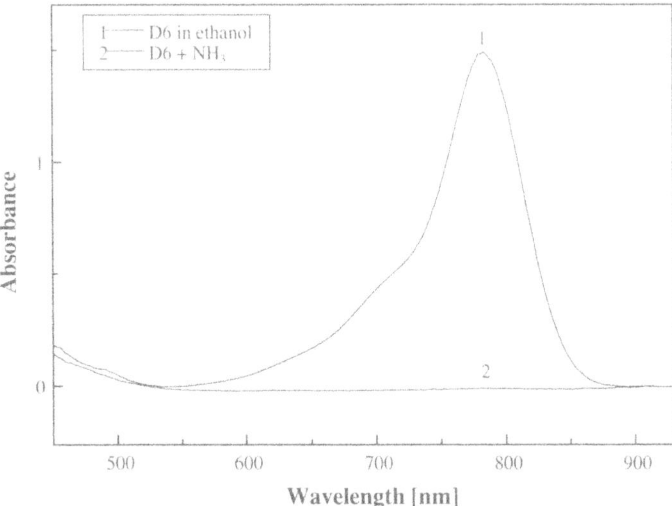

Fig. 5. Absorption spectra of the dye D6 in ethanolic solution before and after exposure to ammonia

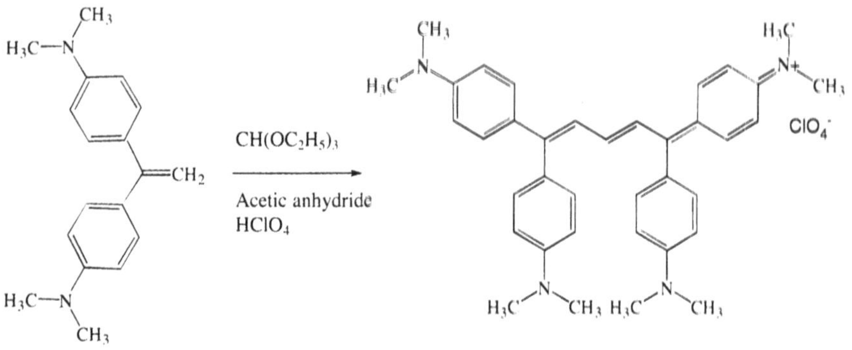

$CH(OC_2H_5)_3$

Acetic anhydride

$HClO_4$

D7

Scheme 6

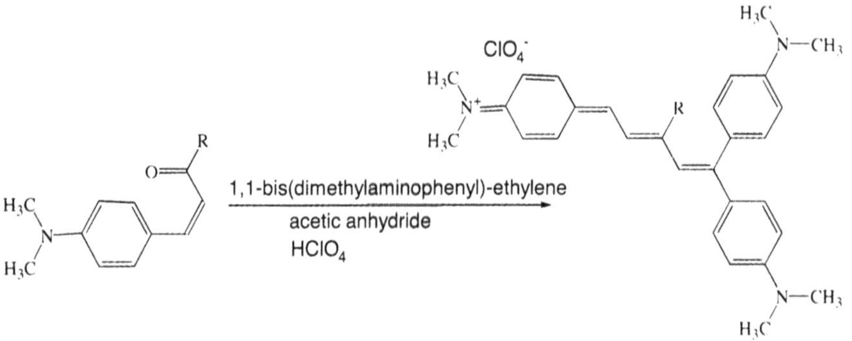

1,1-bis(dimethylaminophenyl)-ethylene

acetic anhydride

$HClO_4$

D8-D11

Scheme 7

The synthesis of D7 was carried out by Tuemmler and Wildi [41], where 1,1-bis(4-dimethylaminophenyl)ethylene was initially synthesised by the same workers [42]. The dye D7 exhibits two absorption peaks at 628 and 800 nm and is reversibly sensitive to ammonia.

Tris(4-dimethylaminophenyl)divinylenes in the form of perchlorates exhibit three absorption peaks, and one of these appears in the NIR region. These dyes were prepared by condensing 1,1-bis(4-dimethylaminophenyl)ethylene with an appropriate ketone, e.g. 4-(dimethylamino)benzalacetone, 1-(4-dimethylaminophenyl)-1-penten-3-one, 4-(dimethylamino)chalcone and 4,4'-bis(dimethyla mino)benzalacetone. The synthesis proceeds according to Scheme 7 [43, 44].

The substituents R, occurring in the dyes, are listed in Table 2. The absorption spectrum of the dye D8 is shown in Fig. 6. The dye is reversibly sensitive to ammonia, and the acid-base properties of this series of dyes have been studied in detail previously [45]. The group of dyes D9–D11 was synthesised with the

Table 2. Substituents and spectral characteristics of the dyes D8–D11

Dye	R	λ_{max} [nm]
D8	CH_3	627, 690, 804
D9	CH_3CH_2	578, 615, 803
D10	phenyl	639, 692, 804
D11	$(CH_3)_2PhCH{=}CH$	629, 690, 804

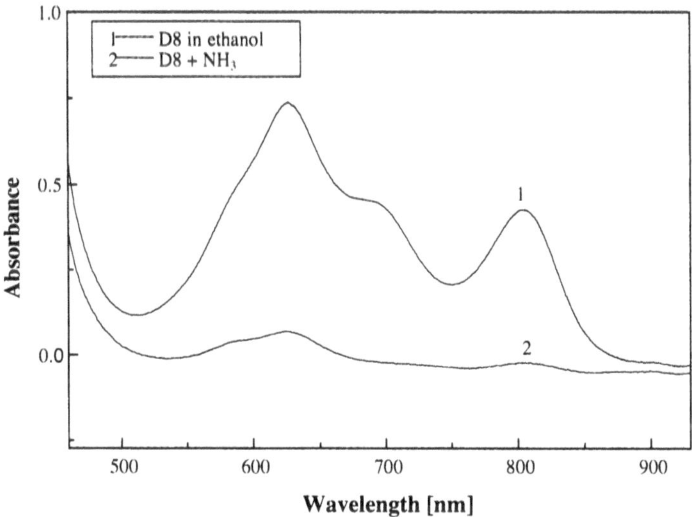

Fig. 6. Absorption spectrum of the dye D8 in ethanol and after its interaction with ammonia

aim of evaluating the effect of various substituents attached to the same framework on the molar absorptivity and maximum absorption wavelength. The bulkier the substituent, the lower the molar absorptivity. The influence of the substituent on the maximum absorption wavelength is small. All compounds are well soluble in ethanol and acetone. These dyes are reversibly sensitive to ammonia. The spectra of the dyes D8–D11 are very similar and their spectral characteristics are listed in Table 2.

It has been shown that the dyes D9–D11 are sensitive not only to ammonia but also to some amines. They may be used as pH indicators in the basic region [45], as it can be seen from Fig. 7. The figure shows that pK_a values of these dyes are approximately 12.

Tetrakis(4-dimethylaminophenyl)hexadienes in the form of diperchlorates were synthesised from 1,1-bis(dimethylaminophenyl)ethylene and an appropriate hydroxy ketone according to Scheme 8 [43, 45]. The substituents R occurring in the dyes are listed in Table 3.

The dyes D12–D14 have also been synthesised with the aim of evaluating the effect of various substituents attached to the same framework on the molar

0.6
0.5
0.4
0.3
0.2
0.1
0.0
Absorbance
2
4
6
8
10
12
pH
D11
D10
D9

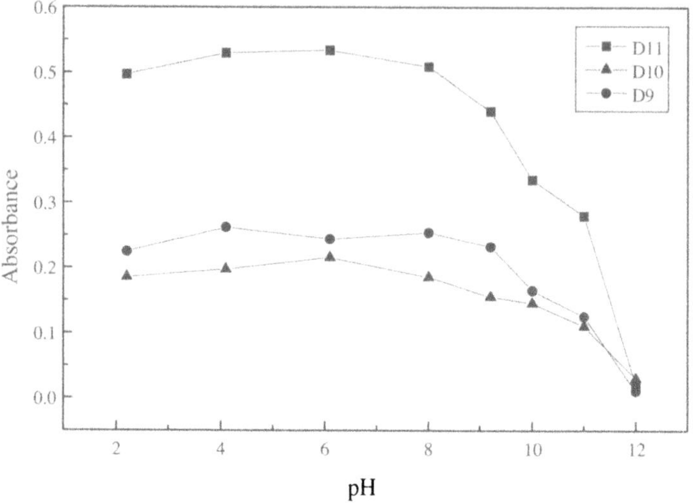

Fig. 7. Absorbance at maximum wavelength as a function of pH for the dyes D9–D11 in water-ethanol solution

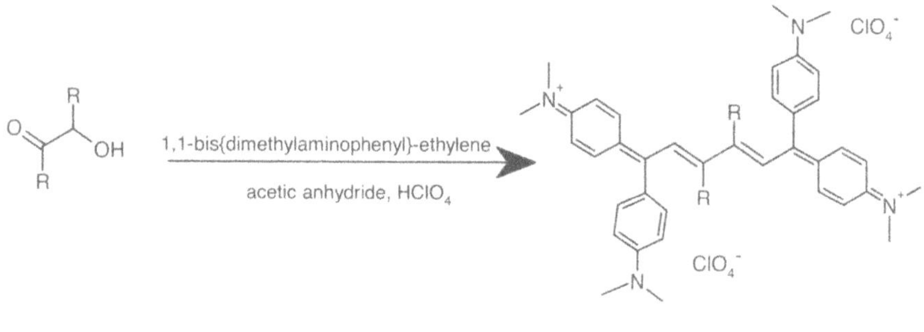

R
O
OH
R
1,1-bis(dimethylaminophenyl)-ethylene
acetic anhydride, HClO4
N
ClO4-
R
R
N
N
N
ClO4-

D12-D14

Scheme 8

Table 3. Substituents and spectral characteristics of the dyes D12–D14

Dye	R	λ_{max} [nm]
D12	CH_3	627, 805
D13	phenyl	595, 642, 802
D14	thienyl	597, 628, 803

absorptivity and peak absorption wavelength. Their maxima of absorption were found at 802–805 nm. The dyes are soluble in acetone and ethanol and are reversibly sensitive to ammonia. The influence of the substituent on both the molar absorptivity and the wavelength of the maximum absorption was found

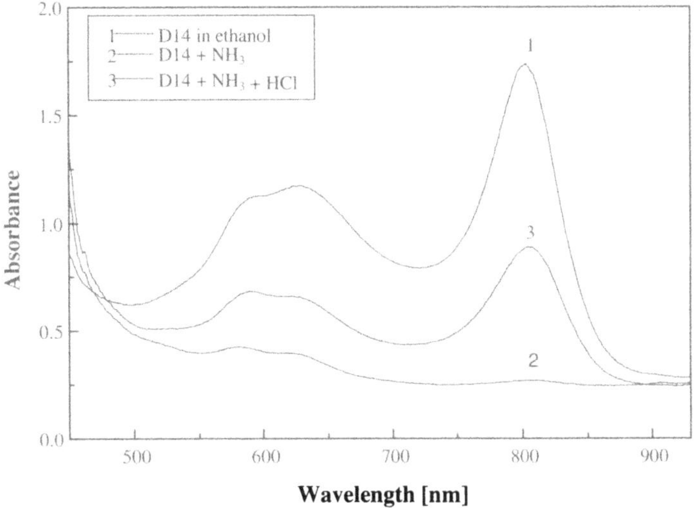

Fig. 8. Absorption spectrum of the dye D14 in ethanol and its reversibility to ammonia

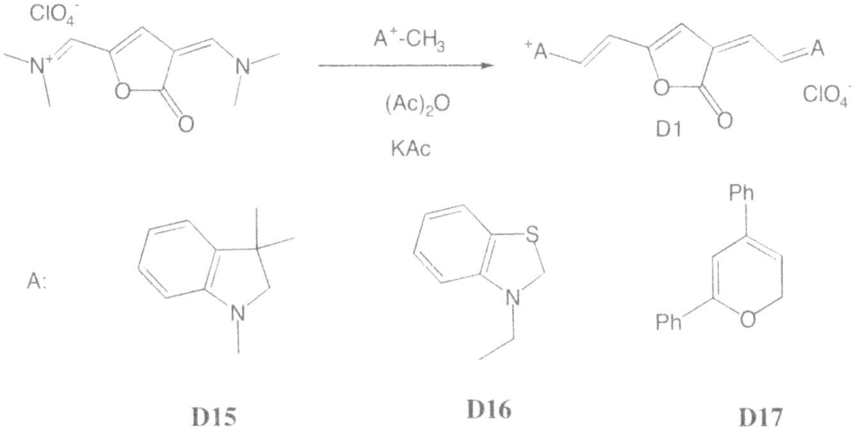

Scheme 9

to be negligible. The absorption spectrum of the dye D14 is shown in Fig. 8; the spectra of the dyes D12 and D13 are very similar and their spectral characteristics are listed in Table 3.

A series of dyes is based on the condensation of 4-(dimethylaminomethylene)-5-oxo-2-furfurylidenedimethyliminium perchlorate with appropriate salts, e.g. 1,2,3,3-tetramethyl(3*H*)indolium iodide, 3-ethyl-2-methylbenzothiazolium iodide and 2-methyl-4,6-diphenylpyrylium perchlorate, according to Scheme 9 [46].

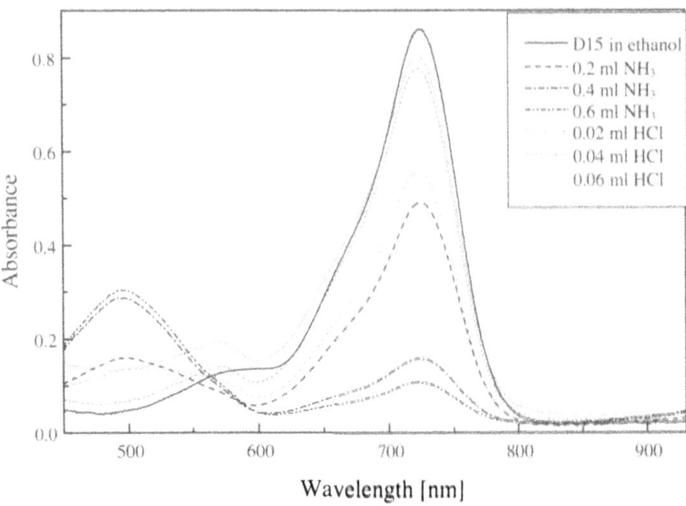

Fig. 9. pH dependence of spectra of the dye D15 in water-ethanol solution

The dyes D15–D17 have the absorption maximum at 720–730 nm. The spectrum of the dye D15 in ethanol and its reversibility to ammonia is shown in Fig. 9. The spectral response to ammonia of the dyes D16 and D17 is similar to that shown in Fig. 9. The response of the dye D16 is not immediate; with this dye, the decrease of absorbance depends on time and the dye-ammonia interaction follows first-order kinetics [46]. Nevertheless, the original absorption spectrum of the dye can be recovered by adding HCl.

Tests of the dye D15 suggest that it is a very promising ammonia transducer. The dye exhibits perfect reversibility and short response time. Unfortunately, its photolability has been observed in several cases.

The dyes D15–D17 can be used also as pH indicators in the basic region. Fig. 10 shows the dependence of their spectra on pH in water-ethanol solutions (1:1 *v*/*v*). It can be seen that pK_a values of these dyes are between 8.8–10.5. The interaction with buffers suggests that ammonia is not the only agent capable of affecting the long wavelength peak in the spectra. The response of the dye D16 is time-dependent also in non-ammonia solutions. Some amines, such as triethylamine and dimethylcyclohexylamine, also interact with the dyes, as is demonstrated in Fig. 11 for the dye D15 [46].

The mechanism of the interaction of the dyes with ammonia is proposed in Scheme 10 for the dyes D15–D17 [46]. The lone electron pair of ammonia is probably attracted by the positive charge delocalised over the π-electron system of the dye. Formation of the bond between ammonia and the carbon skeleton breaks the conjugation and this is manifested by the decrease of the peak at maximum wavelength. This mechanism accounts for the fact that also other electron donors, such as OH^- ion or amines, would interact with the dyes. All the dyes D6–D19 have a positive charge delocalised over the π-electron system. For all these dyes, the absorption peak decreases on interaction with ammonia. The

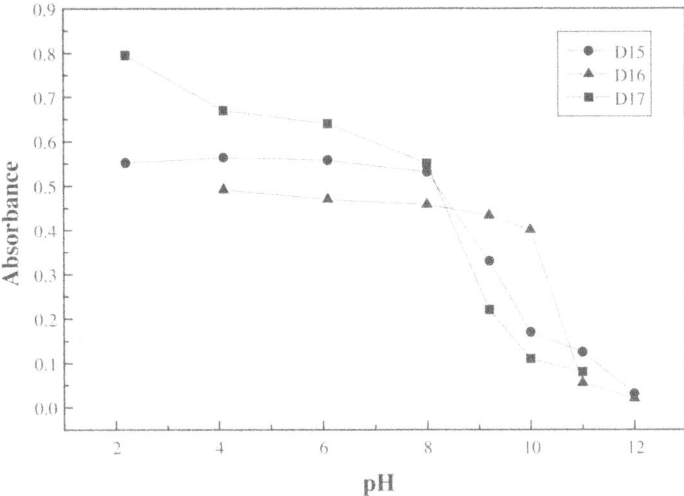

Fig. 10. Absorbance at maximum wavelength as a function of pH for the dyes D15–D17 in water-ethanol solution

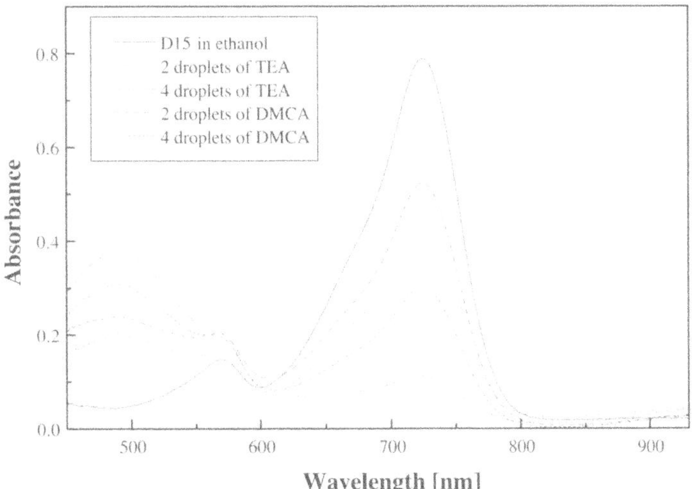

Fig. 11. Sensitivity of the dye D15 to triethylamine (TEA) and dimethyl-cyclohexylamine (DMCA). Volume of the dye solution was 3 mL

mechanism of the interaction suggested in Scheme 10 is likely to take place in the case of all these dyes.

Another series of dyes D18-D20 was prepared by condensing 4-(dimethylaminomethylene)-5-oxo-2-furfurylidenedimethyliminium perchlorate with an appropriate salt, e.g. 1,4-dimethylpyridinium iodide, 1-ethyl-4-methyl-quinolinium iodide and 1-ethyl-2-methylquinolinium iodide. Generally, the conden-

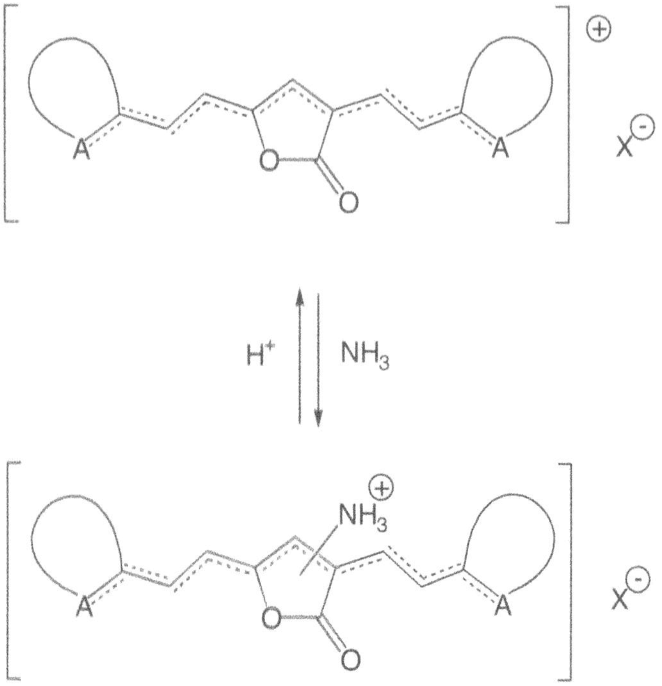

Scheme 10

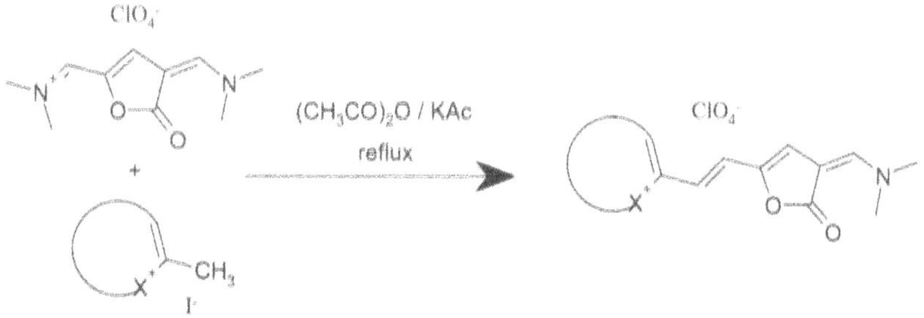

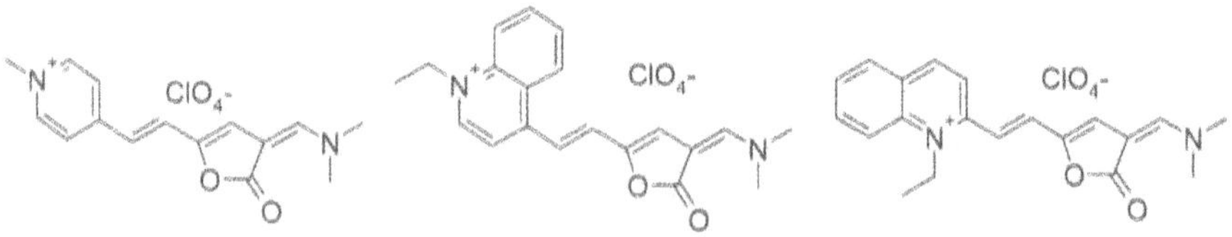

Scheme 11

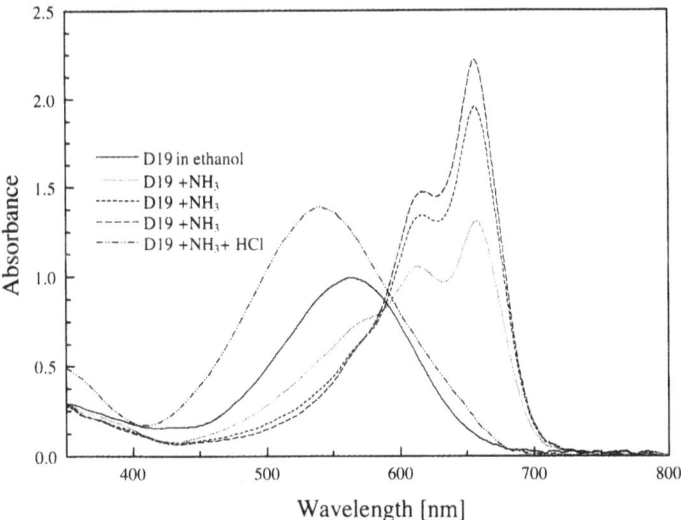

Fig. 12. Solution spectra of the dye D19 in the presence of ammonia and reversibility by addition of HCl

sation reaction was carried out under the same conditions as in the case of the dyes D15–D17; however, in this case the reaction afforded only monocondensation products [47, 48] according to Scheme 11.

The sensitivity of the representative of this group, e.g., the dye D19, to ammonia is shown in Fig. 12. As it can be seen, multiple addition of ammonia led to an increase in the absorbance. Addition of HCl removes free ammonia from the solution and the peak of D19 recovers which shows that the reaction with ammonia is fully reversible in solution.

The original dyes D18–D20 exhibit absorption maxima at 492, 556 and 520 nm, respectively. Under the influence of ammonia, peaks arise at 580, 660 and 620 nm [48]. The edge of the peak for the dye D19 lies in the near infrared region. According to our experience, the absorption maxima of dyes in solid matrix are shifted towards higher values when compared to the spectra measured in solution. Thus, it can be reasonably expected that the peak arising from the exposure to ammonia will have a maximum at over 700 nm if the dye D19 is incorporated in a solid matrix. In contrast to the dyes D1–D17, after the exposure to ammonia the dyes D18–D20 exhibit a shift of the absorption peak towards higher wavelengths. The mechanism of this shift is not fully clear. It may be a formation of highly polar chromophores, as has been suggested by Szablewski et al. [49], or a formation of a charge-transfer type complex.

From other groups of dyes, 3,3-bis[1,1-bis(dimethylaminophenyl)ethylene-2-yl]-substituted phthalides synthesised according to Dai and Peng [50] have been tested. This group of dyes has an absorption peak at about 805 nm; however, they exhibit only partial reversibility to ammonia in solution [37]. Squarilium dyes SQ2 and SQ4 have their maximum absorption slightly above 800 nm, their molar absorptivities are high and they are well soluble in chloroform and acetone, and soluble in ethanol. SQ2 dissolved in ethanol is not sensitive to ammonia, but it is sensitive when dissolved in acetone. The selectivity is thus solvent-dependent. The reaction with ammonia in acetone is not reversible so that these dyes appeared to be unstable in solution under the influence of ammonia [37].

2.2 Structure and Tests of NIR pH Transducers

In the previous section it has been shown that only very limited information on the sensitivity and reversibility of NIR dyes to ammonia is available in the literature. In contrast to this, papers dealing with the synthesis of acid-base indicators absorbing in the NIR region have been published over the past ten years.

Boyer et al. [51] derivatized near-infrared dyes with appropriate functional groups for use as analytical probes. They evaluated the sensitivity to pH of a bis-carboxylic acid derivative of a near infrared dye to illustrate its potential as a probe for determining pH. The dye had an absorption maximum at 795 nm in aqueous solution.

Patonay et al. [52] evaluated an aminodienone dye demonstrating pH dependent absorption and fluorescence spectra as well as solvent polarity dependence. The absorption maximum was at 535 nm in neutral or alkaline solutions in methanol. The absorption spectra underwent a strong bathochromic shift in the presence of acids (λ_{max} = 709 nm) with a concomitant change in the fluorescence spectra. Mason et al. [53] synthesised and tested the stable bis(aminodien)one system with λ_{max} = 648 nm in methanol. The dye underwent protonation at the oxygen atom to give a cyanine chromophore with λ_{max} = 932 nm. The transition was fully reversible and depended solely on pH conditions.

Casay et al. [54] developed a near-infrared fibre optic probe for the determination of caustic soda using a non-commercially available tetrasubstituted chloroaluminum naphthalocyanine NIR dye, susceptible to pH changes in solutions. The maximum wavelength of the dyes was above 700 nm. The probe consisted of a semiconductor laser diode (λ_{max} = 780 nm), a NIR dye and a detector. The probe, made of poly(methyl methacrylate), served as a support for a permeable polymer that was used for the entrapment of the NIR dye. The permeable polymers used were Nafion and Gelatin. As hydrogen ions diffused through the permeable polymer, a complex was formed with the dye, accompanied by changes in its spectral characteristics. The probe showed good reproducibility.

Lindauer et al. [55] synthesized two new acidochromic styrylcyanines and one new benzopyrylotrimethinium dye. Their VIS and NIR absorption spectra were recorded as a function of pH in fluid solution and embedded in polymer layers. The layers proved to be useful as pH sensors. They also synthesized [56] a series of new acidochromic 9-(4-dialkylaminostyryl)-acridines. The colour change in the presence of perchloric acid and the pK_a was investigated and their photosensitizing ability was tested by oxygen consumption measurements. In a further paper [57], they reported the synthesis of a series of new acidochromic dyes absorbing in the NIR spectral region (up to 900 nm) with a systematic variation of the underlying nonsymmetric pyrylium trimethinium acridine structure. The dyes exhibited pK_a values between 5.7 and 6.7. Semiempirical MO-calculations were performed to elucidate the essential characteristics of the chromophores.

Lehmann et al. [58] presented an absorption-based, fibre optic pH meter for the physiological pH range. It employed a sensor membrane based on NIR proton-carrier dye and laser diodes as light sources. It was shown that an optical absorption sensor with a sensing dye placed on the tip of a bifurcated fibre required special provisions to receive sufficient light back from the sensor head to the detector. In view of the availability of cheap and intensive light sources in the NIR, it was advantageous to use NIR dyes for chemical sensor membranes. A new type of acidochromic dye immobilized in a PVC matrix was presented for use with a two-wavelength, single-beam fibre optic photometer. Values of pH measured with the fibre optic pH meter were compared with the results obtained with a glass electrode.

Czerney and Grummt [59] synthesized a series of new near-infrared-absorbing non-symmetric and symmetric trimethinecyanine dyes containing a 7-dialkylamino-1-benzopyrylium end group. The dyes were shown to be useful as pH sensor dyes. They also synthesised [60] acidochromic nonsymmetric pyrylotrimethines NIR dyes, tested their absorption behaviour and showed that they are suitable to be employed as sensor dyes in pH sensing. They also presented new stable acidochromic near-infrared absorbing dyes useful for the design of optodes [61]. The dyes belonged to the classes of 2-(4-hydroxyphenyl)-1-benzopyrylium salts, 9-[3-(pyranylidene)propenyl]-7-diethylamino-1-benzopyrylium salts, 9-[3-(thiopyranylidene)- and 9-[3-(1-benzopyranylidene)-propenyl]acridinium salts, and 9-(4-diallylaminostyryl)acridinium salts. Absorption maxima of the long-wavelength absorbing forms spanned the region from 600 to 900 nm.

Miltsov et al. [62] presented a one step synthesis of new acidochromic indolenine-based cyanine dyes. The dyes had absorbance maxima at 780–805 nm and exhibited spectral changes in pH range from 6 to 2. They also suggested a new one-step synthesis of ketocyanine dyes [63]. The dyes obtained exhibited spectral changes in pH range from 1.7 to 4.3 and their protonated forms absorbed at 715–750 nm. They described the use of ketocyanine dyes [64] as indicators that can operate in bulk optode membranes as the chemically active region was near 780 mm. Their spectral characteristics, acid-base properties, chemical stability, and solubility in the membrane phase were discussed. The response characteristics were first tested in a conventional absorbance/transmittance flow cell. The dyes offered a wide range of pK_a's in PVC membranes, good sensitivity as a result of their high molar absorptivities, excellent solubility in the plasticizer, and short response times. They presented good chemical stability under common laboratory conditions when stored in the dark, and the absence of leaching guaranteed a long lifetime. Membranes were finally applied as the sensing region of an integrated waveguide optode, demonstrating the extraordinary sensitivity improvement while preserving the remaining analytical features. Response times lower than 2 min were obtained.

A series of dyes has been synthesised by condensing 4-(dimethyl-aminomethylene)-5-oxo-2-furfurylidenedimethyliminium perchlorate with an

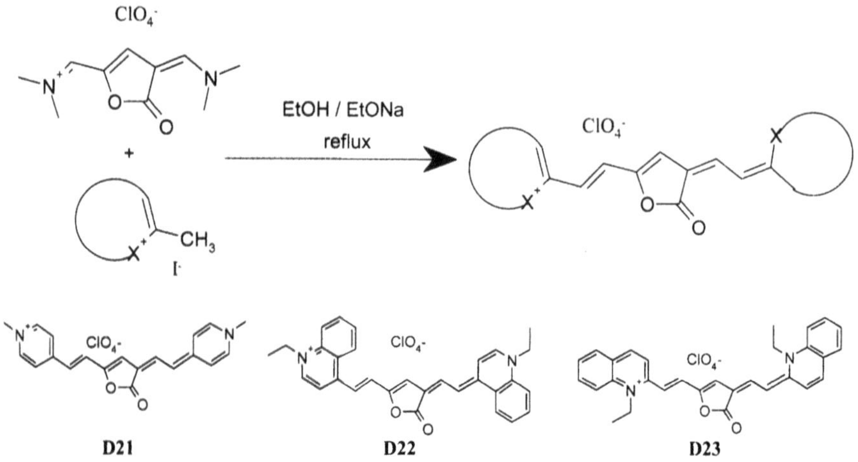

ClO_4^-

EtOH / EtONa
reflux

D21 D22 D23

Scheme 12

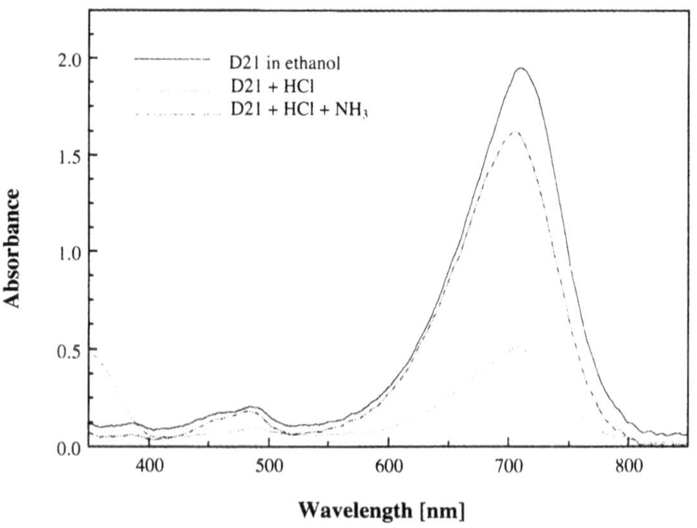

Fig. 13. Solution spectra of the dye D21 in the presence of HCl and reversibility by addition of ammonia

appropriate salt, e.g. 1,4-dimethylpyridinium iodide, 1-ethyl-4-methylquinolinium iodide, and 1-ethyl-2-methylquinolinium iodide in ethanol under reflux with the addition of sodium ethoxide as a catalyst [48]. The procedure is shown in Scheme 12.

The spectral responses of the dyes D21–D23 are very similar and all have their absorption maxima in the near infrared region. Fig. 13 shows the spectra of D21 and its sensitivity to HCl. After addition of HCl to an ethanol solution of the dye, an immediate decrease of peak intensity is observed. The peak recovers after addition of ammonia to the solution. The series of dyes D21–D23 can thus be used as NIR pH indicators in acidic pH range.

3 Quantum-chemical Calculations and General Rules

Absorption of light by a NIR dye brings about a transition of an electron from its ground state to the excited state, thus producing electronic spectra. Since the light absorption depends on the electronic structure of the molecule, the quantum-chemical calculations represent a basic method for deeper insight into the relation between the dye structure and its spectral properties. The excitation energy can be evaluated as the difference in energy of the highest occupied (HOMO) and lowest unoccupied (LUMO) molecular orbital [65].

The first method, using molecular orbitals expressed as a combination of atomic orbitals, was the Hückel molecular orbital (HMO) method. Approaches to the design and synthesis of NIR absorbing dyes have been concisely reviewed in the literature [5]. A book on infrared absorbing dyes edited by Matsuoka [1] contains a chapter on the design of such colorants using the Pople-Parr-Pariser molecular orbital (PPP MO) method. A review of software based on a modified PPP MO method was subsequently published [66]. Although HMO and PPP MO methods take into account only $\pi\rightarrow\pi^*$ transitions, primarily the PPP MO method appeared to be suitable for the strategic design of chromophoric systems [1].

The above-mentioned $\pi\rightarrow\pi^*$ transitions are, however, also affected by σ-electron distribution. This is taken into account in the semi-empirical all valence electron methods. We have used the standard semi-empirical AM1 (Austin Model) method of quantum chemistry (AMPAC program package) [67, 68] in order to find the optimal geometries and to calculate the corresponding electronic structures as well as electronic spectra. As this method is not especially parameterised to produce exact electron spectra, a series of similar compounds was investigated in order to obtain correct trends in peak positions. All calculations were performed in higher precision using the Davidon-Fletcher-Powell optimisation procedure [69, 70]. The electronic structure characteristics were evaluated in terms of charges of individual atoms and bond indices of individual bonds [71]. The systems were characterised by arithmetic mean values and root mean square deviations of the bond distances, atomic charges and bond indices over the carbon skeleton of individual chromophores. The perturbations of the planar π-bond framework were described by the deviations from planarity (dihedral angles). From the results obtained some general rules can be deduced.

From the calculations for tris(4-dimethylaminophenyl)divinylenes (dyes D8–D11) it follows that the calculated λ_{max} values exhibit a systematic shift towards lower wavelengths in comparison with experimental data [44]. This may be explained by neglecting all external interactions in isolated model systems as the higher energy difference between the ground and excited electronic state of the compound in real solutions is due to their different electrostatic interactions with solvent molecules. Chromophore planarity plays the key role. The (dimethylamino)phenyl rings bonded to the same carbon atom cannot be co-planar with the pentamethine chain and their perpendicularity causes the decrease of λ_{max}. On the other hand, the perpendicularity of the single (dimethylamino)phenyl ring increases λ_{max}. All these deformations cause significant changes in the electronic structure of individual chromophores. The highest λ_{max} values are obtained for the highest similarity of all statistical parameters of the benzene rings with the aliphatic divinylene chains and extended divinylene chains including additional chain-to-phenyl bonds. This similarity condition seems to be more important than the maximal planarity condition. The small changes of λ_{max} with different R substituents may be explained by the small changes in the electronic structure of the (dimethylamino)phenyl chromophores (despite significant changes of the divinylene chain system). From this point of view, the use of substituents with higher electron donor or acceptor character might not lead to a λ_{max} increase. A systematic shift of λ_{max} to higher wavelengths with an increase of positive charge could be used to mimic some solvent effects (such as polarisation due to the solvent permittivity) in real systems. The protonation removes the chromophore planarity and brings about a decrease of λ_{max} in accordance with experimental data.

The results of calculations for tetrakis(4-dimethylaminophenyl)hexadienes (dyes D12–D14) [45] are quite similar to those for the dyes D8–D11. Also, a systematic shift towards shorter wavelengths compared to the experimental data is observed, which can be explained by the neglect of all external interactions in the isolated model systems. The mean atomic charges indicate a relatively uniform distribution of the additional charges among individual chromophores. Small changes of λ_{max} with variation in R substituents can be explained by small changes in the electronic structure of (dimethylamino)phenyl chromophores, and substituents with higher electron donor or acceptor character may not lead to a significant λ_{max} increase.

The influence of the central chromophore ring on the spectra of NIR polymethine dyes has been studied recently [72]. The trends in spectral data of two series of dyes containing either two benzothiazole or four dimethylaminophenyl side chromophores connected by a 7-membered aliphatic chain in the central chromophore implemented by cyclohexene, furanone or cyclopentene ring were compared. In both series, the effect of the central chromophore ring variation on the maximum wavelength λ_{max} was studied using the statistical treatment of the chromophores data obtained by quantum-chemical calculations. The significance of the factors influencing λ_{max} is arranged in the following order: the chromophore planarity, mutual angles of individual chromophore planes, the similarity of individual chromophores in the dye and substituents on individual chromophores. The importance of maximal similarity of bond lengths, atomic charges and bond indices between the central and side chromophores (including inter-chromophore bonds) has been also confirmed in agreement with previous publications [44, 45].

From the quantum-chemical calculations for the dyes D18–D23 it follows that the dyes exhibit a very intense single peak at λ_{max} [48]. Deviations from planarity bring about a decrease of λ_{max} due to weaker mutual π-interactions of chromophores. Furane rings in D18 and D21 dyes are coplanar with pyridine ones in contrast to their non-planarity with quinoline rings, especially in D19 (deviation of 16°), D20 (deviation of 28°) and D23 (deviations of 12° and 20°). The polar solvent influences the dyes' non-planarity in real solutions and this might be the origin of the different trends in measured and calculated peaks of absorption spectra.

The quantum-chemistry studies are useful in gaining a detailed understanding of the modifications in geometry as well as electronic structure that are induced by chemical modifications of the dyes. The calculations are related to isolated systems and, consequently, the calculated and experimental values may differ due to interactions with the matrix. This difference can be up to 100 nm or even more. Nonetheless, the calculations can be very helpful when one thinks of varying substituents on a chromophore chain. In such a case, the calculations can reliably predict trends in λ_{max} changes.

4 Influence of Matrix Quality on the Band Shape and Maximum Wavelength

The shape of the absorption peak and the position of its maximum depends much on the matrix quality. In solution, the maximum wavelength of the cationic dyes D6–D23 depends on the relative permeability of the solvent. For the metal-complex dyes D1–D5, the solvent may enter the ligand sphere of the metal ion and the solution of the dye cannot be prepared. For example, the solution of the dye D1 in dimethyl formamide exhibits a peak of the free ligand, not the peak of the dye D1 [26].

From the point of view of using a dye as a sensing material, its immobilization in solid matrices is of key importance. According to our experience, after immobilizing the dye in a solid matrix, the absorption peak becomes broader and the maximum wavelength is shifted towards higher values. As shown in Table 4, this is true not only for the NIR dyes but also for "standard" indicators in the visible region.

As seen from Table 4, the maximum wavelength of the NIR dyes increases by more than 100 nm when immobilised in the silicon rubber. The reason for the broadening and shift to higher wavelengths in the solid matrix obviously lies in the interaction between the matrix and the dye. In the previous paragraph it has

Table 4. Maximum wavelength of selected indicators and NIR dyes in ethanol solutions and immobilized in silicon rubber [26]

Dye	λ_{max}(ethanol) [nm]	λ_{max}(silicon rubber) [nm]
Thymol blue	440	600
Fluorescein	450	460
D1	738	850
D2	740	790
D6	780	890

been calculated by quantum-chemical methods that deviations from planarity often take place. The calculations are carried out for the conditions of isolated molecules which is a model quite close to solution. Due to the van der Waals intermolecular interactions, deviations from planarity may decrease in solid matrix, so that it can be reasonably expected that the maximum absorption wavelength of the dyes will be much higher in solid matrix than in solution. It is very likely that the distribution of the strength of the solid matrix-dye interactions is much broader than of the solvent-dye interaction. Consequently, the solid matrix-dye interaction also brings about the broadening of the absorption bands.

The response time of sensors on exposure to gases and their reversal will be determined by the thickness of the film in which the dye is immobilized, and on the permeability of the film for the target chemical [73]. As demonstrated previously, for example [31], the magnitude of the sensor response depends very often on the relative humidity.

5
Fibre-optic Distributed Sensors

Distributed fibre optic chemical sensing systems can be constructed by permitting target chemicals to permeate into the cladding of optical fibres and to interact with the evanescent field close to the core-cladding interface. In principle, the presence of target chemical can be determined from the attenuation characteristics of the optical fibre. In practice this approach is not very useful since it requires light sources that are matched to the absorption features of the target chemicals and this requirement can only be satisfied for a very small group of chemicals of interest. A more promising approach is to make use of chemical reagents that change their spectral features when exposed to target chemicals. Thus by incorporating such material into the cladding of optical fibres it is possible to produce distributed fibre optic chemical sensor. The interaction regions can be pinpointed using optical time domain reflectometry technique.

There are three possible ways of implementing a distributed sensing system with optical fibres [74]: (i) the multi-branch system where a number of single-point fibre optic sensors are connected to a common optical fibre via fibre optic couplers; (ii) the discrete serial system where the fibre optic sensing sections are joined in-line via a common fibre optic link; (iii) the continuous system which exhibits lowest light losses and thus would give the maximum number of sensing locations over the greatest length of optical fibre. The latter configuration is best suited for the detection and location of chemical leaks.

Only a few fibre optic distributed sensing systems have been reported in the literature. One system uses D-shaped single-mode optical fibre that had thinned cladding for direct detection of methane by direct absorption of the evanescent field [75]. Another system uses a 600-μm core diameter multimode fibre cladded with a polymer containing a visible chemical indicator to detect the presence of water [76]. Both these publications reported the measurement of small attenuation of the light due to exposure of the whole fibre to the target chemical and both techniques required light sources that are incompatible with current

OTDR instrumentation. Another system exploited swelling of polymer claddings of commercial fibres by pure liquid hydrocarbons to demonstrate distributed sensing under laboratory conditions [76], but operation under realistic industrial conditions has yet to be achieved.

At present, the continuous system offers the fastest route to the commercial distributed fibre optic sensing system. In such a system the sensing optical fibre would be plastic-clad silica fibre which has a porous cladding that contains a chemical indicator whose spectral characteristics are sensitive to a particular target chemical or group of chemicals. The operation of this sensing system has to be restricted to the near infrared spectral region where the light losses in optical fibres are minimum. The requirement to operate in the near infrared spectral region means that new chemical indicators must be developed and combined with cladding materials to yield sensing optical fibres. This is because the research into chemical sensing with optical fibres has to date been focused at the development of primarily point sensors that are based on visible chemical indicators that cannot be used for the construction of economical distributed fibre optic sensing systems.

There is a large range of polymeric materials that could be used for cladding purposes. Although many of these are well known, highly stable and produced in large quantities for other purposes, very little research has been carried out into their use as porous support matrices that can accommodate optical chemical indicators and still function as claddings for optical fibres. The situation with near infrared chemical indicators is much more complex. There is a large body of knowledge on near infrared absorbing materials developed for various applications, such as, for example, compact disc media and printers [77]. However, these near infrared absorbers have never been used in chemical sensors, and many of the materials have been synthesised specifically to be stable in adverse environments. Furthermore, new indicators must be compatible with the optical fibre manufacturing technologies to permit their incorporation into the cladding of the sensing fibres.

The performances of many of these new near infrared indicators are yet to be fully established and there is some doubt about their long-term stability. However, results from one COPERNICUS project indicated that fully-reversible response of a sensing fibre to ammonia is achievable over a period of several months [28].

In the field of distributed fibre optic chemical sensing the research efforts are likely to be focused on the development of sensing fibres based on plastic-clad silica structures. New chemical indicators for the near infrared will be required as well as cladding materials and methodologies for incorporating new chemical indicators into claddings. Also methodologies for the production of optical fibres with such claddings and designs of new cable structures to accommodate sensing fibres will be needed. The developments in chemical sensing fibres will need to be accompanied by developments in instrumentation and data processing systems that need to be optimised for the new sensing applications to provide the users with useful information. The first systems are likely to be targeted at leak detection where the concentration of the target chemical is relatively high. It is expected that, as the technology of these systems matures, more sensitive systems will become available for general monitoring of industrial pollutants.

The NIR chemical transducers for the continuous configuration of distributed fibre-optic sensors have to possess another property. Not only do they have to absorb in NIR region and to change reversibly their optical properties after interacting with the target chemical, but, in addition, the absorption peak in the NIR region should only appear after exposure to the target chemical. In that case the attenuation of the optical fibre prior to exposure to the target chemical would be minimal. This is a very restrictive requirements and the number of indicators satisfying this latest condition is severely limited.

For the detection of ammonia, the dyes D18–D20 exhibit an absorption peak around 650 nm after the exposure. These dyes are reversibly sensitive to ammonia in ethanol solutions and are promising for the construction of distributed ammonia sensors since the peak appears under the influence of ammonia. As a chemical transducer, the most promising seems to be the dye D19 since in the solid matrix this peak is expected to shift further into the NIR spectra region. Also the acid-base NIR indicators such as D21–D23 with absorption maxima in the near infrared region could be considered for use as chemical transducers for ammonia. These dyes themselves are not sensitive to ammonia; however, they are reversibly sensitive to hydrogen chloride. After addition of HCl to the ethanol solution of the dye, an immediate decrease of the peak intensity is observed. The peak recovers after addition of ammonia to the solution. Hence these dyes could be used for distributed ammonia sensors provided that the cladding containing the dye could be made to provide an acidic environment. In that case, these dyes would not absorb light in the NIR region in the absence of ammonia. The absorption peaks would only appear on the interaction of the cladding with ammonia. For the sensing of HCl, the acid-base indicators such as D8–D14 could be used provided that the cladding containing the dye could be made to provide a basic environment. Also, the dye presented by Mason et al. [53] could be suitable for fibre optic distributed sensors. In all cases, the interaction dye-target agent should be reversible and the response should be fast.

6 Conclusions

In this chapter, more than twenty NIR dyes, suitable for use as chemical transducers for ammonia and HCl, have been presented. The dyes belong to two basic groups. Also, the sensing mechanisms are suggested. Although many other dyes have been synthesised and tested [37], this chapter has concentrated only on the dyes that are suitable for ammonia, acid and pH sensing.

Distributed fibre optic chemical sensors are expected to play a great role in safety systems of the future. However, the route to their development will be long and difficult since they may need very special transducers. The development of these transducers represents a highly interdisciplinary problem and it is a challenge for future.

Acknowledgements. The developments and advances in the field of chemical sensing with near infrared dyes described in this article could not have been carried out without assistance from many of our colleagues around Europe. The authors wish to express their special thanks to the following co-workers: Ivan Babušík, Roy Blore, Martin Breza, Luis Norena-Franco, Ladislav Kalvoda, Marian Landl, Rudolf Lukáš, Petra Lukášová, Brian MacCraith, Chris Malins, Pavla Nekvindová and Stanislav Sekretár.

References

1. Matsuoka M (ed.) (1990) Infrared Absorbing Dyes. Plenum Press, New York
2. Matsuoka M (1990) Absorption Spectra of Dyes for Diode Laser. Bunshin, Tokyo
3. Fabian J, Kakazumi H, Matsuoka M (1992) Chem Rev 92:1197
4. Matsuoka M (1995) Near IR Spectroscopy. In: Peters AT, Freeman HS (eds) Advances in Color Chemistry Series, vol.2. Blackie Academic & Professional, London
5. Griffiths J (1995) Approaches to the Design and Synthesis of Near-Infrared Absorbing Dyes. In: Peters AT, Freeman HS (eds) Advances in Color Chemistry Series, vol.3. Blackie Academic & Professional, London.
6. Health and Safety Executive (1993) Guidance Note EH40/93 Occupational Exposure Limits. HMSO, London
7. Health and Safety Executive (1988) Control of Substances Hazardous to Health Regulations. HMSO, London
8. Giuliani JF, Wohltjen H, Jarvis NL (1983) Opt Lett 8:54
9. Reichert J, Sellien W, Ache HJ (1991) Sens Actuators A-Phys 25-27:481
10. Kostov YV (1992) Sens Actuators B-Chem 8:99
11. Sadaoka Y, Sakai Y, Murata Y (1993) Sens Actuators B-Chem 13-14:420
12. Klein R, Voges E (1993) Sens Actuators B-Chem 11:221
13. Potyrailo RA, Mikheenko LA, Borsuk PS, Golubkov SP, Talanchuk PM (1994) Sens Actuators B-Chem 21:65
14. Bondarenko DB, Dolotov SM, Koldunov MF, Ponomarenko EP, Sitnikov NM, Startsev AV, Tulaikova TV (1994) Chim Fizika 13:116
15. Grady T, Butler T, MacCraith BD, Diamond D, McKervey MA (1997) Analyst 122:803
16. Potyrailo RA, Hjeftje GM (1998) Appl Spectrosc 52:1092
17. Malins C, Butler IM, MacCraith BD (2000) Thin Solid Films 368:105
18. Raimundo IM, Narayanaswamy R (2000) Quim Anal 19:127
19. Qi ZM, Yimit A, Itoh K, Murabayashi M, Matsuda N, Takatsu A, Kato K (2001) Opt Lett 26:629
20. Raimundo IM, Narayanaswamy R (2001) Sens Actuators B-Chem 74:60
21. Malins C, Doyle A, MacCraith BD, Kvasnik F, Landl M, Šimon P, Kalvoda L, Lukáš R, Pufler K, Babušík I (1999) J Environ Monitor 1:417
22. Lennie AR (1989) Fibre optic sensing utilising the evanescent field, MSc Dissertation. Department of Instrumentation and Analytical Science, UMIST, UK
23. Kvasnik F, Hortin N, Norena-Franco LE (1993) Proc Soc Photo-Opt Instrum Eng 2366
24. Norena-Franco LE, Kvasnik F (1996) Analyst 121:1115
25. Boisdé G, Harmer A (1996) Chemical and Biochemical Sensing with Optical Fibres and Waveguides. Artech House, Boston
26. Kvasnik F (1996) Spectrochemical Index of Optical Sensing Media for the Detection of Chemicals, Research report. DIAS, UMIST, Manchester
27. Norena-Franco LN (1997) Chemically sensitive cladding materials for distributed optical fibre sensors, PhD Thesis. Department of Instrumentation and Analytical Science, UMIST, Manchester
28. Kvasnik F (coordinator) (1997) Copernicus Joint Research Project CIPA CT94-0206: Rapid Detection and Location of Ammonia Leaks, Final Report. Manchester.

29. Kubo Y, Sasaki K, Kataoka H, Yoshida K (1989) J Chem Soc Perk T 1 1469
30. Šimon P, Sekretár S, MacCraith BD, Kvasnik F (1997) Sens Actuators B-Chem 38-39: 252
31. Malins C, Landl M, Šimon P, MacCraith BD (1998) Sens Actuators B-Chem 51:359
32. Kubo Y, Kataoka H, Ikezawa M, Yoshida K, (1990) J Chem Soc Perk T 1, 585
33. Šimon P, Landl M, Sekretár S, Breza M, Budoš A (1996) Synthesis, Spectral Properties and Stabilities of Selected Near Infrared Dyes, Final report to the Copernicus Joint Research Project CIPA CT94-0206: Rapid Detection and Location of Ammonia Leaks. STU, Bratislava
34. Kalvoda L, Dlouhá M, Vratislav S, Lukáš R, Lukášová P, Landl M, Šimon P (2001) Fibre Optic Sensors for Detection of Metal Ions in Aquesous Solutions, Proceedings of Workshop 2001, part A, p. 440. CTU Publishing House, Prague
35. Kim SH, Matsuoka M, Yomoto M, Tsuchiya Y, Kitao T (1987) Dyes Pigments 8:381
36. Balch AL, Holm RH (1966) J Amer Chem Soc 88:5201
37. Šimon P, Landl M, Breza M, Prónayová N, Budoš A (1999) Synthesis and Spectral Properties of Selected Near Infrared Dyes and their Sensitivity to Various Agents, Final report to the Inco Copernicus Joint Research Project ERBIC15CT960819: Development of Personal Safety Monitors Based on New Sensing Reagents. STU, Bratislava
38. CI Acid Green1 (Basacid Green 970) (1996) BASF
39. Norena-Franco LE, Kvasnik F (1998) Analyst 123:2185
40. Schmidt H, Wizinger R (1959) Liebigs Ann Chem 623:204
41. Tuemmler WB, Wildi BS (1958) J Amer Chem Soc 80:3772
42. Tuemmler WB, Wildi BS (1958) J Org Chem 23:1056
43. Landl M, Šimon P (1997) Chem Listy 91:743
44. Landl M, Šimon P, Breza M (1998) Dyes Pigments 40:43
45. Šimon P, Landl M, Breza M (1999) Dyes Pigments 43:227
46. Landl M, Šimon P, Kvasnik F (1998) Sens Actuators B-Chem 51:114
47. Landl M, Šimon P, & Breza M (2000) Synthesis and chemical properties of new type NIR cyanine dyes, 8th Biannual International Conference on Dyes and Pigments COLOR-CHEM 2000. Špindleruv Mlýn, Conference booklet, Poster P14.
48. Šimon P, Landl M, Breza M, Kvasnik F (2003) Sens Actuators B-Chem, 90:9
49. Szablewski M, Thomas PR, Thornton A, Bloor D, Cross GH, Cole JM, Howard JAK, Malagoli M, Meyers F, Bredas JL, Wenseleers W, Goovaerts E (1997) J Amer Chem Soc 119:3144
50. Dai ZF, Peng BX (1998) Dyes Pigments 36:169
51. Boyer AE, Devanathan S, Hamilton D, Patonay G (1992) Talanta 39:505
52. Patonay G, Casay GA, Lipowska M, Strekowski L (1993) Talanta 40:935
53. Mason JC, Patonay G, Strekowski L (1997) Heterocyclic Commun. 3:409
54. Casay GA, Meadows F, Daniels N, Roberson A, Patonay G (1995) Spectrosc Lett 28:301
55. Lindauer H, Czerney P, Mohr GJ, Grummt UW (1994) Dyes Pigments 26:229
56. Lindauer H, Czerney P, Grummt UW (1994) J Prakt Chem-Chem Ztg 336:521
57. Lindauer H, Czerney P, Grummt UW (1995) J Prakt Chem-Chem Ztg 337:216
58. Lehmann H, Schwotzer G, Czerney P, Mohr GJ (1995) Sens Actuators B-Chem 29:392
59. Czerney P, Grummt UW (1996) J Chem Res S 173
60. Czerney P, Grummt UW (1996) J Inform Rec 23:159
61. Czerney P, Grummt UW (1997) Sens Actuators B-Chem 39:395
62. Miltsov S, Encinas C, Alonso JN (1998) Tetrahedron Lett 39:9253
63. Miltsov S, Encinas C, Alonso J (2001) Tetrahedron Lett 42:6129
64. Puyol M, Miltsov S, Salinas I, Alonso J (2002) Anal Chem 74:570
65. Zollinger H (1991) Color Chemistry. VCH Verlagsgeselschaft, Weinheim
66. Hutchings MG (1995) Dyes Pigments 29:95
67. Dewar MJS, Thiel W (1986) AMPAC, Austin Method 1 Package. University of Texas, Austin
68. Dewar MJS, Zoebisch EG, Healy EF, Stewart JJP (1985) J Amer Chem Soc 107:3902

69. Fletcher R, Powell MJD (1963) Comput J 6:163
70. Davidon WC (1968) Comput J 10:406
71. Armstrong DR, Perkins PG, Stewart JP (1972) J Chem Soc Dalton 1972
72. Breza M, Šimon P, Landl M (2001) Chem Papers 55:86
73. Crank J, Park GS (eds) (1968) Diffusion in Polymers. Academic Press, London
74. Šimon P, Kvasnik F, Landl M (1998) Indication of gas leaks and fire alarms with distributed fibre optic sensors, 25th international Conference of Slovak Society of Chemical Engineering, Jasná, Slovak Republic
75. Culshaw B, Muhammad F, Stewart G, Murray S, Pinchbeck D, Norris J, Cassidy S, Wilkinson M, Williams D, Crisp I, Vanewyk R, McGhee A (1992) Electron Lett 28:2232
76. Lieberman RA, Mendoza E, Ferrell DJ, Schmidlin E, Syracuse S, Khalil A, Dergevokian A, Sun Z, Gunther M (1993) SPIE Proceedings No.2068: Chemical, Biochemical, and Environmental Fiber Sensors V, pp.192-201. Boston, Massachusetts
77. Gregory P (1991) High-technology Applications of Organic Colorants, Topics in Applied Chemistry. Plenum Press, NewYork

CHAPTER 9

Piezo-Optical Dosimeters for Occupational and Environmental Monitoring

Kelly R. Bearman, David C. Blackmore, Timothy J.N. Carter, Florence Colin, Steven A. Ross, John D. Wright

1 Introduction

Occupational and environmental monitoring, driven by growing awareness of health hazards and increasingly stringent national and international standards, is becoming focussed on the need for data relating to ever smaller geographical regions, and particularly for personal exposure data. Such personal data are not only the best way to demonstrate compliance with regulations, but also provide the basis for individual exposure/health correlations upon which improved standards can be securely based. The requirements for measurement systems to be used in such local and personal monitoring on a large scale are demanding. They include low cost, small size, reliability, ease of use (preferably by unskilled operators) and the ability to record time-weighted-average exposures over short (e.g. 15 min) and long (e.g. 8 h) periods. Ideally the system should use similar monitoring devices, measured using the same generic low-cost equipment, for a wide range of analytes. Preferably the monitoring device should be capable of measuring exposures to several analytes simultaneously, and the measurement system should include data logging to guarantee proper exposure records. Until recently very few, if any, systems could satisfy these requirements.

Environmental monitoring is now conducted regularly on a large scale using fixed monitoring stations incorporating standard instrumentation such as chemiluminescent analysers for nitrogen dioxide and ultraviolet spectrophotometric analysers for ozone. For example, both the UK and France have networks of these stations, whose data are widely available on the internet [1, 2]. Although these are reliable, accurate and economical over their working lifetime, they cannot provide personal monitoring data and they require substantial initial capital investment in instruments that are dedicated to monitoring a single analyte. Similarly, although occupational monitoring requirements that are generic to particular industries have been tackled by the use of multi-point sampling networks feeding into central analytical instrumentation facilities, these are complex and expensive to install and run. One example was the joint development by several major polymer manufacturers of systems based on gas chromatography and infra-red spectrophotometry coupled with pumped sampling networks to monitor vinyl chloride [3]. Diffusive sampling tubes, in which many different air pollutants are absorbed on a range of materials and subsequently analysed by thermal desorption or solvent extraction followed by chromatographic separation with spectroscopic or mass spectrometric quantification, have also been extensively used. These provide time-weighted-average data for direct comparison with occupational or environmental standards. They are also small enough to be used for personal exposure monitoring and, with care, can be used to measure several pollutants simultaneously. However, the analytical instrumentation required is expensive and demands skilled operators, and the total cost of the tubes themselves can be substantial in large-scale surveys. Diffusive sampling rates for a given sampler tube can be influenced by temperature, nature and concentration of pollutant, and several other variables such as humidity and face velocity and direction of sampled air. Comparative diffusive sampling rates for similar materials (e.g. benzene, toluene and xylene) have been measured in different laboratories and shown to have a standard deviation from the mean value of about 10% [4]. Uncontrolled environmental variables mean that overall accuracy in real environments will be worse than this figure. Nevertheless, given the difficulties of establishing exposure standards (e.g. variations in individual sensitivities, and absence of comprehensive exposure data to all the pollutants that may be present together in any individual's environment), diffusive sampling is commonly accepted as a valid technique for indicating compliance or otherwise with exposure standards. This chapter reviews 10 years of development of the novel PiezOptic dosimeter system, which is now able to fulfil all of the above requirements.

The system consists of a badge that may be used for many different analytes, and a reader unit that can be programmed to work for any combination of analyte badges. The badge is shown in Figs. 1 and 2. Five spots of colorimetric reagent are deposited on a polyvinylidene fluoride (PVDF) piezoelectric film coated with a thin optically-transparent conductive layer of indium-tin oxide. The PVDF film is a laminate of two oppositely-poled PVDF layers. This eliminates effects of stress, acoustic interference and local temperature gradients. The film is supported on both sides by sponge pads in which holes have been die-cut in order to allow both analyte and light to reach the spots. The two sides of the film are contacted at one end by conductive sponge pads.

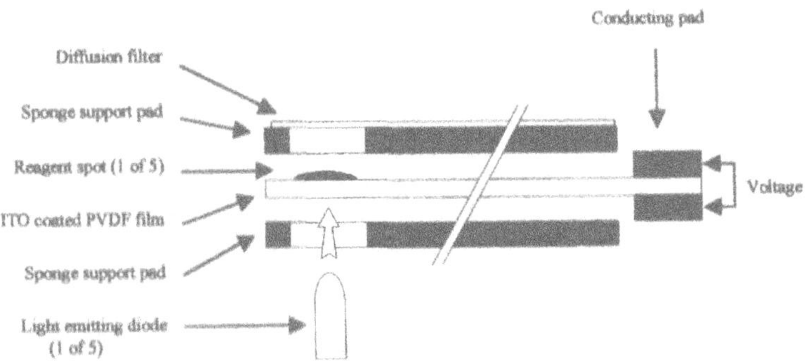

Fig. 1. Schematic diagram of the badge

Fig. 2. The badge

All these components are tightly held inside a polymer casing that gives protection to the reactive spots, allows vapours access to the spots (through a diffusion filter if appropriate), fits inside a slot in the reader and offers apertures for illuminating the spots. The five spots may all be the same, permitting multiple analyses of the same analyte, or they may be different, permitting analysis of several analytes or the inclusion of compensation for interfering substances or for the effects of variations of temperature and humidity.

The operating principle is as follows. When the reagent spot is exposed to the target analyte, a colour change occurs. The quantity of reagent converted to the new colour is directly proportional to the integrated dose (integral of concentration v. time curve over the standard exposure time) of the target analyte received by the spot. When the exposed spot is illuminated using a light emitting diode (LED) that emits in the wavelength range of the colour change, heat is produced in the spot by non-radiative decay of the excited states. The total heat thus produced is directly proportional to the quantity of reagent that has been converted to the new colour, and hence to the analyte dose. This heat expands the spot, stressing the underlying PVDF and thus creating a piezoelectric charge. The charge is collected by the conductive pads and measured by the reader electronics. This principle is essentially a form of photoacoustic spectroscopy. The signal that is measured by the reader depends on many variables. These include LED intensity, optical extinction coefficient of the new coloured form of the reagent in the LED output wavelength region, thickness and illuminated area of reagent spot, spatial distribution of coloured form of the reagent within the spot thickness, heat transfer characteristics of the reagent spot, thermal expansion coefficient of the relevant materials, piezoelectric coefficient of the PVDF strip, and amplifier gain. It is impractical to model all these variables to produce a theoretical response equation. Instead, the operating conditions are controlled to ensure that their values are constant (or, in the case of LED

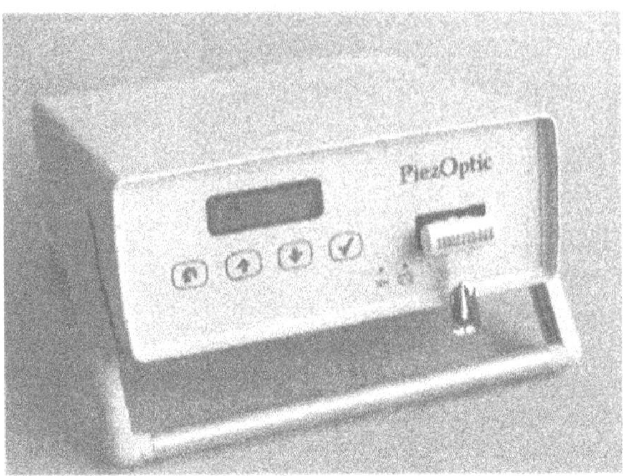

Fig. 3. The PiezOptic Reader Unit

intensity and amplifier gain, compensated by use of a reference as described below). (Some more detailed aspects of this transduction process are considered later in Section 12.)

The PiezOptic reader, shown in Fig. 3, is designed specifically to analyse PiezOptic badges. The badge is inserted into the front aperture and the reagent spots align with the five LEDs inside the reader. The electronics command the flashing of the LEDs and measure the charge thus produced, in the form of a voltage. The measurement is performed using a lock-in amplifier that is phase-locked to the signal.

The badge is read before and after the monitoring period, which can be 15-min STEL (Short-Term Exposure Limit) or 8-h TWA (Time Weighted Average). The readings are corrected for variations in the light output intensity between the five LEDs, using normalisation data obtained using a reference badge that is screen printed with a black ink so as to give signals directly proportional to the light outputs of each of the five LEDs. The initial reading is stored in the reader memory and retrieved during the after-exposure reading, when the built-in algorithm, based on calibration data, transforms the output difference into a time-weighted average exposure in appropriate units. The final result appears on the liquid crystal display and is also stored electronically. The reader also incorporates a bar-code reader that records the individual bar code on each badge for data logging purposes. The whole system is of low cost (reader less than €2000, badge less than €20) and can easily be operated by unskilled staff.

In this chapter, the procedures used to calibrate and evaluate new badges are described, including a general discussion of the performance standards of the system, and then the development of badges for formaldehyde, glutaraldehyde, chlorine dioxide, ozone, nitrogen dioxide, styrene and ammonia are reviewed. Examples of multi-analyte badges are then discussed, together with approaches to minimisation of interferences. Finally the results of a study of the factors that determine the evolution of the piezoelectric signal from each spot are summarised together with the implications of this for future applications of the system.

2 Calibration and Evaluation of New Badges

For optimum performance of this system the badges must be calibrated over the full concentration range to be monitored, in realistic temperature and humidity conditions. The effects of operating in lower and higher ranges of both temperature and humidity must also be explored. Depending on the magnitude of such effects, two of the five reagent spots on the badge may need to be used for automatic temperature and humidity compensation or simple tables of correction factors may suffice if the operating environment has approximately constant temperature and humidity over the measurement period.

The reagent spots themselves generally consist of a dispersion of the reagent adsorbed on silica powder or entrapped in a porous matrix such as a silica sol-gel. This material is then ground and bound into a thin film with a polymer solution. The polymer matrix (commonly polyisobutylene) is flexible and adheres well to the PVDF strip, and may be varied to control ingress of the analyte and/or restrict ingress of interfering substances, e.g. water. The nature and concentration of the colour reagent, and the entrapment matrix and polymer binder used in spot fabrication, must be optimised with respect to sensitivity and shelf life. Furthermore, the use of diffusion filters may be necessary to permit the use of sensitive reagents for monitoring over 8-hour periods. Both the filter and the composition of the reagent spot itself may require further chemical modification in order to minimise cross-sensitivity to additional interfering pollutants present in the atmosphere to be monitored. A further benefit of the presence of five reagent spots on each badge is the possibility of simultaneous monitoring of two or more different pollutants. The effects of cross-sensitivities of reagents to the different pollutants may be eliminated accurately, provided all of the different reagent spots have been calibrated for all of the pollutants. The number of reagent spots with different sensitivity patterns must in practice at least equal the number of different pollutants to be monitored. This is a particularly important facility of the system because colour reagents are rarely totally specific. It is made possible by the powerful microprocessor facility used in the reader, which can implement complex correction algorithms based on calibration data for every spot in every analyte.

The gas rigs used for optimisation and calibration are identical to those commonly used in gas sensor evaluation. A typical installation is shown in Fig. 4.

Particular attention is paid to the control and monitoring of the temperature and humidity of the system, since the badges are passive devices. Pollutant concentrations are generated by a variety of methods including nebulising into a gas flow (e.g. for aldehydes and chlorine dioxide), controlled evaporation of a liquid maintained at constant temperature (e.g. for styrene), bubbling air through an aqueous solution of the analyte (e.g. aldehydes), in-situ generation of the pollutant in the air flow (e.g. generation of ozone using UV light) or from standard certified gas mixtures (e.g. nitrogen dioxide). Materials for the tubing of the rigs are selected to avoid possible reaction or adsorption of the target analyte. Typically glass, polyurethane and PTFE are used. The concentration of the gas or vapour arriving at the test chamber is determined directly by

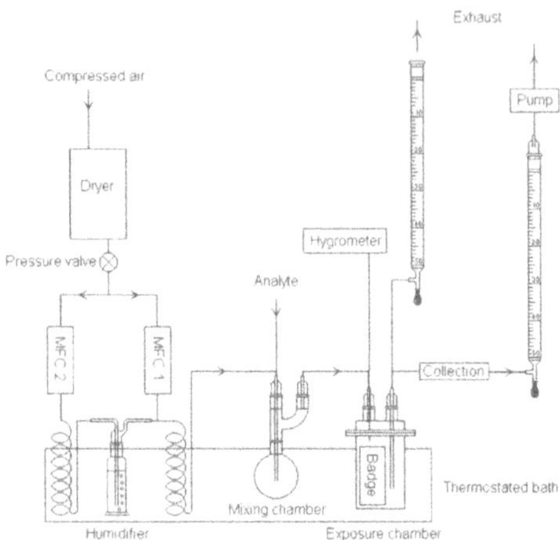

Fig. 4. Gas rig used for badge calibration

passing a known volume through a filter or bubbler and analysing the trapped material chemically, spectrophotometrically or by HPLC. In the case of nitrogen dioxide and ozone, their concentrations have been analysed by sampling from the exposure chamber directly into chemiluminescent and UV analyser instruments, respectively. Such procedures are extremely important since even with careful choice of materials the compositions of many vapours do change as they pass through test rigs.

To quantify the colour change of the reagent, the spots are consecutively illuminated using flashing LEDs with emission wavelength selected to match the optical absorption spectrum of the reagent system. Although thermal drift may lead to changes in LED output and amplifier gain between initial measurement of the standard black badge and the sample, repeat measurements of the standard badge can be normalised to within ±2%. Further errors arise from variations in size and reactivity of individual reagent spots, accuracy of the calibration system, effects of changes in humidity, temperature variations and interfering reactions of other related chemicals with the reagent spots. Effects of variations between different reagent spots are minimised by averaging the normalised readings from several spots deposited on a single PVDF strip. Typically readings of each spot are taken during 16 flashes of the LEDs, the first two being discarded as atypical due to effects of the LEDs and circuitry equilibrating after the initialisation of the measurement cycle. The remainder are normalised, the highest and lowest readings are ignored and the remaining 12 readings averaged. The total error for a typical reagent system such as that for nitrogen dioxide has been estimated [5] as ±10% provided carefully-designed calibration rigs are used.

Before any new badge reagent system is released, careful studies are made to optimise storage life. Many colour reagents undergo slow changes due to reactions with components of the storage environment. These components may be pollutants in the residual air in the storage container, or chemical species emitted from the container itself or even from the plastic used for the badge components. A range of strategies is available for minimising such problems. The simplest is to store the badges in sealed metallised bags in a freezer. The sealed environment minimises contamination, and the low temperature reduces the rates of any diffusion processes or chemical reactions. However, this poses problems during transport of the products to users, so efforts are made to ensure that all badges have a room-temperature storage life of at least one week. The storage life is defined for this purpose as the time at which the background signal of the badge has changed so much that the remaining available signal change when used for monitoring is less than that expected for exposure to the permitted exposure limit dose. Strategies to increase room-temperature storage life include varying reagent concentration, heat-treatment of the plastic badge bodies and the storage bags, inclusion of scavenger packs inside the storage bags and modification of the reagent itself and its entrapment matrix.

3
Badges for Formaldehyde Monitoring

Formaldehyde is a commonly-used biocide. It is toxic and excessive exposure causes respiratory problems. It is also the most common indoor air pollutant. Sources are synthetic materials based on formaldehyde resins, and cigarette smoke. Indoor concentrations average ≈ 0.1 ppm and may exceed 1 ppm in adverse conditions, whereas typical urban outdoor concentrations are around 0.01 ppm. The UK maximum exposure limit is currently 2 ppm [6].

The strong reducing power of formaldehyde is a useful property on which to base detection methods. For example, it is capable of reducing silver salts to a brown silver deposit that absorbs light over a wide range across the visible spectrum.

$$2Ag^+ + HCHO + 2OH^- \rightarrow 2Ag^\circ + HCOOH + H_2O$$

The reagent is made by entrapping silver nitrate in a silica sol-gel matrix prepared by mixing for 90 s 1.375 mL tetramethoxysilane, 1 mL of 0.05 M aqueous silver nitrate solution and 2 mL 0.1 M aqueous NaOH solution, and drying for 2 days at room temperature. This base-catalysed process generates large pores in the silica matrix which allow ready access of the analyte vapour to the reagent. The resulting solid is ground to a powder and bound into 5 µL spots with a solution of polyisobutylene. Fig. 5 shows the calibration curve for a badge optimised for use over 15-minute exposure periods. The formaldehyde concentrations were validated by reaction of known vapour volumes with glass fibre filters soaked in 2,4-dinitrophenylhydrazine. The exposed filters were then solvent-extracted and analysed by HPLC [7]. The formaldehyde vapour was generated by nebulising methanol-stabilised formalin solutions, creating a vapour of mean molar mass 49.8 whose composition is 38% HCHO + 62% $MeOCH_2OH$ [8]. Since the sol-gel entrapped reagent is a diffusive sampler, and diffusion rates are inversely proportional to the square root of mean molar mass, the calibration data require a correction if used to detect pure formaldehyde. Since the

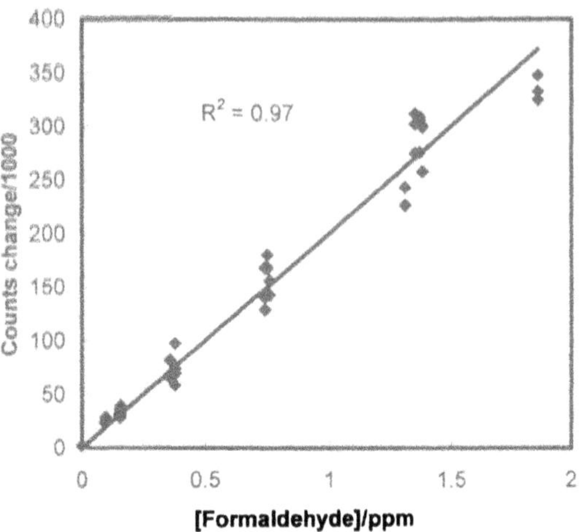

Fig. 5. Calibration curve for the 15-minute formaldehyde badge

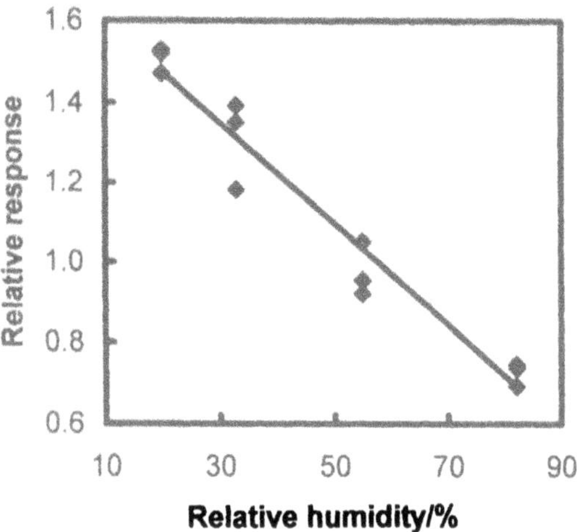

Fig. 6. Humidity effects on the 15-min formaldehyde badge

pure material has a significantly lower molar mass, the effective sampling rate will be higher and the badge response derived from the formalin calibration data must then be reduced by a factor of $\sqrt{(30.0/49.8)} = 0.737$.

However, for many biocidal applications the vapour will originate from such formalin solutions and the calibration curve of Fig. 5 is valid without correction in such cases. This illustrates a very important general principle, that reliable occupational or environmental monitoring demands that the composition of the pollutant in the actual operating environment must be independently established and calibration corrected appropriately depending on the actual calibration vapour used.

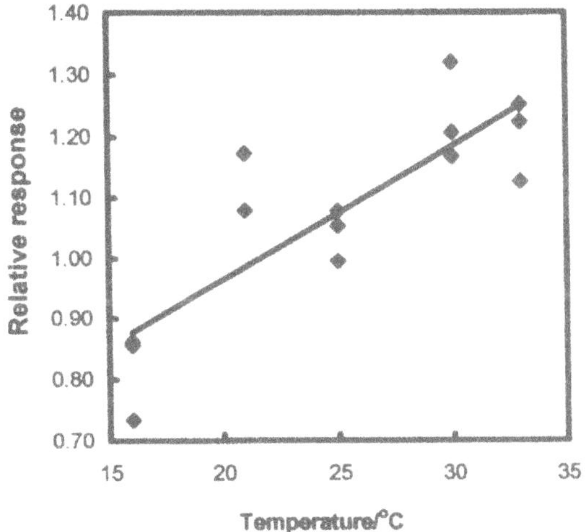

Fig. 7. Temperature effects on the 15-min formaldehyde badge

Since the rates of diffusion processes and chemical reactions are dependent on temperature, and humidity variations may also affect the badge responses, the magnitude of these effects for variations from the standard calibration conditions of 25°C and 50% RH have also been determined. Figures 6 and 7 show the results for this badge. Fortunately the response to glutaraldehyde concentrations as high as double the MEL is less than 10,000 counts, which is negligible in relation to the formaldehyde responses shown in the calibration curve of Fig. 5.

4 Badges for Glutaraldehyde Monitoring

Glutaraldehyde is a biocide which is widely used for sterilizing equipment such as endoscopes in hospitals. It is highly toxic and can also sensitize exposed subjects to asthma. The UK maximum exposure limit is 50 ppb [6], and for hospital use it is necessary to measure the 15-minute time-weighted average exposures of individual workers.

The reaction of glutaraldehyde with 2,7-diaminofluorene produces a yellow-coloured derivative that can be interrogated with a blue LED with emission centred at 470 nm.

Reagent spots are prepared by evaporating a polyisobutylene solution containing a suspension of a powder made by soaking Davisil 644 silica in an acetic acid solution of 2,7-diaminofluorene dihydrochloride and drying the filtered product. By varying the reagent concentration and using a glass fibre diffusion barrier filter as appropriate, the badge response may be adjusted to give readings up to the 50 ppb MEL value in either 15 min or 4 or 8 h as required. Calibration curves for these badges are shown in Figs. 8 and 9.

These calibrations are validated by determining the actual glutaraldehyde concentrations by reaction of known vapour volumes with glass fibre filters soaked in 2,4-dinitrophenylhydrazine. The exposed filters are then solvent-extracted and analysed by HPLC [7]. The response of these badges to formaldehyde is approximately 10% of that to the same concentrations of glutaraldehyde. Since typical formaldehyde concentrations are commonly substantially higher than those of glutaraldehyde as pointed out above, some compensation for this effect is required in atmospheres where both of these biocides may be present

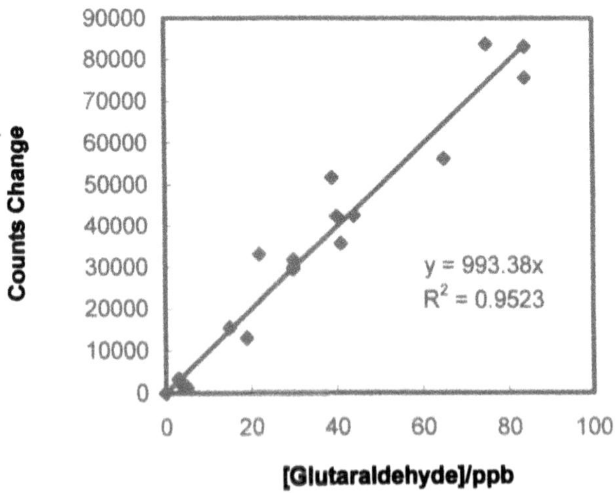

Fig. 8. Calibration curve for the 15-minute glutaraldehyde badge

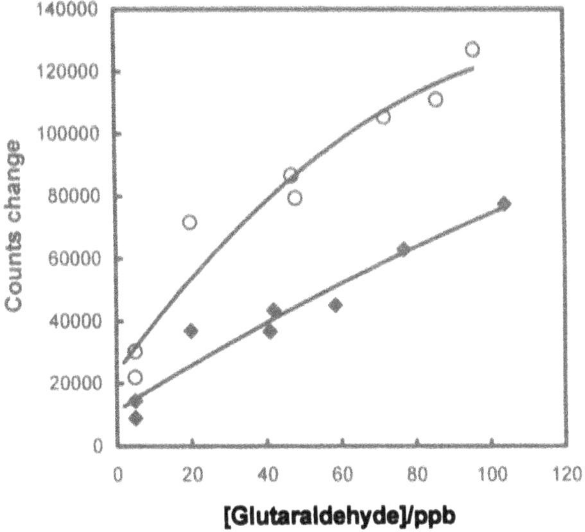

Fig. 9. Calibration curves for the glutaraldehyde badge over 4-hour (diamonds) and 8-hour (circles) exposure periods

together. This is easily achieved since the formaldehyde reagent shows good selectivity against glutaraldehyde, as already mentioned.

However, the diaminofluorene reagent is susceptible to interference from strong oxidising agents (such as atmospheric nitrogen dioxide) which generate a blue quinoneimine product that absorbs at 450 and 620 nm, and thus also gives a response with the 470 nm LED.

H_2N … NH_2 $\xrightarrow[-H_2O]{+[O]}$ HN= … =NH

Two methods have been devised to deal with this problem. In the first of these, the reagent spot is covered with a glass fibre filter soaked in dinitrophenyl-hydrazine. This reacts with any glutaraldehyde, and thus any colouration of the underlying reagent spot is due solely to the interfering oxidants. The reading from this spot, corrected if necessary for the presence of the diffusion barrier, is subtracted from the reading of an identical but unprotected spot to give the response due only to glutaraldehyde. This method is effective, but somewhat complex to implement due to the need to compensate for the effect of the diffusion barrier presented by the filter. A simpler method is to incorporate an antioxidant into the reagent spot itself. When ascorbic acid was tried in this role, it was found to react with glutaraldehyde, giving very low badge responses, in the following reaction:

glutaraldehyde + 2 L-ascorbic acid →

Use of a protected ascorbic acid 6-palmitate eliminated this problem and its effectiveness is demonstrated in the calibration graphs in Fig. 10.

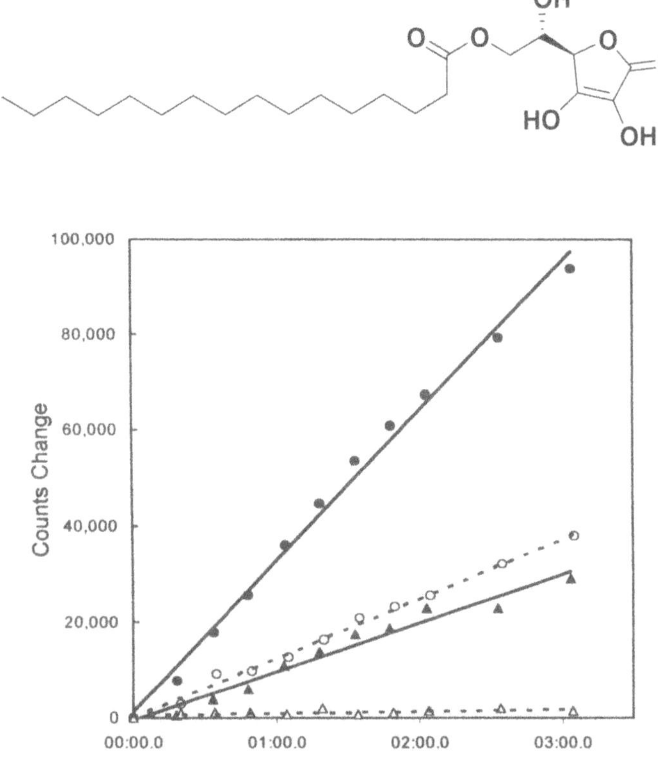

Fig. 10. Response of the 15-min glutaraldehyde badge in natural air (containing traces of NO_2) in the presence of 33 ppm glutaraldehyde (closed symbols) and in the absence of glutaraldehyde (open symbols). Circles denote the original diaminofluorene. Triangles denote the reagent incorporating antioxidant

Effects of humidity and temperature on the badge response have been investigated. The effects of temperature are approximately twice as large as those for the formaldehyde badge, possibly reflecting a higher activation energy for the reaction, but the humidity effect is negligible between 25 and 90% RH. Only in very dry conditions (<5%RH) is an approximate doubling of sensitivity observed.

5
Badge for Monitoring Chlorine Dioxide

Chlorine dioxide is a biocide that is increasingly used as one of the alternatives to the aldehydes. The UK Occupational Exposure Standard limits for this gas are 300 ppb over 15 min or 100 ppb over 8 h [6]. 2,2'-Azino bis(3-ethylbenzthiazoline)-6-sulfonic acid diammonium salt (ABTS) is capable of selecting for ClO_2 against other potentially oxidising species. The colourless ABTS is oxidised by ClO_2 to a stable green-coloured radical cation [9].

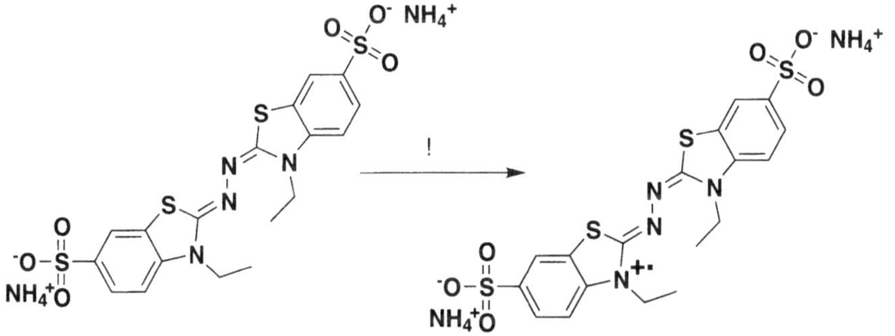

Reagent spots are fabricated from this reagent following the same pattern as for formaldehyde, i.e. entrapment in a base-catalysed silica sol-gel followed by drying, grinding and binding with a polyisobutylene solution. Since ClO_2 is an unstable gas, vapour concentrations cannot be generated exactly from standard sources. 0–500 ppb concentrations of ClO_2 were generated by nebulising freshly-prepared aqueous solutions into air flows, and quantified by sampling from the test chamber at the location of the badge being tested. The ClO_2 solutions were prepared from $NaClO_2 + H_2SO_4$. The sampled gas was absorbed in aqueous ABTS solutions and monitored by spectrophotometry. Fig. 11 shows the calibration curve obtained for this badge.

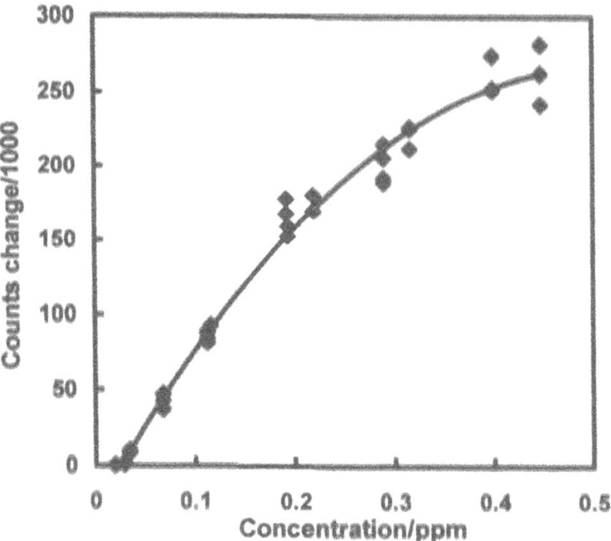

Fig. 11. Calibration curve for the 15-min chlorine dioxide badge

6 Badge for Monitoring Ozone

Ozone is used extensively in water treatment and for odour elimination in buildings, but is also present in the environment as a consequence of the ultraviolet component of sunlight interacting with oxygen in air. Since it is an irritant, and has a role in the generation of photochemical smog, monitoring is necessary. Although there are several well-established standard methods for its determination, notably UV absorption and chemiluminescent techniques, a reliable and cheap method for measuring personal exposures for susceptible individuals (e.g. asthmatic children) is highly desirable. We have therefore developed an ozone badge based on the bleaching action of ozone on indigo disulfonate [10–13]:

NaO_3S–(indigo)–SO_3Na $\xrightarrow{+ O_3 + H_2O}$ 2 NaO_3S–(isatin) $+ H_2O_2$

The reagent spot is prepared using a method analogous to that described above for glutaraldehyde, i.e. adsorption on silica powder followed by binding with a polymer solution. In the case of the indigo reagent, the nature of the polymer binder was found to be critical. When polyisobutylene or ethyl cellulose was used, no bleaching of the reagent occurred on exposure to ozone, but a significant effect was observed when polyethyleneglycol was used. This is believed to be an indication that water is required for the reaction to occur. Using badges

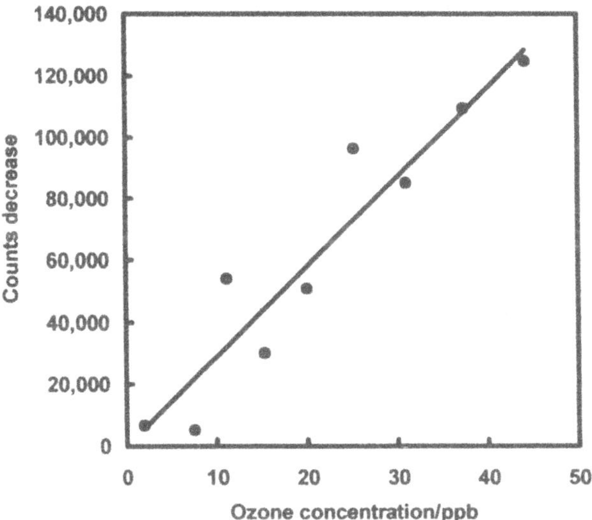

Fig. 12. Calibration curve for the 8-h ozone badge

based on such spots, the calibration curve shown in Fig. 12 was obtained. Exposure of these badges to nitrogen dioxide concentrations of 20 ppb showed no interference from this gas.

7
Badge for Monitoring Nitrogen Dioxide

OES limits for NO_2 are 3 ppm for the 8-hour period and 5 ppm for the 15-minute period. The reagent selected for measuring nitrogen dioxide was *o*-tolidine [14]. The reaction between a primary aromatic amine and nitrous acid HNO_2 leads to a diazonium salt. *o*-Tolidine further reacts with the diazonium salt to form a coloured azo compound. This reaction requires the presence of water for the formation of nitrous acid. In dilute acid, the actual attacking species is N_2O_3, which acts as a carrier of NO^+. Since water is needed for the reaction, the badges showed increased reactivities at higher relative humidities. *o*-Tolidine is also readily oxidised, for example by chlorine and ozone which show cross-sensitivity with the NO_2 badge, and some contribution to the observed colouration of the spots by NO_2 is also likely to arise from formation of quinoneimine oxidation products analogous to that discussed above from the effect of NO_2 on diaminofluorene.

H_2N—(o-tolidine structure)—NH_2

o tolidine

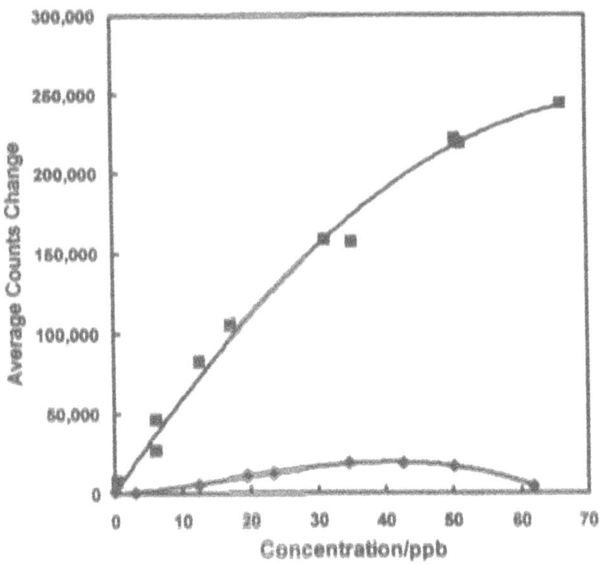

Fig. 13. Calibration of the 8-h nitrogen dioxide badge in nitrogen dioxide (■) and ozone (♦)

o-Tolidine was entrapped in a sol-gel matrix under acid catalysis, and the ground product was bound into spots with a polyisobutylene solution. The reacted reagent powder had a broad absorbance across the visible spectrum, making possible the use of either blue or red LEDs. By varying the amount of reagent, badges capable of covering environmental (10 to1000 ppb) and occupational (2 to 10 ppm) exposures were obtained. Calibration, using standard gas mixtures of NO_2 in air, is difficult due to the high reactivity of NO_2 and the high polarity that gives it a strong tendency to adsorb onto the tubing in the gas rigs. Great care was needed to measure the concentrations actually present at the location of the badges under calibration, using standard chemiluminescent analyser instruments. A calibration curve for a badge designed for environmental application is shown in Fig. 13, together with the interference produced by ozone. Since the ozone badge is unaffected by NO_2 it is possible to determine the concentrations of both of these gases in mixtures by using these two reagent spots together in a single badge.

8
Badge for Monitoring Styrene

Occupational exposure to styrene monomer is widespread. It is used in the preparation of rubber tyres, glass-reinforced plastics, carpet coatings and packaging materials. Styrene causes irritation of the eyes and mucous membranes at 50–100 ppm [15]. It is a central nervous system depressant and may cause dermatitis after repeated exposures. In the United States, the occupational maximum exposure limits for styrene vapour are 20 ppm as an 8-h time weighted

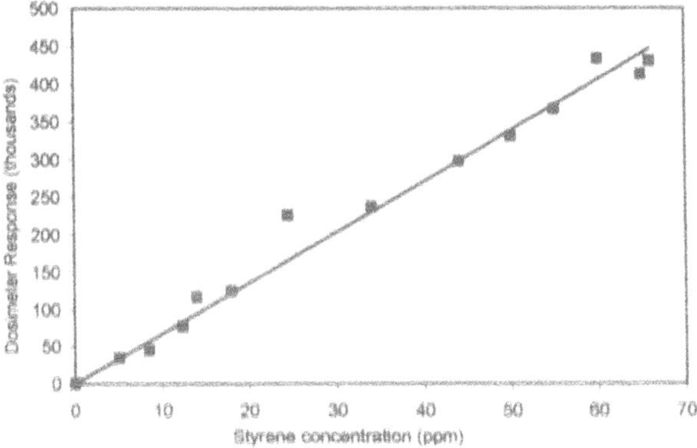

Fig. 14. Calibration curve for the 15-min styrene badge

average (TWA), or 40 ppm as a 15-min short term exposure limit (STEL). The corresponding values in the United Kingdom are 100 ppm (TWA) and 250 ppm (STEL).

Decolorisation of bromine water is the classical spot test for alkenes. Initial attempts to develop a styrene-sensitive reagent using bromine trapped in either a molecular sieve or a sol-gel were unsuccessful. Attention was then turned to polymer-supported tribromide reagents, which were originally developed as convenient solid-phase bromination agents [16, 17]. Polyvinylpyridinium tribromide and Amberlyst®-supported tribromide only showed a small response in a saturated atmosphere of styrene. A large variety of encapsulating matrices (soluble polymers), co-reagents, humectants and solvents were used to try and enhance the reactivity of the bromide reagent. The highest response was obtained using a 10,000 MW polyethylene glycol matrix, containing a small amount of water. Figure 14 shows the response curve for the calibration of the 15-min styrene dosimeter sensor in a controlled atmosphere at 298 K and 50% relative humidity [18].

The reactivity of this reagent makes it an ideal sensor for alkenes in general. Although such lack of specificity is in general a disadvantage, in this case it is in fact an advantage since in industrial applications pollution from a single alkene is the most common situation. Thus, provided full calibration is carried out for the target alkene, the system has wide applicability. We have already shown that the dosimeter can be used in environments which may contain methyl methacrylate (used in plastics production and bone cement), ethyl cyanoacrylate (used in "superglue" fuming to visualise latent fingerprints) and vinyl chloride monomer (PVC production).

9 Badge for Monitoring Ammonia

Although ammonia emissions are widespread in the environment, most arise naturally from biological decay (approximately 100 million tons/annum) while anthropogenic sources (mainly waste treatment) account for less than 5% of this amount. However, measurement of ammonia in exhaled breath of humans and animals is becoming an important diagnostic tool. We have reported two systems that are feasible for use in ammonia badges [5]. One of these uses the basic properties of ammonia, with the pH indicator Rose Bengal, whose pK_a is 4.5, while the other uses the reduction of silver ions by Mn^{2+} to silver. The latter reaction only occurs in the basic conditions produced by ammonia.

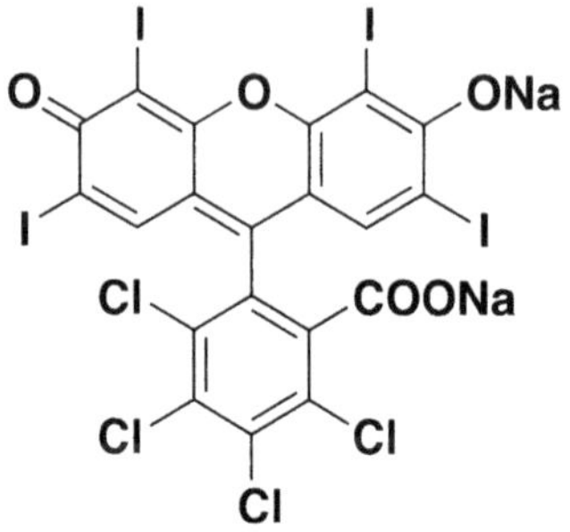

Rose Bengal

Both reagents are entrapped in a silica sol-gel matrix prepared by *p*-toluenesulphonic acid catalysis. The sensitivity of the reagents is dependent on the pH of the sol-gel precursor mixture, and is optimised at pH 1.57 for the Rose Bengal system. Other workers have shown that sol-gel entrapment of pH indicators can have the effect of widening the pH range over which the indicator changes colour. This useful effect extends the dosimeter capacity of the system, and is believed to be a due to a range of pore sizes in the gel which provide different local environments for the entrapped indicator molecules, each with its own characteristic effective pK_a. Similar effects on chemical equilibria have been observed for the complexometric reagent Eriochrome Cyanine R, which is selective for copper(II) ions when entrapped in a sol-gel as opposed to aluminium(III) when free in aqueous solution [19]. The sensitivity of this reagent may also be modified by using a *p*-toluenesulphonic acid-impregnated filter layer above the reagent spot.

Linear calibration curves have been obtained over the range 0–60 ppm ammonia. The response is strongly dependent on humidity and temperature, decreasing linearly with humidity and increasing with temperature. However, the responses are rapid, and exposure of the silver/manganese reagent system to steady-state and equivalent-time-weighted-average doses of varying concentrations of ammonia have been shown to give identical overall badge readings. These results are an important verification of the ability of this diffusive sampling system to record accurate time-weighted average doses in conditions where transient peaks of high pollution levels occur.

10 Multi-analyte Badges and the Minimisation of Interference

The above discussion has included several examples where the use of different reagent spots on a single badge can compensate for interferences. For example: the use of glutaraldehyde and formaldehyde reagent spots to analyse mixtures of these two aldehydes despite some cross sensitivity; the use of a glutaraldehyde reagent spot covered by a dinitrophenyl-hydrazine-coated filter to provide compensation for NO_2 interference in glutaraldehyde badges; and the use of ozone and NO_2 spots in a single badge to analyse mixtures of these two gases despite some mutual interference. The presence of five reagent spots on a single badge, together with the intelligent signal processing provided by the reader microprocessor, gives this system unique advantages for multi-analyte analysis. This is particularly valuable in providing automatic compensation for the widespread effects of variations of temperature and humidity on the response of colour reagents.

Reagent systems capable of measuring temperature and humidity have been developed [20]. The humidity system uses the well-known colour change from blue to pink of silica-gel-entrapped cobalt chloride, as the cobalt(II) coordination changes from the tetrahedral chloro complex to the octahedral aquo complex. Although this reaction is very rapid in finely-ground sol-gel material, by suitable choice of entrapment polymer matrix and use of polymer diffusion barrier filters it has been possible to develop spots suitable for monitoring humidity over 15-min and 8-h periods. Similarly, spots for monitoring temperature have been prepared by sol-gel entrapment of pyrogallol, which is oxidised to a brown product at a rate determined largely by the ambient temperature in air with normal oxygen content. The response of the humidity spot is also dependent on the temperature, while the rate of reaction of the temperature spot is also influenced by humidity. However, if both spots are present together (or if either the temperature or humidity are known) these cross-effects may be compensated by the reader electronics operating with suitable algorithms devised from full cross-calibrations over ranges of temperature and humidity. Such a system works well in conditions where temperature and humidity, although not those of the standard calibration environment of 50% RH and 25 °C, are approximately constant. This is the case in many indoor monitoring environments, where such automatic compensation for effects of non-standard operating environments can significantly improve accuracy especially during use by non-skilled staff who may be unaware of the potential effects of such conditions. Where the temperature and humidity do change during the measurement period, precise compensation becomes more difficult for any system, since time-weighted average readings give no indication of the actual time-profile of, for example, temperature and humidity. Periods of high pollution may be associated with low or high temperature or humidity and the relationship between exposure dose and device response is, in principle, continuously varying. In practice, however, such variations are unlikely to be significant over 15-min periods so wherever operating conditions are likely to vary substantially over time it is advisable to use 15-min badges rather than, for example, 8-h badges.

Overall, the opportunity for measuring several analytes simultaneously, with compensation for mutual interference and for humidity and temperature variations, has many advantages. The accuracy of measurements is improved, the permissible range of operating environments in which monitoring can be carried out within a specified degree of accuracy is increased, and the cost of monitoring is reduced.

11 Fundamentals of the Piezo-optical Measurement

The piezo-optical measurement principle used in this system is a form of photoacoustic spectroscopy. Light passes through the transparent PVDF transducer film and is absorbed in the reagent spot within a distance known as the optical absorption length, *d*. Heat is generated in this region by non-radiative decay of the optically excited states. That part of the heat which is generated within the thermal diffusion length (μ) of the interface between the spot and the underlying PVDF film is able to reach the interface and create thermal expansion and thus stress, which generates the electrical charge signal that is measured. The observed signal therefore depends on the thickness of the reagent spot, the way in which the colour change from reaction with the analyte is distributed through the spot, and the intensity of the colour. If the colour change is uniformly distributed through the sample, very low analyte concentrations may lead to weak optical absorption, so that increasing the spot thickness will give a larger signal. However, if the spot thickness exceeds the thermal diffusion length, no further increase in signal will be observed. If the colour change is non-uniform, with outer layers reacting first, thick spots may lead to very low sensitivity as the colour, and hence the heat, is produced further than the thermal diffusion length from the active interface. In both cases, as the colour intensity increases, more heat is generated close to the interface and the signal rises while the time delay between the flashing of the LED and the generation of the piezoelectric response decreases. Both the signal magnitude and the phase lag between the signal and the exciting light therefore contain information relevant to the optimisation of spot thickness and reagent concentration. A study of these factors using several model systems has been reported [21].

The PiezOptic reader unit has an adjustable phase lag (known as the correlation delay) in the lock-in amplifier section, and the optimum value of this parameter is determined for typical analyte concentrations for each new system that is developed. Although in principle the phase lag should be reduced as the analyte dose increases, in practice a satisfactory response can be achieved across the desired concentration ranges using a single value. Similarly, for most of the developed systems reported above, comparison of the waveforms of the exciting light pulse and the piezoelectric response pulse shows that the spots are sufficiently thin to avoid problems of significant phase lag.

Model systems also show that varying the flashing frequency of the LED can have a significant effect on the magnitude of the observed piezoelectric signal. Since the signal is generated by a change in the stress applied to the PVDF film, very low flashing frequencies lead to lower signals because the stress becomes nearly constant during the long constant on or off periods. However, if the flashing frequency is too high, very accurate phase locking is necessary to ensure that the short sample time actually correlates with the period of maximum stress. The actual flashing frequency used is 8 Hz, which is a compromise between these two extremes.

Although satisfactory responses have been obtained for all of the systems reported in this chapter using the standard reader unit without detailed optimisation of the signal waveform, there is considerable scope for such optimisation in systems where sensitivity needs to be pushed to new limits.

12
Future Development Prospects

This generic monitoring system is capable of application to a very wide range of analytes, and future developments are limited mainly by consumer requirements. As toxic compounds are replaced by less harmful alternatives, monitoring must change accordingly. Furthermore, the availability of cheap versatile personal monitoring systems such as this can greatly assist the identification of associations between chemical exposure and health problems and thus contribute to the setting of new standards.

Although this chapter has focussed on the development of badges giving 15-min or 8-h time-weighted-average gas monitoring data, it has recently been shown [22] that this system can also be applied to continuous monitoring of gas concentrations. Carbon dioxide was selected as a simple analyte for testing this application. Carbon dioxide, in low to medium concentrations, can cause respiratory stimulation and affect blood circulation. It can build up inside confined spaces such as tanks and grain silos, or in carbonated beverage production units. In the UK, the Short-Term Exposure Limit (15-min STEL) is 15000 ppm (i.e. 1.5% or 27400 mg m^{-3}) [6]. Being odourless, it cannot be perceived and rapid continuous monitoring methods are therefore required to ensure the safety of workers. pH indicators can be used to follow the concentration of carbonic acid formed from carbon dioxide and water, and in this study *m*-Cresol Purple was used. Spots, prepared from an ethylcellulose-bound bicarbonate-buffered solution of *m*-Cresol Purple containing tetraoctylammonium hydroxide, were deposited on small pieces of the usual PVDF material placed in the specially-designed single-spot holder shown in Fig. 15.

The tetraoctylammonium hydroxide is a phase-transfer reagent which serves to solubilise the blue anionic form of the indicator in the matrix as well as assisting the incorporation and retention of the water molecules necessary to establish the carbon dioxide/carbonic acid equilibrium. This system gave a large decrease in reader signal on exposure to low concentrations of carbon dioxide when illuminated with an amber LED (592 nm, close to the absorbance maximum of 600 nm of the blue form of the indicator). The 90% response times were 3–4 min, with reversal in clean air somewhat slower. It was also shown that the initial rates of response over the first few seconds of exposure were proportional to the carbon dioxide concentration, so that the device could be used in this mode to provide rapid warning of dangerous increases in carbon dioxide con-

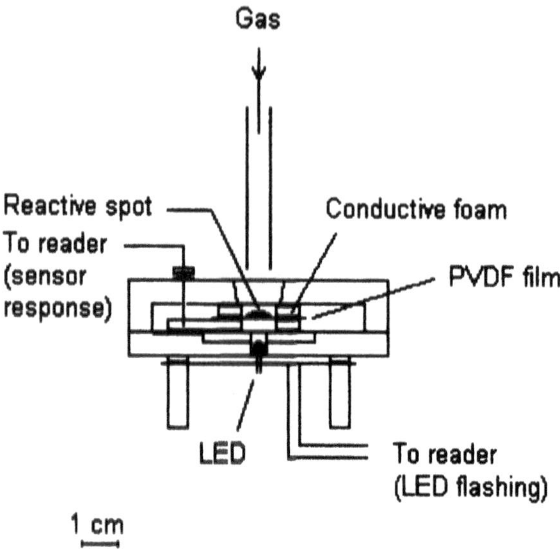

Fig. 15. The single-spot holder used for the continuous monitoring feasibility study

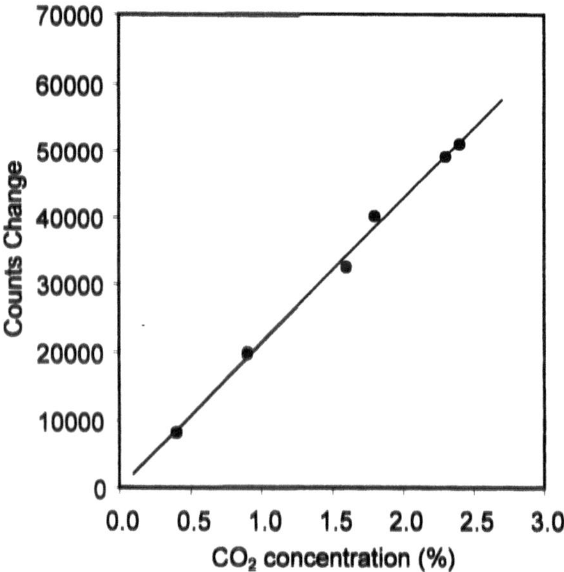

Fig. 16. Calibration curve for the continuous monitoring carbon dioxide sensor

centration. Figure 16 shows the calibration curve for the device as operated in its steady state mode.

This study shows that the system described in this chapter can be used for continuous monitoring as well as for dosimetry applications. Modern solid state electronics technology combined with new surface-mount LEDs render the design of suitable miniaturised low-power signal processing and display circuitry for such a system entirely feasible.

It has also been shown that it is possible to use the PiezOptic system for measuring analytes in solution. For example, spots made from a polymer-bound composite with ground particles of the sol-gel-entrapped Eriochrome Cyanine R copper-selective reagent mentioned earlier in this Chapter [19] can be used to detect 1–40 ppm concentrations of copper(II) in water [23]. The effects of both bulk water and water entrapped within the pores of the spot matrix on the heat-transfer processes within the spot have been shown to influence the size of the observed signals. However, reproducible responses have been obtained in a system where the spot is immersed in the test solution for one minute, before removal of surface moisture with an absorbent pad and drying in air for 3 min before measurement in the normal way. As with the continuous monitoring system, the scope and performance of the system for monitoring species in solution can certainly be improved considerably by further optimisation of the spot and by using a redesigned badge and measurement system optimised for this application.

Acknowledgements. This work has been supported by the UK Teaching Company Directorate, and by the European Community Erasmus, Leonardo da Vinci and Interreg programmes.

References

1. http://www.aeat.co.uk/netcen/airqual/bulletins
2. http://www.ademe.fr/jda/Indice.htm
3. Thain W (1976) The Determination of Vinyl Chloride: a Plant Manual. Chemical Industries Association, London. See also method 1007, NIOSH Manual of Analytical Methods (1994) 4th edn US Government Printing Office, Washington
4. Brown RH (1999) J Environ Monit 1:115
5. Wright JD, Colin F, Stöckle RM, Shepherd PD, Labayen T, Carter TJN (1998) Sens Actuators B 51:121
6. Copyright Unit HMSO (2001) EH40/2001 Occupational Exposure Limits HMSO, London
7. Cuthbert J, Groves J (1995) Ann Occup Hyg 39:223
8. Pengelly I, Groves JA, Levin JO Lindahl R (1996) Ann Occup Hyg 40:556
9. Pinkernell U, Nowack B, Gallard H, von Gunten U (2000) Wat Res 34:4343
10. Bader H, Hoigné J (1981) Wat Res 15:449
11. Grosjean D, Hisham MWM (1992) J Air Waste Manage Assoc 42:169
12. Dorta-Schaeppi Y, Treadwell WD (1949) Helv Chim Acta 32:356
13. Wright JD, von Bultzingslöwen C, Carter TJN, Colin F, Shepherd PD, Oliver JV, Holder SJ, Nolte RJM (2000) J Mater Chem 10:175
14. Colin F, Shepherd PD, Carter TJN, Wright JD (1998) Sens Actuators B 51:244
15. Ede L (1976) (ed.) Proceedings of NIOSH Styrene-Butadiene Briefing, US DHEW/PHS/CDC/NIOSH, Government Printing Office, Washington
16. Parlow JJ (1995) Tetrahedron Lett 36:1395
17. Smith K, James DM, Matthews I, Bye MR (1992) J Chem Soc, Perkin Trans I 1877
18. Bearman KR, Blackmore DC, Carter TJN, Colin F, Wright JD, Ross SA (2002) Chem Commun 980

19. Sommerdijk NAJM, Poppe A, Gibson C, Wright JD (1998) J Mater Chem 8:565
20. Colin F (2002) PhD thesis, University of Kent
21. Gibson CA, Carter TJN, Shepherd PD, Wright JD (1998) Sens Actuators B 51:238
22. Colin F, Carter TJN, Wright JD (2003) Sens Actuators B, 90:216
23. Higginson NAC (2001) PhD thesis, University of Kent

CHAPTER 10

Interferometric Biosensors for Environmental Pollution Detection

L. M. Lechuga, F. Prieto, B. Sepúlveda

1 Background of Interferometer Biosensors

One important step in the development of biosensors is the design and fabrication of a highly sensitive physical transducer, that is, a device capable of transforming efficiently a chemical or biological reaction into a measurable signal. There are several physical methods to obtain this transducing signal such as those based on amperometric, potentiometric or acoustic systems. However, transducers that make use of optical principles offer more attractive characteristics such as immunity to electromagnetic interference, possible use in aggressive environments and, in general, a higher sensitivity.

Generally, in optical transducers, the chemical or biological stimulus produces changes in the characteristics of the medium in contact with the light path, like a variation in its emission properties (luminescence), in the absorption coefficients or in the refractive index. This variation will induce a change in the propagation properties of light (wavelength, intensity, polarisation, phase velocity). Depending on how this change in the light characteristics is measured, several sensor schemes have been developed. Several techniques have been proposed to measure the induced change in the propagation properties of light. The more important one among these is the Surface Plasmon Resonance sensor (SPR) [1, 2]. This method has been improved greatly and nowadays there exist several commercial devices (such as BIAcore [3], Spreeta [4, 5], or Quantech [6]). Other important measurement methods are the grating coupler [7] (commercialised by Artificial Sensing Instruments [8]) and the resonant mirror [9] (commercialised by Affinity Sensors [10]). Another sensing scheme that exists, but is not so far commercially developed, and shows a higher sensitivity is the interferometer. This transducer system will be treated in more detail in this chapter.

All interferometric sensors make use of optical waveguides as the basic element of their structure for light propagation and are based on the same operation principle, *evanescent field sensing*. An optical waveguide is formed with a core layer of material of certain refractive index surrounded by two other media (cladding) with lower refractive indices. Light is confined within the core layer by successive total internal reflections at the core-cladding media interfaces. However, although light travels confined within the core layer, there is a part of the guided light (evanescent field) that travels through a region that extends outward, around hundreds of nanometers, into the media surrounding the waveguide. When there is a change in the optical characteristics of the outer medium (i.e. refractive index change), a modification in the optical properties of the guided wave (phase velocity) is induced via the evanescent field. Sensors based on this operation principle have been fabricated using optical fibres and integrated optical waveguides. However, fibre-based sensors [11] cannot compete with integrated optics with respect to robustness, compact optical circuitry that enables a higher complexity, design flexibility with respect the geometry as well as the choice and combination of materials, ease of access to the optical path in evanescent field sensing or the potential integration with microelectronics and micromechanics systems.

Optical evanescent wave biosensing techniques allow direct monitoring of small changes in the optical properties and are particularly useful in the direct affinity detection of biomolecular interaction as they can record the binding and dissociation in real time. The direct detection method is not as sensitive as indirect ones (i.e. fluorescence, radiolabelling or enzyme amplification) but it generally requires no prior sample preparation and can be used in real time evaluations allowing the determination of concentration, kinetic constants and binding specificity of biomolecules.

The interferometric arrangement for biosensing is highly sensitive and is the only one that provides with an internal reference for compensation of refractive-index fluctuations and non-specific adsorption. Several interferometric devices have been described such as the difference interferometer, the Mach-Zehnder or the Young interferometer. Interferometric sensors have a broader dynamic range than most other types of sensors and show higher sensitivity as compared to other integrated scheme, as shown in Table 1, where a comparison of the different sensor technologies as a function of the limit of detection (in pg/mm^2) is presented. Due to the high sensitivity of the interferometric sensor the direct detection of environmental pollutants (where 0.1 ng/mL must be detected) would be possible with this device. The detection limit is generally limited by electronic and mechanical noise, thermal drift, light source instabilities and chemical noise. But interferometric devices have an intrinsic reference channel which offers the possibility of reducing common mode effects like temperature

Table 1. Comparision of sensitivities for different integrated optical biosensors

Sensing principle	Limit of detection (pg/mm^2)
SPR	2–5
Waveguide-SPR	2
Resonant mirror	5
Grating coupler	1–10
Mach-Zehnder interferometer	0.1
Differential mode interferometer	1
Young interferometer	0.7
Reflectometric interference spectroscopy (RifS)	1–5

drifts and non-specific adsorptions. Detection limits of 10^{-7} in refractive index (or better) can be achieved with these devices, which opens the possibility of development of highly sensitive devices for in-situ pollutant detection.

2 Optical Waveguides

The basic and main structure used for the development of optical interferometer sensors is the optical waveguide. It consists, basically, in a core layer with a refractive index higher than the index of the surrounding media. Light propagates through the structure by successive total internal reflections at the core-cladding interfaces. However, only rays incident with an angle greater than the critical angle will experience Total Internal Reflection (TIR), while the others will be partially reflected at the film boundaries and will leave the waveguide after some reflections (Fig. 1).

In addition to this condition, rays will propagate through the core only if the incident angles satisfy the resonant condition [12, 13]:

$$n_C \cdot k_O \cdot d_C \cdot \cos\theta - \Phi_O - \Phi_S = m \cdot \pi \qquad m = 0,1,2,3\ldots \tag{1}$$

where n_C is the core refractive index, d_C is the core thickness, $k_O = 2\pi/\lambda$, Φ_O and Φ_S are the phase shifts originated at the total internal reflection (see Fig. 1):

$$\Phi_i = \arctan\left[\left(\frac{n_C^2}{n_i^2}\right)^{\rho} \frac{\sqrt{n_C^2 \sin^2\theta - n_i^2}}{n_C \cos\theta}\right] \tag{2}$$

with i = o, s (o=outer medium, s=substrate) and ρ = 0, 1 for the TE and TM polarization, respectively. This characteristic equation defines a discrete set of propagation angles, each associated with an electromagnetic field distribution in the waveguide structure, known as "guided modes". A mode could be under-

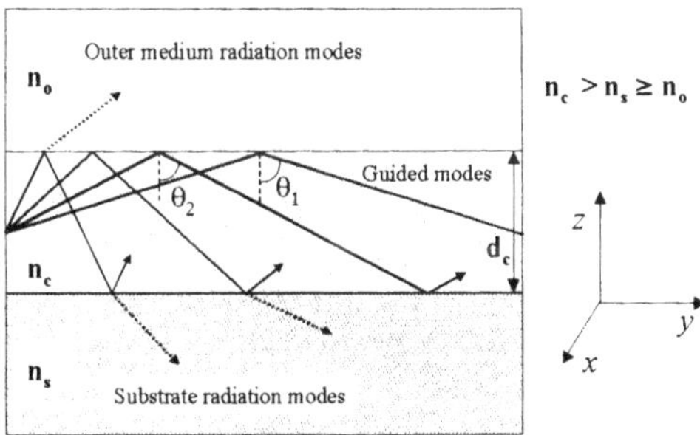

Fig. 1. Slab waveguide. Core layer has a refractive index of n_C and thickness d_C. n_S and n_O are the refractive indices of substrate and outer medium, respectively

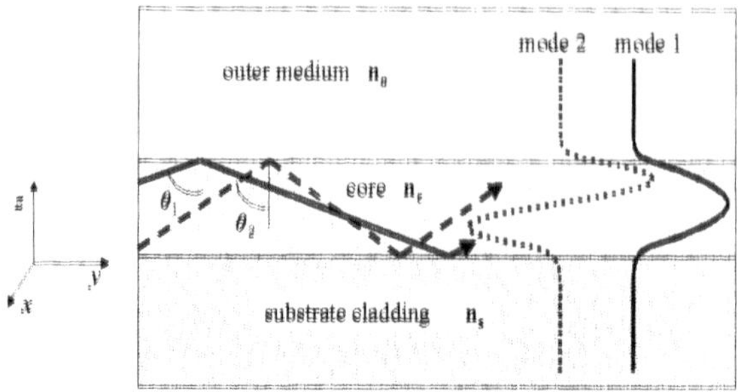

Fig. 2. The two first guided modes are shown

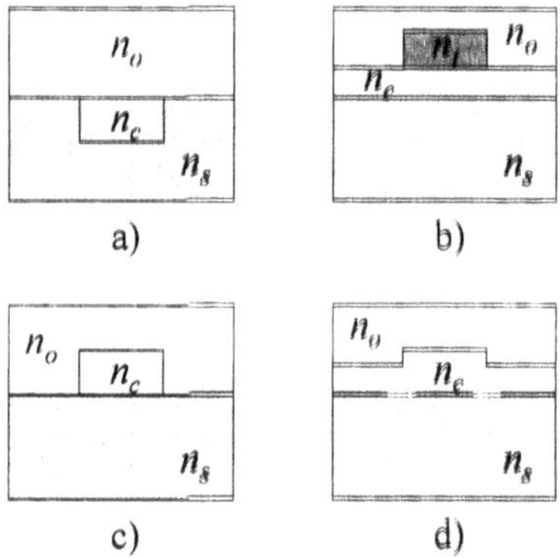

Fig. 3. Optical waveguides with lateral confinement: a) embedded strip waveguide ($n_c>n_s>n_o$); b) ridge waveguide ($n_c>n_t\geqq n_s\gg n_0$); c) and d) rib waveguide ($n_c\gg n_s\gg n_0$)

stood as an energy distribution inside the waveguide and two different modal profiles, as shown in Fig. 2, are analytical solutions of the Maxwell equations. In a waveguide both transverse electric (TE) and transverse magnetic (TM) modes can propagate. Although light is confined inside the waveguide, a part of it (evanescent field) travels through a region that extends outward, around hundreds of nanometers, into the medium surrounding the waveguide. This *evanescent field* is fundamental to assure the interaction of the waveguide modes with the external media, which is the principle of the sensing action.

Optical waveguides with lateral confinement of light (x-direction in Fig. 2) are used for the development of integrated optical devices (lasers, modulators, couplers, etc.). Therefore, the strip geometry, or the so-called channel waveguide, is generally used. This lateral limitation in the light propagation can be achieved with an increment of the thickness or the refractive index of the core with respect to the adjacent media, as shown in Fig. 3, where some of the different types of such waveguides are presented. Fig. 3a depicts the case where the core (with refractive index n_c) is embedded in a substrate with lower refractive index, n_s. Light can also be confined by depositing a raised strip (Fig. 3b) with a refractive index, n_t, lower than the core index, n_c. Figure 3c shows the case where the lateral regions adjacent to the core are completely eliminated, while in Fig. 3d the removal of the surrounding film is incomplete (rib waveguide).

However for the use of the channel or planar waveguides in interferometric biosensor applications, two main conditions must be satisfied: Monomode behaviour and high surface sensitivity.

2.1 Monomode Behaviour

Due to the evanescent sensing principle employed in the sensors, the optical waveguides may be monomode. If several modes were propagated through the structure, then each mode would detect the variations in the characteristics of the outer medium and the information carried by all the modes would be subject to mutual interference. In channel waveguides, monomode behaviour depends on the core thickness, on the width and depth of the channel and on the index contrast, which gives the value of the core refractive index compared to that of the cladding as:

$$\Delta n = \frac{n_c^2 - n_s^2}{n_c^2} \tag{3}$$

We have two different cases:

(a) If the difference of the core-cladding refractive index is very small (Δn lower than 0.5%), it is possible to have monomode behaviour for core thicknesses in the order of several micrometers, dimensions comparable to the core of some commercial optical fibres (between 4 and 10 μm depending on the design wavelength), which implies low insertion losses. However, the cladding is required to be thick enough to reduce propagation losses due to the penetration of the evanescent field into the substrate (Fig. 4). These require a longer fabrication process and the need to control accurately the refractive index. Depending on the thickness of the different layers, a cracking of the

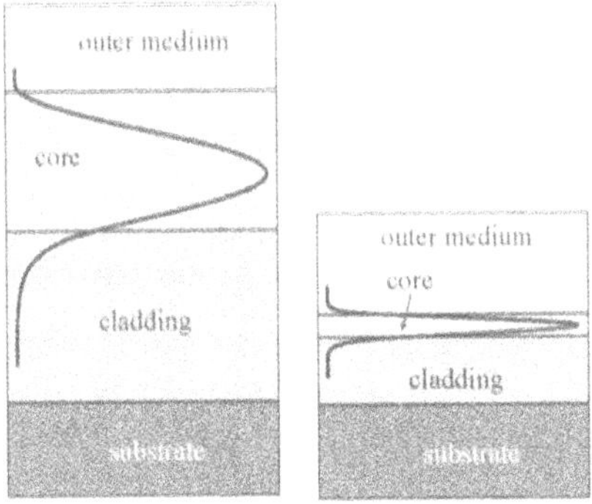

Fig. 4. Monomode optical waveguides: a) low index contrast ($<5\times10^{-3}$); b) high index contrast ($>10^{-1}$)

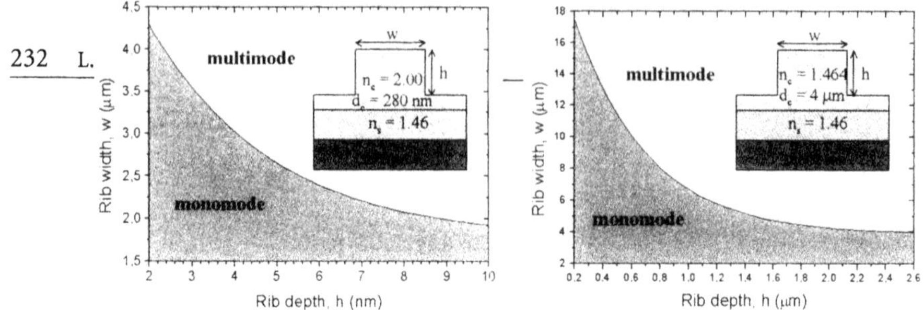

Fig. 5. Modal behaviour as a function of the height (h) and width (w) of the channel for $\lambda = 632.8$ nm and TE polarization. a) low index contrast; b) high index contrast (see inset for the waveguide parameter details)

structure can occur if the stress of each deposited layer is not controlled. For single-mode waveguides, the rib depth is of the order of a few micrometers, as shown in Fig. 5 (a) for a TIR waveguide with a core thickness of $d_c = 4$ μm and an index contrast of $\Delta n = 0.5\%$.

(b) If the difference of the core-cladding refractive index is large (Δn higher than 10%), monomode behaviour is achieved with core thicknesses of hundreds of nanometers. Now, the cladding thickness can be decreased to a few micrometers due to the small penetration of the evanescent field into the cladding (Fig. 4). However, the rib depth must be around several nanometers for single-mode waveguides, as shown in Fig. 5 (b) for a TIR waveguide with a core thickness of $d_c = 280$ nm and an index contrast of $\Delta n = 40\%$. The main disadvantage of these types of waveguides is that the dimension of the rib depth can be of the same order of magnitude as the surface roughness, which implies a technological drawback in the fabrication process.

A new alternative guiding structure has been proposed that offers single-mode behaviour with core dimensions of several micrometers (1–4 μm), rib parameters of a few micrometers and cladding thickness even smaller than the core thickness [14]. This guiding structure, called ARROW (AntiResonant Reflecting Optical Waveguide), is a special multilayer waveguide where light is confined within the core by total internal reflection at the core-outer medium interface and by an antiresonant reflection (with a very high reflectivity of 99.96%) at the two interference cladding layers underneath the core. The refractive indices and thickness of these layers are designed in such a way that they behave as a Fabry-Perot resonator operating at its antiresonant wavelengths [15]. That is, for a given wavelength, rays reflected at the cladding layers (Fig. 6) interfere constructively, leading to a high reflectivity coefficient.

This structure is a leaky waveguide (it does not support guided modes) that has an effective single-mode behaviour, i.e., higher order modes are filtered out by loss discrimination due to the low reflectance of the interference cladding. Other important features of this structure are that it presents low losses for the fundamental mode, has polarization selective characteristics, that is, TM-polar-

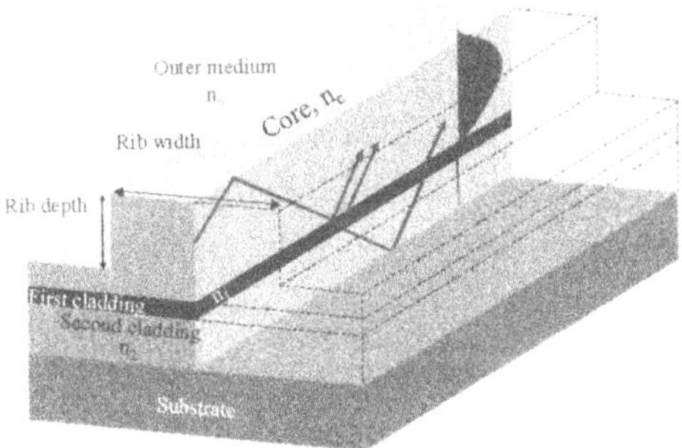

Fig. 6. Channel ARROW structure

ized light presents higher losses than TE-polarized light and there is a large tolerance in the selection of the refractive indices and thickness of the cladding layers.

ARROW structures, based on different materials, have been developed for several applications, mainly in the telecommunications and sensing fields [16–22].

2.2 Surface Sensitivity

For sensing purposes, the optical waveguide may be designed to assure high surface sensitivity which means that the sensor response for changes in the optical properties of the outer medium must be as high as possible (Fig. 7). Surface sensitivity for processes that involve the adsorption of molecules is defined as the rate of change of the effective refractive index of the guided mode, N, as the thickness of the homogeneous molecular adlayer, d_ℓ, varies [23,24]. This sensitivity is related to the squared field magnitude of the guided mode at the core-outer medium interface, considering a homogeneous adsorbed layer of refractive index n_ℓ and thickness d_ℓ (Fig. 7). Surface sensitivity can be expressed as a function of the power fraction in the evanescent field:

$$S_{\text{sup}} \equiv \frac{\partial N}{\partial d_\ell} \approx \frac{\gamma_o}{N} \cdot \left[\left(\frac{n_\ell}{n_o} \right)^{2\varrho} \gamma_o^2 - \left(\frac{n_o}{n_\ell} \right)^{2\varrho} \gamma_\ell^2 \right] \cdot \frac{P_o}{P_T} \tag{4}$$

with $\gamma_i = k_o \sqrt{N^2 - n_i^2}$ (i = o, ℓ), where n_o is the refractive index of the outer medium, n_ℓ is the refractive index of the adsorbed layer, N is the effective refractive index of the guided mode, P_0/P_T is the power fraction of the guided mode at the cover medium and ϱ is 0 for TE mode and 1 for TM mode.

Therefore, for sensing applications the thickness and refractive indices of the waveguide layers must be designed in such a way that a large fraction of the guided mode travels through the outer medium. For TIR waveguides, maximum surface sensitivity is achieved for high refractive index contrast and core thick-

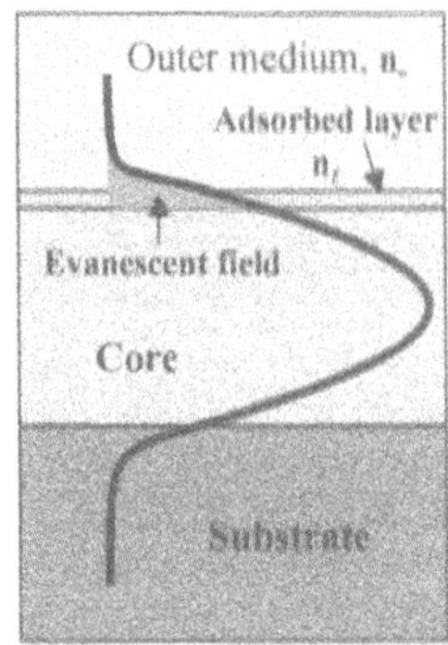

Fig. 7. Surface sensitivity: profile of the fundamental mode for a TIR waveguide. The fraction of the mode that travels through the outer medium (evanescent field) is indicated in the figure

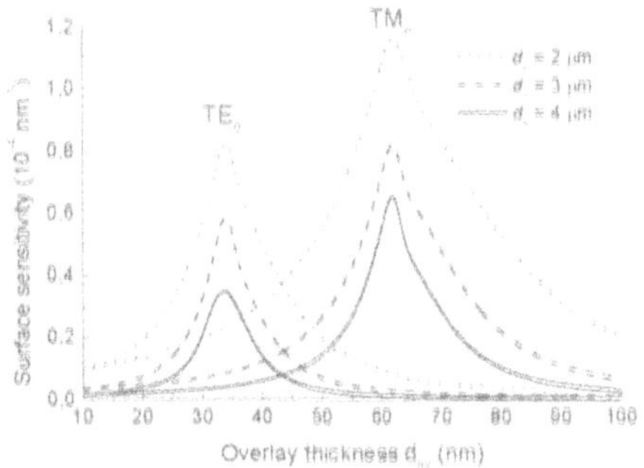

Fig. 8. Sensitivity enhancement for the ARROW structure as a function of the overlay thickness

ness of hundreds of nanometers [23]. Up to now, the sensor with the highest surface sensitivity developed is based on an interferometric arrangement fabricated using microelectronics technology [25], with a core refractive index of 2.00 and a thickness of 100 nm. The corresponding value of the theoretical sensitivity for this structure is $S_s = 3.4\times10^{-4}$ nm^{-1} for TE polarization.

In the case of ARROW structures, due to the high confinement of light within the core layer, surface sensitivity is expected to be lower than sensitivity for TIR waveguides. Although ARROW surface sensitivity can be increased by diminishing the values of the core refractive index and thickness, the maximum values obtained differ between a factor 40 and 60 with the corresponding values of sensitivity for the TIR waveguides [26]. However sensitivity for ARROW structures can be increased by overcoating the surface of the sensor with a high refractive index thin layer [27, 28]. In Fig. 8, surface sensitivity is shown as a function of the high index layer thickness (with a refractive index of 2.00) for an ARROW structure with three values of core thicknesses and refractive index of 1.485. As can be deduced from Fig. 8, sensitivity is enhanced for an overlay thickness around 35 nm for TE polarization and 65 nm for TM polarization (considering that the refractive index of the outer medium is 1.33). Surface sensitivity is enhanced more than one order of magnitude with respect the ARROW without overlay. Maximum values of sensitivity for these ARROW structures only differ a factor 4 compared to the values obtained for TIR waveguides.

3
Principle of Operation of Interferometric Sensors

In the interferometric arrangement (Fig. 9) two light beams of equal intensity are made to travel across two areas of a waveguide (one is the sensor and the other is the reference) and finally they are combined, creating an interference pattern of dark and light fringes [29,30]. When a chemical or biochemical reaction takes place in the sensor area, only the light that travels through this arm will experience a change in its effective refractive index. At the sensor output, the intensity (I) of the light passing through from both arms will interfere, showing a sinusoidal variation that depends on the difference $N_{s,r}$ of the effective refractive index of the sensor ($N_{eff,S}$) and reference arms ($N_{eff,R}$) and on the interaction length (L) :

$$\phi_{s,r} = \frac{2\pi}{\lambda} \cdot N_{s,r} \cdot L \tag{5}$$

where λ is the wavelength. This sinusoidal variation can be directly related to the concentration of the analyte to be measured.

Interferometric sensors have a broader dynamic range than most other types of sensors and also show higher sensitivity than other integrated schemes [2]. Several of these have been described in the literature [29–37]. The type of interferometer that is most commonly employed for biosensing is the Mach-Zehnder device. An attractive aspect of this device is the possibility of using long interaction lengths, thus increasing the sensitivity of the device. Interferometric devices offering a refractive-index resolution of 10^{-7} or even better, have been described [31].

A theoretical study shows that the Mach-Zehnder interferometer sensor seems to be one of the more promising concepts [30] for detection of low concentrations of small molecules without labels (10^{-12} M or even lower).The main problem in the development and possible commercialization of the MZI device is the complexity of the design, fabrication and optical adjustments. The overall

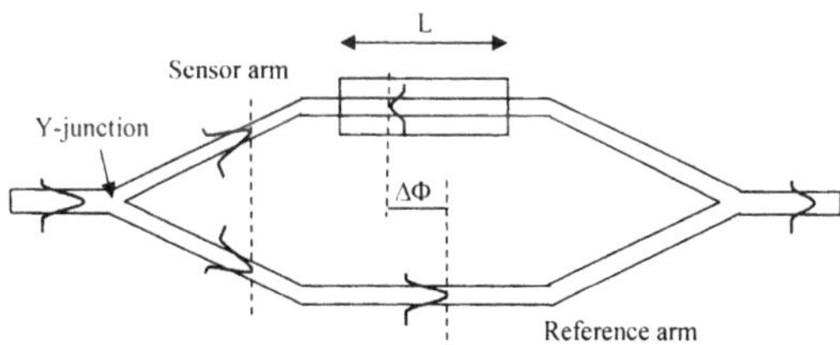

Fig. 9. Mach-Zehnder interferometer configuration

procedure for MZI fabrication is rather laborious and monomode waveguides are required increasing, even more, the complexity of the technology.

Using a conventional total internal reflection planar waveguide structure a Mach-Zehnder interferometer sensor has been developed for immunological purposes with a sensitivity of 10^{-3} nm in the thickness of the adsorbed layer (which correspond to an adsorbed molecular layer of 1 pg.mm^{-2}). Using integrated channel waveguides and for monomode behaviour the height of the waveguide (rib) has to be less than 3 nm, making the fabrication and testing of this device quite complicated.

3.1 Technology of Fabrication

The materials more commonly used for the fabrication of the waveguides, the main structure of the interferometric devices, are: glass, silicon and related materials, polymers, lithium niobate and III–V compounds. Mainly glass and silicon have been used as substrate materials. Glass has been widely used because it is inexpensive and allows simple techniques of fabrication such as ion-exchange that gives low-loss optical waveguides. In the ion-exchange technique the metallic ions that compose the glass are replaced by other ions, like K^+, Ag^+ or Cs^+. The exchange process take place in a molten saline solution, at temperatures between 200 and 300 °C, where the glass substrate is soaked. The optical waveguides are defined using metallic masks deposited on the glass surface, which are removed after the diffusion process. The refraction index could be modified between 0.01 and 0.1 units using the appropriate combination of ions. The technique described here makes use only of heat to produce the ion exchange, thus it is a very slow process. To enhance the exchange rate an electrical field is used as activator of the exchange process.

Silicon is an unique material as substrate due to its mechanical, chemical and electronic characteristics. Both silicon and its compounds have been applied to obtain components. Silicon oxide and silicon nitride are the most widely used materials in integrated optics, due to their transparency in the working wavelengths range and to their amorphousness, being structurally homogeneous. Using thin film techniques and combination of all these materials, different waveguide structures have been obtained which have been used for simple optical components with good optical performance.

To obtain a structure that is able to confine and transmit a light wave, a sequence of steps involving deposition or growth of a thin film on the substrate surface, outlining of the component pattern, and, finally, transfer of the pattern into the growing layer, defines the basic processes that are used in any optical technology for interferometer fabrication, i.e. deposition, photolithography and etching. Among the techniques usually used for waveguide fabrication are ion-exchange, spin or dip-coating, chemical vapour deposition and plasma polymerisation [30].

In the silicon-based technology, fabrication of the devices is based on standard microelectronics silicon technology such as wet and dry etching, photolithography and chemical vapor deposition (CVD). The CVD techniques allow a high control of the waveguide layer homogeneity, both in thickness and in refractive index profile. Several materials can be used as core and cladding, such as silicon nitride, phosphor-doped silicon dioxide, non-stochiometric silicon dioxide or silicon oxynitride. Silicon technology also offers the possibility of integrating sources, detectors and electronics on the same chip (hybrid integration) and the possibility of batch-wise mass-production at low cost. For the processing of the waveguides usually special installations are required (mainly Clean Room facilities) which makes the access to working with this type of sensor more complicated [30].

4 Types of Interferometer Devices: State-of-the-Art

The waveguides described up to now have been used for the development of interferometric sensors. The interferometric devices can be divided into two groups, the multiple-beam and the two-beam interferometers. One aspect common for all the interferometric devices is the problem of light coupling [29] into the waveguide for performing any biosensing process. There are three common methods for light coupling: prism-coupling, end-fired and grating coupling [30]. The advantages and disadvantages of each method depend on the application but end-fired is the most widely used one, mainly when integrated devices are used. On the other hand, for planar interferometric devices, grating coupling is the preferred method [29].

4.1 Fabry-Perot Interferometer

The first group (multiple-beam) is represented by the Fabry-Perot interferometer (FPI). The integrated optical FPI is formed by a straight monomode channel waveguide. The waveguide end-faces form the resonant mirrors, although dielectric coatings can be deposited to increase the reflectivity. The guided light is reflected forwards and backwards and at every reflection, part of the guided light is coupled out (Fig. 10).

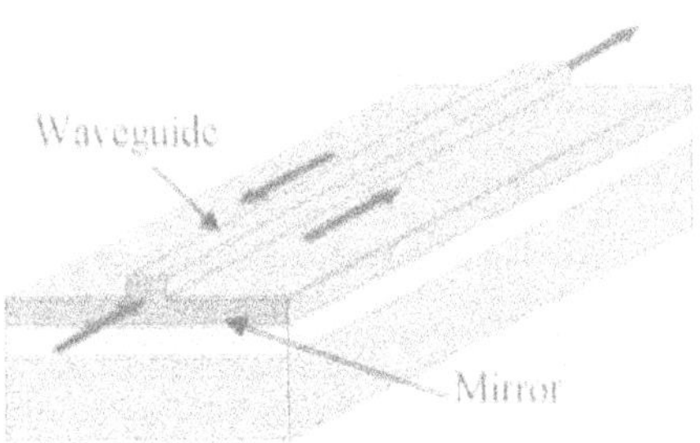

Fig. 10. Fabry-Perot interferometer

The resulting transmission curve of the device is given by the Airy function:

$$\frac{P_T}{P_O} = \frac{(1-R)^2}{\left[(1-R)^2 + 4 \cdot R \cdot \sin^2 \Phi\right]} \tag{6}$$

where P_O and P_T are the incoming the reflected beams and R is the reflectivity of the mirrors. The interferometer phase is given by:

$$\Phi = \frac{4\pi}{\lambda} N_{eff} \cdot L \tag{7}$$

where L is the waveguide length, λ is the wavelength and N_{eff} is the effective refractive index of the guided mode. The main disadvantage of this sensor is the high sensitivity towards temperature changes due to thermal expansion of the substrate. However, a reference FPI can be integrated close to the sensing FPI for the correction of temperature effects. An interferometer of this type has been proposed as refractometer [38].

4.2 Mach-Zehnder Interferometer

The most widely used two-beam interferometer is the Mach-Zehnder interferometer (MZI). In a Mach-Zehnder interferometer (MZI) device the light from a laser beam is split by a Y-junction into two identical beams that travel along the two MZI arms (sensor and reference areas) and are recombined into a monomode channel waveguide giving a signal which is dependent on the phase difference between the two beams. Any change in the sensor area (in the region of the evanescent field) produces a phase difference (and therein a change of the effective refractive index of the waveguide) between the reference and the sensor beam and thus in the intensity of the outcoupled light.

4.3 Planar Versions

In the planar versions of the MZI, a single light beam is used to address multiple sensing elements on a microfabricated optical chip. Several types of planar MZI have been described and evaluated [31–35]. This simplified device is formed by a slab waveguide where the sensing and the reference areas are defined by lithography. Depending on the version the light is coupled to the device by endfire [31, 32] or by gratings [29, 34, 35] (Fig. 11). The beams at the output are recombined using a lens and a beam-splitter. A microscope objective is used to expand the pattern onto the detector.

In the different types of designs, the basic optical chip structure is made with silicon technology. In the MZI of Heideman [30, 33] the planar monomode waveguide is fabricated with standard ($Si/SiO_2/Si_3N_4/SiO_2$) technology, where the thickness of the Si_3N_4 waveguide is 90 nm, with a width of 1 mm and a length

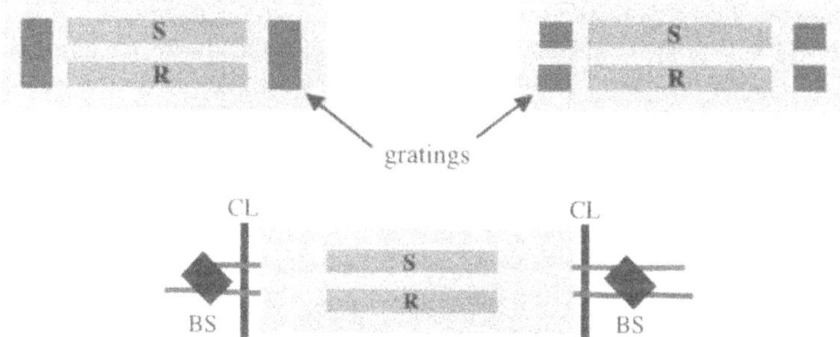

Fig 11. Planar configurations of MZI biosensor devices with (*top*) two grating coupling configurations (*bottom*) end-fire coupling configuration

of 10 mm. Two versions of the device with end-fire and with grating for light coupling have been fabricated and evaluated. With this sensor, a detection limit close to 0.1 pg/mm^2 has been achieved.

In the MZI of Hartman [34], the waveguide is made of Si_3N_4 (140 nm thick) deposited onto a BK-7 glass. Grating couplers were used for light in- and out-coupling. The sensor and reference areas are 15 mm long and 1 mm wide, separated by a 1-mm gap. With this sensor, a detection limit close to 1–5 pg/mm^2 has been achieved.

Another interesting development is the planar polarisation interferometer [36, 37]. In this device, instead of having two waveguiding arms (the sensor and the reference), two orthogonal TM and TE modes, propagating through the waveguide, are employed. The two polarisations are coupled by rotating the laser by about 45°. Both polarisations propagate across the waveguide, but the TM mode experiences a larger phase change than the TE mode, and thus the TE mode is used as reference.

The response of the device is the phase shift between TM and TE modes, as the phase difference between the two modes depends on the interaction of the evanescent wave with the outer medium. At the signal output, both modes are separated using beam splitters, Wollaston prisms or polarisers. The final signal can be acquired using photodetectors or a CCD camera. The ellipticity of the polarised light at the output is related to the chemical or biochemical interaction on the sensor surface, and this effect can be exploited for their monitoring. With this device mechanical and thermal noise are minimised. In order to attain a high sensitivity the waveguides must be designed to maximize the difference in sensitivity between the two modes.

In the Zeeman polarimetric interferometer [37], the device is excited by using a Zeeman laser and thus two frequencies are generated (usually separated 250 kHz). The phase difference which is generated when there is a biomolecular interaction is measured at the output. The 250 MHz frequency sine wave, generated in the recombination of the two modes, is measured and compared with the reference one from the laser.

The main disadvantage of this type of polarimetric device is that the TE mode (used as reference) is not completely insensitive to the specific binding events and the magnitude of the response between both modes is decreased. In some cases, this decrease can be as high as 50%. On the other hand the polarimetric biosensor cannot distinguish between a bulk refractive index change and a specific interaction event. Some attempts have been made to overcome all these problems [29]. A detection limit of 2 pg/mm^2 can be achieved with this type of device.

Another interesting device is the side-by-side or multianalyte planar MZI [29, 35]. This design has thirteen interferometers on a single chip (1×2 cm). Light from a laser is launched into the chip using a broad grating, all interferometers being excited at once. The waveguides are made of Si_3N_4 deposited on glass. The channels are defined by using patterned thick SiO_2. The channels are also defined by the receptors immobilised onto the waveguide surface rather than by the beams themselves. The output beams are recombined and the resulting interference pattern is coupled out using another broad grating and analysed using an array detector.

4.4 Integrated Versions

One of the major disadvantage of planar interferometers is the need for multiple optical components, which have made these devices available only as laboratory prototypes. For further development as portable devices, integrated schemes must be introduced. Integrated optical interferometric devices combine high sensitivity with compactness, simple instrumentation and high degree of batch-production. These structures are well suited for integration of optical and electrical functions onto one substrate (chip).

An integrated version of a MZI can be seen in Fig. 12. If the optical waveguides that form the MZI are monomode and the light source is monochromatic, then the amplitude of the optical field at the divisor output becames $\sqrt{\alpha_1 \cdot k_1 \cdot E_0 \cdot \cos(\omega_0 t)}$ for one of the branches and $\sqrt{\alpha_1 \cdot (1-k_1) \cdot E_0 \cdot \cos(\omega_0 t)}$ for the other, where E_0 is the input optical field, α_1 is the transition loss at the Y-junction and k_1 is the coupling coefficient of the divisor. Light propagates with certain optical losses β_s and β_r, associated with the sensor and reference paths, respectively. Over this distance, light will experience a phase shift ϕ_s and ϕ_r in each one of the arms, where

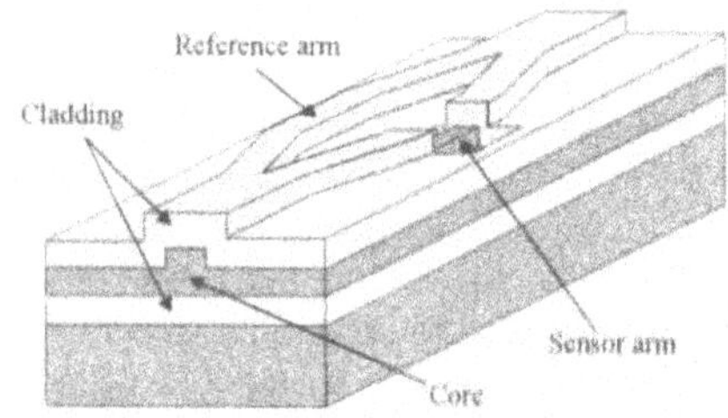

Fig. 12. Integrated Mach-Zehnder interferometer

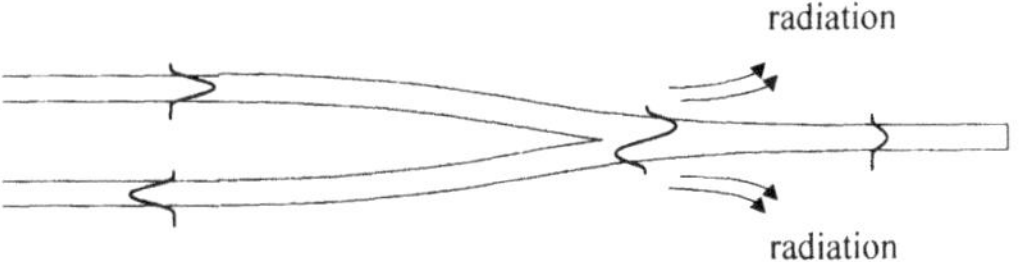

Fig. 13. Combination of two out-of-phase modes at the Y-junction output

$$\phi_{s,r} = \frac{2\pi}{\lambda} \cdot N_{s,r} \cdot L \tag{5}$$

with $N_{s,r}$ the effective refractive index of the guided mode, λ the wavelength and L the interaction length.

When both beams recombine in the output coupled (Fig. 13), the combination of two out-of-phase beams will excite higher order modes in the output optical waveguide of the device. However, as the waveguides have been fabricated as monomode, the modes higher than the fundamental will be radiated out of the output coupler.

Knowing the coupling coefficients of the divisors, the Y-junction losses and the propagation losses of the interferometer arms, it is possible to establish a direct relation between the output signal variation and the phase shift produced in one of the branches with respect to the other. The fringe visibility (V) of the interferometer is defined as:

$$V = \frac{I_{MAX} - I_{MIN}}{I_{MAX} + I_{MIN}} \tag{6}$$

where I_{MAX} and I_{MIN} are the maximun and minimun detected intensity. In the design of MZ interferometers it is very important to have maximum contrast in the output signal, which means that the visibility factor should be close to unity. If we consider the case where both Y-junctions are symmetric and they have the same coupling factors and the same losses, equation 6 will take the form

$$V = \frac{2 \cdot k \cdot (1-k)\sqrt{\beta_r \cdot \beta_s}}{k^2 \cdot \beta_r + (1-k)^2 \cdot \beta_s} \tag{7}$$

If the divisors are designed symmetrically, the coupling factor and the Y-junction losses only depend on the fabrication processes of the devices. The development of most of the interferometric biosensors is based on microfabrication techniques, which implies that with an appropriate control of these processes and considering the small dimensions of the devices (usually in the order of a few millimetres), the couplers are fabricated under the same conditions and their characteristics can be considered to be equal. With these considerations, the visibility factor gets its maximum value ($V = 1$) for a coupling factor $k = 0.5$ or 3 dB (that is, the intensity at the two coupler outputs is the same). Figure 14 shows the transmission curve of a device for different visibility factors.

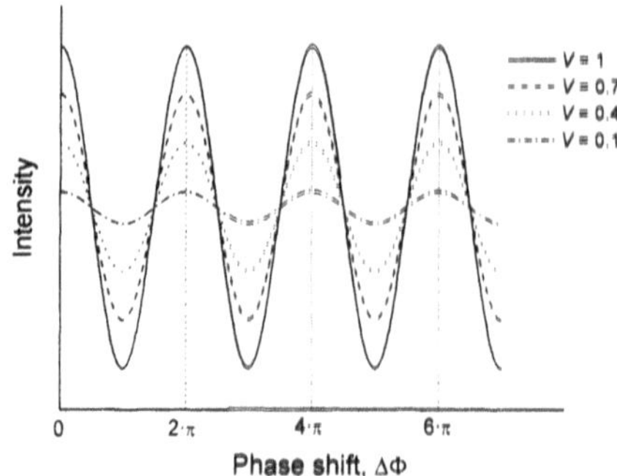

Fig. 14. Transmission curve of the Mach-Zehnder interferometer for different visibility factors

Moreover, V depends on the propagation losses at the two interferometer branches and, considering the real case of a non-monochromatic source, it will also depend on the spectral width of the source (Δf) and on the phase shift between the two interferometer arms ($\Delta\Phi = \Phi_r - \Phi_s$) [30, 31]:

$$V \propto \exp(\Delta f \cdot \Delta\Phi) \tag{8}$$

Therefore, to obtain a maximum visibility factor, it is necessary to have a well-balanced MZI sensor, which is achieved with the design of identical interferometer branches.

Due to the periodicity of the output MZI signal, the detected intensity variation is not proportional to the measurand and, therefore, it is necessary to develop a signal processing scheme to overcome this disadvantage. There are three main problems associated with the non-linearity of the output signal:

1. ambiguity of the signal: it is not possible to determine accurately the phase shift from the output intensity because it is the same for integer multiples of π. For the same reason, it is not possible to deduce the direction of the phase change.
2. intensity fluctuations of the light source may be misinterpreted as phase changes of the interferometer.
3. signal fading: depending on the initial phase shift value between the two interferometer branches, the output MZI signal will be at a certain point of the transmission curve. If the interferometer is tuned close to one of the extreme values of the transmission curve, small phase changes will generate low intensity variations. However, the interferometer sensitivity reaches its maximum value if the interferometer is tuned close to the quadrature condition (Fig. 15). In this point, the MZI response also varies with the visibility factor because the slope of the transmission curve increases with the value of V.

The problems with the periodicity of the interferometer signal output can be solved by generating several intensity values associated with different transmis-

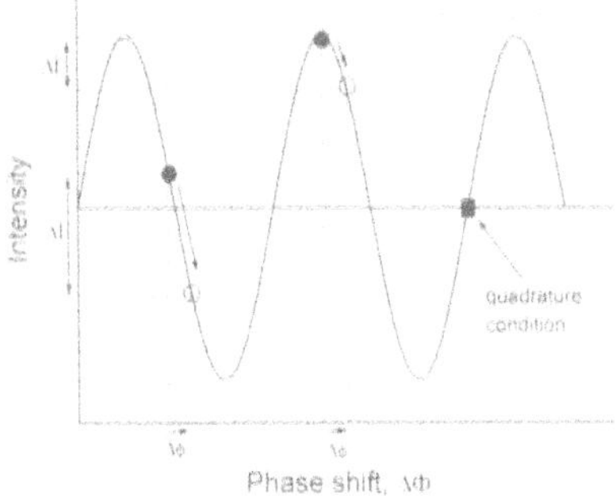

Fig. 15. Signal fading at extreme values of the transmission curve

Fig. 16. Some details of one MZI sensor. (*left*) Y-junction of R= 5mm and (*right*) sensor and reference areas

sion curves, shifted by constant phase values with respect to one another. By the use of simple arithmetical operations, the phase is derived correctly within a single interference period. In the MZ interferometer, the generation of more signals requires a modification of the structure, substituting the output divisor by a 3×3-coupler. The three output waveguides provide signals with a phase difference of 120° with respect to each other [39].

Another alternative consists of tuning the interferometer phase by incorporating a modulation system. Using different operation principles (electro-optic effect, thermo-optic effect, etc.) a phase change is induced in one of the MZI branches with respect to the other by applying an external signal in such a way that the interferometer can be tuned to the quadrature condition [40, 41].

Several configurations of integrated MZI have been developed [22, 25, 39, 42–48]. Most of them are based on silicon or silicon-on-glass technology. An AntiResonant Reflecting Optical Waveguide (ARROW) has been also used for the fabrication of an integrated MZI sensor. The attractive characteristics of the ARROW technology were discussed above. Some details of one of the ARROW MZI interferometers fabricated are shown in Fig. 16.

4.5 Young Interferometer

The Young interferometer is a type of two-beam interferometer. It uses an integrated optical Y-junction acting as a beam splitter, as shown in Fig. 17 [49, 50].

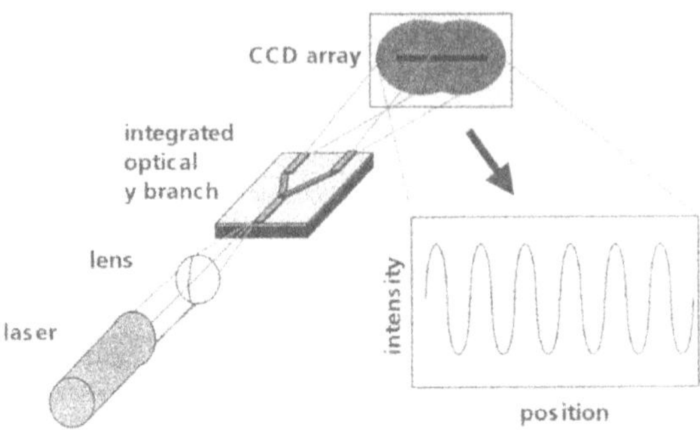

Fig.17. Young interferometer

It is a variant of the integrated Mach-Zehnder interferometer as the sensor and the reference arms are not recombined inside the structure but the light coming out from the two arms is made to interfere at the output. Light is end-coupled in the device. At the output of the sensor, the light emitted in the form of two cones, which superimpose and produce an interference pattern. The intensity distribution is detected by a detector array.

The light coupled out of the two branches generates an interference pattern on a screen or CCD detector with a cosine intensity distribution function. The phase difference of the two interfering rays is given by:

$$\Phi = \frac{2\pi}{\lambda}\left(\frac{d \cdot x}{f} - \left(N_{\text{eff,S}} - N_{\text{eff,R}}\right) \cdot L\right) \tag{12}$$

where d is the distance between the two branches, f is the distance between the output sensor and the screen and x denotes the position on the screen. In one of the arms (sensor arm) and during a certain interaction length L, a change in the optical characteristics of the outer medium is induced, which produces a variation in the effective refractive index in one arm respect to the other ($N_{\text{eff,S}}$ - $N_{\text{eff, R}}$). Under the influence of the adsorption of biomolecules on the sensor branch the fringe pattern moves laterally. One disadvantage of the Young device is the distance from the output to the detector required for maximum resolution. The advantages of this type of interferometer includes the simplicity of the arrangement, the detection of complete intensity distribution and the identical length of the arms which avoids side effects arising from temperature and wavelength drifts. The interferometer must be designed in a symmetric fashion.

A Young interferometer biosensor [49] has been fabricated using resistive waveguides, made by an optimised PECVD process. The films are made of silicon oxynitride, with refractive index of 1.57 and thickness of 400 nm. Lateral structuring has been carried out by wet chemical etching. The waveguide width is about 3.5μm. With this device a theoretical detection limit of 9×10^{-8} in the effective refractive index can be achieved and an experimental limit of 50 ng/mL of proteins has been measured.

A comparison of the different interferometric devices developed until now and their sensitivities is summarised in Table 2. In this Table, the factor Γ is the mass coverage of the surface, which is a measure of the adsorbed mass per unit surface area.

Table 2. Interferometric devices developed to date

Device	Limit of Detection		Technology	Reference
	ΔN_{eff}	Mass coverage (Γ)		
Planar-MZI (with grating)	4×10^{-8}	1 pg/mm^2	Si	33
Planar MZI (end-fire)	$\approx10^{-8}$	0.1 pg/mm^2	Si	31,32
Planar MZI (double grating)	$\approx10^{-8}$	0.5–1 pg/mm^2	Glass/Si	34
Planar MZI (multi-grating)	$\approx10^{-8}$	-	Glass/Si	29
Planar polarisation interferometer	5×10^{-8}	2 pg/mm^2	Glass/Si	36
Integrated MZI (end-fire)	5×10^{-5}	0.1 ng/mm^2	Si (rib = 2 nm)	25
Integrated MZI (3×3 coupler)	4×10^{-7}	4 pg/mm^2	Ion-exchange glass	39
Integrated Young's Interferometer	9×10^{-8}	0.7 pg/mm^2	Si	49,50
Integrated-MZI	$\approx10^{-7}$	10–20 pg/mm^2	Si (rib = 55 nm)	44,46
Integrated-MZI ARROW	4×10^{-7}	15 pg/mm^2	Si (rib = 2.5 μm)	22
Integrated MZI TIR	7×10^{-8}	-	Si (rib = 2–3 nm)	42
Integrated MZI	$\approx10^{-7}$	-	III–V Replicated polymer	47
Integrated MZI	5×10^{-8}	-	Si (rib = 0.7–2 nm)	48

Γ is calculated according to the following equation:

$$\Gamma = \frac{(n_l - n_o)}{\left(\frac{dn}{d[C]}\right)\left(\frac{\partial N_{eff}}{\partial d_l}\right)} \Delta N_{eff}$$

5 Surface Functionalization for Biosensing

For sensing purposes, a layer of receptor molecules capable of binding the analyte molecules in a selective way, has to be previously immobilised at the sensor surface of the interferometer biosensor. Complementary analytes flowing over the surface can be directly recognised by the receptor through a change in the optical properties of the waveguide. In this way, the interacting components do not need to be labelled and complex samples can be analysed without purification.

The immobilisation of the receptor molecule on the sensor surface is a key factor for the performance of the sensor. The chosen immobilisation method must retain the stability and activity of the bound biological receptor. Generally, direct adsorption is inadequate, causing significant losses in biological activity and random orientation of the receptors. Despite these difficulties direct adsorption is widely employed since it is simple, fast and does not require special reagents.

Surface chemistries for stable and defined binding of the molecular receptors have been developed based on affinity immobilisation and/or covalent bonding. In the affinity bonding, a high affinity-capture ligand is non-reversibly immobilised on the sensor surface; for example streptavidin monolayers using then biotinylated biomolecules for recognition. Another approach is to form a Self-Assembled Monolayer (SAM) of alkylsilanes and then the receptor can be coupled using the end of the SAM via a functional group (-NH_2, -COOH,...).

The use of a polymer matrix maximizes the interaction volume probed by the evanescent field, increasing greatly the surface capacity and therefore the sensitivity of the device. The carboxymethyldextran hydrogel approach has been widely used [30]. The receptor molecules are attached to flexible dextran chains and are freely accessible in a three-dimensional space, thus minimizing steric hindrance and increasing the sensitivity. A variety of surface activation chemistries can be used to couple the receptor to the hydrogel via amine, thiol, disulphide or aldehyde groups.

After the immobilisation of the receptor layer, the surface must be backfilled with an inert protein in order to prevent non-specific binding as much as possible. This is not a simple task and different procedures have been employed. The object is to make the reference channel as much equivalent to the sensor area for sensing only the specific binding event for which the interferometer has been designed.

6 Environmental Applications

Although all the developments in MZI devices have been made with the aim of achieving direct detection of small molecules, such as environmental pollutants, until now only two interferometric sensors have been used for environmental applications [51, 52].

A planar MZI using end-fire [51] was used in a competitive assay for atrazine, with a detection limit of 0.1 ppb. An atrazine hapten is synthesised and cova-

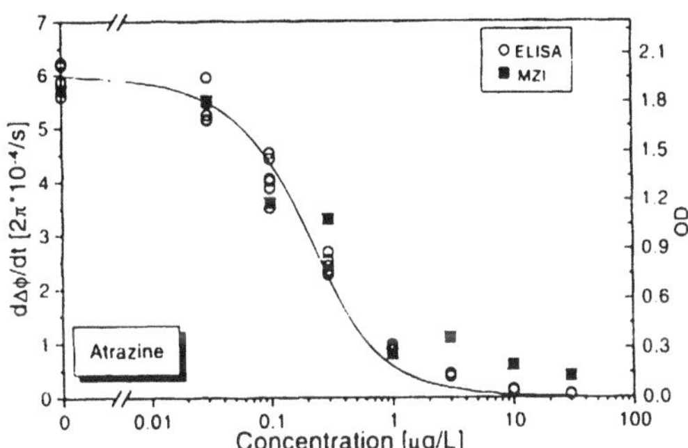

Fig. 18. Atrazine MZI measurements compared with a parallel ELISA. Pesticides samples were premixed for one hour with 1×10^{-7} M monoclonal antibodies specific to atrazine. A Langmuir model was fitted to the results

lently attached to the sensor surface. The liquid sample is introduced over the sensor using a flow system. Then, the atrazine antibody is introduced. The dissolved atrazine competes with the bound atrazine hapten for the antibodies. If there is no atrazine in the sample, then all the antibody will be attached to the atrazine hapten and the signal will be high. When atrazine is present in the sample the signal will decrease for increasing concentrations. In Fig. 18 atrazine measurements with the MZI device are shown. This sensor detects down to 0.1 μg/L or 0.1 ppb of atrazine. The sensor surface can be regenerated without loss of activity and re-used for several analyses.

The integrated sensor with 3×3 coupler has been also used for simazine determination [52] through a competitive assay, where the limit of detection was found to be 0.1 ppb.

7 Future Trends

There is still a need for sensors that are able to detect very low levels of a great number of chemicals and biochemicals in the areas of environmental monitoring, industrial and food processing, health care, biomedical technology, clinical analysis, etc. In the environmental field the demand for new sensors is increasing continuously due to the strict legislation and control being introduced to improve living standards through waste management and remediation programs, for example.

Ideally, evanescent wave sensors must have a very high sensitivity and selectivity, a broad dynamic range, immunity to matrix effects, must be capable of simultaneous multianalyte determination and must be fast, reversible, stable, simple to operate, robust, cheap and of small size (for making a portable system for spatial mapping over large or remote areas).

To reach all the objectives described above, future research and development in optical evanescent wave interferometric biosensors will need to focus on the following key points:

(1) *Integration*: Mass-production of sensors will be possible using integrated optics with the fabrication of miniaturised devices integrating the electronics and optics (sensor-on-a-chip in which the light source, photodiodes and sensor waveguides are combined on a single semiconductor package), the flow system and the reagent deposition (by ink-jet, screen-printing or other technology). A complete system fabricated with integrated optics will offer low complexity, robustness, a standardised device, portability.

(2) *New receptors*: Biological receptors can include chemically and genetically modified enzymes, new types of antibodies with high affinity and selectivity for small molecules, natural or artificial receptors or complex biological recognition elements.

(3) *Multianalyte detection*: Multianalyte capability is required in environmental screening, with thousands of samples per year to be analysed. Direct optical detection with evanescent wave sensors could be a possibility but a parallel detection of as many sites as possible is necessary.

(4) *High sensitivity*. Improved limit detection must be achieved for most of the applications as for example in the environmental field.

References

1. Raether H, (1977) Surface Plasmon oscillations and their applications, in Physics of Thin Films, vol 9, Academic Press, Florida, pp 145–262
2. Lechuga LM, Calle A, Prieto F (2000) Optical sensors based on evanescent field sensing. Part 1: Surface plasmon resonance sensors. Quim Anal 19: 7–13
3. <http://www.biacore.com>
4. Melendez J, Carr R, Bartholomew DU, Kukanskis K, Elkind J, Yee S, Furlong C, Woodbury R (1996) A commercial solution for surface plasmon sensing. Sens Actuators B35-36: 212–216
5. <http://www.ti.com/spr>
6. <http://www.biosensor.com>
7. Tiefenthaler K, Lukosz W (1989) Sensitivity of grating couplers as integrated-optical chemical sensors. J Opt Soc Am B6: 209–220
8. <http://www.microvacuum.com>.
9. Cush R, Cronin JM, Stewart WJ, Maule CH, Molloy J, Goddard NJ (1993) The resonant mirror: a novel optical biosensor for direct sensing of biomolecular interactions. Part I: principle of operation and associated instrumentation. Biosen Bioelectron 8: 347-353
10. <http://www.affinity-sensors.com>.
11. Giallorenzi TG, Bucaro JA, Dandridge A, Siegel GH, Cole JH, Rashleigh SC, Priest RG (1982) Optical Fiber Sensor Technology. IEEE J Quant Electron 18: 626–665
12. Marcuse D (1974) Theory of dielectric optical waveguides. Academic Press, New York
13. Tamir T (1988) Guided-wave optoelectronics. Springer, Berlin
14. Duguay MA, Kokubun Y, Koch TL (1986) Antiresonant reflecting optical waveguides in SiO_2-Si multilayer structures. Appl Phys Lett 49: 13–15
15. Born M, Wolf E (1993) Principles of optics. Pergamon Press, Oxford
16. Mawst LJ, Yang H, Nesnidal M, Al-Muhanna A, Botez D, Vang TA, Alvarez FD, Johnson R (1998) High power single mode Al free InGaAs(P)/InGaP/GaAs distributed feedback diode lasers. J Crystal Growth 195: 609–616
17. Kokubun Y, Asokawa A (1993) ARROW-type polarizer utilizing form birefringence in multilayer first cladding. IEEE Photon Technol Lett 5: 1418–1420

18. Sato S, Pan W, Chen ST, Endo S, Suzuki S, Kokubun Y (1999) 59-nm trimming of centre wavelength of ARROW-type vertical coupler filter by UV irradiation. IEEE Photon Technol Lett 11: 358-360
19. Yamada Y, Sugito A, Moriwaki K, Ogawa I, Hashimoto T (1994) An application of a silica-on-terraced-silicon platform to hybrid Mach-Zehnder interferometric circuits consisting of silica waveguides and $LiNbO_3$ phase-shifter. IEEE Photon Technol Lett 6: 822-824
20. Nathan A, Bhatnagar YK, Benaissa K, Huang W (1995) Micromechanical Mach-Zehnder interferometer compatible with silicon integrated circuit and micromachined technologies. Sens Mater 7: 105–109
21. Garcés I, Villuendas F, Subías J, Alonso J, del Valle M, Domínguez C, Bartolomé E (1997) Bidimensional planar micro-optics for optochemical absorbance sensing. Opt Lett 23: 223–227
22. Prieto F, Lechuga LM, Calle A, Llobera A, Domínguez C (2001) Optimised silicon antiresonant reflecting optical waveguides for sensing applications. J Lightwave Technol, Enero
23. Tiefenthaler K, Lukosz W (1989) Sensitivity of grating couplers as integrated-optical chemical sensors. J Opt Soc Am B 6: 209–220
24. Parriaux O, Veldhuis GJ (1998) Normalized analysis for the sensitivity optimization of integrated optical evanescent-wave sensors. J Lightwave Technol 16: 573–582
25. Shipper EF, Brugman AM, Domínguez C, Lechuga LM, Kooyman RPH, Greve J (1997) The realisation of an integrated Mach-Zehnder waveguide immunosensor in silicon technology. Sens Actuators B 40: 147–153
26. Prieto F, Llobera A, Jiménez D, Domínguez C, Calle A, Lechuga LM (2000) Design and Analysis of Silicon Antiresonant Reflecting Optical Waveguides for Evanescent Field Sensors. J Lightwave Technol 18: 966–972
27. Muhammad FA, Stewart G, Jin W (1993) Sensitivity enhancement of D-fibre methane gas sensor using high-index overlay. Proc Inst Electr Eng Part J 140: 115-118
28. Quigley GR, Harris RD, Wilkinson JS (1999) Sensitivity enhancement of integrated optical sensors by use of thin high-index films. Appl Optics 38: 6036–6039
29. Campbell DP, McCloskey CJ (2002) Interferometric Biosensors. In: Optical Biosensors: present and future. Ch. 9, 277–304. Ed. Ligler FS, Taitt AR. Elsevier Science BV. ISBN: 0-444-50974-7
30. Kooyman RPH, Lechuga LM (1997) Immunosensors based on Total Internal Reflectance". Ch 8, 169–196. In „Handbook of Biosensors: Medicine, Food and the Environment", Ed. Kress-Rogers, E. CRC Press, Boca Raton, Florida
31. Schipper E (1996) Waveguide immunosensing of small molecules. Thesis, University of Twente
32. Lechuga LM, Lenferink ATM, Kooyman RPH, Greve J (1995) Feasibility of evanescent wave interferometer immunosensors for direct detection of pesticides: Chemical aspects. Sens Actuators B 24: 762-765
33. Heideman RG, Kooyman RPH, Greve J (1993) Biosens: Performance of a highly sensitive optical waveguide Mach-Zehnder biosensor. Sens Actuators B 10: 209–217
34. Scheneider BH, Dickinson EL, Vach MD, Hoijer JV, Howard LV (2000) Highly sensitive optical chip immunoassays in human serum. Biosens Bioelec 15: 13–22
35. Campbell DP, Gottfried DS, Roberts DW, Caspall JJ (2002) Proc. Europtrode VI (Sixth European Conference on Optical Chemical Sensors and Biosensors). P. 327. UMIST, Manchester, April
36. Shirshov Y, Snopok BA, Samoylov AV, Kiyanovskij AP, Venger EF, Nabok AV, Ray AK (2001) Analysis of the response of planar polarisation interferometer to molecular layer formation: fibrinogen adsorption on silicon nitride surface. Biosens Bioelec 16: 381–390

37. Ayräst P, Honkanent S, Gracet KM, Shrouf K, Katila P, Leppihalme M, Tervonen A, Yang X, Swanson B, Peyghambariam N (1998) Thin-film chemical sensors with waveguide Zeeman interferometry. Pure Appl Opt 1261–1271
38. Konz W, Brandenburg A, Edelhäuser R, Ott W, Wölfelshneider H (1989) A refractometer with fully packaged integrated optical sensor head, in: Arditty HJ, Dakin JP, Kersten RTh Optical fiber sensors. Springer Verlag, Berlin, pp 443-447
39. Luff BJ, Wilkinson JS, Piehler J, Hollenbach U, Ingenhoff J, Fabricius N (1998) Integrated optical Mach-Zehnder biosensor. J Lightwave Technol 16: 583–592
40. Ikkink TJ (1998) Interferometric interrogation concepts for integrated electro-optical sensor systems, Thesis, University of Twente
41. Johnson GW, Leiner DC, Moore DT (1977) Phase-locked interferometry. Proc SPIE vol 126, pp 152–160
42. Sepúlveda B, Prieto F, Calle A, Llobera A, Domínguez C, Lechuga LM (2002) Integrated Optical Interferometric Biosensors based on Microelectronics Technology for Immunosensing. Proc Biosensors 2002, Kyoto, Japan, May 2002
43. Weisser M, Tovar G, Mittler S, Knoll W, Brosinger F, Freimuth H, Lacher M, Ehrfeld W (1999) Specific bio-recognition reactions observed with an integrated Mach-Zehnder interferometer. Biosens Bioelec 14: 405-411
44. Brosinger F, Freimuth H, Lacher M, Ehrfeld W, Gedig E, Katerkamp A, Spener F, Cammann K (1997) A label-free affinity sensor with compensation of unspecific protein interaction by a highly sensitive integrated optical Mach-Zehnder interferometer on silicon. Sens Actuators B 44: 350–355
45. Busse S, Scheumann V, Menges B, Mittler S (2002) Sensitivity studies for specific binding reactions using the biotin/streptavidin system by evanescent optical methods. Biosens Bioelec 17: 704–710
46. Busse S, DePaoli M, Wenz G, Mittler S (2001) An integrated optical Mach-Zehnder interferometer functionalised by β-cyclodextrin to monitor binding reaction. Sens Actuators B 80: 116–124
47. Kunz RE (1999) Integrated optics in sensors: advances toward miniaturised systems for chemical and biochemical sensing. In Integrated Optical Circuits and components: design and applications, Ed. Murphy EJ. Marcel Dekker, Inc, New York. ISBN: 0-8247-7577-5
48. Heideman RG, Lambeck PV (1999) Remote opto-chemical sensing with extreme sensitivity: design, fabrication and performance of a pigtailed integrated optical phase-modulated Mach-Zehnder interferometer system". Sens Actuators B 61: 100–127
49. Brandenburg A, Krauter R, Künzel C, Stefan M, Schulte H (2000) Interferometric sensor for detection of surface-bound bioreactions. Applied Optics 39: 6396–6405
50. Brynda E, Houska M, Brandenburg A, Wikerstal A (2002) Optical biosensors for real-time measurement of analytes in blood plasma. Biosens Bioelec 17: 665–675
51. Shipper EF, Bergevoet AJH, Kooyman RPH, Greve J (1997) New detection method for atrazine pesticides with the optical waveguide Mach-Zehnder immunosensor. Anal Chim Acta 341: 171–176
52. Drapp B, Piehler J, Brecht A, Gauglitz G, Luff BJ, Wilkinson JS, Ingenhoff J (1997) Integrated optical Mach-Zehnder interferometers as simazine immunoprobes. Sens Actuators B 38–39: 277–282

CHAPTER 11

Fibre Optic Sensors for Humidity Monitoring

MARIA C. MORENO-BONDI, GUILLERMO ORELLANA, MAXIMINO BEDOYA

1 Introduction

Humidity is a term that refers to water vapour, *i.e.* water in gaseous form [1]. It plays an important role in the maintenance of human comfort as well as in many technological applications, agriculture, manufacture of moisture-sensitive products, storage areas, meteorology, automobiles, medical control, and so on, among other activities, some of which are represented in Table 1.

Table 1. Applications of humidity sensors [1–5]

Industry	Application areas	Temperature range (°C)	Relative humidity range (%)
Domestic electric appliances	Air conditioners	5–40	40–70
	Cooking control	5–200	0–100
	Drying of clothing	80	0–40
Medical equipment	Respiratory	20–30	80–100
	Sterilizers	>100	0–100
	Incubators	10–30	25–30
	Pharmaceuticals	20–25	20–40
Industry	Paper manufacture	10–30	50–100
	Textiles	10–30	50–100
	Printing	20–25	90
	Ceramic powders	5–100	0–50
	Electronic parts	5–40	0–50
Alimentary	Dried foodstuffs	50–100	0–50
	Fruit storage	-1–1	75–85
	Chocolate covering	16–17	50–55
Agriculture	Greenhouse	5–40	0–100
	Cereal stocking	15–20	0–45
	Protection of plantations	-10–60	50–100
Automobile	Car-window demisters	-20–80	50–100
	Motor assembly line	17–25	40–55
Others	Soil humidity	5–30	0–90
	Book storage	17–20	38–50
	Photography	20–25	50–70

Water vapour is a part of the earth's atmosphere so that the effect of humidity in the aforementioned areas needs to be monitored and, in many cases, controlled to avoid undesirable processes.

2 Definitions

Humidity can be defined as the concentration of water molecules in the atmosphere [2]. This concentration is usually low and the evaluation of the water vapour content of gaseous atmospheres, i.e. hygrometry (from the Greek "hygros", meaning moist), may become a complex task and "no one solution will meet all the requirements at all times and in all places" [2]. One of the main difficulties associated with the determination of this parameter in the environment lies in the wide range of conditions in which the measurement has to be done. The water vapour content in the atmosphere may change from values in the order of several percent at the earth's surface to parts per million in the stratosphere. At the same time, the ambient pressure drops several orders of magnitude and the temperature varies within a range of 100 K in the atmosphere.

Several parameters have been used to evaluate the humidity of a sample [1, 3, 4].

The absolute humidity (g m^{-3}) or water vapour density, is expressed as the ratio of the mass of water vapour contained in a given volume of air. The specific humidity, q (g kg^{-1}), is defined as the ratio of the mass of water vapour (m_W) to the total mass ($m_W + m_a$) (m_a = mass of air and $W = m_W/m_a$) (Eq. 1):

$$q = \frac{m_W}{m_w + m_a} = \frac{W}{W+1} \tag{1}$$

The water vapour content can also be expressed as parts per million by volume, ppm_v, representing the volume of water vapour per total volume of gas (for an ideal gas). This parameter is also known as the mixing ratio by volume, or the volume ratio.

The parts per million by weight or mixing ratio (ppm_W) can be obtained by multiplying ppm_v by the ratio of the molecular weight of water to that of air (18/29). Sometimes the ppm_W is used to express the amount (mass) of water vapour relative to the total dry gas and, in other cases, it indicates the amount relative to the total moist gas. The mixing ratio and the specific humidity are the measured units when water vapour is an impurity, or a defined component of a gas mixture, in a manufacturing process.

The vapour pressure is another measurement of the water vapour content of air expressed in terms of the partial pressure of the water vapour in air.

The relative humidity, r_h, is defined as the ratio of the partial pressure of water vapour present in the gas, p_w, and the saturation vapour pressure of the gas at a given temperature, p_s. It is expressed as a percentage (Eq. 2).

$$r_h(\%) = \frac{p_W}{p_s} \times 100 \tag{2}$$

Relative humidity is the parameter most commonly measured in hygrometry, mainly in applications related to human comfort (indoor air quality) and outdoor air issues.

The dew point represents the temperature (above 0 °C), T_d (°C), to which the air must be cooled (at constant pressure) to become saturated with respect to liquid water. The frost point is the temperature (below 0 °C), T_F (°C), to which a volume of gas must be cooled to become saturated with respect to ice. The dew point is used as an indicator of the water vapour content at high temperatures and both the dew and the frost points are evaluated when the dryness of a gas is important.

The wet bulb temperature is another parameter used for humidity evaluation. When a stream of unsaturated air passes over the surface a wet thermometer bulb, an equilibrium temperature is attained at which the rate of heat transfer from the air to the wet surface, by convection and conduction, equals the rate of heat lost by the wet surface in the form of latent heat of vaporization.

3 Measurement of Humidity

Commercially available hygrometers can be classified in different ways [6]. Depending on the evaluated parameter (relative humidity, dew/frost point temperatures, or trace moisture determinations), and the transduction mechanism the most widely used humidity sensors can be divided into the following groups [1–4, 7, 8]:

3.1 Relative Humidity Monitoring

Relative humidity measurements can be carried out using psychrometers, mechanical or electric hygrometers. None of these devices can be considered perfect for all applications. On the other hand, since there are not many calibration devices accurate enough for humidity generation and measurement it is sometimes very complicated to confirm the accuracy of a humidity determination [1].

3.1.1 Psychrometers

A psychrometer is a device based on the change in temperature due to the cooling effect of water evaporating into the air. A typical psychrometer consists of two thermometers, one of them with the bulb covered with a cotton wick saturated with water, placed in a reservoir (the "wet-bulb", Fig. 1) and the other in contact with the ambient air (the "dry-bulb"). In modern psychrometers, the wet-bulb is made of a porous ceramic. During operation, a gas stream passes over the wick at an appropriate flow rate and the water evaporation cools the wetted thermometer, the temperature of which will be lower than ambient (measured by the dry-bulb). The humidity value will be obtained by comparing

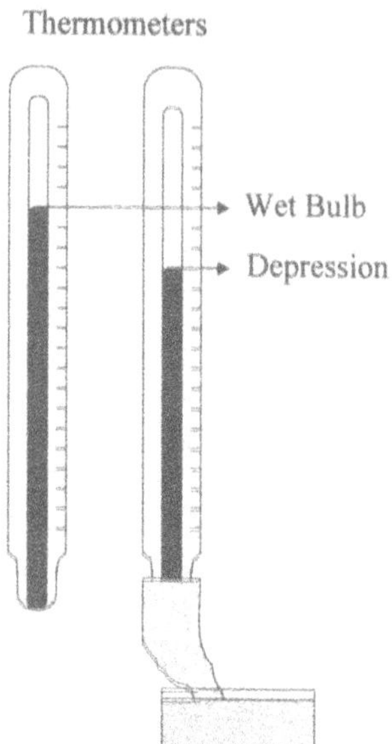

Fig. 1. Psychrometer

the wet and dry bulb temperatures with the aid of a psychometric chart. If all the relevant parameters are known, the temperature depression will be directly related to the heat of evaporation of water, and an absolute determination of the water vapour pressure of the gas will be obtained.

The temperature measurement can be performed with mercury or resistance thermometers, thermocouples, bimetal thermometers and thermistors. For atmospheric humidity measurements a simple device can be used which consists of a wet and dry-bulb thermometers mounted on a sling (sling pyschrometer) that is manually whirled to attain the desired velocity of the air across the bulb. One of the main limitations associated with the use of this device is the difficulty of maintaining the appropriate evaporation rate. In order to solve this problem, in the Assman psychrometer a spring-powered fan provides a constant airflow rate past the bulbs.

3.1.2
Mechanical (Displacement) Hygrometers

This group includes those sensors which are based on a mechanical transduction principle. This is probably the oldest method applied to humidity monitoring (Saussare, XVIIth century). In this case a strain gauge or other mechanism is used to evaluate the expansion or contraction of a material as a function of humidity changes. Materials such as human hair, cellulose or nylon are used for this purpose (Fig. 2).

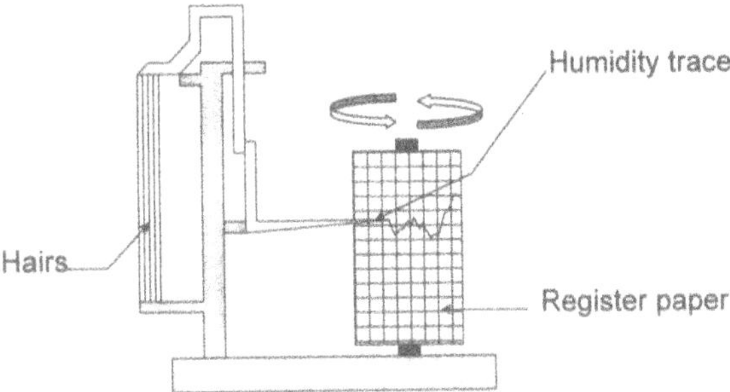

Fig. 2. Hair hygrometer. Humidity is measured by recording the changes in the length of hair (human, horse..). The hair expands as humidity rises and contracts when it decreases

3.1.3
Electric Hygrometers

This group include sensors that measure the change in electrical properties (capacitance, resistance) of a porous layer as a function of the relative humidity. These devices must be calibrated against a primary standard (gravimetric hygrometer) or a fundamental system. The materials used to manufacture these sensors can be divided into three groups: electrolytes, organic polymers and porous ceramics [8,9]. Capacitive sensors are based on the dielectric changes of a thin film when the material absorbs or desorbs moisture from the environment. The resulting change causes a capacitance variation that in turn provides an impedance that will vary with humidity. The nature of the sensitive material and the geometry of the sensor have a considerable impact on the properties of the device. The former is usually very thin to allow maximum sensitivity and short response and recovery times. Resistive humidity sensors are devices that also transduce air humidity into an impedance change that can be measured as a current, a voltage or a resistance. These sensors can be further subdivided into ionic and electronic conduction types.

Electric hygrometers are secondary devices, since they do not measure any fundamental property of the water vapour. They are prone to signal drifts due, for instance, to contamination or temperature effects and require frequent recalibration or even sensor replacement. Nevertheless, they are broadly applied nowadays due to their simplicity and low cost.

3.2
Dew Point Sensors

Dew point detection can be accomplished by visual, optoelectrical, electrical, ratiometric or gravimetric techniques. Dew point hygrometers based on lithium chloride, ceramic or silicone-based materials have been reported. In these cases, a sensitive material showing a sudden change in its properties above 90% relative humidity should be employed [2].

3.2.1
Chilled Mirror (Optical Condensation) Hygrometers

This device provides the most accurate method for dew point measurements and it is used as a primary standard. However, it requires maintenance by skilled personnel and is an expensive instrument. In some cases these instruments are referred to as "optical" hygrometers; nevertheless these devices are not based on a spectral absorption type principle (see below). The instrument consists basically of a mirror-like metallic surface artificially cooled. The cooling of the surface may be accomplished by evaporation of a solvent of low boiling point, such as ether, by vaporisation of a condensed permanent gas (e.g. carbon dioxide or liquid air) or by a stream of water at controlled temperature. The mirror is illuminated with a light source and the water condensation is optically detected by monitoring the reflectivity. When dew forms on the mirror surface the light will be scattered, decreasing the signal at the detector. The temperature of the mirror is carefully controlled to maintain equilibrium between condensation and evaporation of water as the humidity changes (T_{dew}). In the cycling chilled mirror dew point hygrometers the mirror temperature is cycled and kept at the dew point for only a short time until the correct formation of dew is detected; after that, the temperature of the mirror is increased and the dew on the surface is evaporated. In this way the mirror contamination is considerably decreased since its surface is kept dry most of the time. This method is especially useful in cold climates where the humidity levels are low (mountain, polar measurements).

3.2.2
Optical Absorption Hygrometers

These devices are based on the measurement of the electromagnetic radiation absorbed by water vapour in the infrared (bands centred at 2.7 and 6.3 μm) or ultraviolet regions. The latter kind of instrument is known as the Lyman-Alpha hygrometer, indicating the absorption of the specific line in the ultraviolet spectrum at 121.56 nm.

These instruments have been used mainly for certain highly specialised environmental measurements and are very useful for tracking fast humidity fluctuations, for instance, on board aircraft carriers, in chimney smoke or during sand storms, but their broad use is still very limited. They are secondary measuring devices and must be periodically calibrated.

3.3
Measurement of Trace Moisture

The aluminium oxide and the silicon oxide capacitive hygrometers have been used for this type of measurements. Other devices specially suitable for this purpose include:

3.3.1
Mass Sensitive Devices (Gravimetric Method)

This method is particularly useful for the evaluation of water vapour in the low ppm region. A known quantity of gas is passed over a desiccant material, such as calcium chloride or phosphorus pentoxide, and the increase of weight is related to the moisture content of the sample. It is accepted as one of the most accurate techniques for humidity measurement although it is time consuming and not useful for day-to-day measurements. The most widely used gravimetric humidity sensor is the quartz microbalance. Such a device consists of thin plates of piezoelectric quartz, coated with a humidity sensitive layer, the resonant frequency of which changes in response to variations in the ambient humidity. A non-coated reference resonator is used as a reference to minimise temperature and pressure variations.

The mass loading effect is also used in the surface acoustic wave (SAW) sensors. In these devices, the velocity of surface waves is affected by the deposition of a definite mass of water vapour. The SAW technique can also be applied to dew point sensors [6].

3.3.2
Coulometric (Electrolytic) Method

This method is based on Faraday's laws of electrolysis to evaluate the water content of a gas sample. The water vapour in an air stream is passed through a cell where it is electrolysed into hydrogen and oxygen. The current consumed in the process is directly related to the amount of water molecules present in the sample. Provided that all the water in the sample is electrolysed in the cell, an absolute measurement of the water content can be obtained with this method. This sensor is used in very dry environments up to a maximum of 100 ppm_v.

3.4
Miscellaneous Humidity Sensors

There are other humidity sensors and instruments that are less well known but can also be found on the market, and are very useful for particular applications. For instance, devices based on the effect of thermal conductivity change in the presence of water vapour are known [10]. Microwave sensors are based on the variation of the electromagnetic properties of some sensitive materials at ultra-high frequencies [11]. The evaluation of the colour change and the variation of the luminescence characteristics of humidity-sensitive dyes have also been used for the development of humidity sensors. Most of these devices are based on the use of waveguides and will be described in more detail in the following sections along with other fibre-optic humidity sensors (not indicator mediated). Some of the advantages and disadvantages of the sensors for humidity monitoring discussed in preceding sections are presented in Table 2.

Table 2. Characteristics of some humidity sensors

Material	Principle	Advantages	Disadvantages	Typical applications	Ref.
Hair, nylon or cellulose	Expansion or contraction of the material as a function of the relative humidity	▪ Inexpensive ▪ Simple	▪ Dust, oil, pollutants, rain and fog contamination decrease the sensitivity and accuracy of the devices ▪ Hysteresis ▪ Increase in the response time with increasing T and r_h ▪ Drifts over time ▪ Different expansion coefficient for each hair in a bundle	Meteorological stations	[1]
Electrolyte (saturated LiCl)	Measurement of the equilibrium temperature for every water vapour pressure in contact with the saturated salt solution, at which the solution takes or does not give off any water	▪ Simple ▪ Low cost ▪ Low sensitivity to contaminants	▪ Slow response time ▪ Limited measuring range ▪ Moderate accuracy ▪ Needs frequent maintenance	Refrigeration controls, dryers, dehumidifiers, air line monitoring, pill coaters	[1, 6, 8]
Ceramics	**Ionic-type** Humidity-dependent variations in the conductivity of ionic-type oxides upon water adsorption **Electric-type** Conductivity changes upon chemisorption of water on semiconductor oxides **Proton-type** Conductivity changes upon adsorption of water on proton conducting ceramic sur-	▪ Resistant to chemical attack ▪ Low frost point measurement ▪ Safe in explosive installations ▪ Thermal and physical stability ▪ Small in size, in situ use	▪ Need for periodic regeneration due to ageing effects ▪ Hysteresis ▪ Contamination	Humidity measurement and control in a wide range of industries. Petrochemical and power industries. Dryer control. Natural gas in pipelines	[1, 4, 5, 8]

Table 2. continued

Material	Principle	Advantages	Disadvantages	Typical applications	Ref.
Organic polymer films	**Resistance-type** Conductivity change of a polymer upon water adsorption **Capacitance-type** Capacitance change of a polymer upon water adsorption **Swelling-type** Conductivity change depending on carbon or metal particle distance after polymer swelling	▪ Low cost ▪ Miniaturisation	▪ Do not operate at high temperatures and humidity ▪ Hysteresis ▪ Slow response time, long-term drift ▪ Degradation upon exposure to some solvents and electrical shocks	Humidity measurement and control in different industries. Meteorological stations. Water activity measurements. Indoor air quality monitoring.	[1, 4, 6, 8]
Reflective surface	**Chilled mirror hygrometer** Dew point detection by cooling a reflective condensation surface (mirror) until water begins to condense	▪ Broad measuring range ▪ High precision and accuracy ▪ No hysteresis ▪ Fundamental measurement	▪ High cost ▪ Contamination (Kelvin and Raoult effect) ▪ Temperature gradients (from water to mirror and from mirror to thermometer). Errors associated with the particular type of temperature sensor	Humidity calibration standard. Industrial applications where high accuracy and traceability are required: liquid cooled electronics, medical air lines, clean room controls, dryers.	[1, 6]

Table 2. continued

Material	Principle	Advantages	Disadvantages	Typical applications	Ref.
Hygroscopic quartz	**Piezoelectric sensor** The sorption of water on a hygroscopic-sensitive quartz crystal increases the mass and affects the resonance frequency of the crystal	▪ Wide operation range ▪ Low sensitivity to contamination ▪ Measurements of very low frost points ▪ Fast response ▪ High accuracy	▪ High cost ▪ Careful temperature control ▪ Not for in situ use	Moisture monitoring in gases containing corrosive materials. High purity gas production. Semiconductor manufacturing.	[1, 4]
–	**Infrared hygrometer** Water vapour absorption in bands centred at 2.7 and 6.3 μm	▪ Non-contact measurements ▪ Unaffected by contaminants ▪ Fast response time ▪ No hysteresis ▪ Broad measuring range	▪ Complex design and expensive instrumentation (high cost) ▪ Deterioration of the optical windows over time	Measurements in areas with heavy contaminants (smoke stacks, oil vapours and dust)	[1]
	UV hygrometer (Lyman-Alpha Hygrometer) Water vapour absorption in a narrow band centred at 121.56 nm	▪ Fast response ▪ Accurate measurements at high altitude ▪ Cloud measurements	▪ Sensitivity to oxygen at low humidity ▪ Require magnesium fluoride windows which degrade with use ▪ High cost	Highly-specialised environmental measurements	[1]

4 Fibre-optic Humidity Sensors

Humidity sensing devices should fulfil several requirements to be applied in a range of applications [4,6]: a) sufficient sensitivity over a wide range of humidity and temperature values; b) short response time with small or null hysteresis; c) long operational life; d) low temperature effect; e) resistance to contaminants; f) low maintenance requirements and cost. In some cases, low weight, low power and compatibility with a microprocessor may also be required.

Nevertheless, introduction of the optical fibre in the development of humidity sensors offers several additional advantages: possibility of remote sensing and continuous monitoring in confined or hazardous environments, insensitivity to electromagnetic interferences, small size and the feasibility of multiplexing the information from different sensors in one optical fibre.

The main optical humidity sensors (optodes) described in the literature are based on the variation of one of the following phenomena with humidity:

a) The absorption characteristics of a colourimetric reagent.
b) The luminescence intensity and/or emission lifetime of certain dyes.
c) The refractive index, due to the physi- or chemisorption of water directly on the surface of the optical fibre or in a thin film attached to it.
d) the reflectivity of thin films.

4.1 Fibre Optic Sensors Based on Absorption Measurements

Winkler [12] was the first to observe in 1864 that the blue colour of anhydrous cobalt chloride changed progressively in the presence of water vapour, and associated this change with a variation of the hydration state of the salt. The initial blue colour turns pink as $CoCl_2$ is converted into the hexahydrate and the absorption band of the salt shows a dramatic hypsochromic shift (Fig. 3) [13].

Based on this property, cobalt chloride is widely used as a qualitative indicator of the presence or absence of moisture. In fact this compound is still used nowadays to evaluate the residual capacity of solid desiccants. This effect has also been observed with hydrates of cobalt, copper and vanadium salts [14]. Such colourimetric reagents can be immobilised in gelatines or cellulose, or directly deposited on a porous optical fibre for sensor development [15–19, 24] and have the advantage of higher sensitivity and quantitative precision compared to visual methods [15].

Russell and Fletcher [16] reported the immobilisation of cobalt chloride on the surface of a 600-μm, 12 cm long, uncladded silica fibre. The sensitive layer was prepared by dipping the bare fibre into an aqueous solution of gelatine/$CoCl_2$. After drying, the absorbance at 680 nm of the sensitive films (100 nm thick) was measured by internal reflection spectroscopy as a function of relative humidity in the 40–80% r_h range. Ballantine and Wohltjen [17] have developed an optical fibre sensor using the same colourimetric reagent/polymer combination reported by Russell and Fletcher but as a coating on the surface of a glass

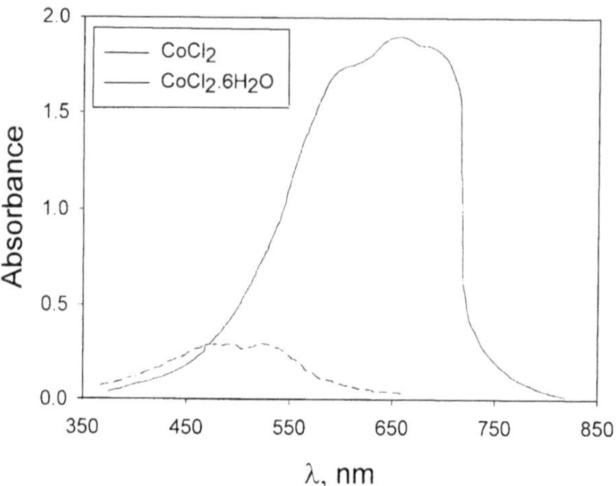

Fig. 3. Absorption spectrum of $CoCl_2$ deposited on a cellulose filter paper as a function of the hydration state of the salt: (—), anhydrous form, (...) hexahydrate

capillary tube. The aim of their work was to develop a portable, inexpensive system for field measurements. The performance of the sensor was similar to that obtained previously [16]. The response follows a S-shaped curve and excellent sensitivity was obtained in the 60–95% r_h range. The main difficulty was related to the preparation of uniform, reproducible films on individual waveguides that would be suitable for a single use device.

In an effort to increase the sensitivity and decrease the size of the previously described sensors, Zhou et al. [18] immobilised $CoCl_2$ directly onto high transparency alkali borosilicate glass fibres (150 to 300 μm in diameter). The large surface area and the direct absorption of light passing through the porous glass sensor segment afforded a device able to measure r_h values down to 0.5% at 25 °C. The measuring range could be extended by changing the concentration of the $CoCl_2 \cdot 6H_2O$ solution used to treat the fibre, although the dynamic interval of the optodes was limited and high r_h values (50% according to the published data) could not be measured.

Boltinghouse and Abel [19] immobilised cobalt chloride directly on cellulose and on acetylated cellulose paper for sensor construction. According to their results, acetylation of cellulose affords a support with reduced interaction with water (compared to the pure cellulose matrix) that allows an important decrease of the light scattering, improves reproducibility of the measurements, shortens the response time and lowers the sensor hysteresis. Nevertheless, the optode response is affected by temperature and the response range is limited, depending on the $CoCl_2$ concentration in the cellulose matrix, so that the use of the sensor was restricted to applications requiring the control of constant r_h levels.

Kharaz and Jones [24] have developed a multi-point distributed humidity-sensing system obtained by depositing a thin film of $CoCl_2$, immobilised in gelatine, on the core of a multimode optical fibre. For relative humidity measurement, the absorbance of the sensor has been evaluated at 670 nm using the absorbance at 850 nm as a reference. Employing this design, a four sensor network was constructed with 200-μm optical fibre and a distance between sensors of 20 m. An optical-time-domain reflectometer (OTDR) with two laser diodes was used to measure the optical attenuation of the different sensors in the 20–80% r_h range, at temperatures between 25 and 50 °C. The resolution of the measurements in the network was limited by the OTDR noise, but might be improved by increasing the averaging time in the instrument.

Other metal salts have also been used for the development of relative humidity optodes. Several authors have reported that in the presence of moisture, Co_3O_4 films [25] obtained by pyrolysis of cobalt 2-ethylhexanoate, and NiO films [30] prepared by plasma oxidation of co-evaporated Ni-C composite films on a glass plate substrate, show humidity-sensitive absorbance changes at room temperature. A drawback of the Co_3O_4 films for sensor development was the fact that their sensitivity increases with decreasing wavelength, being maximum in the UV region. NiO films display the largest absorbance changes in the VIS/NIR region with better sensitivity than the Co_3O_4 films at the measuring wavelength. The response times were in the order of minutes in both cases demonstrating the potential of these sensitive layers for optical humidity detection.

As an alternative to metal dyes Wang et al. [20] describe the use of trifluoroacetophenones immobilised in plasticised poly(vinyl chloride) (PVC) for relative humidity monitoring. Upon hydration, the absorption spectra of these molecules in the UV region changes drastically due to formation of a ketal. It is possible to measure in the range 1–100% r_h using two isologous ligands, with high stability, good reproducibility and short response times. Nevertheless, ethanol behaves as an important interferent of the optode, although other acid and basic gases such as acetic acid (2000 ppm), ammonia (100 ppm), SO_2 (10 ppm) and NO_2 (10 ppm) gave no response. One important limitation to the broad application of these indicator dyes for humidity sensing is associated with their low absorption wavelength maxima (253–261 nm). In order to overcome this drawback, Mohr and Spichiger-Keller [27] have prepared a new indicator, *N*,*N*-dioctylaminophenyl-4'-trifluoroacetyl-azobenzene (ETH^T 4001), with a similar sensitivity and selectivity to the previous trifluoroacetophenones but with the absorption maximum shifted by more than 200 nm into the visible region. For optode development the reagent has been immobilised in polyurethane to afford a broad response range with highest sensitivity in the 5–40% r_h region. However, the response time of the membranes is high (in the order of hours without a preconditioning step in water), preventing their use for continuous monitoring applications.

Sadaoka et al. [22] have observed that when a poly(methyl methacrylate) (PMMA) or poly(ethylene oxide) (PEO) films containing Reichardt's dye is exposed to moist environments, its absorption spectrum in the 400 to 1000-nm range changes considerably and the intensity of light reflected at 750 nm can be used to monitor r_h. The same authors [21] have also prepared composite films of hydrolysed Nafion with various triphenylmethane or cyanine dyes containing terminal *N*-phenyl groups working in the reflecting mode. Dyes included in the study were crystal violet, ethyl violet, malachite green, aizen cathilon pink FGH, aizen cathilon red 6BH, aizen cathilon brilliant red 4GH and aizen athilon yellow 3GLH. The long term stability of the membranes was limited depending on the nature of the dye; crystal violet and malachite green are the best ones. The use of thicker Nafion membranes (0.2 mm) allowed improvement of sensor stability although some hysteresis was observed.

It is clear that the polymer used to immobilise the moisture-sensitive dye is a key component of the optode since it will play an important role in the sensitivity and selectivity of the final device. The choice of polymer is usually governed by its structural characteristics for reagent immobilisation, its stability, its permeability to water vapour and its compatibility for interfacing with fibre optics. Therefore, some authors have proposed preparation of sensing layers without the moisture-indicator dyes, in order to avoid the photochemical degradation that they may suffer over long illumination periods. In this way, Otsuki and Adachi [23] have evaluated the reversible opacity changes, at 250 nm, of hydroxypropyl cellulose (HPC) films, cast from ethanol or acetone onto a quartz plate, as a function of the relative humidity. The response has been attributed to the presence of small liquid crystalline phases in the films that cause light scattering weakened by sorbed water.

A similar optical humidity sensor was developed by Fong and Hieftje [35] taking advantage of the interaction of near-infrared radiation with a silica gel adsorbent layer. Adsorbed water was evaluated both by ordinary absorption of radiation by water molecules and the increased scattering of NIR radiation by the silica gel. The detection limit of water vapor in air at room temperature was found to be 240 ppm. A simplified humidity sensor operating in the Herschel near IR wavelength region (700–1100 nm) has been constructed. A slight hysteresis was observed with this device and the authors propose the application of multivariate, multiwavelength calibration models to overcome the effect of interfering species that can also adsorb on the silica gel layer.

In a later paper Otsuki et al. [28] deposited HPC films labelled with rhodamine B on an uncladded optical fibre. The fibre was curved at an angle of 180°, with all the curved region coated with the sensing layer (4 mm long) instead of the polymer cladding. Another optical fibre was used as a reference to evaluate the transmittance of the sensor head at 552 nm. The sensor response was in the order of 1.0 to 2.3 min for 0–95%, and vice versa, with r_h changes showing a little hysteresis. Following the findings of Muto et al. [36] on the use of phenol-red doped plastic fibres for moisture monitoring, Gupta and Ratnanjali [32] have described the development of a fibre optic sensor prepared by depositing PMMA films containing phenol red over the uncladded core of a multimode silica fibre. A U-shaped design has also been applied to increase the transmission of light from the core to the film and the sensitivity of the optode. The optical measurements were carried out with a He-Ne laser operating at 544 nm, with a response time in the order of seconds and a dynamic range limited to 20–80% r_h.

Narayanaswamy et al. [26, 29, 37, 38] have prepared films of crystal violet immobilised on Nafion for relative humidity monitoring. The composition of the sensitive layers has been optimised [29]; the reflectance at 630 nm of the films prepared with a Nafion/crystal violet molar ratio of 10:1 and treated with methanol gave good sensitivity. However, the coloured layers showed degradation when exposed to ambient air after a couple of days, although they could be stored in a dry desiccator for one month without changing their sensitivity. The response was linear from 30 to 70% r_h with short response times (30 s to > 2 min for direct and reverse changes, respectively) and little hysteresis. Cross reactivity with HCl, NO_2 are described in atmospheres with less than 50% r_h. The effect of NH_3 depends on the water content of the film and could be avoided by short measuring times (< 30 s). In order to extend the linear response range of the optode or to use it for the simultaneous determination of ammonia and relative humidity in air, the authors have proposed the application of artificial neural networks (ANN) [26,38]. The relative standard errors of prediction were lower when the ANN was fed with kinetic data instead of spectral data, after 2 min exposure, allowing a better precision and accuracy in the determination of NH_3. This behaviour may be attributed to the fact that the ammonia reaction takes place at a lower rate in the presence of humidity. In fact, response to this gas is irreversible and new films had to be used for each measurement to complete the study.

Alternatively, Somani et al. [39] described the use of crystal violet incorporated in solid polymer electrolytes (SPE) such as poly(vinyl alcohol) (PVA) doped with H_3PO_4, for humidity monitoring. The films change their colour from yellow to violet as a result of the formation of association-dissociation complexes between the water and the acid molecules, with subsequent protonation-deprotonation of crystal violet. The sensing layers display a linear response between 20–80% r_h, an advantage over the Nafion-crystal violet sensing layers. However, since PVA is water soluble, a sensor fabricated with these membranes cannot be used for a long time under high humidity conditions or in the presence of dew. Other dyes such as methylene blue, rhodamine B and rhodamine 6G have also been incorporated in SPE showing a similar humido-chromic effect that can also be exploited for humidity sensing [34].

Poly(4,4'-diphenylamine-p-heptyloxy)benzylidene (PDPAHB) and poly(4,4'-diphenylimine-p-hetyloxy)benzylidene (PDPIHB) have been applied by Kondratowicz et al. [33] to humidity monitoring by evaluating the absorbance changes of thin films using a flow cell coupled to an optical fibre bundle. In this case, the authors propose that the analytical signal is due to polymer swelling in the presence of water, as no changes are observed in their absorption spectra. The response of the membranes can be reversed using dry air, provided that no ammonia is present in the sample since this molecule accumulates irreversibly on the polymer surface, modifying the absorption spectrum.

Erythrosine B incorporated in a thin sol-gel film (100 nm) has been used to coat a single-mode planar integrated optical waveguide (IOW) for water vapour measurements in the low ppm range [31]. The sensor response was evaluated from the outcoupled light at 514.5 nm, obtaining a linear calibration plot in the 1–6% r_h range, with a limit of detection equal to 115 ppm (3×10^{-3} au) of water vapour and response times in the order of 16–52 s. Lewis bases stronger than water, such as ammonia, can be potential interferences for this sensor. In order to allow the use of this device for practical measurements, the authors have proposed the use of grating couplers instead of prism coupling. This will avoid the use of laser sources and increase the reproducibility of the measurements.

Table 3. Optical fibre sensors based on uv–visible measurements described for relative humidity measurements

Analytical signal	Sensitive layer Dye / Support	Dynamic range (%)	Reported interferents	T (°C)	Response time (min)	Sensor life	Hysteresis	Ref.
Vis absorption	$CoCl_2$ / gelatine	40–80		25	0.5–1	>4 months		[16]
Vis absorption	$CoCl_2$ / poly(vinylpyrrolidone)	60–95		22	0.8–0.2			[17]
Vis absorption	$CoCl_2$ / in a porous optical fibre	0.5–10 10–30 25–50		25	2–3			[18]
Vis absorption	$CoCl_2$ / Cellulose	4–60		≈25	5–20		Yes	[19]
UV absorption	Trifluoroacetophenones / PVC	1–53 5–100	Molecules able to form hemiacetals and acetals	20.5	0.08	10 days a 10% drift		[20]
Vis absorption	CV, EV, MG, ACP, ACR, ACBR, ACY[1] / Nafion	0–83	NH_3, HCl and HNO_3	30	0.2–1	Low	Yes	[21]
Vis absorption	Reichardt's dye / poly(methyl methacrylate) or poly(ethylene oxide)	0–71	Ammonia	30	2			[22]
UV absorption	Hydroxypropylcellulose film	0–91		30.0 ± 0.2	5–17			[23]
Vis absorption	$CoCl_2$ / gelatine	20–80		25 – 30				[24]
Vis absorption	Co_3O_4 / glass plate	10–90		25	<5			[25]
Vis absorption	Crystal violet / Nafion	40–55 40–82[2]						[26]
Vis absorption	*N,N*-Dioctylaminophenyl-4'-trifluoroacetyl-azobenzene/ PVC	0.5–100		24 ± 1	15–50	>4 months		[27]
Vis absorption	Rhodamine B / Hydroxypropylcellulose	0–95		25.0 ± 0.2	1–2.3		No	[28]

Table 3. continued

Analytical signal	Sensitive layer Dye / Support	Dynamic range (%)	Reported interferents	T (°C)	Response time (min)	Sensor life	Hysteresis	Ref.
Vis reflectance	Crystal violet / Nafion	30–70	NH_3, HCl and NO_2	22 ± 0.1	>2.3	1 month	Very little	[29]
Vis absorption	Plasma-oxidized NiO films	0–90		25	3		No	[30]
Vis absorption	Erythrosine / sol-gel	1–70	Lewis bases	Ambient	0.3–1			[31]
Vis absorption	Phenol red / Poly(methylmethacrylate)	20–80		–	0.1			[32]
Vis absorption	Poly(4,4'-diphenylamine-p-heptyloxy)benzylidene or poly(4,4'-diphenylimine-p-heptyloxy)benzylidene	50–100[2]	NH_3	23	1-5	2 days		[33]
Vis absorption	Crystal violet or methylene blue / PVC-H_3PO_4	20–80		Several	10		No	[34]

[1] CV: Crystal Violet; EV: Ethyl violet; MG: Malachite green; ACP: Aizen cathilon pink FGH; ACR: Aizen cathilon red 6BH; ACBR: Aizen cathilon brilliant red 4GH; ACY: Aizen cathilon yellow 3GLH.
[2] Using Artificial Neural Networks.

A fibre-optic relative humidity sensor to quantify capillary pore pressure in unsaturated soils has been developed at US Sandia National Laboratory [40]. The sensing layer consists of a porous polymer material that is placed between two standard optical fibres. One of them is used to transmit the reference and the analytical wavelengths (565 and 660 nm respectively); the second fibre drives the light that passes through the sensitive tip back to the optoelectronic interface. The optode is calibrated to measure r_h at the very high values (between 98 and 100%) required to calculate pore pressure for arid soils. This sensor has been incorporated in a cone penetrometer to allow for real-time measurements that help field personnel to predict the movement of water and solutes during field analysis. Table 3 summarises the analytical characteristics of the fiber optic sensors based on absorption measurements described previously.

4.2 Fibre Optic Sensors Based on Luminescent Reagents

Various optical humidity sensors have been developed which rely on luminescence measurements (Table 4). Most of them are based on absolute emission intensity or intensity-based ratios since they usually provide a better sensitivity than absorption devices [41, 42]. However, lifetime-based sensors are becoming increasingly popular for in situ measurements. Various commercial fluorescent molecules such as umbelliferone [43], rhodamine 6G [44, 52], perylene dibutyrate and various *N*-substituted derivatives of perylenetetracarboxylic acid bis-imides (PTCABs) [46] (Fig. 4) immobilised in various matrices, have been employed for relative humidity monitoring. The developed optodes showed in some cases long recovery times (> 30 min, [46]) and other gases, such as molecular oxygen or gaseous ammonia, may cause serious interference. The emission characteristics of such fluorescent dyes are quite sensitive to their environment. In fact, these fluorophores show a strong fluorescence in organic media but emit very weakly in water. This is also the case with the 5-dimethylaminonaphthalene-1-sulphonic acid (DNSA) that has been immobilised by Otsuki and Adachi [47] in a HPC film for humidity measurements. The sensing membranes display a red-wavelength shift of the fluorescence maximum from 435 to 465 nm, as well as a decrease in the emission intensity and a increase in the bandwidth, on going from 0 to 80% r_h. The response time of these layers was relatively long (> 12 min for step changes) and, whereas the λ_{max} values are not affected by repeated humidification and desiccation cycles, the emission intensity decreases slightly, indicating a progressive association of the fluorophore in the sensing films.

The fluorescence of some dyes whose excited state has a strong intramolecular charge-transfer character is strongly perturbed by association with water molecules if the donor group has an unshared electron pair (typically hydroxy, amino or substituted relatives). This is the case with 5-(4'-aminophenyl)-2-(2'-pyrazinyl)-1,3-oxazole (appzox) [45] that has been immobilised in HPC [55] for the development of a relative humidity optode. The fluorescent sensor is able to measure in the 1.68–100% r_h range with response times in the order of 1–2 min and is insensitive to oxygen and other typical organic vapours that are interferents of commercial capacitive humidity sensors. The device has

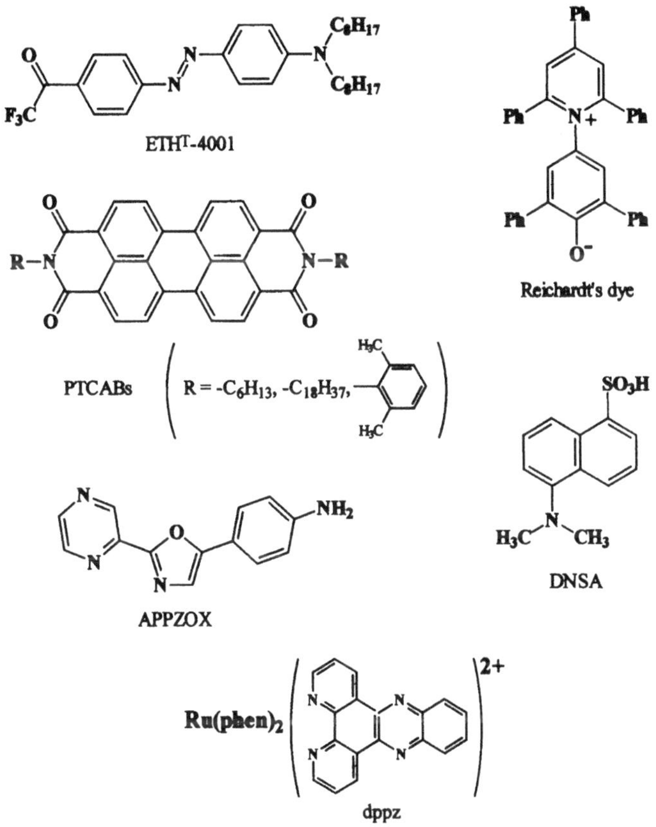

Fig. 4. Chemical structure of some humidity sensitive dyes

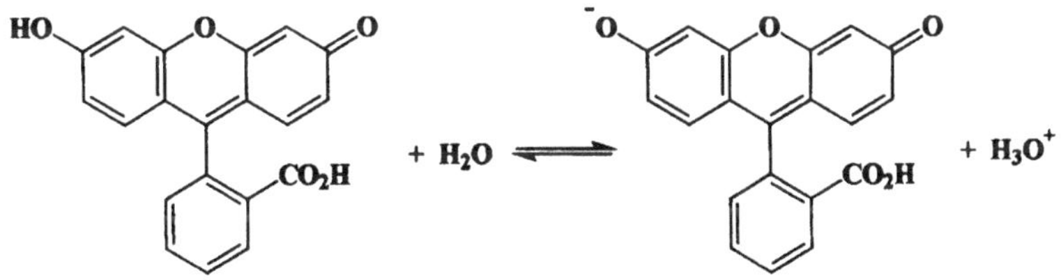

Fig. 5. Effect of water on the ionisation state of fluorescein

been applied to ambient humidity monitoring and the measured data are statistically comparable to those provided by a capacitive sensor.

Muto et al. [49] have applied an optical humidity sensor to breathing-condition monitoring. The device is based on the ionisation reaction that takes place in the presence of water (vapour) and a fluorescein-doped PMMA film (< 10 μm thick), (Fig. 5). The sensitive layer acted as the cladding of a PMMA or polycarbonate core fibre (1 mm diameter and 1–10 cm long). The experimental set

Table 4. Optical fibre sensors for humidity measurements based on luminescent dyes

Analytical signal	Sensitive layer (Dye / Support)	Dynamic range (%)	Reported interferents	*T* (°C)	Response time (min)	Sensor life	Ref.
Fluorescence intensity	Perylenedibutyrate or *N*-substituted derivatives of perylenetetracarboxylic acid bis-imides / Silica gel or filter paper	0–50 20–100	Oxygen and ammonia	Not reported	5–30	>1 month	[46]
Fluorescence intensity	5-Dimethylamino phthalene-1-sulfonic acid / Hydroxypropylcellulose	0–87	Not reported	30.0 ± 0.2	11.8–14.4	Drifts after wet-dry steps	[47]
Phosphorescence lifetime	Pt- and Pd-porphyrins / Langmuir-Blodgett	35–100	Oxygen	20	10 s	6% drift per hour	[48]
Fluorescence intensity	Fluorescein / Poly(methyl methacrylate)	5–85			<0.01 s	>2 months	[49]
Phosphorescence intensity	Tris(8-hydroxy-7-iodo-5-quinolinesulfonic acid)aluminium(III)/ Sol-gel, Dowex 1X2-200 or Bio-Rex AG 1-X8	2.1–100 1.4–100 0.09–30 0.35–80	Oxygen	20 ± 2	4–5		[50]
Fluorescence intensity	Rhodamine B / Nafion or Resin beads	5–80				>6 months	[51]
Fluorescence intensity	Rhodamine 6G / Gelatine	10.2–100	Acetone toluene ethanol chloroform acetic acid and SO_2	20	1	>1½ months	[52]
Fluorescence intensity	Sulforhodamine 101 / Gelatine	35–100	Organic vapours	21	0.3–0.6	>1½ months	[53]
Phosphorescence intensity	2',7'-dibromo-5'-(hydroxymercuri)fluorescein / Sol-gel or Dowex 1X2-200	0.6–40 0.11–50	Oxygen	Room T.	2.6–3.2	>2 months	[54]
Fluorescence intensity	5-(4-aminophenyl)-2-(2-pirazinyl)-1,3-oxazol / HPC	0.54–100	Acetic acid	25			[55]
Luminescence intensity	$[Ru(phen)_2dppz]^{2+}$/Nafion	0–100	Not indicated	Not indicated	>1	5 days	[56]

up includes a fluorescent optical fibre with a waterproof layer as a compensator to eliminate the influence of the fluorescent decay caused by thermal quenching. The maximum sensitivity was obtained for a dye concentration of 10–11 wt% cm^{-1}; under these conditions the fluorescence intensity increased linearly with increasing humidity over the 5 to 85% r_h range.

Sulforhodamine 101 entrapped in gelatine-containing microemulsion organogels composed of isooctane/bis(2-ethylhexyl)sulfosuccinate/water has been optimised by Choi and Shuang [53] for fabrication of optode membranes for humidity monitoring. The strong fluorescence of the dye was effectively quenched by moisture as a result of a membrane swelling along with an increase in flexibility of the medium upon water uptake. The sensing layer shows linear calibration in the 35 to 100% r_h range with response and recovery times in the order of 1 min. The response of the membranes decreases with increasing temperatures, leading to a saturation of the response and a reduction in the working range. Nevertheless, unlike other gelatine-based sensing layers developed by the same authors using rhodamine 6G as indicator [52], no interference was observed in the presence of organic solvent vapours such as acetic acid (0.18% *v*/*v*), acetone (1.04% *v*/*v*), chloroform (0.65% *v*/*v*), ethanol (1.43% *v*/*v*), toluene (0.11% *v*/*v*) or the air pollutant NO_x (0.003% *v*/*v*). The optode has been applied to real sample analyses and the results validated using a commercial hygrometer.

Some optical humidity sensors exploit the phenomenon that water molecules are able to quench efficiently the phosphorescence emission of certain compounds [48, 50, 54]. The use of phosphorescent dyes instead of fluorescent ones offers some advantages for moisture monitoring. The longer excited-state lifetimes provide significant sensitivity and the large Stoke's shifts allow a facile discrimination between the excitation and emission wavelengths. However, oxygen interference can be a serious problem with these devices in some applications. Papkovsky et al. [48] have prepared humidity-sensitive coatings based on Langmuir-Blodgett films containing phosphorescent water-soluble Pt- and Pd porphyrins. The membranes show a bright long-lived phosphorescence in dry form, that could be fitted to a single exponential decay law. The luminescence of the films was quenched by molecular oxygen but it was found that it could be modulated by the environmental water vapour content, so that the effect was applied to the development of fibre sensors for r_h monitoring based on emission lifetime measurements. The films showed some photobleaching; a intensity signal drift of 6% per hour was measured in air.

Costa-Fernández et al. [50, 54] have worked in the optimisation of air moisture sensing materials for the development of room temperature phosphorescence (RTP) sensors. A metal chelate, namely 8-hydroxy-7-iodo-5-quinolinesulfonic acid (ferron) with Al(III), bound to anion-exchange materials (Dowex 1X2-200 or Bio-Rex AG 1-X8) or a sol-gel matrix, and mercurochrome (2',7'-dibromo-5'-[hydroxymercuri]fluorescein) immobilised in Dowex 1X2-200 or in a sol-gel material have been evaluated for relative humidity sensing. The authors point out that the RTP intensities and triplet state lifetimes of both indicators are strongly influenced by the nature of the solid support used for the dye immobilisation. The highest potential for RTP quantification of humidity was obtained with the sol-gel matrices, as they impart a higher rigidity and quencher protection to the indicator dyes. Slightly lower detection limits were achieved using an Al-ferron sensing layer (0.09% r_h) along with a higher precision (3.2% for 8% r_h, $n = 5$). Response times were in the order of 4 min for the sol-gel/Al-ferron combination and 2.6 min for mercurochrome. An increase in the air flow rate decreased the response time but the sensitivity to water was decreased too. In order to enhance both the sample frequency and the linear dynamic range of the sensing phase a flow injection-like gas introduction system has been optimised and applied to the analysis of air samples of various r_h. The results of the mercurochrome based sensor were successfully validated using a commercial hygrometer. Once again, the main limitation of RTP measurements for moisture sensing is oxygen quenching, although the authors have proposed [50] the application of chemometric techniques to monitor both parameters simultaneously. The use of a chelate of morin and aluminium for humidity optosensing based on fluorescence quenching measurements was also proposed by Pedersen et al. [57].

Ruthenium complexes have also been described as indicator dyes for humidity monitoring. Glenn et al. [56] immobilised [(dipyrido[3,2-*a*:2',3'-*c*]-phenazine)bis(1,10-phenanthroline)]ruthenium(II), $[Ru(phen)_2(dppz)]^{2+}$, in a Nafion membrane for optode development based on luminescence lifetime measurements. Dipyridophenazine (dppz) complexes of ruthenium(II) have been described as molecular "light switches" in DNA [58] and in micellar solutions [59] as they display luminescence when the phenazine nitrogen atoms are shielded from water [60]. Ru(II) complexes of dppz display negligible photoluminescence in water due to very fast deactivation of their (emissive) ^{3}MLCT excited state to a short-lived lower lying ^{3}MLCT' state [58]. Details of the photophysics of Ru(II) polypyridyl complexes may be found in Chapter 13 of this book . In non-protogenic organic solvents the latter is destabilised, so that excited state crossing no longer occurs and the typical luminescence of Ru(II) polypyridyls turns on. Unfortunately, the prepared optodes showed hysteresis and degraded after 5 days of use, yet regeneration was possible after storage in a drying environment.

A thorough study of the behaviour of Ru-dppz complexes in acetonitrile-water mixtures (0–10%) has been undertaken [61] using both luminescence intensity and lifetime measurements. Static quenching by water molecules has been evidenced by the strongly non-linear Stern-Volmer plots. Immobilization of the indicator dye onto a poly-fluorinated polymer has allow fabrication of a very robust r_h optode [61,62]. Using a luminescence phase-sensitive optoelectronic system (Grupo Interlab S.A. OPTOSEN® [63]), the humidity sensor has been applied to in situ atmosphere r_h measurements in the 0–100% range with no hysteresis. Validation has been performed against a NIST (National Institute of Standards and Technology) traceable humidity chamber. The optode shows neither O_2 nor alcohol or organic vapour interference. Using the same instrument, the sensor has been applied to humidity determinations in arid soils (validated against a gravimetric method) and to measurements of the water activity (a_w) in foods (validated against a commercial hygrometer). Values of a_w as disparate as 0.2 (biscuits) and 0.96 (plums) have been successfully determined with the fibre-optic sensor.

4.3
Optical Sensors Based on Variations of the Refractive Index

Kunz [64] has reported the use of dielectric waveguides with a spatially varying measurand-dependent effective refractive index distribution to develop humidity sensors based on the variable deflection of a laser beam. Using a porous Ta_2O_5 waveguide that shows a humidity-dependent ordinary refractive index, beam shifts of about 1.4 mm were measured for 0 to 45% r_h. Ronot et al. [65] have tested the performance of different heteropolysiloxane polymers as selective claddings for the detection of chemical vapours using intrinsic optical fibre sensors. The polymer cladding must be non-volatile and allow the diffusion of the analytes to and from the specific interaction sites; its refractive index must be slightly lower than that of the fibre core and must change as a function of the concentration of the selected analyte. Finally, it should also be transparent to the selected wavelength and very homogeneous to avoid light scattering. For water vapour analysis the greatest sensitivities were obtained by coating the bare core multimode optical fibre with amino-propyl siloxane. Nevertheless, the selectivity of the polymeric cladding is not high enough and the combined response of several optical fibres with different claddings would provide better information for complex samples. The effect of humidity on other polymers, such as poly(methyl methacrylate) or fluorinated polyimine, has also been tested for their application in optical sensing [66, 67].

Several attempts have been made to develop fibre optic sensors, based on standard telecommunication grade single-mode fibres, to detect humidity changes at selected points in an optical communications network, as the ingress of water may affect the system performance [68, 69]. In-line sensors are usually an advantageous alternative for this application as they can be installed as a serial array and interrogated by OTDR. In this type of sensor the interaction between the light propagating in the core of the fibre and the analyte can occur in different ways. For instance, as it has been described above, a part of the cladding can be polished away and replaced by an overlay whose refractive index depends on relative humidity. This configuration has been optimised by Bownass et al. [68] for the detection of high humidity in passive optical networks. The authors embedded the fibre in a quartz block (estimated radius of curvature 10–15 cm) and polished the assembly, including the cladding, to produce a half-coupler. A polyethylene oxide (PEO) sensing film was then applied to the polished surface of the half coupler as shown in Fig. 6.

The swelling and hence the refractive index (n) of the PEO film depends on r_h. If the film is dry its n is greater than the effective n of the fibre ($n_{overlay} > n_{fibre}$) and a fraction of the power in the fibre, that will increase as the polishing approaches the core, will be lost in the overlay. With increasing humidity $n_{overlay}$ will decrease, influencing the insertion loss of the sensor that will be negligible when $n_{overlay} < n_{fibre}$. A four-sensor serial-array has been developed and interrogated using OTDR. The devices were optimised to show a loss of 2.5 dB at a 1.31 µm, for a switch between their high-zero loss states at ca. 80% r_h. The need for a precise polishing of the fibre as well as the stability of the sensing film would be limiting factors for this type of sensor.

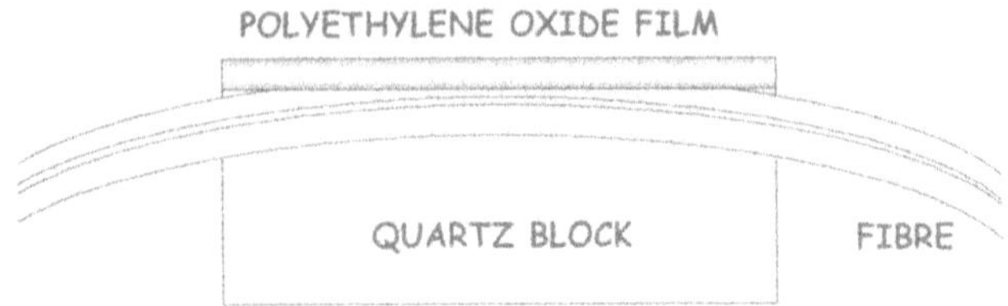

Fig. 6. Schematic diagram of humidity sensitive fibre sensor developed for use in a serial array for network monitoring (adapted from ref. [68])

Alternatively, the fibre-optic sensor can be based on the refractive index changes of a sensing film applied to the cladding of a bent single-mode fibre [69]. The bend generates a whispering gallery (WG) mode to propagate in the cladding and interact with the interface between the fibre and the sensing layer. If the refractive index of the film is lower than that of the cladding, a total internal reflection of the WG occurs at the cladding-overlay interface, causing a strong interference with the core-guided mode that will produce oscillations in the transmitted power as a function of the wavelength. If the refractive index of the film is larger than that of the cladding, the Fresnel reflection of the WG mode will be very weak and the interference with the core guide mode negligible. The selected bend radius and arc length determine the spectral response of the sensor whereas the refractive index of the sensing layer, related to the relative humidity value, is linked to the strength of the WG interactions. In this respect, the sensors fabricated with gelatine layers showed a response at $r_h > 62\%$ whereas for PEO films the point at which the oscillation appeared was about 80% r_h (at 22–24 °C). The main limitations of this type of sensors are associated with the long-term mechanical reliability of the fibre and the ability to fabricate sensors with identical spectral characteristics.

Obviously these sensors, as well as the one described previously, will act as humidity detectors more than as continuous relative humidity monitors, which represent a limitation for some applications. The effect of temperature and the interfering species that could interact with the sensing film on the performance of these sensors should also require further evaluation.

In a different approach, Bariáin et al. [70] deposited an agarose gel surrounding the thinner zone of a biconically tapered single-mode optical fibre. The geometrical parameters corresponding to the sensing area are depicted in Fig. 7.

The sensing mechanism was based on the variation of the optical power transmitted through the taper as a function of the refractive index changes of the agarose gel at different relative humidity values. For each sample, 2 m of a 1.3-μm single-mode standard communication optical fibre were used. After stripping off 10 cm of the fibre cover, the uncovered region was stretched using an splicing unit to obtain the taper depicted in Fig. 7.

Variations of 6.5 dB of the transmitted optical power were obtained for changes between 30–80% r_h with a response time of less than 1 min and low hysteresis. The material was checked for more than half a year and showed no degradation; however, the effect of interfering agents has not been tested.

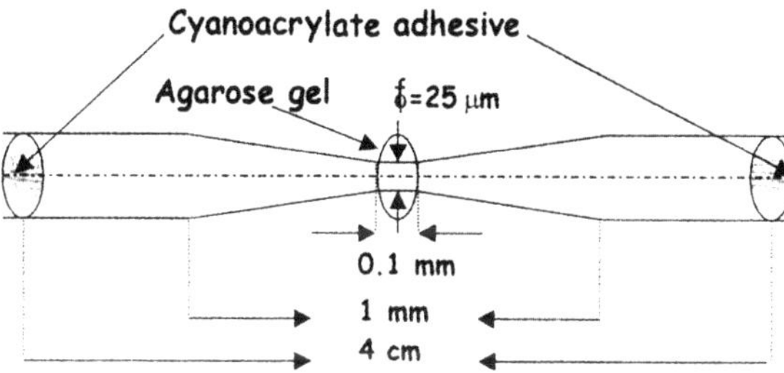

Fig. 7. Geometrical parameters of a biconically tapered single-mode optical fibre used for relative humidity monitoring (adapted from reference [70])

A Mach-Zender interferometer has also been commercialised by Mierij Meteo Instruments (NL) for relative humidity monitoring in the atmosphere [71]. It allows *on-line* measurement using the refractive index change of the water vapour-specific absorbent covering one of the two branches of the interferometer in an integrated optical circuit. The traditional problems of this kind of technology (namely the stability of the polymer adsorbents and chip integration) have been reported as being solved. In order to compensate for various errors, the secondary branch of the interferometer (also called the "reference") is also coated but is not in contact with the ambient air. No temperature correction is required; this fact, together with a fast response (ca. 1 Hz), increases the attractiveness of this sensor.

4.4 Fibre-optic Sensors Based on Changes in the Reflectivity of Thin Films

Mitschke [72] has reported the development of an optode that incorporates a humidity-sensitive porous thin-film interferometer on the tip of an optical fibre (Fig. 8).

The light is guided through a fibre to the moisture-sensitive terminal that consists of a series of alternative layers of SiO_2 and TiO_2 forming a Fabry-Perot filter. The resonance frequency of the spectrum will be given by Eq. 3:

$$\Delta f_{FP} = \frac{c}{2nd} \tag{3}$$

Where c is the speed of light, d the thickness of the spacer and n the refractive index. Upon water adsorption the refractive index of the layers will change and the resonance frequency will be shifted as a function of the ambient relative humidity. A fibre-optic analyser was commercialised using this type of device [1]. The instrument includes the sensor probe and an optical polychromator unit for the detection of the reflection spectrum of the filter as a function of the relative humidity. The sensor sensitivity could be varied from close to 0 to 98% by changing the number and sizes of the micropores in the optical layers. This device has potential applications, for instance, for the measurement of r_h in con-

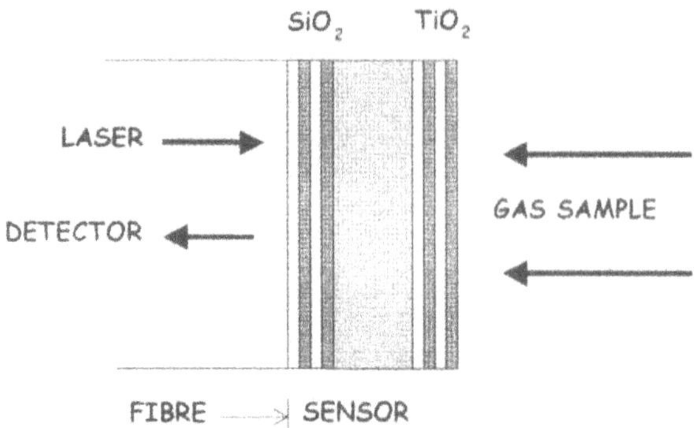

Fig. 8. Fabry-Perot interferometric humidity sensor developed by Mitschke (adapted from reference [72])

taminated areas, natural gas pipes, microwave and induction furnaces or vacuum chambers although it has not been used on a large scale.

Similar principles have also been described by other authors [73–76]. During the past years Lukosz et al. [73–75] have worked on the development of difference interferometers for the analysis of various parameters including relative humidity. In a difference interferometer, the TE_0 and the TM_0 modes in a planar (or rib) waveguide are coherently excited, propagate in a common path and interact with the sample. The sensor response is measured as the time-dependent phase difference $\Delta\Phi(t)$ between the TE_0 and the TM_0 modes induced by the polarization dependent interaction. Different methods have been tested and optimised to measure this phase difference and to achieve multichannel (multianalyte) sensing. For humidity monitoring purposes microporous SiO_2-TiO_2 waveguides ($n_F = 1.75$, $d_F = 198$ nm) were used, as they show large effective refractive index changes in the presence of water, attributed mainly to the sorption of molecules in the micropores of the waveguiding film. However, it has been observed that changes in the temperature of the waveguide, at a constant humidity, also induce a phase difference, complicating the calibration of these sensors for r_h monitoring.

In an attempt to reduce the response time of some of the existing humidity sensors Arregui et al. [77] proposed the use of a nano Fabry-Perot cavity formed by the ionic self-assembly monolayer technique. The method is based on the deposition of successive monolayers of cationic and anionic polymers on the tip of a standard communication optical fibre, previously treated to create a charged surface. In their work, a solution of poly(sodium 4-styrenesulfonate) (PSS^-) was used as the anionic electrolyte and a solution of gold colloids protected by poly(diallyldimethylammonium chloride) ($PDDA^+$) was used to form the cationic electrolyte. The bilayer structure $[Au{:}PDDA^+/PSS^-]_n$, where n indicates the total number of bilayers, was built up at the distal end of the optical fibre and behaves as an homogeneous optical medium, so that local variations of the refractive index occur over lengths much shorter than an optical wavelength. The sensor shows an operation range of 11.3 to 100% r_h, with a maximum variation of the reflected optical power of 4.44 dB (65 bilayers), and response times in the order of seconds that allowed its use for monitoring human breath samples.

5 Calibration of Humidity Sensors

Calibration is a fundamental aspect for application of the sensors described in this chapter. For humidity measurements it is usually accepted [1] that "a standard is a system or device which can either produce a gas stream of known humidity by reference to fundamental base units, such as temperature, mass and pressure, or an instrument that can measure humidity in a gas in a fundamental way, using similar base units". Standards used to calibrate humidity sensors have been classified into three groups [1]:

- *Primary standards.* Rely on fundamental principles and base units of measurement. This is the case with gravimetric hygrometers.
- *Transfer standards.* These instruments are based on fundamental principles but when they are not correctly used can provide incorrect values. The chilled mirror hygrometer, the electrolytic hygrometer and the psychrometer fall into this category.
- *Secondary devices.* These are not fundamental devices and must be calibrated against a transfer standard or other fundamental system. The electric and fibre optical sensors described in this chapter can be included in this group.

Different primary calibration standards can be found in the marketplace (though not inexpensive) that have been extensively used for this purpose (for instance, the dew point or the relative humidity calibration chamber). Another calibration method utilises saturated salt solutions, at a fixed temperature, to generate a known humidity environment [78]. This inexpensive method is widely used for sensor calibration; however, any error in the preparation of these solutions, or in the control of the temperature during the test duration, can generate erroneous calibration data which may be difficult to detect.

When humidity sensors are to be used in field applications there are some variables that may significantly affect the response of these devices and are, in most cases, not considered in their calibration. Among these variables, temperature changes may exert a great influence on the sensor precision and accuracy. In fact, the effect of this parameter has not been evaluated in many of the optical fibre sensors described in this chapter and may be a limiting factor for their broad application for real-time measurements outside the laboratory environment. The sample pressure should also be controlled, as this parameter affects the water vapour pressure of the monitored gas. Flow rate is another essential parameter since pressure gradients or non-equilibrated sampling systems can be generated and affect the measurement accuracy. The electronic instrumentation can also be a problem mainly for unattended long-term measurements where temperature may be difficult to control.

One of the main limitations of many humidity sensors are errors due to contaminants that can be present in the tested samples. As pointed out in the Introduction, the application of these devices is as broad as the possible number of contaminants that may affect their performance. The main interferences can be particulates, ammonia, inorganic salts and organic compounds. Obviously the effect of these species depends on the particular sensing mechanism.

6 Conclusions

It is clear that the broad application range of humidity sensors has diversified the working principles of the instruments and sensors available for the detection of this analyte. The application of fibre sensors for this purpose offers particular advantages over non-optical techniques, such as the possibility of remote monitoring, immunity to electromagnetic interference, safety in explosive environments, small size or the possibility of multiplexing the information from different sensors in the same optical fibre. However, the application of these devices on a large scale is still very limited and restricted to a few applications mainly due to the cost of extensive validation both in the laboratory and in situ, together with the price of the laboratory equipment used to interrogate the reported fibre optic sensors. Considering the benefits of the latter for environmental and industrial humidity monitoring, more effort should be directed to these issues to bring about commercial success.

References

1. Wiederhold PR (1997) Water vapor measurement. Methods and instrumentation, Marcel Dekker, New York
2. Penman HP (1995) Humidity, Chapman and Hall, London
3. Perry RH, Green DW (1997) (eds.) Perry's Chemical Engineers' Handbook, 7th edn. Mc Graw Hill, New York
4. Arai H, Seiyama T (1991) Humidity sensors. In: Göpel W, Hesse J, Zemel JN (eds.) Sensors: A comprehensive survey, vol. 3, part II, VCH, Weinheim, p 981
5. Traversa E (1995) Sens Actuators B 23:135
6. Rittersma ZM (2002) Sens Actuators A 96:196
7. http://www.sensorsmag.com/articles/0997/humidity/main.html
8. Yamazos N, Shimizu Y (1986) Sens Actuators 10:379
9. Sakai Y, Sadaoka Y, Matsuguchi M (1996) Sens Actuators B 35-36:85
10. Kimura M (1996) Sens Actuators A 55:7
11. Bernou C, Rebiere D, Pistre J (2000) Sens Actuators B 68:88
12. Winkler CJ (1964) Prakt Chem 91:209
13. Katzin LI, Ferraro JR (1952) J Am Chem Soc 74:2752
14. King Jr WH. (1965) Humidity and Moisture, vol. 11, Reinhold, New York, p 578
15. Taiichi A (1968) Anal Chem 40:648
16. Russell AP, Fletcher KS (1985) Anal Chim Acta 170:209
17. Ballantine DS, Wohltjen H (1986) Anal Chem 58:2883
18. Zhou Q, Shahriari MR, Kritz D, Sigel Jr. GH (1988) Anal Chem 60:2317
19. Boltinghouse F, Abel K (1989) Anal Chem 61:1863
20. Wang K, Seiler K, Haug JP, Lehmann B, West S, Hartman K, Simon W (1991) Anal Chem 63:970
21. Sadaoka Y, Matsuguchi M, Sakai Y, Murata Y-U (1992) J Mater Sci 27:5095
22. Sadaoka Y, Sakai Y, Murata Y (1992) Talanta 39:1675

23. Otsuki S, Adachi K (1993) Anal Sciences 9:299
24. Kharaz A, Jones BE (1995) Sens Actuators B 46-47:491
25. Ando M, Kobayashi T, Haruta M (1996) Sens Actuators B 32:157
26. Brook TE, Taib MN, Narayanaswamy R (1997) Sens Actuators B 38-39:272
27. Mohr GJ, Spichiger-Keller UE (1998) Mikrochim Acta 130:29
28. Otsuki S, Adachi K, Taguchi T (1998) Sens Actuators B 53:91
29. Raimundo Jr. IM, Narayanaswamy R (1999) Analyst 124:1623
30. Ando M, Sato Y, Tamura S, Kobayashi T (1999) Solid State Ionics 121:307
31. Skrdla PJ, Saavedra SS, Amstrong NR, Mendes SB, Peyghembarian N (1999) Anal Chem 71:1332
32. Gupta BD (2001) Sens Actuators B 80:132
33. Kondratowicz B, Narayanaswamy R, Persaud KC (2001) Sens Actuators B 74:138
34. Somani PR, Viswanath AK, Aiyer RC, Radhakrishnan S (2001) Sens Actuators B 80:141
35. Fong A, Hieftje GM (1995) Anal Chem 67:1139
36. Muto S, Fukasawa A, Osawa T, Morisawa M, Ito H (1990) Jpn J Appl Phys 29:L1023
37. Brook TE, Narayanaswamy R (1998) Sens Actuators B 51:77
38. Raimundo Jr IM, Narayanaswamy R (2001) Sens Actuators B 74:60
39. Somani PR, Viswanath AK, Aiyer RC, Radhakrishnan S (2001) Organic Electronics 2: 83
40. http://www.sandia.gov/Subsurface/factshts/ert/fibreopt.pdf
41. Demas JN, Degraff BA, Coleman PB (1999) Anal Chem 71:793A
42. Wolfbeis OS (1993) (ed.) Fluorescence Spectroscopy New Methods and Applications, Springer, Berlin, p 79
43. Muto S, Fukasawa A, Kamimura M, Shinmura F, Ito H (1989) Jpn J Appl Phys, part 2, 18: 1065
44. Zhu C, Bright FV, Wyatt WA, Hieftje GM (1989) J Electrochem Soc 136:567
45. Orellana G, Gómez-Carneros AM, de Dios C, García-Martínez AA, Moreno-Bondi MC (1995) Anal Chem 67:2231
46. Posch HE, Wolfbeis OS (1988) Sens Actuators 15:77
47. Otsuki S, Adachi K (1993) J Photochem Photobiol A: Chem 71:16
48. Papkovsky DB, Ponomarev GV, Chernov SF, Ovchinnikov AN, Kurochkin IN (1994) Sens Actuators B 22:57
49. Muto S, Sato H, Hosoka T (1994) Jpn J Appl Phys 33:6060
50. Costa-Fernández JM, Díaz-García ME, Sanz-Medel A (1997) Sens Actuators B 38-39: 103
51. Cabrero V, Díaz García ME (1999) Quím Anal 18:117
52. Choi MMF, Tse OL (1999) Anal Chim Acta 378:127
53. Choi MMF, Shuang S (2000) Analyst 125:301
54. Costa-Fernández JM, Sanz-Medel A (2000) Anal Chim Acta 407:61
55. Bedoya M, Moreno-Bondi MC, Orellana G (2001) Helv Chim Acta 84:2628
56. Glenn SJ, Cullum BM, Nair RB, Nivens DA, Murphy CJ, Angel SM (2001) Anal Chim Acta 448:1
57. Pedersen H, Alex T, Ling Chu H, Chung WJ, Sigel Jr. GH (1992) Proc SPIE 1796:135
58. Olson EJC, Hu D, Hörmann A, Jonkman AM, Arkin MR, Stemp EDA, Barton JK, Barton PF (1997) J Am Chem Soc 119:11458
59. Chambron JC, Sauvage JP (1991) Chem Phys Lett 182:603
60. Nair RB, Cullum BM, Murphy CJ (1997) Inorg Chem 36:962
61. Bedoya M (2002) PhD thesis, Complutense University of Madrid
62. Moreno-Bondi MC, Díez Barrio MT, Orellana G, Bedoya M (2002) Proc. Europt(r)ode VII, Manchester, United Kingdom p. 37
63. http://www.interlab.es
64. Kunz RE (1993) Sens Actuators B 11:167
65. Ronot C, Archenault M, Gagnaire H, Goure JP, Jaffrezic-Renault N, Pichery T (1993) Sens Actuators B 11:375

66. Pinceti JC, Naylor DL (1994) Appl Optics 33:1090
67. Wagner FD, Kleckers T, Reuter R, Rohitkumar HV, Blech BA (1993) Appl Optics 32: 2927
68. Bownass DC, Barton JS, Jones JDC (1997) Optics Lett 22:346
69. Bownass DC, Barton JS, Jones JDC (1998) Optics Comm 146:90
70. Bariáin C, Matías IR, Arregui FJ, López del Amo M (2000) Sens Actuators B 69:127
71. http://www. Mierijmeteo.demon.nl/mierij/research/sios.htm
72. Mitschke F (1989) Opt Lett 14:967
73. Lukosz W, Stamm Ch, Moser HR, Ryf R, Dübendorfer J (1997) Sens Actuators B 38-39: 316
74. Lukosz W, Stamm Ch (1991) Sens Actuators A 25:185
75. Stamm Ch, Lukosz W (1996) Sens Actuators B 33:203
76. McMurtry S, Wright JD, Jackson DA (2000) Sens Actuators B 67:52
77. Arregui FJ, Liu Y, Matias IR, Claus RO (1999) Sens Actuators B 59:54
78. Greenspan L (1997) J Res Nat Bureau of Standards 89:81A

CHAPTER 12

Optical Sensing of pH in Low Ionic Strength Waters

BEN R. SWINDLEHURST, RAMAIER NARAYANASWAMY

1 Introduction

The measurement of pH is probably the most commonly performed analytical measurement in a wide range of sciences, including chemistry, biochemistry and environmental studies. This measurement of pH is so important because most chemical reactions are dependent on it in some way. Usually a chemical reaction can only take place within a specific band of pH.

Electrochemical methods have been traditionally used for pH analysis with the glass electrode being by far the most commonly used devise. The glass electrode offers rapid measurement with good precision. However, at low ionic strengths the performance of the glass electrode deteriorates. Errors associated with electrochemical pH measurement at low ionic strengths include streaming potential caused by liquid flowing past the electrode, liquid junction potentials due to differences between the ionic strengths of the calibration solutions and the analyte solutions as well as contamination of poorly buffered solutions by the carryover of material between solutions on the electrode [1].

The spectroscopic measurement of pH involves the use of an indicator species. The spectroscopic properties of such an indicator must be dependent on solution pH. The chemical transduction of pH may rely on a change in the absorption of light, or a change in the luminescent properties of the indicator species. Such acid-base indicators exist in an acid form and a base form over a certain pH range, each form having different spectroscopic properties. The Brönsted-Lowry definition of acids and bases states that an acid or base and its conjugate pair, exists in equilibrium (Eqs. 1 and 2).

$$HA \Leftrightarrow H^+ + A^- \tag{1}$$

where HA is an acid indicator, H^+ is a proton and A^- is the conjugate base of the indicator. For a basic indicator species the equilibrium can be written as:

$$BOH \Leftrightarrow OH^- + B^+ \tag{2}$$

where BOH is a basic indicator, B^+ is this indicator's conjugate acid and OH^- is the hydroxyl ion. Using the example of an acid indicator and applying the law of mass action provides Eq. 3:

$$K_c = \frac{[A^-]\cdot[H^+]}{[HA]} \tag{3}$$

where K_c is the concentration constant of the indicator, and [HA], $[A^-]$ and $[H^+]$ are the concentrations of the indicator, its conjugate base and protons respectively. The concentration constant is related to the thermodynamic constant K_a by the activity coefficients f_x of the individual components (Eq. 4).

$$K_a = K_c \cdot \frac{f_{A^-} \cdot f_{H^+}}{f_{HA}} \tag{4}$$

This can be expressed logarithmically to give the Henderson-Hasselbalch equation:

$$\mathrm{pH} = \mathrm{p}K_a + \log\frac{[A^-]}{[HA]} + \log\frac{f_{A^-}}{f_{HA}} \tag{5}$$

where $\mathrm{p}K_a = -\log K_a$.

As optical pH measurements are dependent on concentration and not activity [8], the Henderson-Hasselbalch equation based on the concentration constant K_c is commonly used.

$$\mathrm{pH} = \mathrm{p}K_c + \log\frac{[A^-]}{[HA]} \tag{6}$$

Ignoring the effect of the activity term in Eq. (5) can only be justified in solutions of very low ionic strength (< 0.1 M).

Given that the concentration constant K_c is related to the ratio of the activity coefficients (Eq. 4), it is dependent on ionic strength and temperature. The activity coefficient of an ion f_x can be related to the ionic strength I by the Debye-Hückel equation:

$$\log f_x = -\frac{A\cdot z_i^2\cdot I^{\frac{1}{2}}}{1 + B\cdot d\cdot I^{\frac{1}{2}}} + C\cdot I \tag{7}$$

where z_i is the charge on species I, A and B are constants which vary with temperature (0.51 and 0.33, respectively, at 25 °C), C is an empirical parameter, and d is the mean ion size parameter in Ångstrom units.

It is because of this relationship to ionic strength that it is important to use buffer solutions of constant ionic strength when calibrating an optical pH sensor. These should also be similar in ionic strength to the measurement solution.

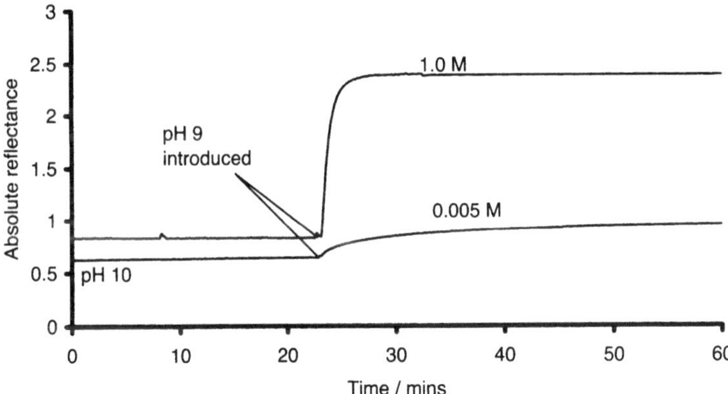

Fig. 1. Effect of ionic strength on response time

The rate constant (k) can be related to ionic strength by the Brönsted-Bjerrum relationship as shown in Eq. 8 [2].

$$\log_{10}\ k = \log_{10}\ k_0 + 2A\ z_1 z_2\ \sqrt{I} \tag{8}$$

where k_0 is the rate constant where the activity coefficients are unity, A is the Debye-Huckel constant, z_1 and z_2 are the charge numbers of the reacting ions and I is the ionic strength. This shows that as the ionic strength decreases so does the rate constant.

The effect of ionic strength on response time is illustrated in Fig. 1. Here the change in response of a physically immobilised pH indicator for a pH transition of 9 to 10 at two ionic strengths is recorded, all other conditions remaining constant. The response time ($t_{95\%}$) of the system at 1.0 M ionic strength was 2.45 minutes, whereas at 0.005 M ionic strength the response was still gradually increasing after 37 minutes.

Diffusion is also an important factor governing the response time of a sensor system. Diffusion can be explained by Fick's first law and is represented in Eq. 9.

$$J = -D\frac{\mathrm{d}N}{\mathrm{d}z} \tag{9}$$

where J is the rate of migration or flux, D is the diffusion coefficient and $\mathrm{d}N/\mathrm{d}z$ is the concentration gradient, with N being the concentration in solution and z being the thickness of the diffusion layer.

2
Optical pH Sensors

Over the past two decades or so, a great deal of work has been carried out in the field of optical pH sensors. Different types of sensing schemes and materials have been employed using a variety of spectroscopic techniques in the transduction mechanism. These are represented in a selection of literature references in Table 1 which lists the methods of immobilisation, optical measurement technique, pH range, detection wavelengths and the reagents utilised for chemical transduction. The techniques of immobilisation of reagents used include physical methods such as adsorption, electrostatic attraction and entrapment, and chemical methods involving covalent attachment on a variety of polymeric matrices, have been widely employed in optical pH sensor development (Table 1). A wide variety of reagent materials has been investigated to cover the entire range of pH, though each reagent system could be applied in a limited dynamic measurement range (ca. 2–3 pH units). A few attempts, however, describe the methods involved in broadening the dynamic pH range [7, 69, 98, 120–123]. Many of the pH sensors studied exhibit a dependence on the ionic strength of the medium, though a few report low or no ionic strength effects [100, 120, 122, 124]. It should be noted that some of the literature reports, particularly recent ones, have described applications of optical pH sensors in industrial and environmental monitoring.

In the following sections, we report an attempt to develop an optical pH sensor for monitoring work in low ionic strength waters. In this study the Mannich reaction is used for the covalent immobilisation of thymol blue onto an aminated controlled-pore glass substrate. Covalent immobilisation in this way reduces problems associated with leaching of the indicator, thus significantly increasing the lifetime of the sensor.

As ionic strength affects the response time of sensing systems and, in general, the rate constant of a reaction decreases with decreasing ionic strength, this sensing system has been designed to remove all barriers to the diffusion of the analyte to the immobilised indicator, in order to reduce reaction times. The removal of diffusion barriers is accomplished by first covalently immobilising the indicator onto a substrate, and then physically immobilising this substrate onto the surface of a film, so that analyte solution can flow directly over the immobilised indicator. This is interrogated with light from the distal end of an optical fibre bundle.

Table 1. Optical pH sensors

pH range	Chemical reagent	λ /nm	Immobilisation method	Measurement principle	Ref.
6–8	PR	557	Entrapment in agarose gel	Absorbance	3
7.04–7.46	PR	560	Indicator solution held in place by a H^+ permeable membrane	Absorbance	4
5.3–7.5	BPB	610	BPB trapped within a cellulose acetate membrane	Absorbance	5
$pK_a = 3.8$	4-Dimethyl-amino-4'-octylazo-benzene	577	Entrapment in cellulose acetate membrane	Absorbance	6
Not given	2-Nitro-4-octyloxy-phenol, 3',3''-dibromo-5-octyl-oxythy-mol phthalein, octylbis(2,4-dini-tro phenyl)acetate	Not given	Entrapment in cellulose acetate membrane	Absorbance	21
4–8	Not given	Not given	Entrapment in a porous polymer fibre	Absorbance	22
pK_as 5.30, 5.65 and 6.35	Styrylcyanine and polymethine dyes	664, 645, 840	Entrapment in a PVC membrane	Absorbance	7
pK_as 5.65 and 6.35	2 unspecified NIR dyes	670 and 826	Entrapment in a PVC membrane	Absorbance	23
2–12	Polyaniline	590, 900, 930	Spin or dip coating onto glass slides	Absorbance	24
6.3–9.8	BCG, BCP, PR	Not given	Entrapment within sol-gel matrix	Absorbance	8
Ranges from pH 0 up to 14	BCG, BPB, BTB, BPR, BCP, CR, CPR, TBPSP	555–625	Indicators electrostatically bonded to a polymer which was covalently bound to a silica fibre	Absorbance	25
1.5–4.5	TTMAPP	440	TTMAPP electrostatically attracted to sulphonated polystyrene	Absorbance	26

Table 1. continued

pH range	Chemical reagent	λ /nm	Immobilisation method	Measurement principle	Ref.
0–5.5	Congo red	511, 577, 620	Covalent binding to a cellulose acetate membrane	Absorbance	18, 27
9.5–13.0 4.0–6.0	Alizarine yellow GG α-naphthyl red	420, 595	Covalent binding to an acetylcellulose membrane	Absorbance	28
3–5.2	Congo red	620	Covalent binding to a cellulose triacetate membrane	Absorbance	29
1–5	Congo red	610	Covalent binding to cellulose acetate	Absorbance	30
0–4, 4–6, 6–10	BCP, BCG, Congo red	580, 615, 600	Not given	Absorbance	31
1.2–3.5, 6–7.5	Congo red, NR, PR	600, 540	Covalent bonding to triacetyl cellulose	Absorbance	32
12–13.5	Thiazole yellow	512	Covalent bonding to a cellulose triacetate membrane	Absorbance	33
Various ranges, between pH 1.1 and 11.8	*N*,*N*-Dimethylaniline, BCP, crystal violet, MR, BCG, phenol phthalein, thymol sulfonephthalein	Not given	Covalent bonding to etched, sintered and controlled pore glasses	Absorbance	19
3–7	BPB	595	Covalent bonding to controlled pore glass	Absorbance	35
1–3.5, 3.5–8	BPB, TB	605, 565	Covalent bonding to controlled pore glass	Absorbance	17
3–8 2–6.8	BTB BCP	562 570	Covalent immobilisation onto fibre tip	Absorbance	36
6.8–7.8	Azo Dye No. 5–91	625	Covalent bonding to a cellophane film	Absorbance	37
6–10	Dye N9	600	Covalent bonding to cellulose acetate	Absorbance	38, 39
10–13	Dye N8	560	Covalent bonding to cellulose acetate	Absorbance	40
Various pK_as ranging from 0.5 to 11.28	A variety of reactive phenolic and amine azo dyes	455–553	Covalent bonding to cellulose membranes	Absorbance	41

Table 1. continued

pH range	Chemical reagent	λ /nm	Immobilisation method	Measurement principle	Ref.
5–10	Nile blue	660	Covalent bonding into a poly(hydroxyethyl methacrylate) matrix	Absorbance	42
1–3	Congo Red	650			
5–9	Prussian blue	720	Covalent bonding in a poly (pyrrolylbenzoic acid) film	Absorbance	43
6–12	Polypyrrole	650	Oxidation of pyrrole onto cuvette	Absorbance	44
Approx. 7–9	PR	565	Indicator in solution	Absorbance	45
3.75–5.5	BCG	616	Indicator in solution	Absorbance	46
3–5, 5–7, 7–9, 9–11, 10.5–12	BPB, BCP, BTB, TB, TP	593	Physical adsorption onto XAD-2 beads	Reflectance	9, 47
Approx. 6–10	PR	560	Physical adsorption onto XAD-2 beads	Reflectance	48
1–4, 10–13, 4–10, 5–11, 4–9, 4.5–7.5, 10–13, 8–10.5, 0–14	TB, BCG, BPB, BTB, PR, NR, MR	525–620	Physical adsorption onto XAD-4 beads	Reflectance	25
7–10	TBPP, xylenol blue, *m*-CP, CR, PR, *o*-cresol-phthalein, BTB, BTR	500–700	Physical adsorption onto XAD-4 beads	Reflectance	11
6–9	BCG	Not given	Physical adsorption onto an XAD-4 bead	Reflectance	50
Various pK_as from 6.17 to 8.86	TBPSP	Not given	Electrostatic attraction onto a variety of ion-exchange resins	Reflectance	12
7–10	PR	Not given	Physical adsorption onto XAD-4 beads	Reflectance	51
pK_a's from 3.69 up to 10.28	BCG, BCP, BTB	580	Either physical adsorption onto XAD-2 beads or electrostatic attraction to IRA-400	Reflectance	13

Table 1. continued

pH range	Chemical reagent	λ /nm	Immobilisation method	Measurement principle	Ref.
A wide range of pK_as from 1.2 up to 12.8	TB, BCG, BPB, BTB, PR, BPR, BCP, CR, *m*-CP	Not given	Physical adsorption onto XAD-4 beads; or Electrostatic attraction onto AG1-X4; AG1-X8; or AG2-X8	Reflectance	14
5.3–9.7	BTB, PR, CPR, Alizarin	550–625	Physical adsorption onto XAD-2 beads	Reflectance	52
6.1–7.1	CR	570	Electrostatic attraction to Dowex 1-X10	Reflectance	15, 53
2.5–5.5, 3.3–4.8, 7.9–11.2	Congo red, BTB, TB	Not given	Entrapment within cellulose acetate membranes	Reflectance	54
7.5–9	BTB	665	Entrapment within a cellulose nitrate membrane	Reflectance	55
6.9–9.5	TB	Not given	Entrapment within a cellulose acetate membrane	Reflectance	56
1–12	TB	450, 550, 640	Electrostatic attraction to an anion-exchange resin followed by entrapment in a cellulose acetate membrane	Reflectance	57
4–9	BTB	540	Electrostatic attraction to an anion-exchange resin followed by entrapment in a copolymer membrane	Reflectance	58
Not given	Polystyrene microspheres	600	Suspension of microspheres within a hydrogel	Swelling dependent reflectance	60
Not given	A variety of pH dependent swelling polymers	830 or 1310	Suspension of microspheres within a hydrogel	Swelling dependent reflectance	61
7–7.4	PR	560	Covalent immobilisation onto polyacrylamide microspheres	Reflectance	16

Table 1. continued

pH range	Chemical reagent	λ /nm	Immobilisation method	Measurement principle	Ref.
7–8.5, 3–5, 8–10	BTB, BPB, PR	Not given	Indicators covalently bound to polyacrylamide microspheres	Reflectance	62
0.5–7	BPB, *m*-CP	520, 590	Covalent immobilisation onto polyacrylamide microspheres	Reflectance	63
0.8–3.2, 3.2–7.0, 10–13	BPB, TB	555, 600, 604	Covalent immobilisation onto polyacrylamide microspheres	Reflectance	64
4–7, 0–14	Commercially available indicator paper	583	Indicator paper	Reflectance	65
3.5–5	Commercially available indicator paper	592	Indicator paper	Reflectance	66
0–14	Commercially available indicator paper	540–646	Indicator paper	Reflectance	67
Several ranges from pH 1 up to 12.5	*N*, *N*-Dimethylaniline, BPB, BCP, PR, *p*-Xylenol blue, TB, *m*-CP	600–690	Covalent bonding to controlled pore glass	Reflectance	20, 68, 69
6.4–8.2	Aminated polystyrene	660	Copolymer attached to the end of a optical fibre	Reflectance	70
Approx. 6–8	Aminated polystyrene	930	Polymerisation reaction onto a chemically treated glass slide	Swelling dependent reflectance	71
4.9–10.5	Poly(*o*-methoxyaniline)	501	Polymeric composite of Poly(*o*-methoxyaniline) doped with *p*-toluene sulphonic acid and cellulose acetate	Reflectance	72
6.2–8	Fluoresceinamine	ex. 430 ex. 485 em. 530	Entrapment in polymer membrane	Fluorescence	73

Table 1. continued

pH range	Chemical reagent	λ /nm	Immobilisation method	Measurement principle	Ref.
3–10	Fluorescein	ex. 488 em. 525 em. 600	Entrapment within a sol-gel matrix	Fluorescence	74
Approx 6.5–7.5	FITC	Not given	Entrapment within a sol-gel	Fluorescence	75,76
$pK_a = 6.3$	FITC	ex. 488 em. 520	Entrapment within a base-catalysed sol-gel	Fluorescence	77
6.5–8.5	HPTS and β-methyl-umbellifer-one	ex. 457 em. 510	Indicators held in place by membrane	Fluorescence	78
5.5–8.0	HPTS	ex. 405 ex. 450 em. 515	Entrapment within an ethylene-vinyl acetate copoly-mer matrix	Fluorescence	79
	HPTS and SR-640	ex. 488 em. 530 em. 610			
9.2–12	Sulphorhodamine	ex. 488 em. 620	Entrapment within a cellulose ether halfester poly-mer	Fluorescence	80
10.0–13.0	BCA	ex. 795 em. 820	Entrapment in Nafion membrane	Fluorescence	81
Appox. 5.0–6.0 and 6.0–9.0	Resorufin, carboxy-SNAFL-2	ex. 442	Entrapment of reagent solution at the fibre tip with dialysis tubing	Fluorescence lifetime measurement	82
5–10.5	carboxy-SNARF-6	ex. 543 em. 660	Entrapment in dextran membrane	Fluorescence lifetime measurement	83

Table 1. continued

pH range	Chemical reagent	λ /nm	Immobilisation method	Measurement principle	Ref.
6.0–8.0	TRH, BTB	ex. 442 em. 600	Sol-gel entrapment	Fluorescence lifeteme. Energy transfer	85
6.8–8.0	SNARF-1C	ex. 535 em. 585 em. 640	Sol-gel entrapment	Fluorescence	86
6–8	SNARF-1	ex. 488 em. 584 em. 630	Entrapment within a poly (hydroxyethyl methacrylate) matrix	Fluorescence	87
7.0–9.0	Ruthenium(II) complex donor and BTB and azo dye acceptors	Not given	Entrapment in a hygrogel	Fluorescence lifetime measurements	88
Not given	Donor/acceptor pairs: Eosin/PR R6G/PR TRH/R6G	ex. em 543/563 580 543/568 600 543/568 600	Entrapment in a hygrogel	Fluorescence energy transfer	89
4.5–8.6	Eosin/PR	546	PR is entrapped and eosin is covalently bound	Fluorescence energy transfer	90
7–8	5 (and 6)-carboxy-naphthofluorescein	ex. 488, 514.5, 598, 632.8 em. 650, 595	Entrapment within a polyacrylamide gel matrix	Fluorescence	91
6–9	HOPSA	ex. 405 em. 470	Electrostatic attraction to anion-exchange resin R-1035	Fluorescence	92
4–8	Fluoresceinamine	ex. 488 em. 530	Covalent bonding to a copolymer which was covalently bound to a silica optical fibre	Fluorescence	93

Table 1. continued

pH range	Chemical reagent	λ /nm	Immobilisation method	Measurement principle	Ref.
5–8	Fluoresceinamine	ex. 488 em. 540	Covalent bonding to a copolymer which was covalently bound to a silica optical fibre	Fluorescence	94
3–6	Fluoresceinamine	ex. 480 em. 520	Either covalent bonding to controlled pore glass or bonding with cellulose acetate	Fluorescence	95
5.5–7.5	Acryloylfluorescein	ex. 430 ex. 490 em. 530	Covalent bonding within a polymer matrix	Fluorescence	96
3–7	FITC	ex. 488 em. 530	Covalent bonding with alkylamine controlled pore glass	Fluorescence	97
0–7	FITC, eosin	ex. 490 em. 540	First FITC is covalently bound on fibrous amino-ethyl cellulose, then eosin produced by bromination of immobilised FITC	Fluorescence	98
0–9	HPTS	ex. 328 em. 520	Covalent bonding to a cellulose membrane	Fluorescence	99
5.0–8.12	HPTS	ex. 465 em. 520	Covalent binding to sintered glass	Fluorescence	100
4.4–9.6	HCC	ex. 410 em. 455			
6.6–7.4	HCC	ex. 410 em. 455	Covalent binding to sintered glass	Fluorescence	101
6–9	5 (and 6)-carboxy-naphthofluorescein	ex. em. a) 620 685 b) 605 682	Either coupling to aminoethyl cellulose (a), or in a sol-gel (b)	Fluorescence	102

Table 1. continued

pH range	Chemical reagent	λ /nm	Immobilisation method	Measurement principle	Ref.
−1–0.6, −0.83–0.14, −1–0.48	PQ, PBQ, PPQ	ex. em. 394 535 394 525 420 560	Covalent bonding to silica optical fibre	Fluorescence	103
6.8–7.8	SNARF-1C isothio-cyanate	ex. 540 em. 650	Covalent bonding with polymer	Fluorescence	104
pK_a 5.45 & 10.7 pK_a 9.5 & 2.8	Acridine 2-naphthol	Not given	Not given	Fluorescence decay time	105
3–9	Methylene blue	632.8	Indicator solution held around fibre	EW attenuation	106
4–9	pH dependent swelling polymer	720	Swelling polymer sandwiched between a metal mirror and a metal island film	Absorption by phase interference	107
Not given	pH dependent swelling polymer hydrogel	Not given	Polymer hydrogel deposited on solid support	Optical time-domain reflectometry	108
3.5–6.5	Fluoresceinamine	ex. 488 em. 530	Entrapment within sol-gel cladding	EW induced fluorescence	109
3–9	BCG and BCP	610	Entrapment within sol-gel cladding	Attenuation of EW	110
HNO_3 conc. from 1 to 10 N	Chromoxane cyanine R	535–550	Sol-gel entrapped reagent replaced optical fibre cladding	Attenuation of EW	111
3–6	BPB	457.9	Entrapment within a sol-gel	Attenuation of EW	112
5.5–8	Polyaniline	1428	Polymer replaced optical fibre cladding	Attenuation of EW	114

Table 1. continued

pH range	Chemical reagent	λ /nm	Immobilisation method	Measurement principle	Ref.
0.5–4	Congo red	1425 cm^{-1}	Covalent immobilisation onto cellulose acetate coated ZnSe internal reflection element	Attenuation of EW	115
4–11	BTB	632.8	Indicator covalently bound to fibre core	Absorption of radiated light	116
2–4.5 2–8 5–9	MR CR 4-Pyridinethiol	647 514.5 647	Coupled with cystamine (MR and CR) and immobilised onto silver films	Surface-enhanced Raman scattering	117
5–7	Poly-(methacrylic acid)	632.8	Polymer replaced optical fibre cladding	Attenuation of EW	118
3–9	Fluorescein	ex. 488 em. 526	Indicator in solution	EW induced fluorescence	119

3 Materials and Methods

3.1 Immobilisation by the Mannich Reaction and Manufacture of Sensing Film

The Mannich reaction was carried out as follows. Thymol blue (30 mg) was dissolved in 2 mL of methanol in a sealable vial. A 1-mL volume of pH 5 buffer solution (pHydrion, Aldrich) was added. CPG particles (200 mg) were then added followed by 3 mL of 1:2 formaldehyde to pH 5 buffer solution (*v*/*v*). The vial was then sealed and submerged in a circulated water bath at 50 °C for 24 h. The particles with immobilised indicator were then rinsed with deionised water and buffer solutions (pH 2–12) until no further leaching of the indicator was observed.

A sensing film was then to be constructed by binding the TB/CPG onto an acetate sheet. Controlling the thickness of the adhesive layer proved difficult, so it was decided that controlling the surface characteristics of the adhesive could result in a reproducible surface for the application of the CPG particles. For this purpose, a commercially available, optically transparent, UV curing polymer was used (Norland optical adhesive 68, refractive index 1.54). This adhesive is a flexible one-part adhesive containing no solvents. The monomer cured in less than 30 min under a low intensity UV lamp.

A drop of monomer was rolled out onto an acetate sheet with a glass rod. This was then pre-cured under the UV lamp for 5 min. CPG particles with immobilised thymol blue were then sprinkled onto the film. The film was then placed under the UV lamp for a further 5 min. The excess particles were then shaken off and the film placed beneath the lamp once again for 30 min for complete curing of the polymer.

The act of pre-curing the adhesive created a surface that remained sticky but not totally set. The CPG particles bound to the surface of the adhesive and did not sink into it significantly. This meant that the thickness of the adhesive layer was not important. The amount of pre-curing controlled the depth to which the CPG particles sank into the adhesive layer. This control of the surface characteristics of the adhesive layer was paramount to the development of a reproducible sensing film.

3.2 Probe Head Design and Flow Cell Construction

This design of the probe head allows the whole probe tip to be replaced as a single unit. The sensing film is at a fixed distance from the distal face of the optical fibre bundle. The probe design, shown in Fig. 2, consists of a vented polypropylene cup that was a push fit onto the end of a fibre bundle. The polypropylene cup was cut from a pipette tip. A series of 1 mm holes were drilled into the side of the cup at the position where the reflectance gap was required. These would allow the analyte solution into the reflectance gap to interact with the sensing film, which was bound to the vented cup. Epoxy adhesive was applied to the rim

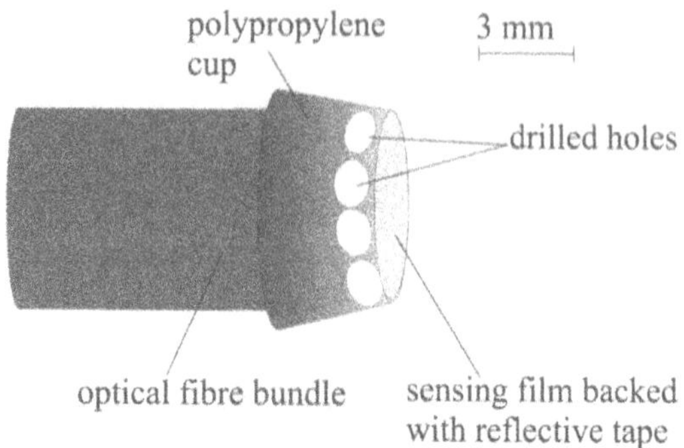

Fig. 2. Vented cup-type probe head

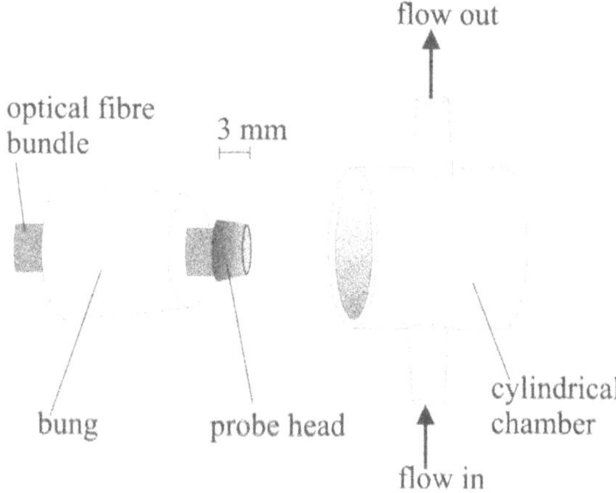

Fig. 3. The probe-compatible flow cell

of the cup and the sensing film was fixed to it. The sensing film was backed with reflective tape to increase the intensity of the reflected light.

In order to simulate a probe functioning in a flowing system, a flow cell was designed where a probe head was held within a flow stream of solution. This probe-compatible flow cell is illustrated in Fig. 3. A cylindrical plastic chamber formed the main body of the flow cell. The cylinder was sealed at one end with a flat plastic disk attached with epoxy adhesive. The other end of the cylinder accepted a plastic bung in a push fit watertight seal. Projecting through the centre of this bung was the distal end of a bifurcated optical fibre bundle. One of a variety of probe head designs could be attached to the end of this fibre bundle. Measurement solutions could enter and exit the flow cell through two openings in the side of the chamber. This arrangement was adjusted so that the probe head was held exactly in line with the inlet and the outlet. A peristaltic pump was used to circulate solutions through the flow cell.

For the characterisation and calibration of this sensing device a series of constant ionic strength buffer solutions were prepared. Buffer solutions of pH

6 to 12 in 0.5 pH increments were prepared. These were of five different ionic strengths (0.005, 0.0525, 0.1, 0.55, and 1 M). A universal buffer stock solution was prepared in accordance with Britton and Robinson [125]. Buffer solutions of ionic strengths 0.005, 0.1 and 1.0 M were made up from this stock solution in accordance with the procedure in the literature [126, 127] by the addition of the appropriate amounts of NaOH solution (0.2 M), KCl and deionised water. Buffer solutions of ionic strengths 0.0525 and 0.55 M were prepared by mixing the appropriate buffer solutions of 0.005 and 0.1 M and of 0.1 and 1.0 M respectively in a 1:1 ratio. All buffer solutions were stored in airtight high-density polyethylene bottles.

4 Instrumentation

The commercially available Ocean Optics S1000 charged coupled device (CCD) based spectrometer was used within a reflectance test rig. The reflectance test rig is shown schematically in Fig. 4. Light from the tungsten-halogen lamp (Ocean Optics LS-1) is coupled into one arm of a bifurcated optical fibre bundle. The light interacts with the sensing element in the flow cell and some is coupled back into the optical fibre bundle. Divergent light emerging from the optical fibre into the spectrometer is collimated by a spherical mirror (SM1). The collimated light is diffracted by a planar grating (G) and is focused onto the 1-dimensional linear CCD array by a second spherical mirror (SM2).

In order to overcome problems of sensing film photobleaching, a band pass interference filter was used in-line with the light source to restrict incident light to the region of interest. The filter was obtained from Coherent and had a 40-nm bandwidth centred at 660 nm with a maximum transmission of 82%.

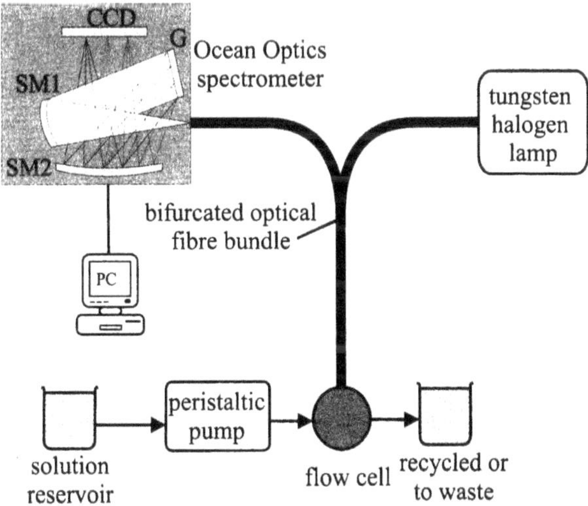

Fig. 4. CCD spectrometer-based reflectance test rig

4.1 Choice of Wavelengths

Three points on a wavelength spectrum were employed in the interpretation of the response of this pH sensing system. First, for correction of detector drift, a wavelength was selected which was distant from areas in the spectra that were dependent on either pH or variations in light source intensity. If the response at a single wavelength was selected this would produce a noisy response as the response from a single CCD pixel is subject to noise. To overcome this, the mean response from all pixels over a wavelength range was utilised. This range is illustrated as range **a** in Fig. 5 and corresponds to a range of approximately 20 nm centred on a wavelength of 610 nm.

For dual wavelength referencing two wavelength ranges were chosen. The first corresponded to the wavelength where the system showed the greatest change in response for a change in pH. For this purpose, a 5-nm range was selected centred on 641 nm. This is illustrated by range **b** in Fig. 5. Ideally, if the system is to exhibit a high sensitivity to the analyte, the second wavelength selected should be at a point where there is no change in response due to pH changes of solution, but where it is still dependent on changes in the light source intensity. Such a point exists at 676 nm, which is an isosbestic point (Fig. 5). However, the steep gradient in response at this point due to the fact that it is at the edge of the transmission profile of the optical filter means that a very small change in the transmission profile of the filter due to temperature effects will result in large changes in the response at this wavelength. This effect results in a very noisy response in this area. Range **c** (Fig. 5) was selected in preference to the isosbestic point. This range is a 5-nm band centred on 668 nm. This range was selected as it was distant from the edge of the filter transmission spectra, and because it exhibited a low sensitivity to pH.

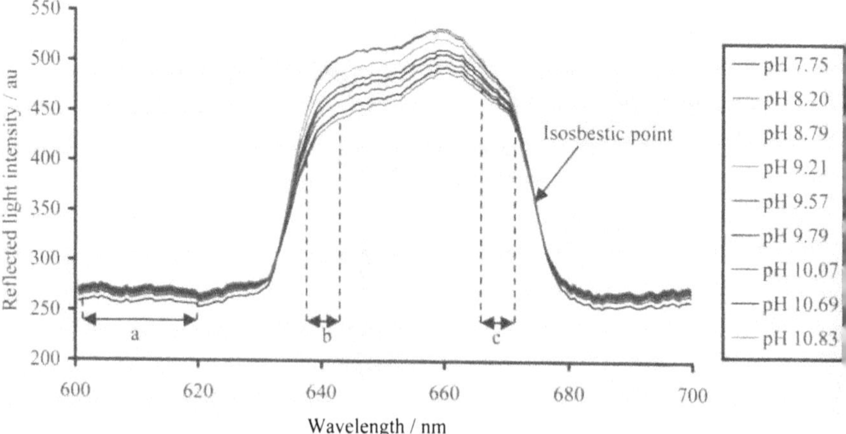

Fig. 5. Example of spectra for the filtered system over a range of pH

Using these wavelength ranges, the referenced response R_{ref} of the system is calculated as follows:

$$R_{ref} = \frac{(\text{mean of range b} - \text{mean of range a})}{(\text{mean of range c} - \text{mean of range a})}$$

This equation may be incorporated into the acquisition software of the spectrometer such that R_{ref} can be displayed in real time. A calibration function may also be incorporated so that the measured pH may be plotted in real time.

5 Results and Discussion

5.1 Variation of System Response between Films

The variation in the response characteristics between different sensing films was investigated. Two separate batches (A and B) of immobilised indicator were prepared; these were then used in the manufacture of probe heads. Two probe heads were manufactured from each batch of immobilised indicator. Evaluation of the response of these four probe heads allowed a comparison between probes made from the same batch and between those of different batches.

Figure 6 shows two calibration plots for both probe heads constructed from batch A of immobilised indicator. A convergence of the plots at a pH of approximately 9.5 represents the pK_a of the immobilised indicator. It can be seen that although all four plots are similar there is a slight difference in gradient of each probe head's plots. This difference is visible as a divergence above pH 9.5. This is caused by probe A2 being slightly less sensitive than probe A1.

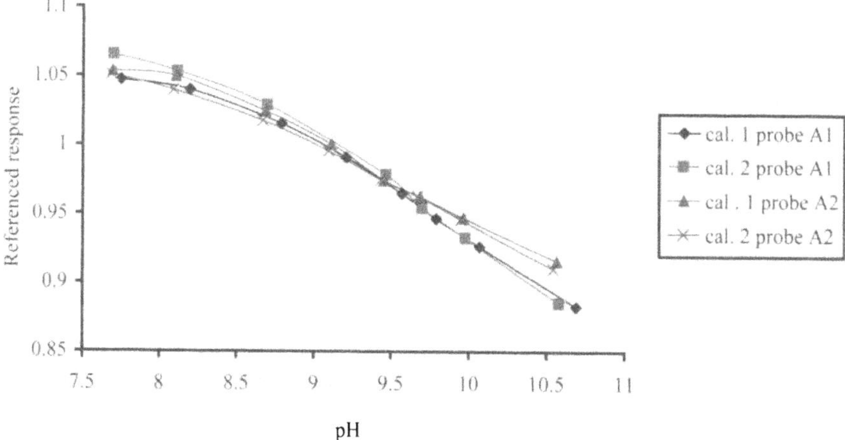

Fig. 6. Calibration plots for two probe heads constructed from batch A of immobilised indicator

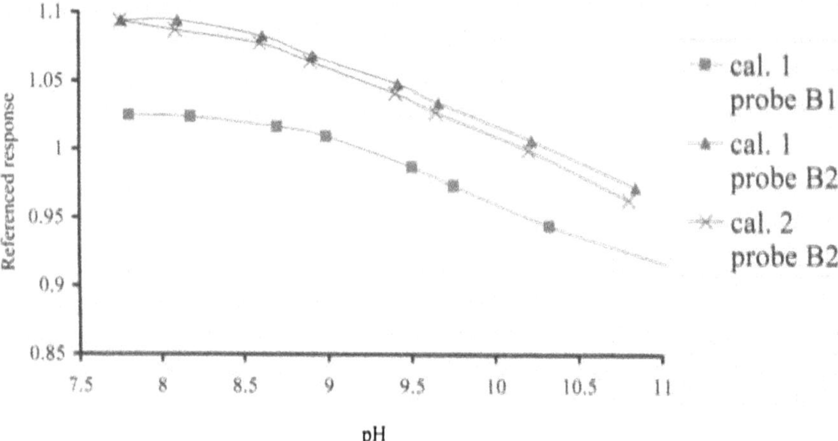

Fig. 7. Calibration plots for two probe heads constructed from batch B of immobilised indicator

The difference between probes is greater with the second batch. Figure 7 shows calibration plots for probe heads constructed using batch B of immobilised indicator (unfortunately, due to problems with the computer's acquisition software, data were lost and only one calibration plot was available for the probe B1). It is apparent that significant differences exist between the response characteristics of individual probes. This being so, a comparison between different batches is meaningless. This means that further work must be done on the reproducible manufacture of the sensing films. However, the use of more reproducible probes will not replace the need for calibration.

Calculation of the precision of each probe from these calibration plots produced a variation of only 8.6 % between all four probes. The average precision of these probes was calculated to be 0.061 pH units ± 8.6 %.

5.2 Temperature Response

Experiments were conducted at room temperature (25±1 °C), 30 ±1 °C, 40 ±1 °C and 50 ±1 °C. Unfortunately, at 50 °C air bubbles were formed within the system. These caused the resulting spectra to become erratic and thus these results were discounted.

Figure 8 shows calibration plots for 25, 30 and 40 °C. Although, there appears to be a steeper gradient to both the 30 and 40 °C plots it should be noted that the room temperature calibration was carried out 2 weeks before the other two. The fact that there is no difference between the plots at 30 °C and at 40 °C indicates that for changes in temperature up to 40 °C there is no significant effect on the response of this system.

The fact that the effect of temperature above 40 °C could not be investigated with this system due to the build-up of air bubbles was regrettable and work is

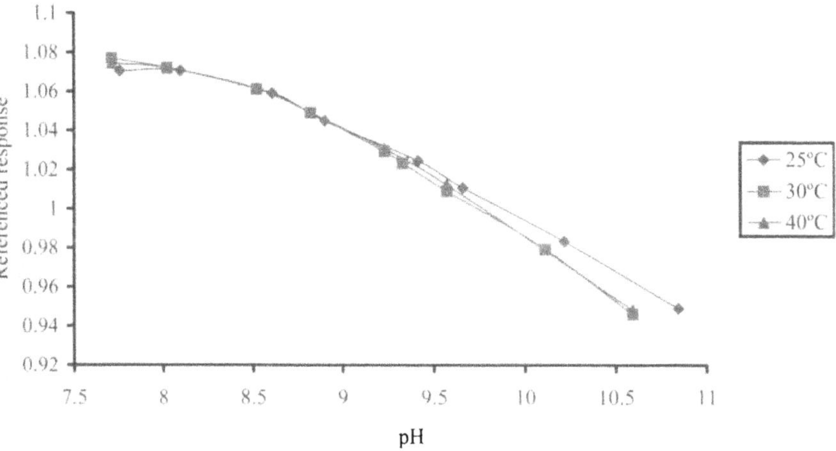

Fig. 8. pH calibration plots for three temperatures

needed to investigate this further. However, were the system to be used industrially, the incorporation of a temperature control unit would be possible.

5.3 Longevity of Sensing Films

The longevity of a sensing film in continuous use has been investigated. Using the LS-1 tungsten-halogen light source with the band-pass filter attached, a sensing film in a vented cup-type probe head was illuminated continuously for 2.5 months. During this period, 0.005 M ionic strength buffer solution, at a pH

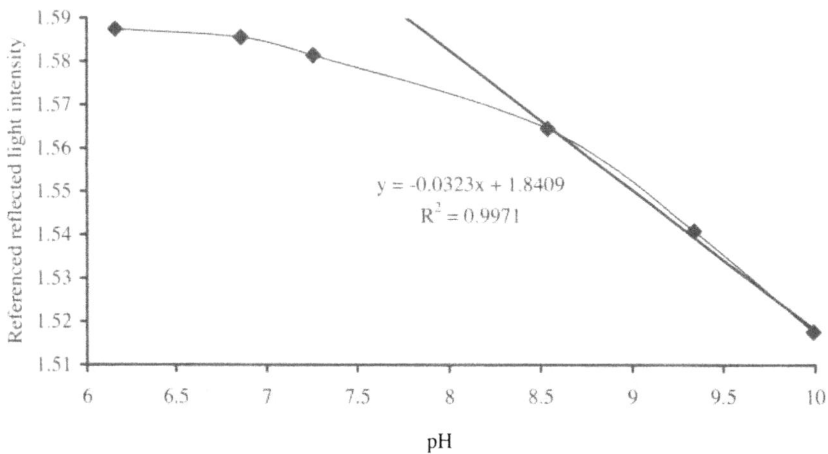

Fig. 9. Calibration of sensing film at 0.005 M ionic strength after 2.5 months of continuous use

of approximately 9, was pumped through the flow cell over the probe head. This arrangement was to simulate 2.5 months of continuous use of the sensing system. After this period, the system was calibrated using 0.005 M ionic strength buffer solutions (Fig. 9).

From the regression equation obtained from Fig. 9, the precision of the measurement was calculated. The precision was found to be 0.08 pH units whereas the precision at the beginning of this study was 0.064 pH units. This means that there was very little loss of sensitivity over 2.5 months. The severe effects of photobleaching have been overcome by the use of an optical filter, which prevents high-energy radiation interacting with the sensing film.

5.4 Effect of Ionic Strength

The effect of different ionic strengths of solutions on the response characteristics was investigated. The pH response of the system was calibrated using buffer solutions of five different ionic strengths: 1.0 M; 0.55 M; 0.1 M; 0.0525 M and 0.005 M.

The calibration plots for all five ionic strengths are shown in Fig. 10. The plots are normalised at pH 6.2 in order to aid comparison. It can be seen that, in general, there is a decrease in the gradient with decreasing ionic strength. This translates into a decrease in sensitivity with decreasing ionic strength.

These data may be used to estimate the precision of this system at lower ionic strengths. If the gradient of each plot, at pH 9, is plotted against ionic strength, it can be seen that there is a logarithmic function that relates the two (Fig. 11). As an example, the ionic strength of a solution with a conductivity of 20 μS was determined as 0.273 μM. Substituting this value into the logarithmic function (Fig. 11) gives a gradient of 0.0392. Taking the average error in referenced response from previous experiments, the precision is calculated to

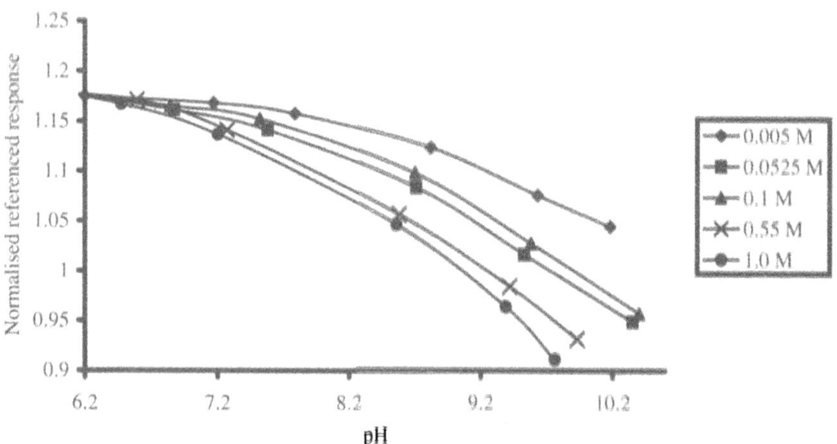

Fig. 10. pH calibration plots for five different ionic strengths

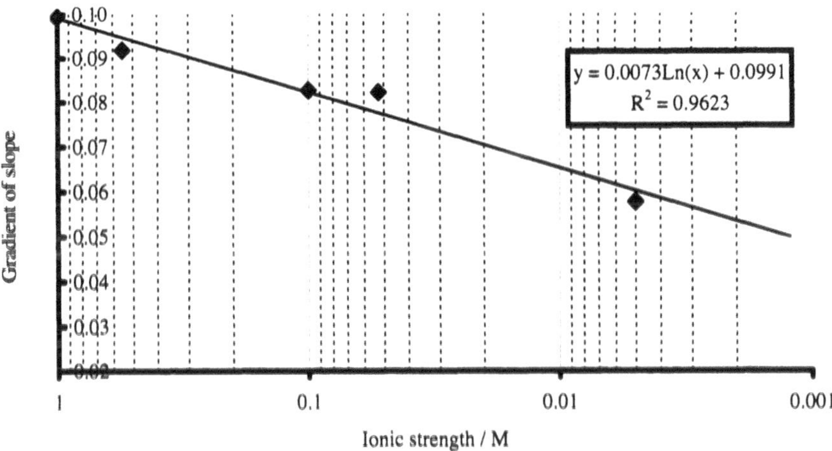

Fig. 11. Gradient of calibration plot (at pH 9) vs the log of solution ionic strength

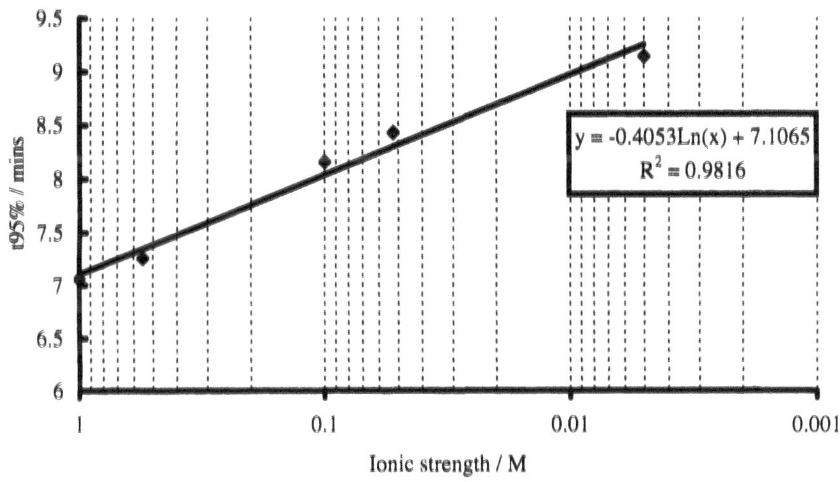

Fig. 12. Plot of the response time ($t_{95\%}$) vs the log of solution ionic strength

be 0.085 pH units. This represents a 28% loss in precision from that obtained at 0.005 M ionic strength. Although this is only a crude estimate, it gives an idea of the order of precision that this system would be expected to achieve at lower ionic strengths.

Studies have been carried out on the effect that ionic strength has on the response time of the system. Transitions between approximately pH 10 and 6 have been recorded over time for all five ionic strengths. The time for 95% of the change to take place ($t_{95\%}$) has been calculated for each (Table 2).

If these values are plotted a logarithmic function can be fitted to the data (Fig. 12). This function can be used to estimate the response time ($t_{95\%}$) for a solution

Table 2. Response times ($t_{95\%}$) for five solution ionic strengths

Ionic strength [M]	$t_{95\%}$ [min]
0.005	9.14
0.0525	8.43
0.1	8.16
0.55	7.26
1.0	7.06

of 0.273 μM ionic strength. Using this value for the ionic strength the response time can be calculated as 10.43 min. This is only a slight increase in response time (from 9.14 to 10.43 min) from 0.005 M ionic strength to the ionic strength of the 20 μS solution.

Abbreviations

AG1	Anion exchange resin with $-N(CH_3)_3^+$ functional group
AG2	Anion exchange resin with $-N(CH_3)_2C_2H_4OH^+$ functional group
BCA	Bis(carboxylic acid)
BCECP	Not specified
BCG	Bromocresol green
BCP	Bromocresol purple
BPB	Bromophenol blue
BPR	Bromophenol red
BTB	Bromothymol blue
BTBss	Bromothymol blue sodium salt
BTR	Bromothymol red
CCD	Charge coupled device
CNF	Carboxynaphthofluorescene
CPG	Controlled pore glass
CPR	Chlorophenol red
CR	Cresol red
DADPA	Diaminodipropylamine
EW	Evanescent wave
m-CP	Metacresol purple
FITC	Fluorescein isothiocyanate
HCC	7-hydroxycoumarin-3-carboxylic acid
HOPSA	8-hydroxyl-1,3,6-pyrenetrisulphonic acid
HPTS	1-hydroxypyrene-3,5,8-trisulphonate
IRA-400	Anion-exchange resin
MR	Methyl red
NR	Neutral red
PBQ	Poly(biphenylquinoline)
PMT	Photomultiplier tube
PPQ	Poly(phenylquinoxyaline)

PQ	Poly(phenylquinoline)
PR	Phenol red
R6G	Rhodamine 6G
SNARF	Seminaphthofluorescein
SNAFL	Seminaphthorhodofluor
SR-640	Sulphorhodamine-640
TB	Thymol blue
TBPP	Tetrabromophenolphthalein
TBPSP	3,4,5,6-tetrabromophenolsulphonephthalein
TRH	Texas red hydrazide
TTMAPP	α,β,χ,δ-tetrakis(4-trimethylaminophenyl)porphine
TP	Thymolphthalein
XAD-2	Cross linked polymer of styrene and divinyl benzene
XAD-4	Cross linked polymer of styrene and divinyl benzene

References

1. McQuaker NR, Kluckner PD, Sandberg DK (1983) Environ Sci Technol 17:431
2. Avery HE, Shaw DJ (1971) Advanced physical chemistry calculations, Butterworths, London.
3. Hao T, Xing X, Liu CC (1993) Sens Actuators B10:155
4. Suidan JS, Young BK, Hetzel FW, Seal HR (1983) Clin Chem 29:1556
5. Kostov YV (1992) Sens Actuators B8:99
6. Cardwell TJ, Cattrall RW, Deady LW, Dorkos M, O'Connel GR (1993) Talanta 40:765
7. Lindauer H, Czerney P, Mohr GJ, Grummet UW (1994) Dyes Pigments 26:229
8. Lin J, Liu D (2000) Anal Chim Acta 408:49
9. Kirkbright GF, Narayanaswamy R, Welti NA (1984) Analyst 109:5
10. Boisdé G, Biatry B, Magny B, Dureault B, Blanc F, Sebille B (1989) SPIE Chemical, Biochemical and Environmental Sensors 1172:239
11. Turner JD (1995) MSc Dissertation Thesis, University of Manchester Institute of Science and Technology
12. Baker MEJ, Narayanaswamy R (1995) Sens Actuators B29:368
13. Andres RT, Sevilla III F (1991) Anal Chim Acta 251:165
14. Motellier S, Toulhoat P (1993) Anal Chim Acta 271:323
15. Moreno MC, Martinez A, Millan P, Camara C (1986) J Mol Struct 143:553
16. Peterson JI, Goldstein SR, Fitzgerald RV, Buckhold DK (1980) Anal Chem 52:864
17. Baldini F, Bechi P, Bracci S, Cosi F, Pucciani F (1995) Sens Actuators B29:164
18. Jones TP, Porter MD (1988) Anal Chem 60:404
19. Harper GB (1975) Anal Chem 47:348
20. Bacci M, Baldini F, Bracci S (1991) Appl Spectrosc 45:1508
21. Cardwell TJ, Cattrall RW, Deady LW, Dorkos M, Kaye AJ, Papanikos A (1995) Aust J Chem 48:1081
22. Tabacco M, Zhou Q, Nelson B (1991) SPIE Chemical, Biochemical and Environmental Fiber Sensors III 1587:271
23. Lehmann H, Schwotzer G, Czerney P, Mohr GJ (1995) Sens Actuators B29:392
24. Grummt UW, Pron A, Zagorska M, Lefrant S (1997) Anal Chim Acta 357:253
25. Boisdé G, Biatry B, Magny B, Dureault B, Blanc F, Sebille B (1989) SPIE Chemical, Biochemical and Environmental Sensors 1172:239
26. Igarashi S, Kuwae K, Yotsuyanagi T (1994) Anal. Sciences 10:821
27. Jones TP, Coldiron SJ, Deninger WJ, Porter MD (1991) Appl Spectrosc 45:1271
28. Ensafi AA, Kazemzadeh A (1999) Microchemical 63:381

29. Kuznetsov VV, Yakunina IV (1994) J Anal Chem 49:847
30. Liu JN, Shahriari MR, Sigel JH (1992) Optics Lett B8:1815
31. Maher MH, Shahriari MR (1993) J Test Eval 21:448
32. Kostov Y, Tzonkov S, Yotova L, Krysteva M (1993) Anal Chim Acta 280:15
33. Safavi A, Abdollahi H (1998) Anal Chim Acta 367:167
34. Baldini F, Bracci S (1993) Sens Actuators B11:353
35. Baldini F, Bracci S, Cosi F, Bechi P, Pucciani (1994) Appl Spectrosc 48:549
36. Sotomayor PT, Raimundo IM, De P, de Oliveira NG, de Oliveira WA (1998) Sens Actuators B51:382
37. Wolthuis R, McCrae D, Saaski E, Hartl J, Mitchell G (1992) IEEE Trans Biomed Eng 39: 531
38. Holobar A, Weigl BH, Trettnak W, Benes R, Lehmann H, Rodriguez NV, Wollschlager A, O'Leary P, Raspor P, Wolfbeis, OS (1993) Sens Actuators B11:425
39. Weigl BH, Holobar A, Trettnak W, Klimant I, Kraus H, O'Leary P, Wolfbeis OS (1994) J Biotech 32:127
40. Werner T, Wolfbeis OS (1993) Fresenius J Anal Chem 346:564
41. Mohr GJ, Wolfbeis OS (1994) Anal Chim Acta 292:41
42. Hisamoto H, Tsubuku M, Enomoto T, Watanabe K, Kawaguchi H, Koike Y, Suzuki K (1996) Anal Chem 68:3871
43. Koncki R, Wolfbeis OS (1998) Sens Actuators B51:355
44. de Marcos S, Wolfbeis OS (1996) Anal Chim Acta 334:149
45. Benaim N, Grattan KTV, Palmer AW (1986) Analyst 111:1095
46. Farquharson S, Swaim PD, Christenson CP, McCloud M, Freiser H (1991) SPIE Chemical, Biochemical and Environmental Fiber Sensors III 1587:232
47. Kirkbright GF, Narayanaswamy R, Welti NA (1984) Analyst 109:1025
48. Serra G, Schirone A, Boniforti R (1990) Anal Chim Acta 232:337
49. Alabbas SH, Ashworth DC, Bezzaa B, Momin SA, Narayanaswamy R (1996) Sens Actuators A51:129
50. Motellier S, Michels MH, Dureault B, Toulhoat P (1993) Sens Actuators B11:467
51. Motellier S, Noire NH, Pitsh H, Dureault B (1995) Sens Actuators B29:345
52. Bacci M, Baldini F, Scheggi AM (1988) Anal Chim Acta 207:343
53. Moreno MC, Jimenez M, Conde CP, Camara C (1990) Anal Chim Acta 230:35
54. Wróblewski W, Roüniecka E, Dybko A, Brzózka Z (1998) Sens Actuators B48:471
55. Edmonds TE Ross ID (1985) Anal Proc 22:206
56. Dybko A, Wróblewski W, Maciejewski J, Romaniuk R, Brzózka Z (1997) SPIE Chemical, Biochemical and Environmental Fiber Sensors IX 3105:361
57. Takai N, Sakuma I, Fukui Y, Kaneko A, Fujie T, Taguchi K, Nagaoka S (1991) Artif Organs 15:86
58. Takai N, Hirai T, Sakuma I, Fukui Y, Kaneko A, Fujie T (1993) Sens Actuators B13–14: 427
59. Takahashi C, Kaneko A, Komata Y, Yokoyama S, Fujie T (1993) Sens Actuators B13–14: 756
60. Rooney MTV, Seitz WR (1999) Anal. Commun 36:267
61. Seitz WR, Rooney MTV, Miele EW, Wang H, Kaval N, Zhang L, Doherty S, Milde S, Lenda J (1999) Anal Chim Acta 400:55
62. Dafu C, Qiang C, Jinghong H, Jinge C, Yating L, Zemin Z (1993) Sens Actuators B12: 29
63. Netto EJ, Peterson JI, Mcshane M, Hampshire VA (1995) Sens Actuators B29:157
64. Boisde G, Blanc F, Perez JJ (1988) Talanta 35:75
65. Zhu J, Jin Y, Xue J (1992) Fres J Anal Chem 342:42
66. Woods BA, Ruzicka J, Christian GD, Rose NJ, Charlson RJ (1988) Analyst 113:301
67. Woods BA, Ruzicka J, Christian GD, Charlson RJ (1986) Anal Chem 58:2496
68. Baldini F, Bracci S, Bacci M (1991) SPIE Chemical, Biochemical and Environmental Fiber Sensors III 1587:67

69. Baldini F, Bracci S, Cosi F (1993) Sens Actuators A37-38:180
70. Shakhsher Z, Seitz WR, Legg KD (1994) Anal Chem 66:1731
71. Zhang L, Langmuir ME, Bai M, Seitz WR (1997) Talanta 44:1691
72. Sotomayor MDPT, De Paoli MA, de Oliveira WA (1997) Anal Chim Acta 353:275
73. Agayn VI, Walt DR (1993) Biotechnology 11:726
74. McCulloch S, Uttamchandani D (1997) IEE Proc-Optoelectron 144:162
75. Badini GE, Gratton KTV, Palmer AW, Tseung ACC (1989) Pringer Proc Physics 44:436
76. Badini GE, Gratton KTV, Tseung ACC (1995) Analyst 120:1025
77. Nivens DA, Zhang Y, Angel SM (1998) Anal Chim Acta 376:235
78. Opitz N, Lubbers DW (1983) Sens Actuators 4:473
79. Luo S, Walt DR (1989) Anal Chem 61:174
80. Liebert TF, Walt DR (1995) J Control Release 35:155
81. Zen JM, Patonay G (1991) Anal Chem 63:2934
82. Thompson RB, Lakowicz JR (1993) Anal Chem 65:853
83. Szmacinski H, Lakowicz JR (1993) Anal Chem 65:1668
84. Lakowicz JR, Szmacinski H (1993) Sens Actuators B11:133
85. Bambot SB, Sipior J, Lakowicz JR, Rao G (1994) Sens. Actuators B22:181
86. Grant SA, Glass RS (1997) Sens Actuators B45:35
87. Ji J, Rosenzweig Z (1999) Anal Chim Acta 397:93
88. Kosch U, Klimant I, Werner T, Wolfbeis OS (1998) Anal Chem 70:3892
89. Lakowicz JR, Szmacinski H, Karakelle M (1993) Anal Chim Acta 272:179
90. Jordan DM, Walt DR, Milanovich FP (1987) Anal Chem 59:437
91. Song A, Parus S, Kopelman R (1997) Anal Chem 69:863
92. Zhujun Z, Seitz WR (1984) Anal Chim Acta 160:47
93. Munkholm C, Walt DR, Milanovich FP, Klainer SM (1986) Anal Chem 58:1427
94. Tan W, Shi ZY, Kopelman R (1992) Anal Chem 64:2985
95. Saari LA, Seitz WR (1982) Anal Chem 54:821
96. Ferguson JA, Healey BG, Bronk KS, Barnard SM, Walt DR (1997) Anal Chim Acta 340: 123
97. Fuh MRS, Burgess LW, Hirshfeld T, Christian GD, Wang F (1987) Analyst 112:1159
98. Posch HE, Leiner MJP, Wolfbeis OS (1989) Fresenius J Anal Chem 334:162
99. Schulman SG, Chen S, Fenglian B, Leiner MJP, Weis L, Wolfbeis OS (1995) Anal Chim Acta 304:165
100. Offenbacher H, Wolfbeis OS, Furlinger E (1986) Sens Actuators 9:73
101. Wolfbeis OS, Offenbacher H (1986) Sens Actuators 9:85
102. Wolfbeis OS, Rodriguez NV, Werner T (1992) Mikrochim Acta 108:133
103. Carey WP, Jorgenson BS (1991) Appl Spectrosc 45:834
104. Parker JW, Laksin O, Yu C, Lau ML, Klima S, Fisher R, Scott I, Atwater BW (1993) Anal Chem 65:2329
105. Draxler S, Lippitsch ME, Leiner MJP (1993) Sens Actuators B11:421
106. Deboux BJC, Lewis E, Scully PJ, Edwards R (1995) J Lightwave Technol 13:1407
107. Schalkhammer T, Lobmaier C, Pittner F, Leitner A, Brunner H, Aussenegg FR (1995) Sens Actuators B25:166
108. Michie WC, Culshaw B, Mckenzie I, Konstantakis M, Graham NB, Moran C, Santos F, Bergqvist E, Carlstrom B (1995) Optics Lett 20:103
109. MacCraith BD, Ruddy V, Potter C, O'Kelly B, McGilp JF (1991) Electron Lett 27:1247
110. Ding ZY, Shahriari MR, Sigel GH (1991) Electron Lett 27:1560
111. Noire MH, Bouzon C, Couston L, Gontier J, Marty P, Pouyat D (1998) Sens Actuators B51:214
112. Lee JE, Saavedra SS (1994) Anal Chim Acta 285:265
113. MacCraith BD (1993) Sens Actuators B11:29
114. Ge Z, Brown CW, Sun L, Yang SC (1993) Anal Chem 65:2335
115. Jones TP, Porter MD (1989) Appl Spectrosc 43:908
116. Rao BS, Puschett JB, Matyjaszewski K (1991) J Appl Polym Sci 43:925

117. Mullen KI, Wang D, Crane LG, Carron KT (1992) Anal Chem 64:939
118. Niwa M, Yamamoto T, Higashi N (1991) J Chem Soc Chem Comm 444
119. Hale ZM, Payne FP (1994) Sens Actuators B17:233
120. Deboux BJC, Lewis E, Scully PJ, Edwards R (1995) Proc SPIE-Int Soc Opt Eng 2542: 167
121. Taib MN, Andres R, Narayanaswamy R (1996) Anal Chim Acta 330:31
122. Sotomayor PT, Raimundo IM, Zarbin AJG, Rohwedder JJR, Neto GG, Alves OL (2001) Sens Atuators B74:157
123. Vishnoi G, Goel TC, Pillai PKC (1999) Proc SPIE-Int Soc Opt Eng 3538:319
124. Boutin P, Mugnier J, Valeur B (1997) J Fluoresc 7:215S
125. Britton HTS, Robinson RA (1931) J Chem Soc 458:1456
126. Coch Frugoni JA (1957) Gazz Chim Ital 87:403
127. http://www.bi.umist.ac.uk/users/mjfrbn/Buffers/Makebuf.asp

CHAPTER 13

Environmental and Industrial Optosensing with Tailored Luminescent Ru(II) Polypyridyl Complexes

GUILLERMO ORELLANA, DAVID GARCÍA-FRESNADILLO

1 Introduction

The sensitivity, specificity and versatility of *optical methods* for chemical determinations have turned spectroscopy into one of the most popular techniques for *environmental* analysis and *process* control [1, 2]. In most cases, however, the very same attractive features have led (so far) to expensive instrumentation and/or complex methods compared to, for instance, the well-established electrochemical sensors. Fibre-optic chemical sensors (also known as „*optodes*") are bound to overcome such limitations provided they use cost-effective optoelectronics and prove to be specific, sensitive and robust enough to fulfill their analytical tasks in air, water and soil quality monitoring as well as in the industrial environment.

Over the last 25 years hundreds of optical fiber sensors have been described for the analysis of chemical parameters but, surprisingly enough, hardly any of them has come onto the market yet [3]. The broadly *different* spectroscopic *properties* of the *many* optical *indicator* dyes [4], the necessity of immobilizing them onto a *suitable* solid support, the required *selectivity* and *sensitivity*, and the tough ambient conditions in environmental and industrial control, among other factors, are responsible for such a gap. Competition from improved alternative sensors, the intrinsic cost of sophisticated optodes and their unavoidable field validation, as well as a frequent failure to recognize the end-user needs, help to explain the slow pace of transforming basic research in this area into industrial devices [5].

The design and synthesis of a *homogeneous* family of tailored luminescent indicators and the exploitation of selected sensing schemes are trying to fill the gap [6, 7]. Luminescent ruthenium(II) complexes have turned out to be the cornerstone of successful optodes for molecular *oxygen* monitoring, the first ones to be marketed by a number of companies [8]. As it will be described below, this fact is due to the recent availability of blue LEDs, the output of which overlaps with the strong absorption of such metal compounds, their well-separated luminescence in the red with μs emission lifetimes and their significant photochemical and thermal stability compared to most purely organic dyes. Moreover, one (or more) of the coordinating ligands can be chemically modified (i) to adjust the net electrical *charge* ($n+$, 0, $n-$) of the overall metal complex; (ii) to provide functional groups to *attach* the indicator dye to the polymer support;

(iii) to impart or enhance the interaction of the indicator with the targeted *analyte* or to prevent response to *interferent* species; (iv) to alter the *solubility* of the dye; (v) to vary the *spectroscopic properties* of the complex. This chapter aims to illustrate how the unique features of this family of luminescent dyes can be employed to fabricate successful luminescent optosensors for on-line environmental and process analysis. Selected examples of the molecularly engineered dyes and photochemistries will be presented in some detail, such as those that have led to oxygen, temperature, pH, carbon dioxide, chloride, sodium ion, biochemical oxygen demand (BOD), selected metabolites and humidity fibre-optic (bio)sensing.

2 Ru(II) Polypyridyl Complexes

Since the first report by Paris and Brandt on the intense luminescence observed from the photoexcited $[Ru(bpy)_3]^{2+}$ cation (bpy = 2,2'-bipyridine) in solution [9], a great deal of attention has been devoted to the study of Ru(II) coordination compounds with polyazaaromatic chelating ligands [10]. Such heterocyclic ring systems contain an α-*diimine* moiety that is responsible for their (mostly) bidentate properties (Fig. 1), so that up to three of them may be incorporated into the pseudo-octahedral coordination sphere of the transition metal (Fig. 1).

The interest in these peculiar complexes was stimulated in the 1970's by their ability to act as electron-transfer photoredox partners [11], and their potential use in devices for the chemical conversion and storage of solar energy through photocatalytic splitting of water. Although the severe limitations of schemes to accomplish the latter aim had already been put forward, active research on Ru(II) *polypyridyl* complexes[1] was far from decreasing due to the broadness and variety of novel applications quickly demonstrated, e.g. quantum counters [12], luminescent probes for micelles and other organised media [13], artificial photonucleases and photochemical reporters of the rich DNA morphology [14], constituents of supramolecular edifices [15], singlet molecular oxygen generators [16], chemiluminescent analytical reagents [17], and optosensing dyes [6], to name just a few. Moreover, the thorough elucidation of the electrochemical, spectroscopic, photophysical and photochemical features of these metal complexes has been a challenging task for the scientific community [18].

The versatility of Ru(II) polypyridyls arises from the fact that both their ground- and excited-state optical and (electro)chemical properties, together with their solubility and charge, can be tuned both coarsely and finely by a judicious selection of the heterocyclic chelating ligands. Particularly, the binomial *structural-electronic* attributes of the latter (namely *ring* size, fusion and substitution; nature, number, and position of the *heteroatoms*; number and type of polyazaaromatic *ligands*) allow the knowledgeable researcher to perform

[1] Although, as will be shown in this chapter, many other azaheterocyclic chelating ligands that do not contain *pyridine* moieties have been used to coordinate the Ru(II) cation in addition to the original 2,2' -bipyridine, the generic name *polypyridyls* pervades the literature and has been maintained to designate this family of related dyes.

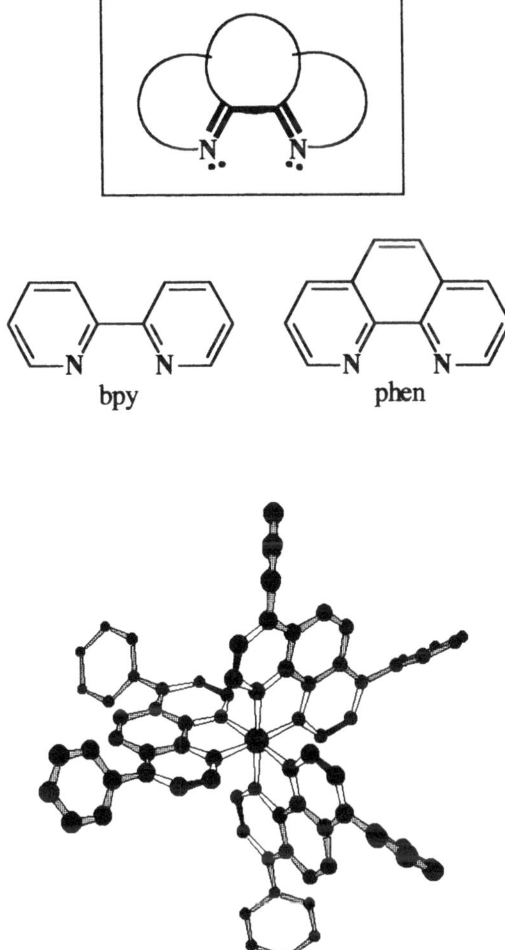

Fig. 1. General structure of the α-diimine chelating ligands and of the two prototypical 2,2'-bipyridine and 1,10-phenanthroline ligands, together with the 3-D structure of the $[Ru(dip)_3]^{2+}$ complex (dip = 4,7-diphenyl-1,10-phenanthroline)

the molecular engineering required to tailor the photoluminescent and photochemical features of the Ru(II) complex to the sought application. Obviously, this extraordinary versatility is very much desirable for sensor design and paves the way for developing a homogeneous array of indicator dyes based on a single family of coordination compounds.

2.1 Light Absorption Features

The UV-VIS absorption spectra of $[Ru(bpy)_3]^{2+}$ cation and its relatives (e.g. $[Ru(phen)_3]^{2+}$, Fig. 2, where phen = 1,10-phenanthroline), and the assignment of the main bands to specific electronic transitions, is now firmly established [10]. The *visible* region of their spectra is dominated by a strong absorption peaking at 440–500 nm (typically $4000 < \varepsilon_{max} < 40\,000$ $dm^3\,mol^{-1}\,cm^{-1}$).[2] On the basis of its intensity, shape and position, such a band has been ascribed to a spin-allowed d-π* *metal-to-ligand* charge transfer (MLCT) transition. Despite some initial controversy, the scarce orbital interaction between the chelating ligands in the complex seems to be established nowadays and, therefore, the ligand-localised features of the photoexcited electron. For instance, in the case of a homoleptic complex (NN stands for any azaheterocyclic chelating ligand) the electronic transition might be accurately represented as:

$$[Ru^{2+}(NN)_3] + h\nu \rightarrow [Ru^{3+}(NN)_2(NN^-)]$$

Nevertheless, electron hopping between the different ligands is possible within the lifetime of the excited state. Excited-state absorption spectra, circular dichroism, pico- and femto-second laser kinetic spectrometry, time-resolved resonance Raman, electrochemical measurements and theoretical calculations

[2] Usually two absorption maxima in the blue region are observed due to the D_3 (or lower) symmetry of the *homoleptic* metal complexes and/or to the presence of different heterocyclic chelating ligands in the coordination sphere of Ru(II) (*heteroleptic* complexes). The lower symmetry of the latter molecules also yields broader absorption bands.

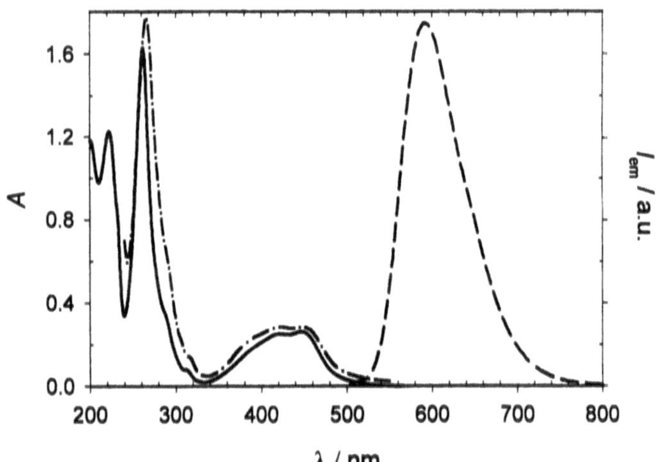

Fig. 2 Absorption (——) excitation (–·–·–·) and emission (– – –) spectra of $[Ru(phen)_3]^{2+}$ in water (the excitation spectrum has been offset for the sake of clarity; phen = 1,10-phenanthroline)

support the ligand-localised model. If the coordinating heterocyclic ligands have very different electronic features, the promoted electron will dwell predominantly in the most electron-withdrawing ligand (i.e. the one with the lowest-lying LUMO).

The tail observed on the low energy side of the MLCT absorption band has been attributed to a spin-forbidden MLCT transition. The Ru(II) polypyridyls also display several bands in the UV region of the electronic spectrum (Fig. 2). The intense absorptions centred at 185–220 nm and at 280–320 nm (typically $40\,000 < \varepsilon_{max} < 100\,000$ $dm^3\,mol^{-1}\,cm^{-1}$) correspond to *intraligand* π-π^* transitions, as follows from their similar position to those of the corresponding free ligands. Other less intense absorptions (typically $\varepsilon_{max} < 30\,000$ $dm^3\,mol^{-1}\,cm^{-1}$) observed between the ligand-centered and the MLCT bands are more difficult to assign but they are generally attributed to higher-energy MLCT or Laporte-forbidden ligand-field (d-d) transitions.

It has been shown that the position of the visible MLCT absorption is determined by both σ- and π-effects [19]. Therefore, a decrease of the energy gap between the metal-based d(π) and the ligand-centered π^* orbitals (i.e. a red shift of the MLCT band) may originate from a rise in the energy level of the former (for example, due to stronger σ-donor ligands) and/or a stabilisation of the latter (for example, caused by electron-withdrawing substituents on the ligands).

As far as the heteroleptic $[Ru(NN)_2(NN')]^{2+}$ complexes are concerned [10], their absorption pattern resembles the weighted electronic spectra of the corresponding component homoleptic chelates (i.e. 2/3 $[Ru(NN)_3]^{2+}$ and 1/3 $[Ru(NN')_3]^{2+}$). Therefore, provided the absorption maxima of the latter are distant enough, two MLCT bands are normally observed in the spectra of the above-mentioned heteroleptic Ru(II) species, lending further support to the ligand-localised model of

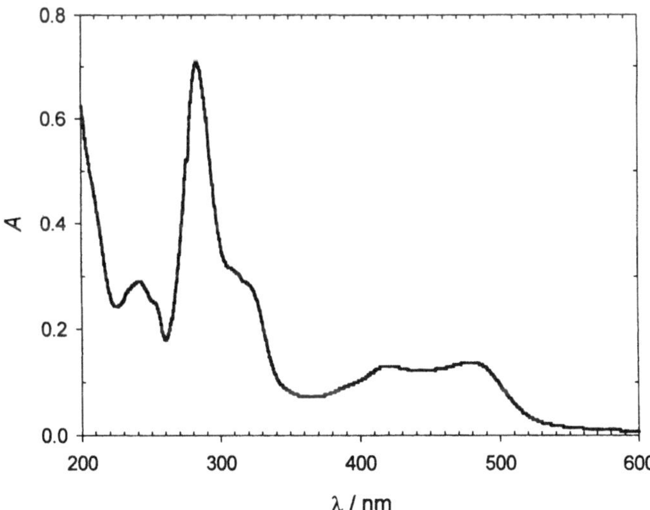

Fig. 3. Absorption spectrum of $[Ru(bpy)_2(pzth)]^{2+}$ in water (bpy = 2,2'-bipyridine; pzth = 2-(1,3-thiazol-2-yl)pyrazine)

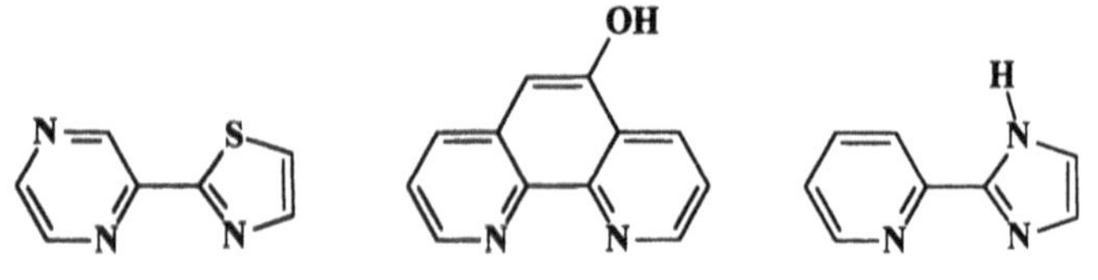

pzth 5-OHp pyim

Fig. 4. Structure of some acidity-sensitive α-diimine chelating ligands; pzth = 2-(1,3-thiazol-2-yl)pyrazine, 5-OHp = 5-hydroxy-1,10-phenanthroline and pyim = 2-(2-pyridyl)-1,3-imidazole

the photoinduced electronic transition. Since the electronic features of the coordinated ligand determine the energy level of its π* orbital (vide supra), individual d→π* absorptions will be displayed by heteroleptic complexes containing azaheterocyclic ligands with very difficult electron withdrawing/releasing properties in the coordination sphere of the metal atom (Fig. 3).

Naturally, *acid-base* equilibria of metal-chelated ligands, involving functional groups or heteroatoms that do not participate in the binding to the Ru(II) core, show up in the UV-VIS absorption spectra of the complex, too. Again the most pronounced changes occur when deprotonation or protonation of the metal complex provokes significant variations in the electronic distribution of the ligand. For instance, ground-state monoprotonation of the pyrazinyl moiety of coordinated 2-(1,3-thiazol-2-yl)pyrazine (pzth, Fig. 4) in $[Ru(pzth)_3]^{2+}$ shifts the MLCT maximum from 465 to 511 nm due to significant stabilisation of the π* orbital of the heteroaromatic acidic ligand [20]. However, the acid-base equilibrium of 5-hydroxy-1,10-phenanthroline (5-OHp, Fig. 4) in $[Ru(tmp)_2(5\text{-}OHp)]^{2+}$ (tmp = 3,4,7,8-tetramethyl-1,10-phenanthroline) does not produce any measurable change in the MLCT blue absorption band of the complex [21].

It must be born in mind that the Ru(II) core may alter dramatically the ground-state acid/base properties of the *coordinated* ligand from those observed for the *free* ligand. For example, the pK_a for protonation of pzth (0.8) drops to −1.9 in $[Ru(pzth)_3]^{2+}$ because of the strong electron-withdrawing effect of the transition metal ion [20]. Spectrophotometric titration yields a pK_a value of 7.9 for the $[Ru(bpy)_2(pyim)]^{2+}$ complex (pyim = 2-(2-pyridyl)-1,3-imidazole, Fig. 4), a 10^5-fold increase in the acidity of the free imidazole ligand [22]. The electronic effect of the metal center will be maximum when the acid/base functional group or atom sits in a conjugated position with respect to the coordinated nitrogen atoms.

2.2 Luminescence Features

After some initial controversy, it is now well established that the orange-red luminescence displayed by photoexcited Ru(II) polypyridyls arises from a radiative π*-dπ electronic transition [10, 18]. The emission quantum yield (Φ_L) of $^*[Ru(bpy)_3]^{2+}$ in oxygen-free aqueous solution at 298 K (very often used as a quantum yield standard) is 0.042 ± 0.002 [23]. However, Φ_L values as high as

0.40 for $[Ru(dip)_3]^{2+}$ (dip = 4,7-diphenyl-1,10-phenanthroline) have been determined in deaerated fluid media at room temperature [24]. Not every Ru(II) complex with polyazaheterocyclic chelating ligands displays luminescence: for instance, $[Ru(pyim)_3]^{2+}$ and $[Ru(Me\text{-}pyim)_3]^{2+}$ (pyim = 2-(2-pyridyl)-1,3-imidazole and Me-pyim = 1-methyl-2-(2-pyridyl)-1,3-imidazole) do not emit at room temperature due to the very low-lying π^* of the diazole moieties (vide infra), yet their orange-red photoluminescence switches on in alcohol medium at 77 K [10, 25].

Although the lowest excited state responsible for the emission is commonly regarded as a *triplet* metal-to-ligand charge transfer (^{3}MLCT) level,[3] generated by fast (and efficient) intersystem crossing from the initial singlet MLCT state reached upon light absorption (Fig. 5) [26],[4] the strong spin-orbit coupling provided by the heavy metal loosens the meaning of spin labels. The same effect determines the excited state lifetimes of ruthenium polypyridyls at room temperature to be typically in the order of 0.1–7 µs, too short for a regular phosphorescence but clearly exceeding those of a fluorescent emission (e.g. the excit-

[3] Due to symmetry reasons, even for homoleptic Ru(II) polypyridyls, the emissive ^{3}MLCT excited state is actually a *manifold* of at least three very close-lying (e.g. within 0.12 kJ mol^{-1} for the bpy complex) $d\pi^*$ levels where the photoexcited electron is in thermal equilibrium. Therefore, they are usually considered as a single excited state [10].

[4] The quantum yield of intersystem crossing from the ^{1}MLCT to the ^{3}MLCT state has been determined to be 1.0 for $Ru(bpy)_3^{2+}$ and several other analogues [10], yet generalisation of this situation to all Ru(II) polypyridyls is very dangerous. For instance, Φ_{ISC} values as low as 0.24 have been measured by laser kinetic spectrometry for Ru(II) complexes containing azole-coordinated moieties [26].

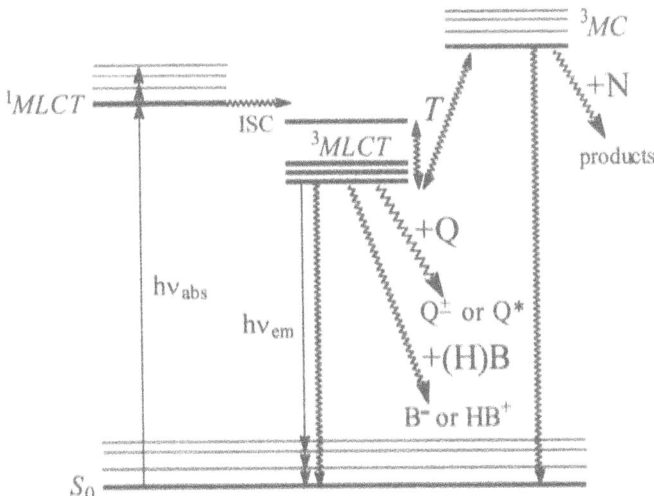

Fig. 5. General Jablonski diagram displaying the excitation of a Ru(II) complex to its ^{3}MLCT excited state and the major non-radiative deactivation processes competing with luminescence

ed state lifetime of $*[Ru(bpy)_3]^{2+}$, $*[Ru(phen)_3]^{2+}$ and $*[Ru(dip)_3]^{2+}$ are 0.61, 1.0 and 4.7 μs, respectively in oxygen-free aqueous solution at 295 K) [10]. While the *radiative* deactivation rate constant from the ^{3}MLCT state is similar for most Ru(II) polypyridyls ($3–6 \times 10^4$ s^{-1}), their *non-radiative* return to the ground state may show markedly different rate constants due to electronic factors ($10–1000 \times 10^4$ s^{-1}).

Deactivation of the ^{3}MLCT state may also proceed by thermally activated crossing to a near higher-lying metal-centred (^{3}MC) level which, in turn, decays to the ground state without emission (Fig. 5). Such (antibonding) MC excited state is also responsible for ligand photosubstitution reactions, i.e. undesirable dye *photodegradation* [27]. The latter effect is particularly dramatic in less polar solvents (or polymer matrices) and in the presence of nucleophilic counteranions or other basic species due to important ion pairing in these media. Photolability is minimized if chelating polyazaheterocycles with high-energy π^* orbitals or caged polypyridyl ligands are selected [27, 28]. Nevertheless, photodecomposition quantum yields of most Ru(II) polypyridyls in water is several orders of magnitude lower that those of purely organic dyes.

The MLCT nature of the electronic transition dramatically enhances the polarity of Ru(II) polypyridyls in their excited state, making their emission sensitive to the microenvironment (e.g. solvent polarity) around the dye. Moreover, deuteriation experiments have shown that the O–H oscillators are efficient quenchers of the ^{3}MLCT state so that the emission lifetime tends to be shorter if the Ru(II) dye is dissolved in hydroxylic solvents or immobilised on OH-containing supports [23].

The emission wavelength of Ru(II) polypyridyls undergoes a significant hypsochromic shift (20–30 nm) upon freezing of their solutions to 77 K. This effect has been attributed to *rigidochromism*: the highly polar ligand-localized excited state is quickly and strongly solvated in fluid solution; in a solid matrix (not necessarily at 77 K) the polar molecules around the photoexcited complex cannot re-orientate so fast and, therefore, a less stabilised excited state is produced [29]. A similar effect has been observed in many cases when the luminescent Ru(II) complex is electrostatically or covalently *immobilized* or *entrapped* in a solid support (Nafion) [30].

The largely triplet nature of the lowest-lying electronic excited state of Ru(II) polypyridyls (vide supra) determines the rather long *emission lifetime* (τ) displayed by these unique dyes. Although luminescence lifetimes ranging from 0.020 to 7 μs have been measured in deaerated fluid solution at room temperature, typical values in air-equilibrated media (including polymer-supported Ru(II) complexes) are found in the 0.1–2.5 μs region. Unlike *triplet* states of purely organic dyes (and some inorganic species), the ^{3}MLCT excited state lives short enough to survive quenching by molecular oxygen that kills phosphorescence at room temperature under most experimental conditions [31]; unlike the short-lived *singlet* excited states of most organic fluorophores, the ^{3}MLCT excited state lasts enough to facilitate the development of optosensors based on luminescence *kinetics* (both time-resolved and phase-sensitive detection) instead of just using the emission *intensity* [32]. Fluorescent dyes with decay profiles in the few nanosecond time window require sub-ns pulsed excitation sources and expensive (fast) transient digitizers for analytical measurements based on time-resolved emission kinetics; if phase-sensitive detection is used, MHz–GHz modulation frequencies are required to achieve the most sensitive analytical response (the emission phase shift, ϕ) since $\tan \phi = 2\pi f \tau$, where f is the modulation frequency of the excitation source [32].

Even with the latest advances in optoelectronics, the less demanding requirements (longer-pulsed light sources, relatively slower signal acquisition or kHz-modulated excitation) imposed by the longer-lived luminescent Ru(II) complexes have made them the dyes of choice for optosensing applications based on τ measurements.[5] The recent availability of inexpensive light-emitting diodes (LED) and laser diode (LD) sources down to the near UV range (370 nm), and particularly the *blue* LEDs and LDs [33], that can be either pulsed with ns width or modulated at the required frequencies, allow a perfectly matched excitation of the Ru(II) polypyridyls and cheap instrument designs. While the modern red-sensitive compact photomultiplier modules that contain power electronics and amplifier in a 20-cm^3 volume, are versatile sensitive detectors for those dyes, solid-state avalanche photodiodes (APDs) are a more rugged and cheaper option for detecting the emission from the most photoluminescent metal complexes.

Recently, Ru(II) polypyridyls covalently linked to polycyclic aromatic hydrocarbons (PAHs) (Fig. 6) have been demonstrated to display emission lifetimes in excess of 100 μs [34]. The energy of the lowest-lying triplet excited state (T_1) of pyrene, anthracene and other PAHs is similar to that of the ^{3}MLCT state of the Ru(II) complexes; as a consequence, an equilibrium is immediately established between ^{3}MLCT and T_1 to yield an extremely long-lived ^{3}MLCT emission (in rigorously oxygen-free solution) due to the ms excited state lifetime of T_1. The dramatic slow down of the ^{3}MLCT deactivation kinetics is a function of the PAH structure (pyrene boosts the ^{3}MLCT emission lifetime more than anthracene or naphthalene due to the more similar T_1 energy level of the former) and of the type of covalent link between the two moieties. The oxygen sensitivity of such conjugates increases dramatically according to their photophysical features.

---

[5] Luminescence *lifetime*-based analytical measurements are intrinsically more robust than *intensity*-based determinations. This is due to the fact that the former are not subject to fluctuations, drift and/or ageing of both the light source and the photon detector, are immune to dye leaching (as far as a minimum amount of it to be measured remains at the sensitive tip), and are not affected by photobleaching of the dye (provided the decomposition products are not luminescent). Advantages of the frequency domain approach are the following [4]: (i) the averaged frequency domain waveforms have less noise when compared to the time domain waveforms; (ii) the frequency domain signal is easier to capture since it expands over much longer time periods; (iii) signals with reduced harmonic content can be determined with fewer samples and, in turn, fewer data points translate into faster data processing speeds, an advantage for real-time implementations; (iv) no deconvolutions are necessary in the frequency domain compared to the time domain measurements, where the need for them adds extra software and computation time and limits the versatility of time domain techniques for real-time applications. Moreover, phase-sensitive luminescence detection is generally regarded as superior compared to time-resolved acquisition due to the cheaper and simpler electronics required to collect the signal in the former.

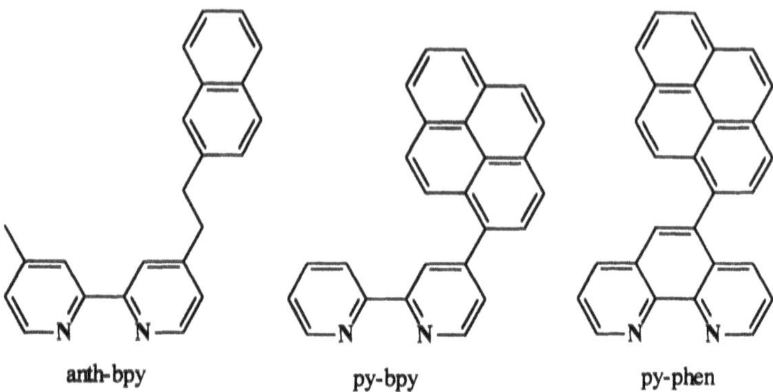

Fig. 6. Some α-diimine chelating ligands incorporating polynuclear aromatic hydrocarbons; anth-bpy = 4-methyl-4'-[2-(2-naphthyl)ethyl]-2,2'-bipyridine, py-bpy = 4-pyren-1-yl-2,2'-bipyridine and py-phen = 5-pyren-1-yl-1,10-phenanthroline

2.3 Redox Features

The existence of a metal center confers a rich redox (photo)chemistry upon Ru(II) complexes with polyazaheteroaromatic ligands [10]. The relative stability of other oxidation states (typically III and I) in this family of dyes, together with their long excited-state lifetimes (see above), makes them efficient partners in many photoinduced electron transfer reactions to and from their ^{3}MLCT level [11]. Occurrence of such processes opens up a competitive (photochemical) deactivation pathway, so that the luminescence intensity and lifetime of Ru(II) polypyridyls diminish accordingly. With the aid of widely available experimental techniques such as cyclic voltamperometry or pulse polarography, up to four redox potentials (namely $E^{3+/2+}$, $E^{2+/+}$, $E^{+/0}$ and $E^{0/-}$) can be routinely measured at room temperature in dry acetonitrile or dimethylformamide. The family leader $[Ru(bpy)_3]^{2+}$ displays ground-state redox potentials of

[6] There is an annoying difficulty in comparing the redox potentials reported in the literature for Ru(II) polypyridyls. Some authors report them vs. the Normal Hydrogen Electrode (NHE) while others prefer to do it vs. the Standard Calomel Electrode (SCE). While the difference between the potential of both reference electrodes is well known, correction of experimental values is not straightforward. Redox potentials for Ru(II) polypyridyls are usually measured in non-aqueous medium (dry acetonitrile) so that the use of a (aqueous) SCE reference leads to an important liquid-junction potential and severe drift due to the incoming organic solvent. Special reference electrodes are available for non-aqueous solvents, particularly the $Ag/AgNO_3$ in acetonitrile for this medium. However, the relationship between the redox potential vs. this electrode and SCE or NHE is ill defined. From our experience, the redox potentials of Ru(II) polypyridyls reported vs. NHE and SCE should be compared *directly* (i.e. without applying the conversion factor of 0.24 V that separates them), due to the fact that a fortunate coincidence roughly equals such a difference to the unavoidable liquid-junction potential (and other possible effects).

+1.26, –1.43, –1.61 and –1.86 V/NHE, respectively.[6] Nevertheless, the presence of *electron-withdrawing* ligands (e.g. pzth, Fig. 4) in the coordination sphere enhances the 3+/2+ redox potential of the corresponding Ru(II) complex: good π-acceptor heterocyclic rings possess low lying (vacant) π* orbitals capable of mixing efficiently with the filled metal d(π) orbitals ("π-back-bonding"), lowering the energy of the latter and therefore increasing the oxidation potential of the metal polypyridyl complex. Introduction of *electron-donating* ligands (e.g. pyim, Fig. 4) raises the d(π) level of the metal complex and provokes a decrease of $E^{3+/2+}$.

Unlike the oxidation potential, $E^{2+/+}$ of these dyes is related to the energy of their lowest-lying π* orbital of the ligand sphere. In this way, the reduction potential becomes less negative as the π-acceptor character of the heteroaromatic ligands increases. Given the variety of known polypyridyl chelating ligands, $E^{3+/2+}$ redox potentials ranging from +0.4 to +2.0 V/NHE and $E^{2+/+}$ redox potentials from –0.7 to –1.8 V/NHE have been determined [10]. Since the lowest excited state of Ru(II) polypyridyls is located ca. 2 eV above their ground state ($E^{0\text{-}0}$), the oxidation and reduction potentials of the former can easily be estimated ($E^{3+/2+*} = E^{3+/2+} - E^{0\text{-}0}$ and $E^{2+*/+} = E^{2+/+} + E^{0\text{-}0}$).

While their tunable redox potentials can make luminescent Ru(II) polypyridyls very attractive indicator dyes for a number of electroactive analyte species, it is also a matter of worry as far as chemical *interference* is concerned. For instance, oxidizing gases such as sulphur dioxide or chlorine severely perturb the response of luminescent oxygen sensors based on $[Ru(dip)_3]^{2+}$ and analogues. Sometimes the interfering species may be kept at bay through a judicious choice of the polymer support of the indicator dye: for instance, embedding $[Ru(dip)_3]^{2+}$ into a gas-permeable membrane (e.g. silicone rubber or polyester) to fabricate sensors for dissolved O_2 avoids interference of water-soluble metal ions or other polar species of appropriate redox potential (see below).

2.4 Preparation

The wide variety of heterocyclic chelating ligands described in the literature has led to multiple synthetic procedures to prepare their coordination complexes with Ru(II) [35]. Nevertheless, agreement on the optimum method of synthesis has not been reached so far. Anderson and Seddon [36] have stated that the best procedure to prepare *homoleptic* Ru(II) polypyridyls is reduction of commercially available $RuCl_3{\cdot}xH_2O$ (in alcohol solution) with dihydrogen (ca. 2 atm) in the presence of platinum black. Subsequent reaction with an excess of ligand leads to „high" chemical yields (ca. 60%) of a very pure metal complex (after column chromatography and recrystallization!) [36]. Heating of $RuCl_3{\cdot}xH_2O$ and the polyazaheteroaromatic ligand in methanol, ethanol or dimethylformamide has also been used. However, according to our experience, one of the best methods for preparation of homoleptic Ru(II) polypyridyls is that described by Meyer and coworkers [37]. Such procedure involves heating at reflux in ethylene glycol (ca. 200 °C) a solution of $RuCl_3{\cdot}xH_2O$ and 20–30% excess of ligand over the stoichiometric amount.[7] The advantages of this method can be summarised as follows:

- *Applicability* of the synthesis is wide enough; polypyridyl ligands with very different electronic properties lead to good yields (typically > 70%) of the corresponding $[Ru(NN)_3]^{2+}$ complex [25].
- *Product isolation* is very efficient after cooling down the reaction mixture to room temperature, dilution with some water, filtration through sintered glass and dropwise addition of a concentrated solution of ammonium hexafluorophosphate. The PF_6^- salt precipitates out immediately and can be recovered easily by filtration, followed by thorough washing with water to remove the alcohol solvent and ether (or halogenated solvents) to remove unreacted ligand. If the water-soluble chloride salt is desired, metathesis of the counteranion can be performed in an anion-exchanger column.
- Thanks to the high temperature (higher still can be achieved using glycerol instead of ethylene glycol), the *reaction time* is relatively short (typically 1–3 hours and very often less than 60 min) and the reaction can frequently be monitored visually: the initial dark brown of the solution ($RuCl_3$) changes quickly to dark green (corresponding to the $[Ru(NN)_3]^{3+}$ complex). Formation of the homoleptic Ru(II) complex is signalled by the orange-red colour change of the reaction mixture. The higher reaction rate of this procedure is a consequence of the reducing power of ethylene glycol (compared to monohydroxyl alcohols) and the high temperature of the synthetic procedure.
- Since no "foreign" reducing agents are used [36], the *purity* of the product is very high, particularly after re-precipitation of the homoleptic complex from diethyl ether/acetonitrile or diethyl ether/acetone mixtures.

It has been discussed above that *heteroleptic* Ru(II) polypyridyls (e.g. Fig. 3) are interesting indicator dyes because the different ligands allow a fine tuning of the electronic, physico-chemical and photochemical features of the complex. Heteroleptic Ru(II) polypyridyls are usually prepared from the precursor bis(chelate)dichloro complex and the stoichiometric amount (or a ca. 10% excess) of the second azaheteroaromatic ligand in a solvent compatible with the solubility of the latter (e.g. dimethylformamide, ethanol, methanol,...). The bis (chelate)dichlororuthenium(II) species is synthesised by heating the commercial $RuCl_3 \cdot xH_2O$ salt and the stoichiometric amount of ligand in the presence of a large excess of LiCl (typically 6–20 mol per mole of Ru(III) salt) in DMF. The reaction does not normally afford the target complex if the corresponding chloride salt of the homoleptic species is very insoluble in the medium used. In this case, the $[Ru(NN)_3]^{2+}$ complex precipitates out and is the major product. The bi s(chelate)dichlororuthenium(II) species is isolated by slow introduction of acetone (or other volatile miscible solvent) in the reaction mixture by placing both

[7] Although the reported procedure does not mention it, the reaction is best carried out under argon or dinitrogen. The inert atmosphere protects the chelating ligand from oxidation at the high temperature used, and facilitates reduction of the initial Ru(III) species to the more stable Ru(II) product.

the reaction mixture and the solvent in a closed container over a few days at room temperature. The acetone vapours progressively dissolve in the DMF solution and promote precipitation of the neutral metal complex.

The most critical issue of the synthesis of heteroleptic species is their essential *purification* from the accompanying reaction products (starting materials, the corresponding homoleptic complexes and other heteroleptic complexes formed by ligand methathesis). There is no general recipe to carry out an adequate purification: the ligand structure, overall charge of the metal complex, and the presence of particular functional groups may render the reported procedures useless. Currently, preparative HPLC is sometimes used but its cost and time-consuming optimisation of the product separation remove part of its appeal. Column chromatography is probably the most general purification technique. However, traditional stationary phases and eluents used in organic synthesis are not appropriate for separating mixtures of Ru(II) complexes due to the extremely high retention in those phases. Ion-exchange column chromatography on anionic or cationic Sephadex®, cellulose or polystyrene resins, together with an aqueous NaCl or HCl concentration gradient, are very convenient separation media. In any case, extremely caution should be exercised to avoid contamination of the luminescent heteroleptic complexes by the usually much more luminescent homoleptic species. Such contamination may ruin the effect of the analyte on the indicator sensitivity or may lead to undesired sensor responses. Actually, careless purification of Ru(II) complexes has sometimes led to literature claims of dual luminescence or multiexponential decay profiles, among other flaws.

Structural *confirmation* of the novel Ru(II) polypyridyls is not an easy task either. Molecular complexity (particularly for species with the lowest symmetry such as heteroleptic ones bearing asymmetric or asymmetrically-substituted chelating ligands), together with the small amounts of luminescent indicator dyes usually prepared after a multi-step synthesis, create such difficulties. Aside from the tedious crystallisation plus X-ray diffraction (XRD) technique, high-field nuclear magnetic resonance (≥ 300 MHz for ^{1}H-NMR) and the modern 2-D NMR methods are the most powerful structural elucidation tools, provided at least a 20–30 mg sample is available and solubility permits [38]. In this regard perdeuteriated dimethyl sulfoxide and acetonitrile are the most common solvents. A thorough analysis of the coordination-induced shifts of the NMR signals, i.e. comparing those of the free and the Ru(II)-chelated ligands, provides a „picture“ of the target molecule that rivals XRD, including structural and coordination isomer assignment, molecular configurations and electronic features of the metal-ligand interaction [38]. The huge chemical shift range of ^{99}Ru-NMR spectroscopy (more than 10,000 ppm!) and its correlation to the metal-centred electronic transitions were promising tools for investigating Ru(II) polypyridyls [39]. However, the small sensitivity of this quadrupolar nucleus, together with its very low resonance frequency ($\Xi = 4.605$ MHz) and wide signals, deter from the practical use of this technique for most complexes.

Mass spectrometry with electrospray ionisation is increasingly used due to wider availability of the specialised apparatus and the extremely small sample size required. Nevertheless, impurities may go unnoticed if they do not generate abundant ions with the same *charge* as the major Ru(II) product obtained Precise determinations of the luminescence lifetime of the novel dye in organic solvent solution, using time-correlated single photon counting (TC-SPC), may help to reveal impurities and even to shed light on their possible chemical identity. Acquisition of a high enough number of counts (typically over 10,000 at the peak channel) and sufficiently different emission lifetimes ($\geq$ 2-fold, e.g. those of a mixture of homoleptic and heteroleptic complexes) are desirable features for the success of this analytical technique. However, on-line monitoring (using fibre optics) in the UV-VIS region of the electronic absorption spectrum is the most convenient and simple way of following the progress of the metal complexation (for instance, from the dichlorobis(chelate) complex to the heteroleptic species, see above). Moreover, careful inspection of UV-VIS absorption features in combination with the above mentioned emission lifetime determinations in solution may be powerful criteria in many instances to assess the purity of the indicator dye.

2.5 Physical Properties

Ru(II) coordination complexes with azaheterocyclic chelating ligands are isolated as salts of various counter anions (PF_6^-, BF_4^-, Cl^-, ClO_4^-). They usually contain one or more solvent molecules after the crystal formation. Therefore, they are solids with high melting point (typically over 300 °C, with decomposition). Salts with the bulkier anions tend to be soluble in polar aprotic solvents (acetonitrile, dimethyl sulfoxide, dimethylformamide,...) and methanol; chlorides

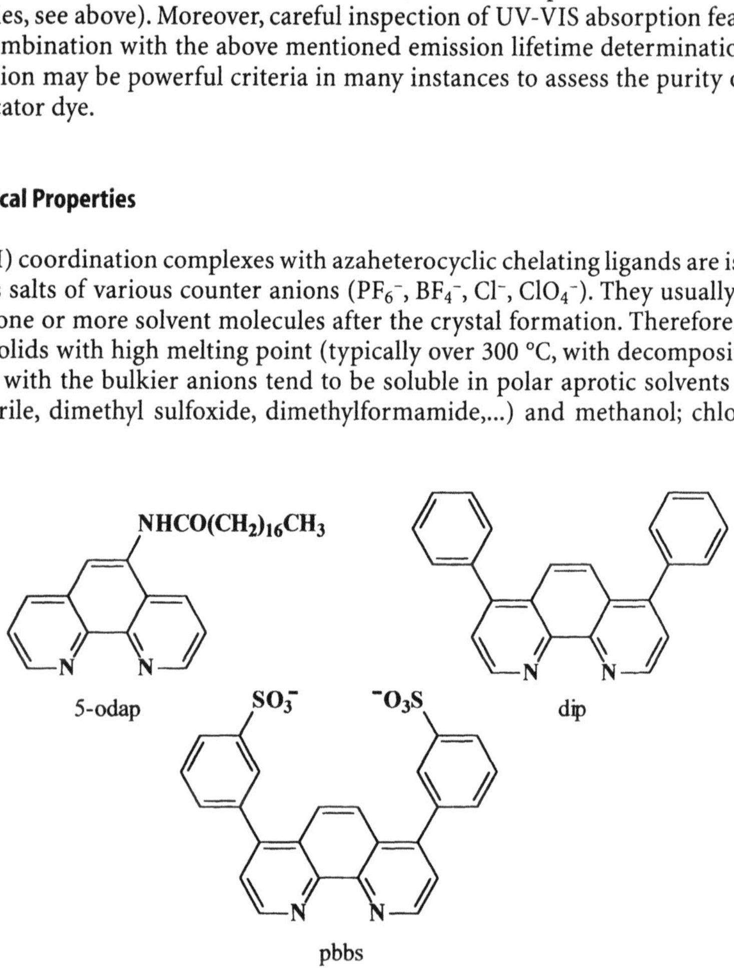

Fig. 7. Examples of structural modification in the 1,10-phenanthroline ligand to alter the physical properties of the corresponding Ru(II) complexes; 5-odap = *N*-1,10-phenanthrolin-5-yloctadecanamide, dip = 4,7-diphenyl-1,10-phenanthroline and pbbs = 1,10-phenanthroline-4,7-diyl)bis(benzenesulfonate)

are best solubilised in water. The solubility of Ru(II) complexes is also highly dependent on the peripheral substituents. For instance, tris(*N*-1,10-phenanthrolin-5-yloctadecanamide)ruthenium(II) (Fig. 7) is significantly soluble in toluene or chloroform [40], and a 10^{-2} M solution of tris(4,7-diphenyl-1,10-phenanthroline)ruthenium(II) (Fig. 7) in dichloromethane may be readily prepared, yet tris[(1,10-phenanthroline-4,7-diyl)bis(benzenesulfonate)]ruthenium(II) requires water to achieve the same concentration. Solubility of Ru(II) polypyridyls bearing acidic or basic functional groups (CO_2H, NH_2) is strongly dependent on pH. Amphiphilic counter-anions (e.g. sodium dodecylsulfate) have also been used to increase the solubility of the metal complexes in hydrophobic media (e.g. polymer matrices) [41]. Therefore, not only may the spectroscopic features of Ru(II) indicator dyes be finely tuned to the target analyte species, but it is also possible to adjust their physical characteristics to the dye immobilisation procedure and to the sample nature.

Introduction of tailored chemical groups in the periphery of the coordination compound is a way of *covalently* tethering the indicator dye to a suitable polymer support. Formation of a chemical bond between the optical indicator molecule and an inorganic or organic substrate is the best way of increasing sensor life, but suitable functional groups and linkers based on non-commercially available ligands are needed and the synthesis and purification procedures are lengthy, affording lower chemical yields. Moreover, it is not infrequent that the photophysical and photochemical properties of the indicator dyes (vide supra) undergo a substantial change upon covalent binding onto a solid support.

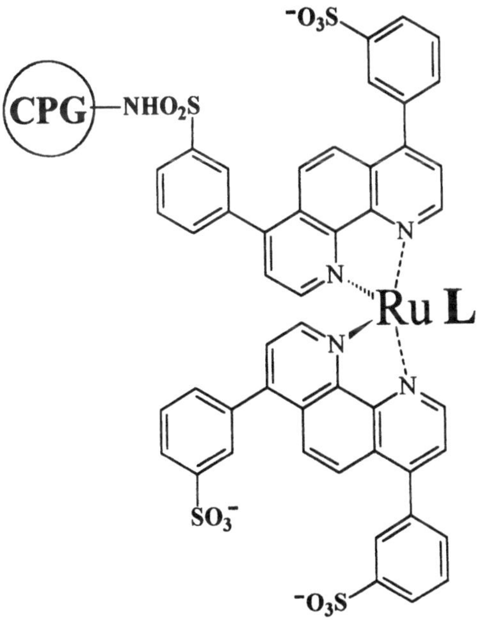

Fig. 8. Heteroleptic Ru(II) complex covalently bound to porous glass via a sulfonamide bond

Nevertheless, a careful selection of the functional groups and of the chemical reaction may yield very useful luminescent sensors. For instance, a Ru(II)-based sensor for measuring molecular oxygen in organic media has been developed by reacting sulfonate-substituted Ru(II) polypyridyls and aminopropyl porous glass (Fig. 8) [42]. Such a sensitive phase avoids the use of silicone or other polymer membranes (typically employed in oxygen optodes and electrodes) to hold the indicator dye that do not withstand many organic solvents such as those commonly found in the chemical industry.

2.6 Photochemistry

Ru(II) coordination complexes with polyazaheteroaromatic chelating ligands display a rich photochemistry due to their long-lived (lowest) excited state, significantly high oxidation and reduction potentials upon illumination, and feasibility of efficient energy transfer (see above) [10, 11]. Moreover, introduction of appropriate substituents into the ligands may provide additional photochemistry (e.g. proton transfer, redox, complexation, isomerisation,...). While the variety of photochemical reactions is excellent for designing a large number of luminescent sensors based on those complexes, it also poses the drawback of multiple interfering chemical species that might prevent, at a first glance, the development of useful sensors based on such materials. Polymer support of the indicator dye, together with additional polymer layers covering the indicator phase, are the best means of keeping off unwanted chemical and biological species.

The Stern-Volmer treatment of luminescence quenching (Eqs. 1 and 2) [43] describes bimolecular reactions involving photoexcited Ru(II) polypyridyls (τ_0, τ, Φ_0 and Φ are their excited state lifetimes and emission quantum yields in the absence and in the presence of quencher, respectively, and k_q is the bimolecular quenching rate constant).

$$\Phi_0/\Phi = 1 + k_q\tau_0\,[\text{quencher}] \tag{1}$$

$$\tau_0/\tau = 1 + k_q\tau_0\,[\text{quencher}] \tag{2}$$

Both equations are coincident if purely *dynamic* deactivation of the excited state of the luminophore occurs. However, if this process competes with *static* quenching (i.e. both species are associated before photoactivation, with a constant K_{as}), the expression for Φ_0/Φ contains higher order terms (e.g. Eq. 3) while Eq. (2) is still valid since the „static" component of the excited state deactivation may be regarded as instantaneous compared to the "dynamic" bimolecular process [44].

$$\Phi_0/\Phi = 1 + (k_q\tau_0 + K_{as})\,[\text{quencher}] + k_q\tau_0\,K_{as}\,[\text{quencher}] \tag{3}$$

Obviously, more complicated behaviours may be found, particularly in (micro)heterogeneous media [45,46]. This is a consequence of the heterogeneous distribution of Ru(II) indicator dyes, quencher molecules and photophysical and photochemical rate constants after incorporation into/onto solid supports. In these cases, the Stern-Volmer plots (Eqs 1 and 2) are no longer linear (even if purely dynamic quenching takes place). Examples of such behavior will be discussed for the oxygen sensors.

A number of *inorganic* species deactivate the ^{3}MLCT excited state of Ru(II) polypyridyls, most of them by photoinduced electron transfer (PET) [11]. Ions such as Fe(III), Cu(II), Eu(III), Tl(III), Ag(I), Hg(II) and $S_2O_8^{2-}$ quench *$[Ru(bpy)_3]^{2+}$ oxidatively to yield the Ru(III) species as the product. With Fe(II), Eu(II), SO_3^{2-} and ascorbate, the excited state of the complex (and luminescent relatives) is photoreduced. Quenching by Cr(III) and Ti(III) has been proposed to occur by energy transfer. Bimolecular deactivation rate constants (k_q) span four orders of magnitude: while *$[Ru(bpy)_3]^{2+}$ quenching by Eu(III) in water is relatively slow ($k_q = 3.6 \times 10^5$ L mol^{-1} s^{-1}), Fe(III) deactivates the same dye at a much faster rate ($k_q = 2.7 \times 10^9$ L mol^{-1} s^{-1}). Oxygen is one of the few molecules that effectively quenches Ru(II) polypyridyls ($k_q = 3.2 \times 10^9$ L mol^{-1} s^{-1} in water) predominantly by energy transfer. Other transition metal complexes also quench the luminescence from these dyes with rate constants in the range of 10^8 to 10^{10} L mol^{-1} s^{-1}: $[Co(NH_3)_6]^{3+}$, $[Ru(NH_3)_6]^{3+}$, $[Fe(C_2O_4)_3]^{3-}$, $[Rh(bpy)_3]^{3+}$, $[Cr(bpy)_3]^{3+}$, $[Fe(phen)_3]^{3+}$, $[Co(phen)_3]^{3+}$, $[Fe(CN)_6]^{4-}$, $[Ni(CN)_4]^{2-}$ and $[Cu(acac)_2]$ are a few examples. Static quenching with oppositely-charged species plays an important role in the quenching processes and ion-pairing constants can be calculated therefrom. *Organic* compounds also quench the luminescent ^{3}MLCT state of Ru(II) polypyridyl dyes [11]. Oxidative deactivation has been demonstrated by laser flash photolysis for nitroaromatics, quinones and bipyridinium ions („viologens“), with rate constants above 10^9 L mol^{-1} s^{-1} (i.e. close to diffusion-control rates) in water and organic solvents. *$[Ru(bpy)_3]^{2+}$ and the like are also efficiently quenched reductively by aromatic amines and dithioanions to afford the Ru(I) complex. The latter are strong reductants that react with numerous species including water, so that these reactions are normally carried out in non-aqueous solvents. Energy transfer to anthracene, pyrene, stilbene and benzophenones has been documented. Organometallics such as ferrocene and related species also quench the excited state of Ru(II) polypyridyls by photoinduced energy transfer.

Particularly interesting for the development of optosensors based on Ru(II) polypyridyl dyes is luminescence *resonance energy transfer* (LRET) to suitable organic dyes. If the donor (Ru complex) and the energy acceptor (typically a colourimetric indicator that absorbs in the red) are within the characteristic Förster distance, and if the acceptor absorption spectrum changes in response to the analyte, then the emission intensity and lifetime of the donor will change in response to the analyte [47]. Consequently, LRET can be used for sensing a wide variety of species (pH, CO_2, ammonia, glucose,...) [47]. Nevertheless, the phenomenon of LRET requires an acceptor concentration of about 1–10 mM to achieve a suitable proximity with the co-dissolved donor. Such concentrations result in high optical densities at the excitation and the emission wavelengths, and inner-filter effects are difficult to overcome, so that absolute intensity measurements are difficult to perform. Moreover, lifetime-based measurements rely on the absorption by the acceptor, so that the system cannot correct for possible bleaching or leaching of the latter.

If a ^{3}MC excited state is populated efficiently (by thermal activation, vide supra) from the ^{3}MLCT state (Fig. 5), then the complex will undergo an additional photochemistry of ligand *dechelation*. This is due to the $\sigma^*(e_g)$ antibonding nature of the orbital where the excited electron dwells. Photodetachment of a bidentate ligand may lead to rebinding and isomerization, or to *photosubstitution* reactions by nucleophiles present in the vicinity of the complex (anions, solvent molecules, functional groups in the polymer support,...), particularly if the Ru(II) polypyridyl is dissolved in a medium with low ionisation power [27]. Such dechelation photochemistry is deleterious for long term monitoring in environmental and industrial processes since it leads to photobleaching and compromises long term stability of the sensor head. Fortunately, most Ru(II) bis(chelate) complexes are scarcely or non luminescent, so that the effect of indicator photobleaching may be avoided using emission lifetime-based measurements (see above).

3
Acidity Sensors

Introduction of proton-donor or proton-acceptor groups (or atoms) into the polypyridyl ligands coordinated to Ru(II) (e.g. those depicted in Fig. 9) leads to luminescent complexes sensitive to the solution acidity. In most transition metal complexes, the coordinated ligand is more acidic (or less basic) than the free ligand. The metal ion acts as an electron-withdrawing atom which is satisfied by σ-donation from the ligand [38]. However, the excited state acid-base features of such dyes may differ significantly from those in their ground state due to important changes in the electron distribution between the two states (see above). A general scheme of the proton transfer processes in Ru(II) polypyridyls is summarised in Fig. 10.

The theoretical basis of proton transfer in the (electronically) excited state has been described many times in the literature [48, 49]. The Förster treatment (or „Förster cycle") allows an estimation of the excited state acidity (pK_a^*) based only on *thermodynamic* grounds, i.e. the indicator dye pK_a value and the excited state energies of the corresponding acid and basic species ($E_{HR}^{0\text{-}0}$ and $E_R^{0\text{-}0}$, respectively, Fig. 10). If entropic changes are neglected and if the protonated complex emits at higher energy than the deprotonated one, then Eq. (4) holds:

$$pK_a^* = pK_a + (E_R^{0\text{-}0} - E_{HR}^{0\text{-}0})/2.303RT \qquad (4)$$

However, accurate values of the 0-0 transitions are required to get meaningful values of the excited state levels. This is difficult to determine for Ru(II) polypyridyls so that the Förster cycle for these complexes should only be considered as a guideline with limited quantitative applicability [49]. Equation (4) tells us that if R* luminesces more to the red than HR*, the former will be a weaker base in its excited state. If the opposite situation is observed (i.e. R* emits at higher

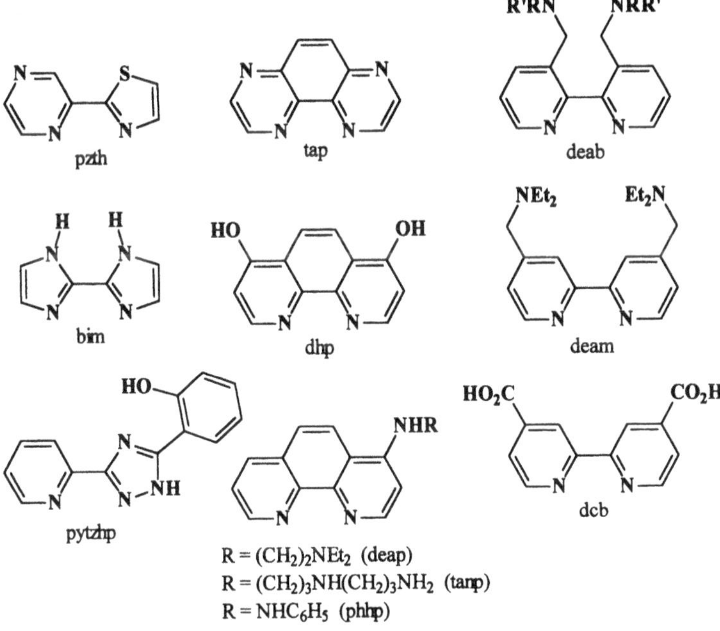

Fig. 9. Polyazaheteroaromatic chelating ligands used to prepare pH-sensitive Ru(II) complexes; pzth = 2-(1,3-thiazol-2-yl)pyrazine, tap = pyrazino[2,3-*f*]quinoxaline, deab = *N*-({3‘-[(diethylamino)methyl]-2,2‘-bipyridin-3-yl}methyl)-*N*,*N*-diethylamine, bim = 1*H*,1‘*H*-2,2‘-biimidazole, dhp = 1,10-phenanthroline-4,7-diol, deam = *N*-({4‘-[(diethylamino)methyl]-2,2‘-bipyridin-4-yl}methyl)-*N*,*N*-diethylamine, pytzhp = 2-(3-pyridin-2-yl-1*H*-1,2,4-triazol-5-yl)phenol, deap = *N*,*N*-diethyl-*N*‘-1,10-phenanthroline-4-ylethane-1,2-diamine, tanp = *N*-[3-(1,10-phenanthrolin-4-ylamino)propyl]propane-1,3-diamine, phhp = *N*-phenyl-1,10-phenanthrolin-4-amine, dcb = 2,2‘-bipyridine-4,4‘-dicarboxylic acid

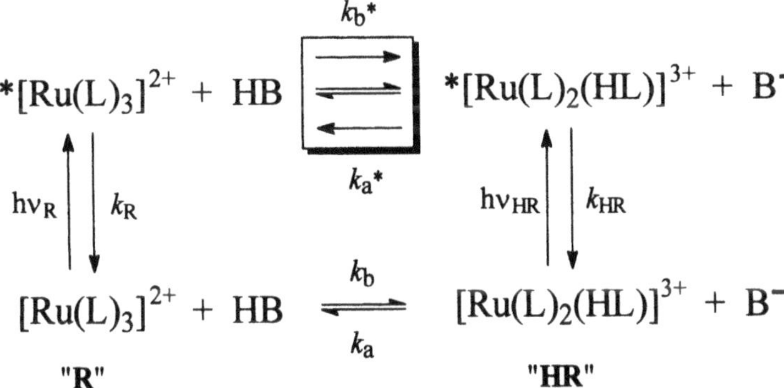

Fig. 10. General scheme of the proton transfer processes in luminescent Ru(II) polypyridyl complexes (a basic one with homoleptic structure has been chosen to illustrate the phenomena, but it can be also applied to heteroleptic or acidic complexes)

energy than HR*), then R* will be a stronger base (HR* a weaker acid) than the corresponding ground state species.

In addition to thermodynamics, the *kinetic* aspects of the excited state acid-base system (Fig. 10) must also be considered. If either R* or HR* are short-lived (i.e. the rate constants k_R or k_{HR} are large enough) then an equilibrium between them will not be established. In such a situation, proton transfer from/to the ^{3}MLCT state will be an irreversible process and the pK_a* calculation will be meaningless. Unfortunately this seems to be the case for many Ru(II) polypyridyl complexes (e.g. $[Ru(pzth)_3]^{2+}$, $[Ru(tap)_3]^{2+}$, etc) [20, 50]. If irreversible proton transfer quenching of the excited state in competition with radiative (and non-radiative) decay pathways occurs, the strength and concentration of the Brönsted acid or base used to test the effect of pH on the emission of the Ru(II) indicator dye will determine the experimental (sigmoidal) curve. This phenomenon has been clearly evidenced by Orellana and coworkers [20, 51] In this case, the luminescent Ru(II) complex is useless as a general pH indicator dye.[8] Unless different buffers have been tested and a thorough photophysical investigation is performed, it is difficult to predict (and discover in the literature reports!) which Ru(II) polypyridyls will display acid-base equilibrium in their lowest excited state.

Orellana and Moreno-Bondi were the first to realise that luminescent Ru(II) complexes bearing at least one pH-sensitive ligand could be employed to monitor the solution acidity, using both emission intensity and lifetime measurements [52]. A brief report of the pH effect on the emission of $[Ru(bpy)_2(deab)]^{2+}$ (Fig. 9) in solution was published shortly afterwards [53]. Inflection points in the intensity pH curves were found at pH 1.9, 2.4, 4.1, 5.9, 7.8 and 8.6, depending on the alkyl and aryl substituents of the tertiary amine groups bound at the 3,3' positions of the bpy system. Similar 4-mono and 4,4'-disubstituted dialkylaminomethylbipyridine (deam) heteroleptic Ru(II) complexes have been described by Murtaza et al. [54] and by Xu et al. [55]. While the former shows a pH effect on its luminescence in the 6.8–8.5 range, the latter undergoes emission changes in the 3–5 and 8–10 pH ranges when immobilized onto a cyclic siloxane/poly(ethylene oxide) support. Heteroleptic Ru(II) chelates with 1,10-phenanthroline-4,7-diol (dhp) have been electrostatically attached to the latter support [56] or to perfluorosulfonated Nafion® membranes [57, 58] for measuring pH in the 2–6 and 1–8 ranges, respectively. Response times of 3 to 6 min have been reported for those sensors. The emission of monosubstituted 4-hydroxy-1,10-phenanthroline complex with phen ancillary ligands shows the same pH behavior than the disubstituted analogue [21]. The pH effect on heteroleptic deap, tanp and phhp (Fig. 9) Ru(II) complexes has been reported, too [59]. For all of them a decrease in their luminescence intensity and lifetime is observed at high

[8] The situation is so dramatic that buffers of approximately the *same* pK_a value (e.g. tris-(hydroxymethyl)aminomethane and phosphate, both with $pK_a \sim 7$) quench $^*[Ru(pzth)_3]^{2+}$ at extremely different rates: while the latter shows $k_q = 5.94 \times 10^8$ dm^3 mol^{-1} s^{-1}, the acidic species of the former displays $k_q \approx 0$ due to its positive charge that prevents competitive excited state bimolecular deactivation.

alkalinity, pH 7–11, 8–12 and 9–12 being their respective variation ranges due to deprotonation of the 4-amino group.

Acidic heterocycles such as the imidazole and triazole moieties of bim and pytzhp of heteroleptic $[RuL_2(bim)]^{2-}$ and $[RuL_2(pytzhp)]^{2+}$ complexes (Fig. 9) have been described as pH sensors in the 4–7, 5–9 (bim; L = pbbs or L = bpds, where bpds stands for 2,2'-bipyridine-4,4'-disulfonate, respectively) [52, 59], and 3–8 range (pytzhp; L = bpy) [60]. The bim complexes have been immobilised onto cationic Nylon membranes and cationic glass beads or covalently attached to porous glass via the ancillary sulfonate groups [61]. The pyridyl-triazole complex was incorporated into a silica xerogel. The same support has been used to entrap $[Ru(bpy)_2(dcb)]^{2+}$ (dcb = 2,2'-bipyridine-4,4'-dicarboxylic acid, Fig. 9) [60]. The fabricated luminescent pH sensor shows response between pH 3 and 7. Luminescence resonance energy transfer from homoleptic $[Ru(bpy)_3]^{2+}$ and $[Ru(phen)_3]^{2+}$ complexes to colourimetric pH indicator dyes such as bromothymol blue, has been employed to fabricate acidity-sensitive terminals upon co-immobilisation in plasticised PVC [62, 63]. The colorimetric indicator controls the pH working range of the luminescent optosensor and both intensity- and lifetime-based measurements can be performed but, in this case, the latter do not correct for indicator bleaching or leaching. On the positive side, LRET sensors are not subject to the problem of absence of excited state acid-base equilibrium. Finally, perturbation in the pH 5–9 range of phase-sensitive detection of the luminescence from $[Ru(bpy)_3]^{2+}$ by the pH-sensitive 6-carboxyfluorescein co-immobilised on poly(vinyl acetate), has been described as a novel sensing principle [62]. Although it is an emission lifetime-based measurement, it depends on the ratio between the fluorescence *intensity* of the organic and the Ru(II) polypyridyl dyes so that the sensor is not immune to indicator bleaching or leaching. In any case, sensors based on the combination of pH-insensitive and pH-sensitive dyes are certainly easier to fabricate, since no specific heterocyclic chelating ligand and the corresponding heteroleptic Ru(II) have to be synthesised (and purified!).

4 Carbon Dioxide Sensors

In a similar fashion to the pH electrode, separation of a pH-sensitive tip from the sample with a gas-permeable membrane allows monitoring of carbon dioxide, due to the hydrolysis it undergoes with the release of hydronium ions [3, 64]. Therefore, a good pH optode based on Ru(II) polypyridyls may be readily adapted to CO_2 measurements, provided it displays the appropriate pK_a^* value for an optimum response to the analyte (≈7). This fact seems to be the reason for the lack of CO_2 sensors based on such dyes reported so far. However, unlike pH measurements, since the sample does not directly contact the pH-sensitive indicator layer, even luminescent Ru(II) polyazaheterocyclic complexes that undergo *irreversible* proton transfer reactions in their excited state may be used to monitor CO_2, both in the gas phase and in solution [52].

Orellana and co-workers have published a number of papers describing the basics and analytical application of fibre-optic CO_2 measurements using

$[Ru(pzth)_3]^{2+}$ (Fig. 9) and both emission intensity- and lifetime-based methods [20, 51, 65, 66]. The indicator dye, immobilized in a polymer gel phase, is separated from the sample by a thin silicone membrane. The working principle follows the scheme depicted in Fig. 10. The ground state complex is completely non-protonated ($pK_a = -1.9$); however, its basicity increases more than 10^6-fold in its ^{3}MLCT excited state due to the high acceptor character (low-lying π^* orbital) of the pyrazine moiety. Therefore, it undergoes efficient (irreversible) proton transfer from suitable Brönsted acids present in the reservoir indicator phase (phosphate, hydrogen phthalate, acetic acid, H_3O^+,...). Thus, the pH value of the buffered (inner) sensitive layer may be adjusted to provide little quenching to $^*[Ru(pzth)_3]^{2+}$ (e.g. 0.1 M phosphate buffer at pH 8). Permeation and subsequent hydrolysis of CO_2 will increase the concentration of the acidic form of the internal buffer and the luminescence from the indicator dye will decrease accordingly (and so will its emission lifetime) (Fig. 10). This mechanism has been clearly demonstrated by the linear Stern-Volmer plot obtained at different CO_2 levels in the gas sample when the corresponding values of the internal hydrogenphthalate concentration are input in the x-axis [51]. In this way, a limit of detection of 3.5×10^{-4} MPa (0.4% v/v) CO_2 (in the gas phase) is achieved with a sensor containing $[Ru(pzth)_3]^{2+}$ dissolved in 0.1 M phosphate pH 8.2 buffer, with a dynamic range up to 0.09 MPa CO_2.

The gel-type nature of the indicator phase makes it suitable only for aqueous measurements and moist gas determinations due to small but non-negligible permeation of water vapor across the gas-permeable membrane. The sensitivity of the device is adequate for blood-gas analysis, food storage and protective atmosphere determinations, but not enough for CO_2 measurements in seawater. A recent application to effluent gas monitoring in *compost processes*, with simultaneous fiber-optic O_2 measurements, has been described [67].

CO_2 optosensors based on energy transfer to pH-sensitive colorimetric indicators from luminescent Ru(II) polypyridyls (see Section on pH sensors) have been reported too [68]. Alternatively, another possibility of developing emission lifetime-based pH/CO_2 sensors is the so-called dual luminophore referencing (DLR) technique [69]. Perturbation of a long-lived luminescence (such as that from ruthenium complexes) by a short-lived fluorophore that absorbs and emits in the same wavelength range provokes an important phase shift in the signal from the former. If only one of the two dyes is sensitive to the analyte the combination indicator layer may be used to fabricate an appropriate sensor using phase-sensitive detection of the luminescence. Actually, it is a clever way of transforming absolute intensity-based signals into lifetime-based measurements, but it is affected by the luminophore's bleaching or leaching. Its application to non-destructive *packaged food analysis* has been very recently described [70]. The sensor is based on a membrane sandwich of the analyte-insensitive $[Ru(dip)_3]^{2+}$ complex and the pH-sensitive 8-hydroxypyrene-1,3,6-trisulphonic acid (HPTS) fluorophore. It is expected that more CO_2 sensors will appear in the near future as more perfect pH optodes based on such metal complexes are developed.

5 Temperature Sensors

Temperature is one of the most important parameters in process control and environmental monitoring. Although thermometry by mechanical or electrical methods is widely employed, quantification of this parameter can be difficult sometimes due to hostile media, restricted accessibility and/or the presence of strong electromagnetic fields that preclude the use of conventional thermometers. Commercially available non-contact IR thermometry may be used under some of those situations provided there is a straight free path between the measuring device and the sample. When this requirement is not met, fibre-optic sensing offers a convenient alternative. Among the various temperature-dependent optical properties that can be used to develop a suitable *optode*, luminescence thermometry is probably the most sensitive and versatile technique [71]. In addition to numerous inorganic materials, temperature sensing with Ru(II) polypyridyls has been reported and patented [6, 7, 72–74]. It seems that the only temperature indicators of this class used so far have been those ubiquitous for oxygen sensing, namely $[Ru(phen)_3]^{2+}$ and $[Ru(dip)_3]^{2+}$. Unfortunately, the same photophysical features that provide a good sensitivity to O_2 [16], make them less sensitive to temperature [75].

The basis of the temperature effect on the emission from the ^{3}MLCT state of Ru(II) polypyridyl complexes has already been described above (Fig. 5). Mathematical derivation of the function that relates the luminescence lifetime to the sample temperature allows to find the maximum response [72]. However, since the thermally-activated ^{3}MC level is a dissociative one, the challenge is to design and prepare photostable indicators of this family showing the maximum temperature sensitivity [76]. This can be done by selecting the chelating ligands along the lines discussed in this chapter. Interference from molecular oxygen and other quenchers present in the sample is excluded by encapsulating the dye in a (transparent) plastic support such as poly(acrylonitrile). In this way, it is possible to achieve a temperature resolution of 0.1 °C in the –20 to 50 °C range. Response times are strongly dependent on the sensitive layer thickness and on the nature of its polymer. Simultaneous measurement of temperature and analyte (typically oxygen) may even be performed using a two-sector sensitive layer at the distal end of a fibre-optic bundle, provided one of them is analyte-permeable [74]. In this way temperature compensation of the response from Ru(II)-based optosensors may be realised.

6 Oxygen Sensing with Luminescent Ru(II) Polypyridyl Dyes

Molecular oxygen is a very important species in nature. It is produced by photosynthesis in the presence of light and consumed via different respiration processes, since it is the final electron acceptor for degradation of organic material. Therefore, oxygen is a central substrate for aerobic organisms. Adequate availability of molecular oxygen is a prerequisite for most fauna and flora in the biosphere, and microbial processes are largely governed by the availability of oxygen as well. In the atmosphere, both the diffusion rate and the concentration of oxygen are sufficiently high to consider that it is always available. However, this is not the case under water due to the low solubility of oxygen in aqueous systems (7.55 mg O_2 per L at 1 atm and 298 K). The amount of available oxygen is, for aquatic organisms, at least as important as the quantity of food, so that the oxygen concentration is an essential parameter of studies in aqueous media. In this way, there are many application fields where oxygen determination is important, for instance, in environmental applications such as biological oxygen demand analysis [77], or in medical applications, where it is of utmost importance to monitor dissolved oxygen levels in the blood, as blood transports oxygen to tissues and cells throughout the body (e.g. physiological and critical care, among other medical studies) [78–82]. Control of the oxygen level is also needed in many processes within biotechnological, chemical and petrochemical industries, such as fermentation, composting, chemical synthesis, and combustion, where concentration of this gas is often a key factor, and in the food preparation industry [83, 84]. Continuous monitoring of the oxygen concentration is required in water treatment plants, drinking water and environmental analysis of continental and sea water, since the amount of dissolved oxygen is an indicator of the water quality and a decrease in this amount usually signals the presence of organic waste [85]. Sensitive devices having different dynamic ranges are required when applications are to be developed. For example, oxygen monitoring in a patient's breath requires an optical sensor that works optimally at 159 Torr whereas a sensor for checking the integrity of an oxygen-free food package would need to perform best at around 1 Torr.

Because of its universal importance numerous techniques for measuring gas-phase and dissolved oxygen have been developed [86]. Winkler iodometric titration has been adopted for oxygen determination for many years and is still considered, to some extent, to be the standard method. However, the time-consuming and cumbersome nature of the titration prevents its application to process monitoring. The amperometric Clark electrode developed in 1956 was a breakthrough since it is easily calibrated, has relatively rapid response and requires relatively inexpensive instrumentation [87].

During the 1970's and the 1980's, the scientific community working on oxygen sensing was aware of the need develop new measurement schemes to replace the widely used Clark electrode in those situations where its limitations show up. The electrochemical oxygen sensor is based on the analyte consumption and suffers from electrode poisoning, flow dependence, calibration drifts, electrical interference and fragility. Moreover, it is difficult to miniaturise and cannot be used to monitor this species in the gas phase, so that alternative techniques must be used (galvanic cells, paramagnetic detectors, zirconia oxide monitors, etc.). Therefore, the development of alternative technologies to monitor oxygen concentrations in many environments, that were comparable in cost and superior in performance to the Clark electrode, was a challenge to scientists working in the area of optical sensing. Although some of the problems of the Clark electrode have been properly addressed during the last years, compact fibre-optic-based oxygen sensors have reached a competitive price and a satisfactory level of performance (low energy consumption, calibration required only once a year, little sensitivity to contamination, biofouling or ageing, easy miniaturization, suitability for long measuring periods and short response time) [8]. Some of them, particularly those in the medical market, have been commercially available for more than ten years [88]. No doubt Ru(II) polypyridyl indicator dyes are responsible for this commercial success.

6.1 Oxygen Optosensors

Molecular oxygen is a well-known quencher of the excited states of luminescent molecules [89–94]. Therefore, many luminophores such as polynuclear aromatic hydrocarbons, some heterocyclic compounds, transition metal coordination compounds and metalloporphyrins, all of them having long emission lifetimes (for best sensitivity), were target indicator dyes to develop oxygen sensors. Several reviews and book chapters about environmental sensing or fluorescence spectroscopy give the main features that any oxygen sensor based on luminescence measurements must fulfil [3, 95–99].

Optical oxygen sensing is usually based on collisional quenching by molecular oxygen of a luminophore embedded in a polymer support [100, 101]. The oxygen quenching process is described by the Stern-Volmer equations for emission intensity or lifetime measurements (Eqs. 1 and 2). The slope of the plot (also known as the Stern-Volmer constant, K_{SV}) is a measure of the oxygen sensitivity of the sensor. Oxygen partial pressure (P_{O_2}) can be substituted for the oxygen concentration in the Stern-Volmer equation for quenching processes in liquid phase if one takes into account Henry's law and the oxygen Henry coefficient in a given solvent. The Stern-Volmer equations (Eqs. 1 and 2) are linear and coincident under these conditions since quenching of a luminophore by O_2 in solution is a dynamic process. However, in (micro)heterogeneous media, Stern-Volmer plots typically display a downward curvature (see below) [40, 42, 102, 103].

In order to fabricate an optical sensor, luminophores are usually *immobilised* on a polymer support, which commonly plays an active (but not always predictable) role in the sensor response, compared to the well-known behavior observed for dissolved indicators. Materials for oxygen optosensing should (i) have a good O_2 permeability for a rapid response; (ii) display a high local quenching constant around the indicator molecule for good sensitivity; (iii) permit an easy and reproducible immobilisation of the indicator dye; (iv) possess good optical properties and high thermal, mechanical and chemical stabilities; (v) be easily and robustly attached to a light-guide system (e.g., an optical fibre) and (vi) provide some antibiofouling features when they are to be used for environmental or in vivo applications. A high loading of indicator in hydrophobic/hydrophilic polymers is often necessary to obtain sufficient signal intensity, specially in the case of micro-optodes, since the sensing area on the fibre is extremely small and the luminescence signal is relatively low. Although it is not a requisite for oxygen sensing in the gas phase, water insolubility of the supported indicator is another essential property in order to minimize dye leaching in sensors for dissolved oxygen measurements.

As far as the indicator molecule is concerned, good oxygen sensing materials require photostable luminescent indicators excitable in the visible region, with moderate to high emission quantum yields (Φ_L) and long excited state lifetimes in deoxygenated media (τ_0), for developing low cost, compact lifetime-based measuring instrumentation. Also high bimolecular oxygen quenching constants (k_q) are required even when the indicator dye is supported on the polymer matrix.

6.2 Luminescent Ru(II) Complexes as Oxygen Indicators

Luminescent transition metal coordination complexes, particularly those of ruthenium containing polyazaheterocyclic chelating ligands, are very attractive oxygen indicator dyes. Their unique (photo)physical and (photo)chemical features (vide supra) make these molecules superior in performance to other O_2 indicators having long-lived excited states, such as metalloporphyrins or polyaromatic hydrocarbons. The importance of Ru(II) coordination compounds is evidenced by the number of scientific publications and patents in which these molecules are used for implementation of oxygen sensing schemes and devices, besides the number of commercial monitors based on these dyes [61, 88, 104]. The most commonly used Ru(II) complexes for fabricating oxygen sensors have been those of the type $[Ru(L)_3]^{2+}$ with L = 2,2'-bipyridine (bpy), 1,10-phenanthroline (phen) and 4,7-diphenyl-1,10-phenanthroline (dip). The long emission lifetime of the latter has made it particularly attractive.

6.3 Polymer Support and Indicator Design

Devising the sensitive phase is one of the critical issues in the development process of any sensor. The indicator layer contains an analyte-sensitive compound supported in a polymeric matrix, normally a thin film, which often determines a selective interaction between the indicator and the analyte, in this case, molecular oxygen. For this reason oxygen-sensitive membranes have to be characterised thoroughly from both the photochemical and analytical points of view [105]. Therefore, in order to fabricate an oxygen indicator material, the luminescent Ru(II) coordination compounds have been immobilised in *organic polymers* such as polystyrene, polyvinyl chloride, polymethacrylates, polyacrylamide, modified celluloses and perfluorinated ionomers [30, 106–109]. *Inorganic polymer* supports (zeolites, silicone, silica gel, porous glass and silica xero-gels) have also been used [101, 102, 110–115]. The indicator dye is supported following different immobilisation procedures (physical entrapment, adsorption, electrostatic interaction or covalent bonding) that usually have a dramatic effect on the indicator luminescence characteristics and on the oxygen response of the optode in terms of sensitivity, long-term stability and linearity of the Stern-Volmer plots. In this way, Ru(II) complexes have often served as benchmarks of theoretical models of collisional quenching by oxygen in heterogeneous phases (see below in this Section).

An important aspect that has to be taken into account is that many commercially available prepolymers (specially silicone precursors) contain non-volatile solvents, fillers, low molecular weight crosslinkers, catalysts and other additives, which can radically affect the performance of the sensor. These proprietary components may cause unexpected problems due to various chemical and physical interactions between the indicator, the analyte and the support of „unknown" structure [102, 116–118].

Among the non-porous materials used for oxygen sensing, room temperature vulcanization (RTV) silicone polymers are the most important matrices because of their extremely high oxygen permeability (367×10^{-13} cm^3 cm/cm^2 s Pa) [119]. This feature imparts very fast responses, good optical transparency and chemical inertness, despite the lack of mechanical robustness. A broad range of materials less permeable to oxygen have also been employed, but they require indicator dyes with higher sensitivity embedded in very thin layers (< 5 μm). In general, chloride or perchlorate salts of Ru(II) coordination compounds such as $[Ru(bpy)_3]^{2+}$ and $[Ru(phen)_3]^{2+}$ are water-soluble species, and even the diphenyl substituted complex $[Ru(dip)_3]^{2+}$ is rather hydrophilic; therefore, they are soluble neither in hydrophobic silicone rubbers nor in most organic polymers, where they tend to aggregate and precipitate producing oxygen-sensitive matrices that show a bad performance, suffer from dye leaching into the aqueous media and/or film fogging due to water uptake. This problem has been solved by two different strategies that have been applied to almost every oxygen optosensor: (i) modification of the polymer matrix with incorporation of additives (at the expense of increasing the heterogeneity of the sensor phase); (ii) modification of the indicator dye via the coordinated ligands or the counter-ion of the metal complex ion.

Pure homogeneous polymers derived from polydimethylsiloxane can be formulated to give acceptable sensor responses even without the addition of a heterogeneous component (filler) such as silica [117]. However, these polymers show highly nonlinear Stern-Volmer plots, while the silica-doped systems are often more linear and display steeper responses. This common observation apparently reflects the fact that even in „homogeneous" (i.e., chemically pure) polymers, the indicator dye still dwells in different microenvironments (e.g. regions of low and high quenching efficiency) that are not time-averaged over the excited state lifetime. Therefore, performance can usually be enhanced by adding a heterogeneous component, which selectively adsorbs the sensitive dye and affords a good response [117].

Addition of a filler material (e.g. hydrophobically- or pyrolytically-produced silica gel), dyed with the adsorbed luminescent indicator, to the silicone pre-polymer before vulcanization has been a common approach because of the better sensitivity, response linearity and ease of preparation of the indicator material it provides [120]. The contribution of nonpolar and electrostatic interactions in the adsorption process of $[Ru(bpy)_3]^{2+}$ to silica has been investigated [121]. In fume silica, it has been argued that oxygen reaches the indicator dye via surface diffusion and quenching is exclusively dynamic. However, the Ru(II) adsorption sites are highly heterogeneous as shown by nonexponential emission decay kinetics both in the presence and absence of oxygen [112, 114, 120, 122]. Oxygen quenching of the metal complex in these supports is completely reversible and highly reproducible. Response times of the corresponding sensors are typically less than 1 min for gas-phase oxygen sensors and longer than 1 min for dissolved oxygen sensors, due to the necessary phase transfer step (in the recovery cycle, response times are even longer compared to pure silicone because molecules such as oxygen are known to adsorb strongly on silica surfaces) [78]. As expected, the response time of the sensitive film depends on its thickness [123]. The silicone prepolymer can also be mixed previously with the filler and then, after vulcanization of the polymer, the membrane is dyed with indicator. In this way, $[Ru(dip)_3](ClO_4)_2$ was incorporated into silicone rubber by using a dichloromethane solution of the complex, since this solvent swells up and penetrates the film [105]. This approach usually yields indicator layers with higher O_2 sensitivity and shorter response times, since accessibility of oxygen to the Ru(II) complex is facilitated; however, fabrication of the sensing phase is more tedious and the mechanical stability of the material decreases (film curling). Independently of the chosen method for dye immobilisation, the heterogeneity around the luminophore is significant since, due to the various interactions between the polymer and the silica filler, there are multiple binding sites for the Ru(II) polypyridyls. Therefore, the sensor response can deviate from that observed for pure silica or silicone supports, and the linearity of the oxygen dose-response curve in the 0–100% range may be seriously affected. This fact makes difficult a two-point sensor calibration procedure, a feature that is highly desirable for the commercial application of any monitoring device.

Indicators placed inside zeolites are promising materials for use in oxygen optical sensors since the high degree of crystallinity of these aluminosilicates provides excellent indicator retention without seriously altering its photophysics. In this way, $[Ru(bpy)_3]^{2+}$ was synthesized inside zeolite Y supercages or adsorbed onto the zeolite surface, and the resulting material was embedded in silicone. The sensor features include fast oxygen response (in the order of seconds), excellent long-term stability, complete lack of dye leaching and, for certain applications, the material can be operated at high temperatures [111].

Copolymers, cross-linkers and plasticizers can be used to tailor the properties of the indicator layer [107, 117, 118, 124]. This strategy has been used, for example, to increase the oxygen sensitivity of some indicator-matrix compositions where the poor solubility of polar indicators in nonpolar polymers affects performance due to modification of their luminescent features or poor local oxygen diffusion or solubility around the luminescent indicator. In this way it is possible to get polar domains into an otherwise nonpolar polymer, yet transition regions cannot be avoided. Variations in the polymer polarity can be controlled by varying the amount and polarity of the additive. However, even though a modified polymer may be a proper matrix for the Ru(II) complex, severe heterogeneity in the resulting sensing layers is demonstrated by heavily nonlinear Stern-Volmer responses. Plasticizers (trimethyl phosphate, tributyl phosphate, dioctyl phthalate,...) are usually included in the formulation for most plastic film gas sensors in order to enhance the O_2 diffusion rate through the film due to increased mobility of the polymer segments. Therefore, changes of the additive also affect markedly the response and recovery times of the oxygen sensors. Additives may also be included in the polymer support to modify its mechanical properties, e.g. to increase or decrease the polymer rigidity, without drastically affecting transparency or, in some cases, sensitivity [118, 124].

Another important aspect that affects long-term performance of the sensor head and, therefore, has to be taken into account for the sensor design, is that polymers containing good coordinating groups such as primary amines or carboxylates may enhance photoinstability by themselves entering into the metal coordination sphere (via photodechelation, Fig. 5) and should probably be avoided for this reason [117]. In addition, since *singlet oxygen* is produced upon energy transfer quenching of the luminescent Ru(II) polypyridyl by oxygen, complexes having phen ligands tend to exhibit a significant photodegradation. This phenomenon is manifested not only by a decrease in their luminescence intensity but also by a reduction in their emission lifetime and specific absorption spectral changes, as has been demonstrated for Ru(II) dyes adsorbed on silica gel or dissolved in polystyrene or PVC [125, 126]. The effects depend strongly on the nature of the dye, the type of polymer matrix and the oxygen and dye concentrations. Singlet oxygen attack on the C5–C6 bridge of the phenanthroline ligands (Fig. 1) is suggested to be responsible for the observed effects on the luminescence features [125, 127].

Other polymer materials such as xerogels have been used in the preparation of oxygen sensing phases. The sol-gel process is a method of material preparation by room temperature reaction of a metal alkoxide, water and a solvent. In this way, tetramethoxy- or tetraethoxysilane undergo hydrolysis and polycondensation to form a three-dimensional silica network. The sol material can be pre-doped with an oxygen-sensitive indicator whose molecules will be entrapped in the subsequent gel cage-like microstructure by the cross-linking action of silicon and oxygen atoms. Drying and heat treatment will then progressively densify the gel by elimination of solvents and water. The densification process and porosity of the resulting material is controlled by an appropriate choice of the starting pH and temperature. Such conditions try to ensure that the large dye molecules cannot leach out but smaller analyte molecules can permeate the interconnected cages. The flexibility of the sol-gel process facilitates adjustment of those coating properties (e.g. thickness) which determine critical sensor parameters such as sensitivity and response time. For example, silica gels prepared at low pH (< 3) are generally less porous than those prepared under more basic conditions (pH 5–7). In addition, monolithic or thin-film sensing structures may be fabricated, although the latter offer the advantage of faster response and permit a better design of the sensor (e.g. via dip- or spin-coating of a decladded portion of an optical fibre, or other appropriate optical waveguiding substrate, by the sol at a suitable stage in the densification process and further curing produces a dye-doped microporous sensing layer on the light guide). Therefore, this approach offers a number of advantages over other methods of sensor fabrication [128].

The possibility of tailoring the properties of xerogel films to optimise quenching in different environments (gas-phase compared to aqueous phase) and for different oxygen concentration ranges depends on the material porosity and on the detailed quenching mechanism in a particular phase. The water/

alkoxysilane ratio was found to be the most important parameter that must be controlled. While for gas-phase oxygen sensing the quenching process is highly non-linear and films show a high sensitivity to oxygen at low oxygen concentrations, in aqueous environments the quenching response of the films is in general lower than in the gas phase and at higher water/precursor ratios the oxygen response decreases. The nature of the xerogel surface is also of importance to operation of the sensor. In particular, enhancement of the surface hydrophobicity increases quenching by dissolved oxygen (in water), since under these conditions the quenching is a gas-phase interaction, as compared to a surface-adsorbed process. This is achieved by using suitable proportions of organically modified precursors of the form $R(OEt)_3Si$, where R is an alkyl group, in the fabrication procedure. It has been shown that by increasing the ratio of modified precursor, the O_2 response in the aqueous phase increases. A very low limit of detection (6 ppb), was obtained for the modified films („ormosils"). Extending the aliphatic group of the modified precursors (e.g. butyltriethoxysilane) further enhances the quenching response in the aqueous phase [129–131].

A different strategy for addressing the problem of low solubility of the indicator complex in the polymeric support has been derivatization of the dye to increase the hydrophobicity of the coordinating ligand(s) of the Ru(II) complex or by covalently bonding dye and polymer, thus avoiding the need for co-immobilisation on silica gel before or after fabrication of the membrane. In this way, Ru(II) complexes with octadecanamide or biphenyl-substituted 1,10-phenanthroline ligands have been prepared and tested as oxygen probes immobilised in silicone matrices [102, 122]. These indicator dyes are soluble in silicone and yield linear Stern-Volmer plots, yet the sensitivity of such indicator layers is lower compared to silica gel-doped silicones. The reduced dimensionality and concentration effect of the oxygen quenching that occurs on the *surface* of the solid particles of the filler probably accounts for the observed difference.

Good solubility and, therefore, linear quenching characteristics are also promoted by alkyl substitution on the ligands and the use of a plasticizer (tributyl phosphate) with marked cation solvating power [132]. New hydrophobic Ru(II) complexes have also been synthesised as potentially useful longwave absorbing luminescent oxygen probes (λ_{max} = 560 nm) [133]. The new Ru(II) complexes comprise bpy or dip ligands and the low field strength dmch or omch ligands (dmch = 5,8-dimethyl-6,7-dihydrodibenzo[*b*,*j*]-1,10-phenanthroline, omch = 5,8-dimethyldibenzo[*b*,*j*]-1,10-phenanthroline). Their absorption bands strongly overlap the emission of existing bright green LEDs but at the expense of a very low Φ_L.

Heteroleptic Ru(II) complexes containing sulfonated diphenylphenanthroline and alkylamide-1,10-phenanthroline derivatives have been covalently linked through a sulfonamide bond to porous glass in order to develop sensors robust enough for long-term monitoring of dissolved oxygen in *organic* solvents. These supports provide good sensitivity but also afford nonlinear responses to oxygen [101, 104].

Ru(II) complexes can be readily solubilized in different polymers by ion-pairing with organophilic anions such as tetraphenylborate [107, 132], although it has been demonstrated that this anion is a quencher of their luminescence.

Such effect has allowed the use of this counterion to rank the polarity of the different polymers by their tendency to promote ion pairing [127]. Organic anions such as aliphatic carboxylates and sulfates or anions containing silicone-like structures make Ru(II) polypyridyls silicone-soluble by converting them into the respective ion pairs [41]. These counterions are known to exert almost no effect on the lifetime and photostability of the cation, while the presence of a viscous or solid solvent may increase the lifetime even further. The dodecyl sulfate counteranion has been found to be the most useful. Acetic acid-releasing one-component RTV silicones provide the best matrix for these oxygen sensitive membranes. The luminescence intensity of the indicator layers was observed to be more than 10 times higher than those of membranes based on silica gel-adsorbed Ru(II) probes having the same film thickness. It is lower, though, by at least a factor of 2 than that of sensing films based on plasticized PVC or polystyrene and having the same indicator concentration. Unlike silicagel-adsorbed $[Ru(dip)_3]^{2+}$, the concentration of the counter-ion modified dyes in silicone can be greatly increased, while $[Ru(dip)_3]^{2+}$ deposited on silica gel tends to undergo self-quenching of luminescence when its concentration exceeds a certain critical value, which strongly depends on the loading capacity of the particular silica gel used. In contrast to what might be expected for a homogeneous dye-in-polymer solution in which no inhomogeneities are visible under the microscope, Stern-Volmer plots are still nonlinear. Strong evidence is presented from quenching experiments that the luminescent ion pairs are present in both a monomolecular and an aggregated (not crystalline) form that have different quenching constants, weighing factors and decay times, resulting in nonlinear quenching plots. The decrease in oxygen sensitivity (K_{SV}) compared to silica-doped $[Ru(dip)_3]^{2+}$ films has been attributed to the effect of a slow transition from the homogeneous to the aggregated (less quenchable) form.

6.4
Luminescence Quenching Models in Heterogeneous Supports

Although the Stern-Volmer plots are linear in homogeneous media, in practice, downward curved graphs are practically the rule with polymer-supported luminescent indicators, as a result of (micro)heterogeneity in the polymers. This effect is unfavorable for routine calibration purposes [134]. Such non-linearity manifests itself in non-exponentiality of the emission decays both in the presence and in the absence of oxygen. Therefore, a great effort has been made to rationalize the complex oxygen quenching behavior exhibited by immobilised Ru(II) polypyridyls. In general, departure from linearity has been attributed to the heterogeneity of the solid support on the microscopic scale (multiple sites or wide distribution of sites for the indicator dye, various degrees of oxygen accessibility to the immobilised luminophore, and/or incomplete solubilization of the dye in the polymer matrix). The combination of such effects makes it difficult to predict the behavior of a novel oxygen sensor.

The first question to be addressed was the nature of the luminescence quenching in the sensing layers. A method for assessing the relative contributions of *static* and *dynamic* quenching in heterogeneous systems has been developed based on a comparison of intensity quenching with lifetime quenching data using a preexponential weighted emission lifetime, τ_M [106, 109, 114, 127, 135]. This parameter is calculated by fitting the observed decay curves to a sum of a relatively small number of exponentials (between 2 and 4) and is defined according to Eq. (5), where B are the preexponential weighting factors of the multiple-exponential fit.

$$\tau_{\mathrm{M}} = \frac{\sum B_{\mathrm{i}} \tau_{\mathrm{i}}}{\sum B_{\mathrm{i}}} \tag{5}$$

If plots of τ_{M0}/τ_M and I_0/I (where I_0 is the luminescence intensity in the absence of quencher and I the luminescence intensity in the presence of quencher) versus the quencher concentration are linear and superimposable, then bimolecular deactivation is purely dynamic. Although this method does not provide direct information on the details of system heterogeneity, it has the advantage of requiring no a priori information on its nature. Oxygen quenching of Ru(II) complexes has been shown to be *exclusively dynamic* in different polymers such as silicone or polystyrene, and simulations confirm that the method works for a wide range of heterogeneous systems [135].

One of the first models developed to explain the downward curved Stern-Volmer intensity plots was surface diffusion quenching in oxygen sensing layers containing adsorbents such as silica gel [114]. In principle, the immobilised indicator could be deactivated by gaseous oxygen and/or adsorbed oxygen. This model considers that adsorption isotherms are expected to provide an additional effect on the curvature of the Stern-Volmer plots. For *adsorbed* oxygen both the Langmuir and Freundlich isotherms may represent the surface concentrations of this gas at a particular partial pressure. The Langmuir isotherm model describes adsorption phenomena up to monolayer coverage. The Freundlich isotherm (Eq. 6) is an extension of the Langmuir isotherm and provides an empirical parameter (n) to account for the variation in the interaction energy between adsorbate and support:

$$\theta = aP^{1/n} \tag{6}$$

where θ is the fractional surface coverage of the adsorbate, a is a collection of constants, P is the equilibrium gas-phase pressure, and n is an empirical parameter related to the intensity of the adsorption. The Langmuir model assumes that all sites have the same binding energy, while the Freundlich model assumes that there will be a distribution of site energies with preferential binding to the most energetic sites.

For multiple emitting sites with different oxygen quenching constants, the Stern-Volmer equation is given by Eq. (7) [114, 127, 135]:

$$I_0 / I = \left(\sum_{\mathrm{i}} \frac{f_{0\mathrm{i}}}{1 + K_{\mathrm{SVi}}[\mathrm{Q}]} \right)^{-1} \tag{7}$$

where f_{0i} is the steady-state fraction of light emitted from the i-th component and K_{SVi} is its corresponding Stern-Volmer quenching constant.

In order to evaluate the adequacy of the surface diffusion quenching model, the authors explored several approaches concerning the nature and distribution of luminescent indicator sites on the adsorbate and the mode of quencher delivery to the sensitizer. The most interesting models were those considering: (a) *two independent binding sites* with both sites quenched by gaseous diffusion with different quenching constants (Eq. 8)

$$I_0 / I = \left(\frac{f_{01}}{1 + K_{SV1}[Q]} + \frac{f_{02}}{1 + K_{SV2}[Q]} \right)^{-1} \tag{8}$$

(b) diffusional quenching by adsorbed oxygen only with oxygen adsorption following a *Freundlich isotherm* (Eq. 9):

$$I_0 / I = 1 + K_{SV}[Q]_{adsM} aP^{1/n} \tag{9}$$

Experimental results point out that dynamic quenching in silica disks can best be described by a Freundlich surface quenching model (b), although gaseous quenching cannot be neglected. On the other hand, oxygen quenching in silica-doped silicone membranes and organic polymers is best fitted by the two-site quenching model (a). This model assumes that the indicator exists in only two sites, each of them having a different unquenched contribution to the emission intensity and a different quenching constant. Although multiple sites would be expected for these membrane systems, a simple two-component model appears sufficient to fit most calibration data and it has been found indeed that the two-site model has an excellent fitting value for oxygen quenching in different solid supports [102, 106, 124, 127, 136–138]. However, because it will quantitatively fit more complex systems, one should use its fitting parameters with great caution for detailed *mechanistic interpretations* [117] (see below).

A modified form of the Stern-Volmer equation, referred as the *non-linear oxygen solubility* model, has been proposed for correlating the luminescence intensity of a Ru(II) complex in a polymer with the analyte concentration [110]. Equation (10) is based on the kinetics of oxygen quenching and the oxygen solubility in the polymer,

$$I_0/I = 1 + A\,P_{O_2} + B\,P_{O_2}/(1 + b\,P_{O_2}) \tag{10}$$

where A and B are empirical parameters combining the oxygen quenching and solubility parameters, and b is a parameter in the solubility equation. This model considers that all the photoexcited molecules in the silicone-rubber film can be quenched by oxygen and it assumes that the deviation of the Stern-Volmer plot from linearity is due to the non-linear solubility of oxygen in the polymer. The above equation has been used to fit successfully the observed data associated with most oxygen optodes [107].

In the case of oxygen sensors based on silicone sensing layers containing additives such as cross-linkers, co-monomers, silica gel, etc., the two-site model was found to be too simplistic [117, 118]. Therefore, the investigated systems were characterized by three types of regions: (i) a largely hydrophobic domain whose major constituent is poly(dimethylsiloxane); (ii) interfacial regions; (iii) insoluble silica particles or, for those silica gel-free polymers, more polar regions originating from the cross-linker or co-monomer.

As mentioned above, the luminescence decay profile of Ru(II) complexes immobilised in polymer supports cannot be fitted to single exponential functions. Many alternative fitting models have been proposed that yield good quality fits (e.g. multiexponential decays, Gaussian distributions, stretched exponentials,...), though they provide no physical information on the reasons for the microheterogeneity and are only mathematical constructs [139]. Recently, a new model that considers the interaction of the fluorophore with the nonuniform environment provided by the polymer has been proposed to describe the nonexponential emission decay of indicator molecules, giving also physical reasons for the experimentally observed nonlinearity of the Stern-Volmer plots [103, 140]. This model has a better physical basis than the usual fits with multiple exponentials or lifetime distributions of arbitrary shape. Oxygen quenching is described by a single parameter which, in favorable cases, depends linearly on the oxygen pressure. This may be of advantage for the calibration of optical oxygen sensors based on luminescence quenching, and it opens the possibility of describing the characteristic dependence of sensor signal on oxygen concentration with a minimum number of variable parameters.

The sensing layers studied include PVC, silicone, ethylcellulose and polystyrene polymers. Most polymers are nonuniform media in a sense that there is no long-range structural order and, to properly describe the deviations from a single exponential decay of the luminophore embedded in a polymer matrix, the influence of the microenvironment on its radiative (k_r) and nonradiative (k_{nr}) deactivation rate constants has to be considered. This effect can be due to spatial nonuniformity in the refractive index of the support, variations in the steric stabilization of the luminescent indicator by the polymer environment, electron transfer or non-resonant energy transfer from the luminophore to acceptor sites in the matrix. The medium surrounding the luminophore should affect predominantly k_{nr} (see Luminescence Features section); this effect depends on the distance between the indicator dye and the nearest interacting polymer site. In the simplest case, such distances are homogeneously distributed and k_{nr} depends on the distance between the luminophore and the interacting site of the polymer according to a power law (r^{-6} for a dipolar interaction). Other relationships may be used depending on the nature of the indicator dye and polymer support. Following this approach, integration of the differential equation describing the time evolution of the excited-state population (luminescence intensity) yields Eq. (11), which deviates from the usual exponential form even in the absence of any external quencher.

$$I(t) = \exp\left[-(1+c)(t/\tau) - a\sqrt{(t/\tau)}\right] \quad (11)$$

In Eq. (11), c is a quenching parameter related to the oxygen concentration and a is a parameter proportional to the density of quenching sites (if no matrix interactions are present, $a = 0$ and the decay remains a single exponential). This model gives a physically more reasonable procedure than multiple-exponential or rate-distribution models do because Eq. (11) relates the form of the decay curve to a certain quenching mechanism. In principle, a single quenching parameter is sufficient for a full description, and in favorable cases, i.e. when the indicator is well dispersed in a single-phase matrix for which Henry's law is valid, this quenching parameter varies linearly with the oxygen pressure. Such favourable conditions, however, prevail only in certain polymer matrices, while a nonlinear dependence of the quenching parameter on the oxygen pressure is found in others.

Finally, attempts have been made to *linearize the calibration plots* of oxygen sensing layers containing Ru(II) complexes as indicator dyes [141]. The simple model also accounts for stray radiation within the system [78, 142, 143]. It assumes that the measured signal (I) is composed of quenchable and non-quenchable radiation. The quenchable light (I_Q) corresponds to luminescence generated from O_2-accessible Ru(II) dye molecules. The oxygen quenching process is considered to be described by a single K_{SV}. Non-quenchable light (I_{NQ}) corresponds to any unremoved excitation radiation, all sources of stray radiation and any non-quenchable luminescence from indicator aggregates within the sensing membrane. The measured intensities in the absence and presence of quencher are given as $I_0 = I_{0Q}+I_{NQ}$ and $I = I_Q+I_{NQ}$, respectively. Rearranging and introducing them into the Stern-Volmer equation gives the final working Eq. (12).

$$I = I_{NQ} + \left(I_0 - I_{NQ}\right)/\left(1 + K_{SV}\left[O_2\right]\right) \tag{12}$$

6.5 Instrumentation Used in Oxygen Sensing with Ru(II) Dyes

Twenty years ago, $[Ru(bpy)_3]^{2+}$ was already recognized to be a promising indicator for oxygen sensing among other blue-excitable indicators which were known to be strongly quenched by oxygen [144]. Subsequently, the first simple optoelectronic sensors for oxygen based on measurement of the luminescence lifetime rather than intensity, using a frequency-modulated light emitting diode (LED) as a light source, were described [145, 146]. Such studies started the development of commercial applications based on fibre-optic oxygen sensing, that has evolved from the first fluorescence spectrometers or fibre-optic photometers with rather limited applicability and high cost, to the modern and less expensive all solid-state instruments that are being marketed nowadays [8].

The number of commercially successful applications is still low because most optical oxygen sensors have never been developed beyond the laboratory stage, and do not fulfil crucial requirements such as long-term stability, ruggedness, cost effectiveness and handling convenience. In order to perform in situ monitoring of oxygen, low-cost, battery-operated portable sensors based on luminescence measurements that make use of solid-state technology, being able to measure low power optical signals with good signal-to-noise ratios, are the desirable instruments. Recently, lifetime-based oxygen monitoring systems suitable for the determination of gas phase or dissolved oxygen, using standard electronics, digital components and low-cost semiconductor devices such as narrow-band powerful blue LEDs and photodiode detectors, have been developed [98, 147]. These instruments, which are compatible with the oxygen-sensitive membranes described above, may be calibrated by a simple two-point procedure with air and oxygen-(saturated) gas (or water in the case of dissolved oxygen sensors). It has been shown that lifetime-based sensors can be stable over years without any need for recalibration (their shelf life is in the order of years), and reproducible results can be achieved using sensor elements from different production lots without adherence to stringent specifications concerning sensor composition. This fact makes quality assurance much easier and time-consuming and costly calibration procedures can be kept to a minimum in the production process. Another advantage for the end user is that there is no need for individual calibration steps during the measurement, which makes application of the sensors much more easy to handle by untrained personnel, saving time and cost [147–151].

The capability of light guides for simultaneous long-range transmission of optical signals has been widely exploited for constructing oxygen optosensors. In addition to the most employed front-face interrogation of the sensitive membrane, placed at the distal end of a bifurcated fibre-optic bundle or a single waveguide [95, 96, 98], other configurations such as evanescent-wave excitation [128] or capillary optosensing [151, 152] have been explored. Fibre-optic oxygen sensors have a number of advantages compared to other optical sensing schemes [3]. Moreover, simultaneous measurement of chemical and physical parameters such as oxygen, pH, carbon dioxide and temperature (multiparametric analysis) is an attractive feature of fiber-optic sensing that was exploited from the very beginning of optical sensing [153, 154].

6.6 Applications

The estimation of the oxygen partial pressure in gaseous samples or the oxygen concentration in solution or biological fluids has very important implications in environmental, industrial, medicinal and analytical chemistry. Therefore, numerous luminescence-based optical sensors have been developed for gas and liquid phase measurements, involving gases and water monitoring [105, 128, 155, 156], clinical analysis of blood [157], through-skin oxygen flux analysis and diagnosis of circulatory disturbances and their consequences [158]. Some interesting oxygen optosensors have recently been developed, specially those concerning areas of environmental and clinical application such as biosensors, microsensors, two-dimensional measurement of the distribution of oxygen in heterogeneous systems (oxygen imaging), multianalysis and sensors based on oxygen transduction. These recent developments will be commented below; former applications may be found in refs. [3–5, 95, 98]. The many applications to blood gas analysis will not be considered here because they lie outside the scope of the present book.

A lifetime-based (pulse-probe) optode for marine oxygen monitoring has been designed and the influence of several oceanographic parameters specific to the marine environment, such as salinity and hydrostatic pressure, were investigated [159]. The sensing cell is a thin cylindrical cavity within two silicone membranes in which $[Ru(phen)_3]^{2+}$ is dissolved in water to avoid problems arising from the immobilisation of the dye in a solid matrix. Salinity appears to have no influence on the sensor response but there is a clear effect of the hydrostatic pressure, the latter attributed to changes in the relative position of the fibres, resulting in variations in the collected luminescence intensity.

MacCraith and co-workers have reported a high performance dissolved oxygen optical sensor based on phase fluorometric detection with a dual LED referencing system [155]. The limit of detection is < 10 ppb, while the long-term stability should enable reliable operation for months. The temperature dependence of the sensor response has been characterised and incorporated in a calibration function. The optode operates satisfactorily over the complete range of dissolved oxygen concentrations and the response is particularly sensitive at low oxygen levels due to the dynamics of the optical quenching process in the heterogeneous environment of the microporous film (a rugged hydrophobic xerogel sensing layer doped with $[Ru(dip)_3]^{2+}$ that can be incorporated in a disposable cap configuration). The sensor head has been designed specifically for waste-water monitoring but the sensor can be repackaged for other potential applications.

An optosensor that withstands sterilisation for oxygen monitoring in bioreactors has been described [160]. Various polymers were investigated under these harsh conditions. Polysulfone and polyetherimide displayed comparable results and excellent characteristics for this application. The sensor is calibrated before use and does not need to be re-calibrated after sterilisation. This is very important for the application in fermentation control because there the sensor cannot be removed after autoclaving due to the danger of contamination of the sterilised bioreactor. The developed sensor allows measurements between 5 and 50 °C and oxygen partial pressures from 0 to 5×10^4 Pa. Within these ranges, the oxygen measurement at a temperature different from the calibration temperature can be compensated with sufficient accuracy.

One of the most significant problems when using sensing devices in the real world is surface *fouling*, that is, the rapid accumulation of adsorbed material on the working surface of sensors. Such foreign material leads to drift and eventually sensor failure as a result of disruption of the sensitive material or prevention of analyte transport into the sensitive reagent layer. A major cause of concern, particularly in environmental, clinical, or bioreactor monitoring, is the formation of protein layers or the adhesion of microorganisms and cells. Sensor biofouling makes long-term monitoring difficult and requires frequent (field) maintenance operations and probe replacement. Deposition of a non-toxic, biocompatible polymer coating based on phosphorylcholine-substituted methacrylate units has been demonstrated to be a suitable technology to impart highly effective anti-biofouling properties to luminescent oxygen optodes

[161]. Nanometer-thick coatings of such materials are shown to reduce significantly the adhesion of marine bacteria and thrombocytes to the surface of $[Ru(dip)_3]^{2+}$-doped silicone layers. The thin coatings perturb the analytical features of these sensors only slightly or not at all and are mechanically stable for more than one year of continuous immersion. Such biomimetic materials will help to extend the operational lifetime of optical sensors in the application fields of environmental monitoring, clinical chemistry, and biotechnology.

The spatial distribution of chemical and physical parameters in most natural environments is heterogeneous. By use of one or several *microsensors*, point measurements or the one-dimensional distribution of chemical parameters can be performed at a high spatial resolution. In this case, multiple-microsensor systems or arrays of sensors are required to monitor the bi-dimensional distribution of chemical parameters [162]. A submicrometer optical fiber oxygen sensor based on the fluorescence quenching of $[Ru(phen)_3]^{2+}$ has been fabricated [163]. The indicator dye has been incorporated into acrylamide and is attached covalently to a silanized optical fiber tip surface by photoinitiated polymerization. Leaching of the sensing reagent from the polymer host matrix has been minimized by optimisation of the acrylamide monomer/*N,N*-methylenebisacrylamide ratio. The sensor is fully reversible and highly reproducible. The sample volume required for measurements is 100 fL and an absolute detection limit of 1×10^{-17} mol of dissolved oxygen has been achieved.

A fibre-optic oxygen multisensor system and its applications in biotechnology has been described [164]. The oxygen optode is based on $[Ru(dip)_3]^{2+}$ adsorbed on silica gel and embedded in silicone or polystyrene polymers, where the fiber tip was dipped. The multisensor setup involves a 4-channel system that was used to monitor the oxygen profile and to determine the diffusion coefficients of oxygen in matrices where microorganisms are cultured, to determine the aerobic/anaerobic transition.

Conventional lifetime imaging requires rather complex and expensive instrumentation, such as lasers and intensified CCD cameras. Recently, a low cost luminescent oxygen sensor based on all-solid-state technology has been described for oxygen imaging based on lifetime measurements. The authors demonstrate that an array of light emitting diodes and a directly gatable CCD camera can be used as an alternative to image oxygen concentrations on a surface covered by an optical sensor [165]. The sensing layer was prepared by adsorbing $[Ru(phen)_3]^{2+}$ on silica gel and embedding the particles in filler-free silicone rubber. The reported results show that for an illuminated sensor area of 5 cm^2 and an aimed spatial resolution of 1 mm, oxygen resolution was better than 0.4 Torr in the absence of oxygen and $\Delta P_{O_2} = 2.5$ Torr at $P_{O_2} = 100$ Torr. Luminescence lifetime imaging of oxygen is of considerable interest, for example, for noninvasive medical applications, which are expected to have a great potential in the fields of microcirculation and cutaneous respiration, critical processes in a great variety of diseases. The use of fiber optic microsensors, whose signals can be recorded simultaneously with a single opto-electronic device, is the best way to measure gradients of physical and chemical parameters with sufficient spatial resolution (< 50 μm). Such micro-optodes have very small tip diameters (5–50 μm) that enable an almost non-invasive measurement to be made in intact biological systems. For instance, the oxygen production or uptake of a sediment layer can vary from hypersaturation to fully anaerobic conditions only in the upper few millimeters and can be calculated from the measured micro-profile in the diffusive boundary layer. Oxygen micro-optodes with moderate sensitivity are therefore required to monitor oxygen within the full dynamic range from 0.1 to 100% oxygen saturation and sufficient spatial resolution. A luminescence lifetime-based portable microoptode array, that enables simultaneous oxygen sensing with eight oxygen microoptodes using a simple optical setup both in the laboratory and under field conditions, has been developed and marketed [162, 166]. The measuring system consists of (i) an optical unit, a special fibre-coupler array, optical filters, lenses, light sources (LEDs) and light detectors (photodiodes, PMT); (ii) an analog signal processing unit (phase-angle detection, filtering) and (iii) a digital signal processing (control, data storage and display). Each sensor signal is sampled in a time-multiplexed mode. This multisensor array system was primarily designed for investigation of the oxygen distribution in biofilms and aquatic sediments. The oxygen concentration is measured with tapered silica-glass fibers (tip diameter 20–30 μm) using $^*[Ru(dip)_3]^{2+}$ immobilised in polystyrene. Coatings made of soft polymers such as silicone or plasticized PVC are not robust enough for measurement in harsh sediments or other complex biological systems that require probes which are not fragile due to their very small size. Organically modified sol-gels (ormosils) also appear to be promising indicator supports. The oxygen-sensitive coating is deposited in this case by dipping the fiber tip into the polymer solution and subsequently evaporating the solvent.

The combination of optical sensing with imaging techniques to obtain two-dimensional microscale measurements of chemical parameters by using planar optodes is also a powerful method to determine the distribution of chemical parameters within foils. It has been successfully applied to 2-D oxygen sensing in sediments with high spatial resolution and negligible disturbance of the microenvironment [166].

Since oxygen holds great significance as an indicator of the metabolic state of living cells, xerogel-based, ratiometric, optical nanoprobes termed PEBBLEs („*probes encapsulated by biologically localized embedding*“) have been demonstrated to enable reliable real time measurements of subcellular molecular oxygen [167]. The radii of these spherical probes range from 50 to 300 nm. They incorporate the oxygen-sensitive luminescent indicator $[Ru(dip)_3]^{2+}$ and an oxygen insensitive fluorescent dye as a reference for the purpose of ratiometric intensity measurements. These probes reportedly have excellent reversibility, dynamic range, and stability to bleaching and photobleaching. Their small size and inert matrix allow them to be inserted into living cells with minimal physical and chemical perturbations to their biological functions. The sol-gel matrix protects the dyes from interference by proteins in cells, enabling reliable in vivo chemical analysis, compared to using free dyes for intracellular measurements. Moreover, the matrix also significantly reduces the toxicity of the indicator and reference dyes to the cells.

The interdependence of pH, carbon dioxide and oxygen during chemical and biochemical processes has driven the need to monitor them simultaneously, continuously and in situ, in order to exert better control. The fabrication and performance of a multi-analyte imaging fiber sensor that allows pH, carbon dioxide and oxygen to be monitored simultaneously with rapid response has been recently described [168]. Sensing elements were fabricated by entrapping $[Ru(dip)_3]^{2+}$ within polymer matrices via photopolymerization to yield distinct regions of analyte-sensitive material at the distal end of the fiber. The working range of the multianalyte sensor is 0–100% for oxygen, 0–10% for carbon dioxide in the 5.5–7.5 pH range (carbon dioxide and pH levels are not monitored with Ru(II) polypyridyls). The sensor was used to monitor changes in these parameters during a beer fermentation. The small size of the sensor and the simplicity of the instrumental design may allow its use in a clinical environment.

A multiparametric environmental monitoring system for remote analysis has been developed [7, 76, 122, 169]. The system comprises up to 8 optical channels which permit the simultaneous determination of dissolved (or gas-phase) oxygen, pH, temperature, CO_2, humidity and biological oxygen demand (BOD), employing luminescent indicators specifically designed for this purpose and waveguides. Alternatively, the same parameter(s) can be monitored in different locations. A simple change of the sensitive tip by screwing different caps allows the user to select the monitored parameter in each channel. The instrument incorporates its own communication features (PSTN, GSM, RS-232) and offers the possibility of integrating conventional (electrical) sensors too. The sensor field validation carried out against the instrumentation employed in the Water Quality Network of different river basins and the National Institute for Meteorology (INM-MMA) has demonstrated the stability of the novel optical sensors. The general analytical features of the optoelectronic system are 0.04 to 35 mg L^{-1} of dissolved O_2, 4 to 9 pH units, –30 to 80 °C, 0.6 to 50 mg L^{-1} BOD, 0.1 to 30% CO_2 in air, and 1–100% relative humidity, but most of them can be customised according to the particular application.

A new solid-state integrated array format called OSAILS („*optical sensor array and integrated light source*") for simultaneous multi-analyte detection has been recently developed [170]. The oxygen sensing device uses microwells machined directly into a LED face and filled with a xerogel doped with the oxygen-sensitive $[Ru(dip)_3]^{2+}$ dye. The resulting luminescence is collected by a microscope objective, filtered and imaged onto a CCD. The OSAILS device is simple to construct, robust and requires low voltages and no optical fibres.

Several *biosensors* for molecules of bioanalytical interest have been developed using *oxygen transduction*, i.e. monitorization of changes in the oxygen concentration due to enzyme-catalysed oxidation reactions of target substrates. In this way, luminescent biosensors for **ethanol** [120, 171], **glucose** [142], **cholesterol** [172] and choline-containing **phospholipids** [173] have been developed. Typically the appropriate enzymes are immobilized on the surface of the oxygen optode and measurements are performed in a flow-through cell.

An optical biosensor for the continuous determination of glucose in beverages based on sol-gel entrapped glucose oxidase has recently been presented [174]. The matrix was ground to a powder form and packed into a flow cell. This mini-reactor was positioned in a spectrofluorimeter and connected to a continuous sample flow system. An oxygen-sensitive optode membrane was fabricated from $[Ru(dip)_3]^{2+}$ adsorbed on silica gel particles and entrapped in a silicone rubber film. The membrane was placed against the wall of the flow cell to sense the depletion of oxygen upon exposure to glucose.

Finally, Wolfbeis et al. have described the development of the first optical fibre microbial biosensor for determination of the *biochemical oxygen demand* (BOD) using oxygen transduction with $[Ru(dip)_3]^{2+}$ [175]. The sensitive tip of a fibre bundle consists of adjacent layers of (a) the oxygen-sensitive luminescent material, (b) *Trichosporon cutaneum* immobilised in poly(vinyl alcohol), and (c) a substrate-permeable polycarbonate membrane to retain the microorganism layer. The layers are placed, in this order, on top of an optically transparent gas-impermeable polyester support. Typical response times are 5–10 min and the dynamic range is 0–110 mg/L BOD when a glucose/glutamate standard is used. The luminescence signal is affected by various parameters, including the thickness of the layers, the cell density of the yeast and the rate at which the substrate is passed through the flow-through cell. BOD values estimated by the optical biosensor correlate well with those determined by the conventional BOD_5 method. The main advantages are (a) a more rapid estimation of BOD (in comparison with the 5-day BOD_5 standard); (b) the possibility of performing in situ monitoring using fibre-optics and (c) the option of designing inexpensive disposable sensor cells.

7 Miscellaneous Sensors and Concluding Remarks

Special Ru(II) polypyridyls have been described also as luminescent indicators for *direct* **humidity** measurements in the atmosphere (0–100% RH) and for water activity determination in food [176] using a multichannel phase-sensitive fibre-optic instrument. The Ru(II) complex contains a dipyridophenazine moiety [177], the excited state of which interacts strongly with water molecules. This interaction yields a new electronically excited state that deactivates via a non-radiative pathway. Therefore an increase of the water vapor concentration decreases the luminescence of the immobilised metal complex, the process being fully reversible. The selected polymer for the indicator support prevents oxygen interference. Humidity sensors are the topic of another chapter in this book.

Resonance energy transfer from Ru(II) complexes to colorimetric pH indicators dyes (typically bromothymol blue, BTB) embedded in plasticised PVC membranes, along with an anion-selective carrier, has been used to develop **chloride** sensors [178, 179]. The response is based on co-extraction of the anion and protons from the aqueous sample into the organic layer, that leads to a large change in the absorbance of BTB. The dynamic range can be adjusted via the sample pH (e.g. 30–180 mM chloride at pH 7.0). A similar procedure can be employed to monitor **potassium** ion in the 1–100 mM range (at pH 8.7) with valinomycin as the analyte-selective carrier and $[Ru(dib)_3]^{2+}$ (dib = 4,4'-diphenyl-2,2'-bipyridine) [179, 180]. Actually, a set of luminescence decay time-based chemical sensors for clinical applications, all of them based on luminescence resonance energy transfer from Ru(II) polypyridyls and suitable for use in a microplate format, has been reported [157]. In addition to the above mentioned sensors for O_2 and plasticised membranes for pH, CO_2, K^+ and Cl^- measurements, additional PVC indicator layers for **Na^+**, **Ca^{2+}**, **ammonia, urea** and **glucose** were fabricated with the metal dye and the appropriate carrier.

Chloride ion has also been measured to provide an estimation of seawater salinity via the dual luminophore referencing method [181]. The sensor uses a composite membrane containing Nafion-immobilised lucigenin, the fluorescence of which is quenched by chloride, and a ruthenium complex entrapped in poly(acrylonitrile) beads.

A highly sensitive detection of **ammonia** in the gas phase can be carried out with a fibre-optic sensor based on a „sandwich" membrane containing a long-lived Ru(II), insensitive to the analyte, and the indicator dye 2,7-dimethyl-2,7-diazoniapyrenium (DAP) [182]. DAP forms strong charge transfer complexes with amines and ammonia in organic media, that display significant red fluorescence ($\Phi_f = 0.05$) and blue absorption. Using phase-sensitive emission detection and a mathematical model to optimise the composite sensitive layers, NH_3 concentrations as low as 9×10^{-4} Torr (1.3 ppmv) were detected.

It is hoped that this chapter has provided a clear account on the feasibility of designing luminescent optosensors based on tailored Ru(II) polypyridyl dyes. Both *direct* and *indirect* schemes can be arranged to monitor a very wide variety of analyte species including gases, ions and metabolites. Nevertheless, new analytes will certainly be introduced among the target species in the near future and more applications outside the laboratory environment will certainly be reported. The subject is ripe for them!

Acknowledgement: The authors would like to express their tanks for the continued support of their sensor research by the Spanish government agency CICYT, the Ministry of Science and Technology, the European Union, the Madrid Community Council, and the UCM during the last 15 years. We are also indebted to the companies Grupo Interlab, TGI, Repsol-YPF and Agilent Technologies which have also funded a large part of our sensor investigations, and have allowed the transfer of our technology to the market.

List of Abbreviations and Symbols

5-odap	*N*-1,10-phenanthrolin-5-yloctadecanamide
5-OHp	5-hydroxy-1,10-phenanthroline
acac	acetylacetonate
anth-bpy	4-methyl-4'-[2-(2-naphthyl)ethyl]-2,2'-bipyridine
APD	avalanche photodiode
B	preexponential weighting factors of a multiple-exponential fit
bim	1*H*,1'*H*-2,2'-biimidazole
BOD	biochemical oxygen demand
bpds	2,2'-bipyridine-4,4'-disulfonate, disodium salt
bpy	2,2'-bipyridine
BTB	bromothymol blue
CCD	charge coupled device

DAP	2,7-dimethyl-2,7-diazoniapyrenium
dcb	2,2'-bipyridine-4,4'-dicarboxylic acid
deab	*N*-({3'-[(diethylamino)methyl]-2,2'-bipyridin-3-yl}methyl)-*N*,*N*-diethylamine
deam	*N*-({4'-[(diethylamino)methyl]-2,2'-bipyridin-4-yl}methyl)-*N*,*N*-diethylamine
deap	*N*,*N*-diethyl-*N*'-1,10-phenanthroline-4-ylethane-1,2-diamine
dhp	1,10-phenanthroline-4,7-diol
dib	4,4'-diphenyl-2,2'-bipyridine
dip	4,7-diphenyl-1,10-phenanthroline
DLR	dual luminophore referencing technique
dmch	5,8-dimethyl-6,7-dihydrodibenzo[*b*,*j*]-1,10-phenanthroline
DMF	dimethylformamide
$E^{0\text{-}0}$	energy difference between the lowest excited state and the ground state
$E^{m+/n+*}$	redox potentials of a Ru(II) complex in its excited state
$E^{m+/n+}$	redox potentials of a Ru(II) complex in its ground state
f	modulation frequency of the excitation source
GSM	global system for mobile communications
HPLC	high performance liquid chromatography
HPTS	8-hydroxypyrene-1,3,6-trisulphonic acid
I	luminescence intensity in the presence of quencher
I_0	luminescence intensity in the absence of quencher
K_{as}	association constant
k_{nr}	nonradiative deactivation rate constant
k_q	bimolecular quenching rate constant
k_r	radiative deactivation rate constant
K_{SV}	Stern-Volmer constant
LD	laser diode
LED	light-emitting diode
LRET	luminescence resonance energy transfer
MC	metal-centered state
Me-pyim	1-methyl-2-(2-pyridyl)-1,3-imidazole
MLCT	metal-to-ligand charge transfer state
NHE	normal hydrogen electrode
NMR	nuclear magnetic resonance
NN	polyazaheterocyclic chelating ligand
omch	5,8-dimethyldibenzo[*b*,*j*]-1,10-phenanthroline
OSAILS	optical sensor array and integrated light source
PAH	polycyclic aromatic hydrocarbon
pbbs	(1,10-phenanthroline-4,7-diyl)bis(benzenesulfonate), disodium salt
PEBBLE	probes encapsulated by biologically localized embedding
PET	photoinduced electron transfer
phen	1,10-phenanthroline
phhp	*N*-phenyl-1,10-phenanthrolin-4-amine
K_a	ground state acidity constant

K_a*	excited state acidity constant
PMT	photomultiplier tube
P_{O_2}	oxygen partial pressure
PSTN	public switched telephone network
PVC	poly(vinyl chloride)
py-bpy	4-pyren-1-yl-2,2'-bipyridine
pyim	2-(2-pyridyl)-1,3-imidazole
py-phen	5-pyren-1-yl-1,10-phenanthroline
pytzhp	2-(3-pyridin-2-yl-1*H*-1,2,4-triazol-5-yl)phenol
pzth	2-(1,3-thiazol-2-yl)pyrazine
RTV	room temperature vulcanization
SCE	standard calomel electrode
T	triplet excited state
tanp	*N*-[3-(1,10-phenanthrolin-4-ylamino)propyl]propane-1,3-diamine
tap	pyrazino[2,3-*f*]quinoxaline
TC-SPC	time-correlated single photon counting
tmp	3,4,7,8-tetramethyl-1,10-phenanthroline
UV-VIS	ultraviolet-visible
XRD	X-ray diffraction
ε_{max}	absorption coefficient at the maximun of the absorption band
ϕ	luminescence phase shift
Φ	luminescence quantum yield in the presence of quencher
Φ_0	luminescence quantum yield in the absence of quencher
Φ_{ISC}	intersystem crossing quantum yield
Φ_L	luminescence quantum yield
τ	luminescence lifetime
τ_0	excited state lifetime in the absence of quencher
τ_M	preexponential weighted emission lifetime
Ξ	nuclear magnetic resonance frequency

References

1. Reeve RN (1994) Environmental analysis, Wiley, Chichester, UK
2. Arpe HJ (1994) Ullmann's encyclopedia of industrial chemistry, vol B6: analytical methods II and process control engineering. Wiley-VCH, Weinheim
3. (i) Wolfbeis OS (ed) (1991) Fiber optic chemical sensors and biosensors vol 1 & 2. CRC Press, Boca Raton; (ii) Cámara C, Pérez-Conde C, Moreno-Bondi MC, Rivas C (1991) Chemical sensing with fiber optic devices. In: Wise D, Wingard L (eds) Biosensors with fiber optics, Humana Press, Clifton, p 29; (iii) Cámara C, Pérez-Conde C, Moreno-Bondi MC, Rivas C (1995) Fiber optical sensors applied to field measurements. In: Quevauviller P, Maier, EA, Griepink B (eds) Quality assurance for environmental analysis, Elsevier, Amsterdam, p 165; (iv) López-Higuera JM (ed) (2002) Handbook of optical fibre sensing technology, Wiley, New York
4. Lakowicz JR (1994) Topics in fluorescence spectroscopy, vol 4: probe design and chemical sensing. Plenum, New York
5. Colin F, Quevauviller P (eds) (1998) Standards, measurements and testing for the monitoring of water quality: the contribution of advanced technologies. Elsevier, Amsterdam

6. Demas JN, DeGraff BA (1991) Anal Chem 63:829A
7. Orellana G, Moreno-Bondi MC, Bustamante N, Hidalgo E, Ruiz-López JL, García-Alonso JL, Delgado J, Bedoya M, Yáñez A, Sicilia JM, (1998) Proc Europt(r)ode IV. Münster, Germany, p 121
8. For example, PreSens Microx, Neuburg a. d. Donau, Germany, for environmental and laboratory oxygen measurements. Ocean Optics FOXY, Dunedin, Fla, USA, for dissolved or gaseous oxygen sensing. Van Essen Instruments Dive, Delft, The Netherlands, for dissolved oxygen in groundwater or surface water. Bioasure DO3000, Boston, MA, USA, to monitor dissolved oxygen during fermentation or cell culture processes. Grupo Interlab Optosen, Madrid, Spain, for in situ environmental/industrial oxygen monitoring. Comte MOPS, Hannover, Germany, for environmental and biotechnological applications
9. Paris JP, Brandt WW (1959) J Am Chem Soc 81:5001
10. Seddon EA, Seddon K (1984) The chemistry of ruthenium. Elsevier, Amsterdam, p 1173; (ii) Juris A, Balzani V, Barigelletti F, Campagna S, Belser P, Von Zelewsky A (1988) Coord Chem Rev 84:85; (iii) Kalyanasundaram K (1992) Photochemistry of polypyridine and porphyrin complexes. Academic Press, London; (iv) Orellana G, Quiroga ML, de Dios C (1993) Trends in Inorganic Chemistry 3:109
11. Hoffman MZ, Bolletta F, Moggi L, Hug GL (1989) J Phys Chem Ref Data 18:219
12. Mandal K, Pearson TDL, Demas JN (1981) Inorg Chem 20:786
13. (i) Kalyanasundaram K (1987) Photochemistry in microheterogeneous systems. Academic Press, Lausanne; (ii) Moreno-Bondi MC, Orellana G, Turro NJ, Tomalia DA (1990) Macromolecules 23:910; (iii) Ottaviani MF, Ghatlia ND, Turro NJ (1992) J Phys Chem 96:6075
14. (i) Pyle AM, Barton JK (1990) Probing nucleic acids with transition metal complexes. In: Lippard SJ (ed) Progress in inorganic chemistry, vol 38: bioinorganic chemistry. Wiley, New York, p 413; (ii) Nordén B, Lincoln P, Akerman B, Tuite E (1996) DNA interactions with substitution-inert transition metal ion complexes. In: Sigel A, Sigel H (eds) Metal ions in biological systems, vol 33: probing of nucleic acids by metal ion complexes of small molecules. Marcel Dekker, New York, p 177; (iii) Moucheron C, Kirsch-De Mesmaeker A, Kelly JM (1998) Structure and Bonding 92:163
15. Lehn JM (1995) Supramolecular chemistry. Wiley-VCH, New York
16. García-Fresnadillo D, Georgiadou Y, Orellana G, Braun AM, Oliveros E (1996) Helv Chim Acta 79:1222
17. Gerardi RD, Barnett NW, Lewis SW (1999) Anal Chim Acta 378:1
18. Riesen H, Wallace L, Krausz E (1997) Int Rev Phys Chem 16:291
19. Caspar JV, Westmoreland TD, Allen GH, Bradley PG, Meyer TJ, Woodruff WH (1984) J Am Chem Soc 106:3492
20. Orellana G, Moreno-Bondi MC, Segovia E, Marazuela MD (1992) Anal Chem 64:2210
21. García Fresnadillo D (1996) PhD thesis, Universidad Complutense de Madrid
22. Haga M-A (1983) Inorg Chim Acta 75:29
23. Van Houten J, Watts RJ (1976) J Am Chem Soc 98:4853
24. Alford PC, Cook MJ, Lewis AP, McAuliffe GSG, Skarda V, Thomson AJ, Glasper JL, Robbins DJ (1985) J Chem Soc Perkin Trans II:705
25. Orellana G, Alvarez Ibarra C, Quiroga ML (1988) Bull Soc Chim Belg 97:731
26. Orellana G, Braun AM (1989) J Photochem Photobiol A 48:277
27. Ross HB, Boldaji M, Rillema DP, Blanton CB, White RP (1989) Inorg Chem 28:1013
28. Barigelletti F, De Cola L, Balzani V, Belser P, von Zelewsky A, Vögtle F, Ebmeyer F, Grammenudi S (1989) J Am Chem Soc 111:4662
29. Milder SJ, Gold JS, Kliger DS (1986) J Phys Chem 90:548
30. Garcia-Fresnadillo D, Marazuela MD, Moreno-Bondi MC, Orellana G (1999) Langmuir 15:6451
31. Vo-Dinh T (1984) Room temperature phosphorimetry for chemical analysis. Wiley, New York

32. Lakowicz JR (1999) Principles of fluorescence spectroscopy. Kluwer Academic/ Plenum
33. Nakamura S, Fasol G (1997) The blue laser diode. Springer, Berlin Heidelberg New York
34. Tyson DS, Henbest KB, Bialecki J, Castellano FN (2001) J Phys Chem A 105:8154
35. Krause RA (1987) Struct Bond 67:1
36. Anderson S, Seddon KR (1979) J Chem Res Synop:74
37. Rillema DP, Allen G, Meyer TJ, Conrad D (1983) Inorg Chem 22:1617
38. Orellana G, Alvarez Ibarra C, Santoro J (1988) Inorg Chem 27:1025
39. Orellana G, Kirsch-De Mesmaeker A, Turro NJ (1990) Inorg Chem 29:882
40. Mingoarranz FJ, Moreno-Bondi MC, García-Fresnadillo D, de Dios C, Orellana G (1995) Mickrochim Acta 121:107
41. Klimant I, Wolfbeis OS (1995) Anal Chem 67:3160
42. Xavier MP, García-Fresnadillo D, Moreno-Bondi MC, Orellana G (1998) Anal Chem 70:5184
43. Stern O, Volmer (1919) Phys Z 20:183
44. Demas JN (1983) Excited state lifetime measurements. Academic Press, New York
45. Green NJB, Pimblott SM, Tachiya M (1993) J Phys Chem 97:196
46. Klafter J, Drake JM (1989) Molecular dynamics in restricted geometries. Wiley, New York
47. Szmacinski H, Lakowicz JR (1993) Lifetime-based sensing using phase-modulation fluorometry. In: Czarnik AW (ed) ACS Symp Ser 538 Fluorescent chemosensor for ion and molecule recognition. ACS, Washington, p 196
48. Nelly RN, Schulman SG (1988) Proton transfer kinetics of electronically excited acids and bases. In: Schulman SG (ed) Molecular luminescence spectroscopy. Methods and applications, part 2. Wiley-Interscience, New York, p 461
49. Vos JG (1992) Polyhedron 11:2285
50. Nazeeruddin MK, Kalyanasundaram K (1989) Inorg Chem 28:4251
51. Marazuela MD, Moreno-Bondi MC, Orellana G (1998) Appl Spectrosc 52:1314
52. Orellana G, Moreno-Bondi MC (1990) Spanish Patent 2 023 593
53. Grigg R, Jamilaprasadh Norbert WDJ (1992) J Chem Soc, Chem Commun 1300
54. Murtaza Z, Chang Q, Rao G, Lin H, Lakowicz JR (1997) Anal Biochem 247:216
55. Xu W, Mehlmann J, Rice J, Collins JE, Frasser CL, Demas JN, DeGraff BA, Bassetti M (1999) Proc SPIE-Int Soc Opt Eng 3534:456
56. Price JM, Xu W, Demas JN, DeGraff BA (1998) Anal Chem 70:265
57. Chan C-M, Fung C-S, Wong K-Y, Lo W (1998) Analyst 123:1843
58. Chan C-M, Lo W , Wong K-Y (2000) Biosens Bioelectron 15:7
59. Ayala D, Navarro F, Bustamante N, Orellana G (2000) Proc Europt(r)ode V. Lyon, France, p 61
60. Malins C, Glever HG, Keyes TE, Vos JG, Dressick WJ, MacCraith BD (2000) Sens Actuators B 67:89
61. Orellana G, García-Fresnadillo D, Moreno-Bondi MC, Xavier Céforo MP (1996) Spanish Patent 2 130 964
62. Lakowicz JR, Castellano FN, Dattelbaum JD, Tolosa L, Rao G, Gryczynski I (1998) Anal Chem 70:5115
63. (i) Kosch U, Klimant I, Werner T, Wolfbeis OS (1998) Anal Chem 70:3892; (ii) Kosch U, Klimant I, Wolfbeis OS (1999) Anal Chem 364:48
64. Leiner MJP (1991) Anal Chim Acta 255:209
65. Orellana G, de Dios C, Moreno-Bondi MC, Marazuela MD (1995) Intensity- and lifetime-based luminescence optosening of carbon dioxide. In: Scheggi AV (ed) Chemical, biochemical and environmental fiber sensors VII. vol 2508 SPIE, Bellingham, p 18
66. Marazuela MD, Moreno-Bondi MC, Orellana G (1995) Sens Actuators B 29:126
67. Xavier MP, Orellana G, Moreno-Bondi MC, Díaz-Puente J (2000) Quim Anal 19:118

68. Neurauter G, Klimant I, Wolfbeis OS (1999) Anal Chim Acta 382:67
69. Klimant I (1999) WO Patent 99/06821
70. Von Bültzingslöwen C, McEvoy AK, MacCraith BD, McDonagh C, Klimant I, Wolfbeis OS (2002) Proc Europt(r)ode VI. Manchester, UK, p 103
71. Grattan KTV, Zhang ZY(1994) Fiber optic fluorescence thermometry. In: Lakowicz JR (ed) Topics in fluorescence spectroscopy, vol 4: probe design and chemical sensing. Plenum, New York, p 335
72. Demas JN, DeGraff BA (1992) Proc SPIE-Int Soc Opt Eng 1796:71
73. Schulze JE (1990) US Patent 4 895 156
74. Klimant I (1997) WO Patent 97/24606
75. Lecomte J-P, Kirsch-DeMesmaeker A, Orellana G (1994) J Phys Chem 98:5382
76. Bustamante Alvarez N (2001) PhD thesis, Universidad Complutense de Madrid
77. Weigl BH, Holobar A, Trettnak W, Klimant I, Kraus H, O'Leary P, Wolfbeis OS (1994) J Biotechnol 32:127
78. Lübbers DW (1992) Advances in Biosensors, vol 2 JAI Press Ltd, p 215
79. Bambot SB, Lakowicz JR, Sipior J, Carter G, Rao G (1995) Optical measurement of bioprocess and clinical analytes using lifetime-based phase fluorimetry. In: Rogers KR, Mulchandami A, Zhow W (eds) Biosensor and chemical sensor technology. American Chemical Society, Washington DC, p 99
80. Cooney CG, Towe BC, Eyster CR (2000) Sens Actuators B 69:183
81. Singer E, Duveneck GL, Ehrat M, Widmer HM (1994) Sens Actuators A 42:542
82. Bambot SB, Rao G, Romauld M, Carter GM, Sipior JH, Terpetching E, Lakowicz JR (1995) Biosens Bioelectron 10:643
83. Xavier MP, Orellana G, Moreno-Bondi MC, Díaz-Puente J (2000) Quim Anal 19:118
84 Ni TQ, Melton LA (1993) Appl Spectrosc 47:773
85. Laws EA (1993) Aquatic pollution. An introductory text. Wiley, New York
86. Glud RN, Gundersen JK, Ramsing NB (2000) Electrochemical and optical oxygen microsensors for in situ measurements. In: Buffle J, Horvai G (eds) In situ monitoring of aquatic systems. Chemical analysis and speciation. IUPAC Series on Analytical and Physical Chemistry of Environmental Systems, vol 6. Wiley, Chichester, p 19
87. (i) Clark LC (1956) Trans Am Artif Intern Organs 2:41; (ii) Clark LC (1959) US Patent 2 913 386
88. For example, the blood gas analyzers Terumo-Cardiovascular Devices CDI System 2000, Irvine, CA, USA. Roche Diagnostics OPTI, Basel, Switzerland. GE-Marquette ABG System, Milwaukee, WI, USA. Diametrics Medical Paratrend 7, St. Paul, MN, USA. Philips Neotrend Sensor, Best, The Netherlands.
89. Kautsky, H (1939) Trans Faraday Soc 35:216
90. Parker CA (1968) Photoluminescence of solutions. Elsevier, Amsterdam
91. Pfeil A (1971) J Am Chem Soc 93:5395
92. Winterle JS, Kliger DS, Hammond GS (1976) J Am Chem Soc 98:3719
93. Lin C-T, Sutin N (1976) J Phys Chem 80:97
94. Demas JN, Harris EW, McBride RP (1977) J Am Chem Soc 99:3547
95. Wolfbeis OS (1997) Chemical sensing using indicator dyes. In: Dakin J, Culshaw B (eds) Optical fiber sensors IV: applications, analysis, and future trends. Artech House, Boston, p 53
96. Gottlieb A, Divers S (1991) In vivo applications of fiberoptic chemical sensors. In: Wise DL, Wingard LB (eds) Biosensors with fiberoptics. Humana Press, Clifton, p 325
97. Szmacinski H, Lakowicz JR (1994) Lifetime-based sensing. In: Lakowicz JR (ed) Topics in fluorescence spectroscopy, vol 4: probe design and chemical sensing. Plenum, New York, p 295
98. Trettnak W, Reininger F (1998) Optochemical sensors in water monitoring. In: Colin F, Quevauviller P (eds) Monitoring of water quality: the contribution of advanced technologies. Elsevier, Amsterdam, p 117
99. Demas JN, DeGraff BA, Coleman PB (1999) Anal Chem 71:793A

100. Bergman I (1968) Nature 218:396
101. Lübbers D, Opitz N (1975) Z Naturforsch C 30:532
102. Sacksteder LA, Demas JN, DeGraff BA (1993) Anal Chem 65:3480
103. Draxler S, Lippitsch ME, Klimant I, Kraus H, Wolfbeis OS (1995) J Phys Chem 99: 3162
104. (i) Lo KP, Groger HP, Luo S, Churchill RJ (2001) US Patent 6 207 961; (ii) Singh R (1994) WO Patent 94/10553; (iii) Shulze JE (1991) US Patent 5 012 809; (iv) Kaneko M, Nakamura H (1989) Spanish Patent 2 007 362; (v) Walt DR, Barnard SM (1994) US Patent 5 244 636
105. Bacon JR, Demas JN (1987) Anal Chem 59:2780
106. Hartmann P, Leiner MJP (1995) Anal Chem 67:88
107. Mills A, Thomas M (1997) Analyst 122:63
108. Di Marco G, Lanza M, Campagna S (1995) Adv Mater 7:468
109. Ishiji T, Kudo K, Kaneko M (1994) Sens Actuators B 22:205
110. Li X-M, Ruan F-C, Wong K-Y (1993) Analyst 118:289
111. Meier B, Werner T, Klimant I, Wolfbeis OS (1995) Sens Actuators B 29:240
112. He H, Fraatz RJ, Leiner MJP, Rehn MM, Tusa JK (1995) Sens Actuators B 29:246
113. Matsui K, Sasaki K, Takahashi N (1991) Langmuir 7:2866
114. Carraway ER, Demas JN, DeGraff BA (1991) Langmuir 7:2991
115. Lee SK, Shin YB, Pyo HB, Park SH (2001) Chem Lett 4:310
116. Klimant I, Leiner MJP (1992) Proc Europt(r)ode I, Graz, Austria
117. Xu W, McDonough RC, Langsdorf B, Demas JN, DeGraff BA (1994) Anal Chem 66: 4133
118. Kneas KA, Xu W, Demas JN, DeGraff BA (1997) Appl Spectrosc 51:1346
119. Brandrup J, Immergut EH (1989) Polymer handbook. Wiley, New York
120. Wolfbeis OS, Leiner MJP, Posch HE (1986) Mikrochim Acta 3:359
121. Huang X, Kovaleski JM, Wirth MJ (1996) Anal Chem 68:4119
122. Delgado Alonso J (2000) PhD thesis, Universidad Complutense de Madrid
123. Chan C-M, Chan M-Y, Zhang M, Lo W, Wong K-Y (1999) Analyst 124:691
124. Mills A, Thomas MD (1998) Analyst 123:1135
125. Hartmann P (2000) Anal Chem 72:2828
126. Roth T (2000) PhD thesis, Eidgenössische Technische Hochschule, Zürich
127. Carraway ER, Demas JN, DeGraff BA, Bacon JR (1991) Anal Chem 63:337
128. MacCraith BD, McDonagh CM, O'Keefe G, Keyes ET, Vos JG, O'Kelly B, McGilp JF (1993) Analyst 118:385
129. McEvoy AK, McDonagh CM, MacCraith BD (1996) Analyst 121:785
130. McEvoy AK, McDonagh C, MacCraith BD (1997) J Sol-Gel Sci Technol 8:1121
131. McDonagh C, MacCraith BD, McEvoy AK (1998) Anal Chem 70:45
132. McMurray HN, Douglas P, Busa C, Garley MS (1994) J Photochem Photobiol A 80:283
133. Klimant I, Belser P, Wolfbeis OS (1994) Talanta 41:985
134. Alcala JR (1994) Real-time chemical sensing employing luminescence techniques. In: Lakowicz JR (ed) Topics in fluorescence spectroscopy, vol 4: probe design and chemical sensing. Plenum, New York, p 255
135. Carraway ER, Demas JN, DeGraff BA (1991) Anal Chem 63:332
136. Demas JN, DeGraff BA, Xu W (1995) Anal Chem 67:1377
137. Hartmann P, Leiner MJP, Lippitsch ME (1995) Sens Actuators B 29:251
138. Choi MMF, Xiao D (2000) Anal Chim Acta 403:57
139. Bossi ML, Daraio ME, Aramendía PF (1999) J Photochem Photobiol A 120:15
140. Draxler S, Lippitsch ME (1996) Anal Chem 68:753
141. Choi MMF, Xiao D (1999) Analyst 124:695
142. Moreno-Bondi MC, Wolfbeis OS, Leiner MJP, Schaffar BPH (1990) Anal Chem 62: 2377
143. Chuang H, Arnold MA (1998) Anal Chim Acta 368:83
144. Wolfbeis OS, Carlini FM (1984) Anal Chim Acta 160:301

145. Wolfbeis OS, Leiner MJP (1988) Proc SPIE-Int Soc Opt Eng 906:42
146. Lippitsch ME, Pusterhofer J, Leiner MJP, Wolfbeis OS (1988) Anal Chim Acta 205:1
147. Lippistsch ME, Draxler S, Kieslinger D (1997) Sens Actuators B 38–39:96
148. Hauser PC, Tan SSS (1993) Analyst 118:991
149. Gruber WR, Klimant I, Wolfbeis OS (1993) Proc SPIE-Int Soc Opt Eng 1885:448
150. Trettnak W, Gruber W, Reininger F, Klimant I (1995) Sens Actuators B 29:219
151. Kieslinger D, Draxler S, Trznadel K, Lippitsch ME (1997) Sens Actuators B 38–39:300
152. Weigl BH, Wolfbeis OS (1994) Anal Chem 66:3323
153. Wolfbeis OS, Weis LJ, Leiner MJP, Ziegler WE (1988) Anal Chem 60:2028
154. Weigl BH, Holobar A, Trettnak W, Klimant I, Kraus H, O'Leary P, Wolfbeis OS (1994) J Biotech 32:127
155. McDonagh C, Kolle C, McEvoy AK, Dowling DL, Cafolla AA, Cullen SJ, MacCraith BD (2001) Sens Actuators B 74:124
156. Goswami K, Klainer SM (1988) Proc SPIE-Int Soc Opt Eng 990:111
157. Wolfbeis OS, Klimant I, Werner T, Huber C, Kosch U, Krause C, Neurauter G, Dürkop A (1998) Sens Actuators B 51:17
158. Holst GA, Köster T, Voges E, Lübbers DW (1995) Sens Actuators B 29:231
159. Gouin JF, Baros F, Birot D, André JC (1997) Sens Actuators B 38-39:401
160. Voraberger HS, Kreimaier H, Biebernik K, Kern W (2001) Sens Actuators B 74:179
161. Navarro-Villoslada F, Orellana G, Moreno-Bondi MC, Vick T, Driver M, Hildebrand G, Kiefeith K (2001) Anal Chem 73:5150
162. Holst G, Glud RN, Kühl M, Kimant I (1997) Sens Actuators B 38–39:122
163. Rosenzweig Z, Kopelman R (1995) Anal Chem 67:2650
164. Kohls O, Scheper Th (2000) Sens Actuators B 70:121
165. Hartmann P, Ziegler W (1996) Anal Chem 68:4512
166. Klimant I, Kühl M, Glud RN, Holst G (1997) Sens Actuators B 38-39:29
167. Xu H, Aylott JW, Kopelman R, Miller TJ, Philbert MA (2001) Anal Chem 73:4124
168. Ferguson JA, Healey BG, Bronk KS, Barnard SM Walt DR (1996) Anal Chim Acta 340: 123
169. (i) Bedoya M, Delgado J, García-Ares E, García-Alonso JL, Orellana G, Moreno-Bondi MC (2000) Proc Europt(r)ode V. Lyon, France, p 221; (ii) Orellana G, Moreno-Bondi MC, Bedoya M, Bustamante N, Delgado J, García E, García-Alonso JL (2001) Proc 1st SENSPOL Workshop: Sensing Technologies for Contaminated Sites and Groundwater. Alcalá de Henares, Spain, p 213; (iii) Orellana G, Moreno-Bondi MC (2002) Proc 15th Optical Fiber Sensors Conference. Portland, OR, p 115
170. Jeong Cho E, Bright FV (2001) Anal Chem 73:3289
171. Wolfbeis OS, Posch HE (1988) Fresenius Z Anal Chem 332:255
172. Marazuela MD, Cuesta B, Moreno-Bondi MC, Quejido A (1997) Biosens Bioelectron 12:233
173. Marazuela MD, Moreno-Bondi MC (1998) Anal Chim Acta 374:19
174. Wu X, Choi MMF, Xiao D (2000) Analyst 125:157
175. Preininger C, Klimant I, Wolfbeis OS (1994) Anal Chem 66:1841
176. Moreno-Bondi MC, Díez-Barrio MT, Orellana G, Bedoya M (2002) Proc Europt(r)ode VI. Manchester, UK, p 37
177. Bedoya M (2002) PhD thesis, Universidad Complutense de Madrid
178. Huber C, Werner T, Krause C, Klimant I, Wolfbeis OS (1998) Anal Chim Acta 364:143
179. Werner T, Klimant I, Huber C, Krause C, Wolfbeis OS (1999) Mikrochim Acta 131:25
180. Krause C, Werner T, Huber C, Klimant I, Wolfbeis OS (1998) Anal Chem 70:3983
181. Huber C, Klimant I, Krause C, Werner T, Mayr T, Wolfbeis OS (2000) Fresenius J Anal Chem 368:196
182. Delgado J, Orellana G (2002) Proc Europt(r)ode VI. Manchester, UK, p 115

CHAPTER 14

TIRF Array Biosensor for Environmental Monitoring

Kim E. Sapsford, Frances S. Ligler

1 Introduction to Biosensors

The field of biosensors has been the subject of an increasing number of books and literature reviews [1–5]. Biosensors consist of two major components: the molecular recognition element and a signal transduction mechanism. The molecular recognition element in biosensor technology takes the form of either a biological molecule (antibodies, enzymes or nucleic acids), a biological system (membranes, tissues or whole cells) or a biomimetic (a species which consists of an active site comparable to that of a naturally produced biomolecule). The increase in sensitivity and specificity, in comparison to chemical sensors, is a direct result of utilizing a biomolecule as the recognition element. The choice of transduction method largely depends on the intrinsic properties of the biomolecule, such as the co-factors found within the structure, and the recognition event measured. Typical transduction techniques used in array biosensor formats include optical and electrochemical; however, mass-mechanical, thermal and acoustic are also possibilities.

Biosensors have numerous applications such as medical diagnostics and healthcare, environmental monitoring of pollutants, process monitoring in the chemical, food and beverage industries and military defense. Despite much research and numerous publications in the area of biosensing, there has been little success in biosensor commercialization due to concerns about stability, sensitivity and quality assurance (during mass production) relative to established and competitive technologies [6, 7]. However, notable exceptions include biosensing devices for glucose [8], biological oxygen demand (BOD) [8, 9] and the home pregnancy and ovulation tests [10], which have made the leap into the commercial market. Much of the initial biosensor research and development has been in the clinical diagnostics market [11]; however, increasingly the use of biosensors for environmental applications is being investigated [7, 12]. Clearly in order to develop commercial biosensors for the environmental market, the final product must offer advantages over existing technology, such as the ability to perform faster, more sensitive, multi-analyte and real-time measurements.

1.1
Biosensors for Environmental Applications

The area encompassing environmental pollutants covers a variety of target analytes considered hazardous to both ecosystems and human health. As a result, there is a requirement to monitor the impact of numerous industrial processes, such as water treatment and chemical production, for potential pollutants, as well as the need to monitor air, food and drinking water for biological warfare (BW) agents. A number of recent reviews have highlighted the competition of biosensors with a number of well established laboratory methods and field test kits for environmental applications [7]. These existing commercial techniques have the ability not only to identify both expected and unexpected analytes but also to map a contaminated region for spatial distribution of a specific species. It appears that the main niche for biosensors in environmental applications is in the area of continuous and rapid field monitoring and on-site mapping of a particular known target analyte, currently not offered by established measurement techniques. Continuous monitoring for a specific analyte would eliminate the need for repeated visits to a particular site and would provide early warning in the case of BW agents, such that appropriate action could be taken. Standard sampling for depth profiling often disrupt the system measured; continuous monitoring would eliminate the stirring inherent in collecting discrete samples.

Concerns over the contamination or degradation of the sample during transport have led to interest in the development of techniques, such as biosensors, for on-site analysis. Such devices must be small, lightweight and portable. The development of microfabrication technologies in the area of miniaturized transducers increases the feasibility of portable biosensors. Advances in the field of lasers and light emitting diodes (LEDs) have led to light sources with a wider range of wavelengths, lower prices and lighter weight. This has resulted in an increased use of optical transduction over the more common electrochemical methods used previously in biosensors. An additional benefit is that reactions which do not involve electroactive species can be employed since electrical contact with the transducer is not required.

Size is not the only issue that must be addressed in the development of a biosensor for environmental monitoring. A number of important factors must be considered such as sensing environment, sensitivity requirements, shelf-life and capability for simultaneous detection of multiple analytes. It is clear that for use in environmental applications, the biosensor must perform in a number of complex sample matrices such as soil, river and sea water and food, preferably with little or no sample pretreatment. Possible detrimental effects of the sample matrix (pH and temperature) on the immobilized biomolecule may be avoided, however, by utilizing a simple sample pretreatment. Whether the biosensor is fixed into position at a site or carried around by an individual has ramifications for the size and operating constraints. If the biosensor is a continuous monitoring device, it must avoid the process of biofouling. Sensitivity requirements such as the concentration range and limit of detection must be considered, as these are different depending on the application. The response and recovery time of the biosensor, as well as how easily the sensing platform is replaced if the device is for one-time use, and the working lifetime and shelf-life of the biosensor are all important considerations. One of the major obstacles to biosensor commercialization is the diversity of potential environmental pollutants ranging from simple organic and inorganic molecules to bacterial spores. The specificity of biosensors, although a huge advantage, also leads to high cost per analyte. One solution to this concern is to produce biosensors capable of simultaneous detection of multiple analytes.

High density arrays are well known in the literature through their application to DNA gene chips [13, 14] and, more recently, antibody chips [15, 16]. These high density arrays have revolutionized the field of genomics and have been used to investigate gene expression as markers for disease and pathogens, drug response and development, and in the study of normal cell functions. Typically, thousands of discrete DNA sequences are printed onto microscope slides using an arrayer. The complementary DNA strands are fluorescently labeled and the resulting hybridization monitored using confocal microscopy. Unfortunately, environmental field monitoring requires small portable devices and confocal microscopy is still very much a laboratory-based technique. The time required for analyte binding to the chips is also a matter of hours. These factors have limited the application of high density arrays so far to the areas of biological, clinical and pharmaceutical screening and discovery, rather than detection using biosensors.

High density arrays themselves share much in common with their low density counterparts which are more typically used in biosensor applications: (1) the chemistry used to immobilize biomolecules onto the planar surfaces without denaturation; (2) the sensitivity attainable; and (3) the need for standardization and quantification of the resulting images [17, 18]. The use of high density arrays for environmental applications has so far been limited by the transduction method, confocal microscopy, used to interrogate them since the arrays themselves are micron-sized. However a recent study by Duveneck and coworkers [19], comparing confocal microscopy and total internal reflection fluorescence (TIRF), demonstrated that TIRF offered a more sensitive measurement. Hence, the future use of high density arrays coupled with TIRF for environmental applications is an exciting prospect. The remainder of this chapter will focus on low density array biosensors based on optical transduction mechanisms.

2 Technical Aspects of Optical Array Biosensors

The development of array biosensors is a relatively new field for both optical and electrochemical transduction and owes much to advances in microfabrication technology which has led to the ability to immobilize arrays of biomolecules in discrete regions on the transducer surface. This chapter will deal with optical transduction, although the field of electrochemical microarray transduction is also generating much interest and research [20–27]. Optical transduction has a number of advantages over electrochemical in as much as electrical contact with the surface is not required and multi-analyte detection is possible provided the various responses occur at different wavelengths or the surface response is spatially mapped via 2D imaging techniques.

2.1 Optical Transduction Used in Array Biosensors

There are a variety of optical behaviors which can be exploited in the development of biosensors including fluorescence, absorbance, bioluminescence, chemiluminescence and refractive index changes. In terms of array biosensors, fluorescence and refractive index change using reflectance transduction are the most popular. Techniques which can be grouped under the principle of reflectance include: attenuated total reflectance (ATR) which monitors alterations in the IR, visible and UV regions; surface plasmon resonance (SPR) [28] and interferometric [29] techniques which measure variations in refractive index; and total internal reflection fluorescence (TIRF) [30] which monitors changes in fluorescence. SPR, interferometry and, to a greater extent, TIRF have been developed as transduction methods to investigate the interactions of arrays of biomolecules immobilized on a sensing surface. The basic experimental arrangement of a system based on the principle of specular reflectance (where $\theta_{in} = \theta_{out}$), as opposed to diffuse reflectance (where $\theta_{in} \neq \theta_{out}$), is illustrated in Fig. 1. All of the optical approaches mentioned above are based on the principle of total internal reflection; however, they harness the resulting evanescent wave in different ways to produce a signal.

2.1.1 *Total Internal Reflection*

At the interface of two media with different refractive indices, described by Eq. 1, (n_x: where n_x is either n_1 or n_2; Fig. 1) incident light from the higher refractive index medium will be partly refracted and partly reflected,

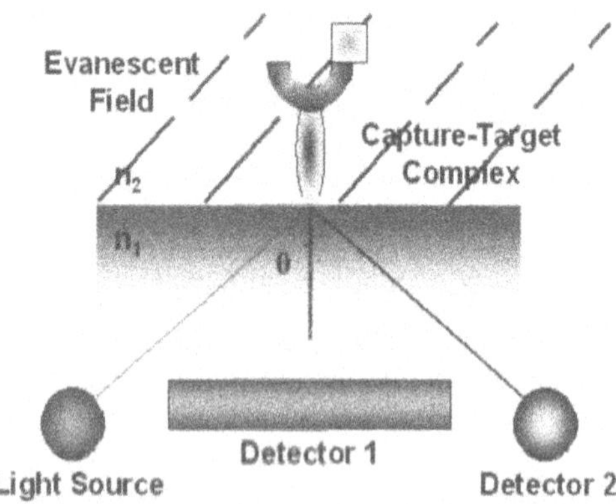

Fig. 1. The basic experimental arrangement of a system based on the principle of specular reflectance

$$n_x = [(\varepsilon_i \mu_i / \varepsilon_o \mu_o)]^{1/2} \quad (1)$$

where ε_i and ε_o are the permittivity of light in the dielectric and a vacuum, respectively, and μ_i and μ_o are the permeability of light in the dielectric and a vacuum, respectively. When the angle of incidence is greater than the critical angle (θ_c), given by Eq. 2, the phenomenon of total internal reflection is observed, whereby all the light is reflected and none refracted.

$$\theta_c = \sin^{-1} (n_2/n_1) \text{ where } n_1 > n_2 \quad (2)$$

Under the condition of total internal reflection, a standing wave of the electromagnetic field is set up at the point of reflection. This standing wave is known as the evanescent wave and penetrates into the lower refractive index medium with an exponential decay. The depth at which the intensity of either the electric or magnetic component of the electromagnetic wave drops to 1/e of its original value is known as the depth of penetration and is described by Eq. 3; where $n_{21} = n_2 / n_1$.

$$d_p = (\lambda_{vac} / 2\pi n_1) [1 / (\sin^2\theta - n_{21}^2)]^{1/2} \quad (3)$$

As shown in Eq. 3, d_p is dependent on the incident angle (θ), the refractive indices of the dielectrics (n_1 and n_2), and the wavelength (λ) of the light. Typical penetration depths range between 100 and 500 nm. It is this evanescent wave, generated at the interface between the two dielectrics that, in TIRF and interferometry, interacts with the surface species immobilized at the interface and in SPR generates surface plasmons in the metal surface.

The intensity of the reflected light will decrease after each reflection due to interactions at the surface of the waveguide. This decrease in the intensity of the reflected light can be measured using detector 2, described in Fig. 1, and is typically used in interferometric techniques and SPR. When the evanescent light excites a fluorophore on the planar waveguide, as with TIRF, the resulting fluorescence emission can be measured either at detector 2 as shown in Fig. 1 or, as is more common, at right angles to the waveguide interface, as represented by detector 1.

2.1.2 Interferometric Techniques

Interferometry is a label-free optical method which utilizes the evanescent field generated under the condition of total internal reflection to directly probe refractive index changes at an interface. The technique is based on the principle that an event such as biological binding will result in a change in the refractive index at the surface and hence change the speed of light through this region. In order to measure this change, a reference beam, which is not exposed to the binding event, is placed adjacent and in close proximity to the sensing beam; the sensing and reference beam are then combined to produce an interference pattern. Reaction specificity is provided by functionalizing the sensing channel

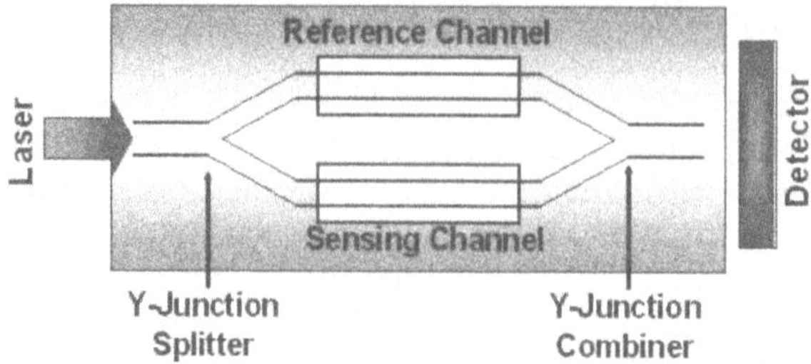

Fig. 2. Typical flow cell arrangement of an interferometric biosensor based on the Mach-Zehnder interferometer

with an analyte-specific capture molecule, such as an antibody. The reference channel is designed to account for background signals arising from thermal differences, mechanical changes and bulk index changes, arising, for example, due to non-specific binding.

Interferometric biosensors are generally based on the Mach-Zehnder interferometer. A typical channel flow cell arrangement which uses a monochromatic light source such as an argon-ion or helium-neon laser is shown in Fig. 2. The laser light is split into two beams using a Y-junction, one for the reference and one for the sensing channel. After passing through the sensing channels, the beams are recombined at a second Y-junction to produce the interference pattern. Waveguides typically take the form of integrated optical waveguides (IOW) and are mono-mode, providing uniform longitudinal excitation of the sensing regions. When the thickness of the waveguide is much greater than the wavelength of the reflected light, the waveguide is referred to as an internal reflection element (IRE). However, if the thickness of the waveguide is decreased such that it approaches the wavelength of the incident light, the pathlength between the points of total internal reflection become increasingly shorter. At the thickness where the standing waves, created at each point of reflection, overlap and interfere with one another, a continuous streak of light appears across the waveguide, and the IRE becomes known as an IOW [31].

IOWs are prepared by depositing an optically transparent thin film of high refractive index onto the surface of a glass substrate. These thin films are typically 80 – 160 nm in thickness and consist of materials such as silicon nitride (Si_3N_4) [32–36]. The IOW thin film is then commonly coated, with the exception of the reference and sensing regions, with a silicon dioxide cladding layer to improve optical contrast. The coupling of light into the waveguide is achieved either by end firing, prism coupling or diffraction gratings. End firing, although the easiest to perform is not as efficient a method of coupling light to thin films as the other two. Prisms on the other hand, although efficient at coupling, often complicate sensor design, due to the intimate contact, pressure and messy index matching fluids required to optically attach the prism to the surface of the waveguide. Gratings therefore are the most commonly used coupling method in interferometry measurements as they are easily fabricated into the waveguide by either embossing or etching the surface. Generally detectors in interferometric measurements take the form of an array of photodiodes.

Interferometry is capable of multi-analyte or multi-sample measurements, as demonstrated for chemical sensors [37–39], by creating arrays of the sensing-reference channel pairs on a single waveguide. However, microfabrication of the Y-junctions used in such array formats has proven to be difficult. Schneider and coworkers [34] overcame this issue by coupling a single broad beam of laser light into the waveguide; the light then interacted with multiple parallel sensing-reference regions, and an array of integrated optic elements combined the signal from adjacent regions producing the interference pattern. This greatly simplified the waveguide fabrication process, leading to miniaturized waveguide surfaces, which are cheap and easy to manufacture consistently.

One of the major disadvantages of interferometry is its inability to measure small molecular weight analytes, a problem shared by SPR. This complicates experimental design, as demonstrated by Schipper and coworkers [35] who developed a biosensor for atrazine which required the synthesis and surface immobilization of a BSA-atrazine complex. The free atrazine could be measured in solution via a competitive assay with the antibody, also in solution. Another important problem is the fabrication of a "real" referencing channel, which takes into account all processes other than the specific binding which occur in the sensing channel. The referencing channel must be designed such that it takes into account bulk refractive index changes of the solution, non-specific binding and temperature fluctuations.

2.1.3 SPR Imaging

SPR biosensing offers a label-free method of investigating biological interactions and was first demonstrated in 1983 [40]; since then it has continued to be the subject of numerous reviews and publications [41–45]. SPR imaging on the other hand is a recent development in SPR technology and is still in its infancy. SPR is the oscillation of free electrons which propagate along the interface between a conductor, typically a metal (e.g. gold, silver or copper) and a dielectric (e.g. air, water or buffer). Incident light, originating from a monochromatic light source such as a laser (e.g. HeNe) or LED, is coupled to the surface plasmons in the metal via the evanescent wave created under the condition of total internal reflection. This coupling is typically done using prisms, such as the Kretschmann configuration; however, IOWs [46] and gratings [47] have also been used. The coupling results in an attenuation of the reflected light intensity measured at detector 2 (Fig. 1) which commonly takes the form of a single or multiple photodiodes, the later is referred to as a "fan" SPR. The angle θ at which this coupling occurs is found to be critically dependent on the dielectric properties at the metal-dielectric surface, including the refractive index and film thickness.

One of the major limitations set by the signal-to-noise ratio of the detection system is the measurement of low molecular weight species, typically less than 5000 Da [48], although this limit of detection is improving. Often experimental design is limited to a system in which the larger of the two binding partner becomes the analyte with the smaller component immobilized as the "capture" element. Also due to the nature of its measurement (refractive index change), SPR is highly sensitive to ambient temperature variations and can be susceptible to bulk matrix effects, such as non-specific binding. There are, however, an increasing number of reports in the literature which address these problems. O'Brien and coworkers [49], for example, developed a two-element SPR array similar in concept to that described earlier for the interferometric measurements, consisting of a reference and sensing region, which was capable of compensating for both thermal and bulk effects.

The simplest approach to multi-analyte or multi-sample SPR biosensing is the production of multichannel, parallel sensing areas on the waveguide surface. Here the laser beam is split to excite surface plasmons in each channel; these channels are then interrogated separately to yield SPR information about each particular channel. An alternative to interrogating each channel separately is to spectrally discriminate the channels by switching to polychromatic light with a wavelength-sensitive detector, such as a spectrograph. This allows the production of a number of SPR curves at different wavelengths either by altering the angle of incidence using a sloping prism [50] or preparing distinct regions in the sensing area via introduction of a high refractive index overlay to part of the surface [51, 52]. SPR imaging, however, holds promise for the development of an SPR biosensor for environmental monitoring due to its potential to carry out both multi-analyte and multi-sample measurements. Research in SPR array biosensing is still very much in the early stages of development, and has mostly concentrated on proof of concept.

O'Brien and coworkers [53] expanded on their previous work to develop an SPR imaging system which combined conventional "fan" type SPR with charge coupled device (CCD) detection. The instrument excited and detected the reflected light intensity at all angles simultaneously using one dimension of the CCD array; the second dimension was used to study all 10 separate PDMS micro-flow channels in real time. The authors investigated antibody-antigen and avidin-biotin type interactions, depending on the immobilized species in a particular channel, demonstrating that such an instrument could be used to carry out either multi-analyte or multi-sample SPR measurements. This is different from the more conventional form of SPR imaging developed by Corn et al. [54–61] in which an expanded HeNe laser beam, reflected off the gold surface at a fixed angle, close to the SPR angle, illuminates the entire surface. Any changes in the index of refraction near the interface will result in changes in the intensity of the reflected light at a fixed angle. SPR imaging measurements were demonstrated with multi-analyte assays using initially 1D then 2D DNA sensing microarrays; multi-sample and multi-analyte measurements would be possible with the 2D sensing array combined with 2D detection format.

SPR imaging is still limited by its inherent inability to detect low molecular weight species. This is further amplified by the fact that the technique measures intensity variation, making it vulnerable to stray light and other noise. SPR imaging techniques based on intensity modulation also possess inferior resolution compared to SPR instruments based on angular or wavelength modulation, which can affect sensitivity. Much of the instrumentation is still very much laboratory-based and large in size and therefore not practical for field applications. The remainder of this chapter will therefore concentrate mainly on the development of TIRF array biosensors for environmental monitoring.

2.1.4
TIRF

As mentioned previously, TIRF is currently the most popular and well developed technique used in the study of 2D arrays of biomolecules immobilized onto the surface of planar waveguides. TIRF differs from SPR in that the evanescent wave, generated under the condition of total internal reflection, interacts with and excites a fluorophore near the surface of the waveguide, and the resulting fluorescence is measured by the detector [31, 62–65]. There has been extensive research into improving the optics and sensitivity of TIRF instrumentation. Most of the final systems described consist of a number of similar components such as the light source and detector and a variety of focusing lenses to improve detector response [66–70]. Golden [69] used a 2-dimensional graded index (GRIN) lens to focus the fluorescence from the planar waveguide onto a CCD. The GRIN lens provided a shorter working distance than a standard lens with a concomitant decrease in overall instrument size. The introduction of bandpass and longpass filters was found to improve the rejection of scattered laser light and hence reduce the background of the system [70].

Coherent light in the form of lasers is typically used as the excitation source in TIRF studies. The exact choice of the laser is dependent upon the fluorescent label used. The most commonly used lasers are the argon-ion (488 nm) laser for fluorescein and a helium-neon (633 nm) or diode laser (635 nm) for the cyanine dye (Cy5). A number of devices have been used in the detection of the resulting fluorescence emission, in particular CCD cameras [70–72], photomultiplier tubes (PMT) [73], photodiodes [74], a single photomultiplier tube [75, 76] and more recently a complementary metal oxide-semiconductor (CMOS) camera [77].

Unfortunately, a side effect of using bulk waveguides and collimated light is the production of sensing "hot spots" along the planar surface which occur where the light beam is reflected, illuminating only discrete regions. These hot spots have been successfully utilized as sensing regions by Brecht and co-workers in the development of an immunofluorescence sensor for water analysis [74, 78]. Feldstein et al. [70, 79] overcame this problem by incorporating a line generator and a cylindrical lens to focus the beam into a multi-mode waveguide which included a propagation and distribution region prior to the sensing surface. This resulted in uniform lateral and longitudinal excitation at the sensing region.

Another method of achieving uniform longitudinal excitation of the sensing region is to use IOWs [67, 72]. These are frequently used in TIRF studies and consist of thin films of inorganic metal oxide compounds such as tin oxide [66], indium tin oxide (ITO) [80], silicon oxynitride [72] and tantalum pentoxide [81, 82]. The light is coupled into these IOWs via a prism or grating arrangement. Studies by Brecht and co-workers [74] compared IRE- and IOW-based waveguides and concluded that the integrated optics significantly improved the sensitivity of the system by a factor of 100 compared to waveguides illuminated in discrete spots.

2.2 The Molecular Recognition Element

The choice of biomolecule used in the development of an array biosensor for a particular analyte is dependent on a number of factors, including the availability of the biomolecule for the analyte of interest and the application required [83]. Enzymes, antibodies and nucleic acids are all extremely specific, selective, sensitive and have fast response times. Whole cell and tissue-based array biosensors contain a number of enzymes resulting in multiple responses and less selectivity. Such systems do give a measure of the bioactivity of a particular toxin; however, they have mostly been studied using electrochemical transduction [27].

The different types of binding events that are typically monitored via TIRF and interferometry include antibody-antigen interactions, nucleic acid hybridization (DNA/RNA) and to a lesser extent receptor-ligand binding. Unlike interferometry, TIRF requires that one of the binding partners generates either intrinsic or extrinsic fluorescence. The introduction of an extrinsic fluorescence probe, such as rhodamine, coumarin, cyanine, or fluorescein dyes is favored over intrinsic fluorescence (which results from amino acid residues or cofactors) as the labels can be bound so that function is unimpaired and selected such that its fluorescence is distinct from the sample matrix background.

2.2.1 *Immunoassays*

To date antibody-antigen binding interactions are the most well characterized systems employed in sensors based on TIRF. The assays carried out using antibody-antigen systems, illustrated in Fig. 3, can be divided into four main categories: direct, competitive, displacement and sandwich immunoassays. As can be seen from Fig. 3, the direct assay (Fig. 3A) is the simplest method to perform; however, it requires that the antigen contain some form of intrinsic fluorescence that can be detected. In the absence of a fluorescent antigen, the preferred formats are competitive and sandwich assays (Fig. 3B, 3D). In the competitive assay (Fig. 3B), a fluorescently labeled antigen competes with a nonlabeled antigen for binding sites on the immobilized antibody [73, 74]. Hence, the resulting fluorescence signal is inversely proportional to the unlabeled antigen concentration. Competitive formats are especially useful in the detection of molecules not large enough to possess two distinct epitopes (e.g., haptens). Sandwich assays (Fig. 3D), on the other hand, require relatively large antigens; the antigen is bound to the immobilized capture antibody at one epitope and is detected by a fluorescent-labeled tracer antibody bound to a different epitope [71–73, 84, 85]. Sandwich and direct assays produce a fluorescence signal that is directly proportion-

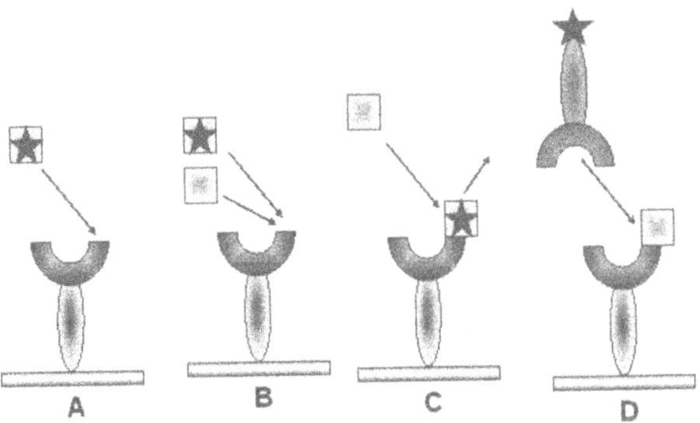

Fig. 3. Schematic representation of the four main categories of solid-phase immunoassays. A) Direct non-competitive assay between the immobilized antibody and the antigen in solution. B) Competitive assay between the fluorescently-labeled (known concentration) antigen and unlabeled (unknown concentration) in solution for binding sites on the immobilized antibody. C) Displacement assay between fluorescently-labeled antigen, previously bound to the immobilized antibody, and unlabeled antigen in solution. D) Sandwich assay, in which the amount of immobilized antigen is determined by passing a second fluorescently-labeled antibody over the surface

al to the amount of bound antigen. The displacement assay format Fig. 3C has only recently been demonstrated in planar waveguide TIRF for measurement of the explosive TNT; here immobilized antibodies are first saturated with a fluorescently labeled antigen. Upon introduction of the unlabeled antigen, displacement of the labeled antigen, which has a lower affinity for the antibody, occurs and is measured as a reduction in the fluorescent intensity [86, 87].

Interferometric techniques have used direct [32, 34], competitive [33, 35] and sandwich assay [36] formats to study antibody-antigen binding interactions. However, due to the low molecular weight limitation interferometry shares with SPR measurements, experimental design plays an important role when planning immunoassays for a particular species. For example Schipper and coworkers [33] were able to measure the herbicide atrazine (MW $\approx$ 250 Da) down to 100 ng L^{-1} by immobilizing a BSA-atrazine derivate onto the sensor surface and undertaking a competitive immunoassay between the surface immobilized BSA-atrazine and unlabeled solution atrazine for binding sites on an anti-atrazine antibody in solution. Schneider et al. [34] were able to detect the binding of human chorionic gonadotropin (hCG) to immobilized anti-hCG antibody via a direct immunoassay; the response was enhanced and sensitivity improved by carrying out a sandwich assay using a second anti-hCG antibody labeled with a gold or latex nanoparticle. However, introduction of a label such as a nanoparticle partially defeats the purpose of using label-free transduction methods in the first place.

2.2.2
DNA and mRNA Analysis

As mentioned previously, high density DNA/RNA microarrays have been extensively used in conjunction with confocal microscopy for clinical diagnostics and screening. There is currently only one example of measuring DNA or mRNA hybridization between complementary strands using interferometry [34]; however, the process has become increasingly studied using planar waveguide TIRF [76, 81, 88]. Duveneck et al. [81] coupled 16-mer oligonucleotides to functionalized tantalum pentoxide waveguides and studied the specific binding of Cy5-labeled complementary oligonucleotides using TIRF. The system demonstrated good stability and excellent signal reproducibility during repeated cycles of binding and regeneration with 50% urea. Budach and co-workers [88] immobilized two different 5'-amino-terminated oligonucleotides in a checkerboard pattern on tantalum pentoxide waveguides using ink-jet printing. One of the Cy5-labeled complementary oligonucleotides was then passed over the waveguide surface and the fluorescent signal intensity of the spots monitored. There was no measurable cross-reactivity between the non-complementary oligonucleotides, and detection limits of 50 fM were achieved for each analyte. The surface was regenerated numerous times, using urea, without any apparent loss in activity. Schuderer et al. [76] opted to immobilize biotinylated 18-mer oligonucleotides onto glass surfaces functionalized with adsorbed avidin using a flow cell assembly; this would allow the immobilization of different capture biomolecules to the same waveguide should it be required. The fluorescein-labeled complementary oligonucleotide was then passed over the surface and the fluorescent signal intensity in each channel monitored using a single photomultiplier, which was scanned over the surface. Detection limits ranged between 3 and 10 fmol with surface regeneration achieved using sodium hydroxide. The main disadvantage to this system was the need to scan the detector over the surface, making it unreliable for the determination of fast binding kinetics.

The use of DNA biosensors for both environmental monitoring and the detection of BW agents, using mainly electrochemical transduction mechanisms, has been reviewed in the literature [20, 83]. However, the optical transduction systems described above have so far looked at simple DNA hybridization, between an immobilized strand and its fluorescently labeled complement, and have not as yet considered environmental applications for such systems.

2.2.3
Membrane Receptor-ligand Interactions

Ligand binding to receptors other than antibodies and DNA has been much less studied in the literature. Currently, only a limited number of studies describing receptor-ligand binding using planar waveguide TIRF have been published. Schmid et al. [89, 90] immobilized the green fluorescent protein (GFP) and serotonin receptor, Rowe-Taitt et al. [91] studied gangliosides as receptors and Pawlak et al. [82] investigated the membrane enzyme Na,K-ATPase in membrane fragments. One major problem with studying receptor-ligand binding has been the immobilization of the receptor protein such that it remains active on the surface. When successful, receptor-ligand binding studies offer applications in the pharmaceutical industry for drug development, for investigating membrane processes and also in environmental monitoring applications, as demonstrated by Rowe-Taitt et al. for cholera toxin [91].

2.3 Immobilization of the Biomolecule to the Transducer

One important prerequisite for all immobilization techniques is that the integrity of the biomolecule be preserved and that the active site remain accessible to the binding partner. There are various methods by which the biological component of a biosensing system can be immobilized onto the surface of the transducer, including physical adsorption, covalent immobilization, and entrapment in polymer matrices [1]. Physical adsorption and covalent binding to functionalized surfaces are the most commonly used in TIRF and interferometry measurements. Adsorption of a biomolecule to a surface under physical processes occurs via dipole-dipole interactions, van der Waals forces or hydrogen bonding, depending on the nature of the substrate surface and the adsorbate. Physical adsorption is the simplest method to perform and has been successfully used in a number of interferometric studies. Schipper et al. [32] adsorbed anti-hCG antibody directly onto an O_2-plasma cleaned Si_3N_4 waveguide for detection of hCG. The same group later used physical adsorption to immobilize a BSA-atrazine complex, but further enhanced the interaction by functionalizing the Si_3N_4 waveguide with dichlorodimethylsilane to increase hydrophobicity [33, 35]. Unfortunately, physical adsorption in general is not only strongly influenced by changes in the ambient conditions, such as pH and the solvent used, but may also be a reversible process. Furthermore, adsorption may not provide as high a density of immobilized biomolecules as covalent immobilization [92]. Physical adsorption is generally nonspecific, random and multi-orientated in nature, often resulting in the inaccessibility of the active binding site.

Covalent immobilization provides surfaces with reproducibly attached biomolecules at relatively high densities (2 ng mm^{-2}) [93]. Most methods of covalent immobilization involve the activation of the surface (e.g., using silane or thiol self-assembled monolayers), followed by covalent linkage of the biomolecule either directly or using a crosslinker. Self-assembled monolayers (SAMs) are extensively used in the construction of artificial biomolecular surfaces due to the simplicity of preparation and adaptability of the resulting surface chemistry. The surface of the SAM can be engineered by incorporating different functionalities into the tail group, such that hydrophobic or hydrophilic surfaces which can undergo further chemical reactions are produced. The use of SAMs in the development of biosensors and biosurfaces has been reviewed by a number of groups [94–96]. Although most research has been concerned with the immobilization of biomolecules onto noble metal surfaces, typically for SPR studies, an increasing number of groups are interested in the use of glass substrates for immobilization [93, 97–100].

There are a number of different planar surfaces used in the immobilization of biomolecules for study with TIRF and interferometry. These includ simple bulk waveguides such as glass, silica and polystyrene, and the slightl more complicated IOW waveguides such as tantalum pentoxide (Ta_2O_5). Ther are, likewise, a variety of different surface chemistries used to modify thes waveguides in order to facilitate biomolecule immobilization. Silanization o the waveguide, whether it be the bulk glass or an integrated optical waveguide is a popular method of functionalizing the surface for further chemistry, wheth er physical [33, 35, 72] or covalent [34, 36, 80] in nature. For example, Plowma et al. [72] functionalized a silcon oxynitride IOW with dimethyldichlorosilan to which an antibody was then directly adsorbed for study with TIRF. Schneide and co-workers [34] modified a silicon nitrite waveguide with 3-glycidoxypro pyldimethylethoxysilane (GOPS). The GOPS was then oxidized to form surfac aldehydes which then underwent reductive amination coupling with either avi din or anti-salmonella antibody and the resulting surfaces studied using inter ferometry. The avidin-biotin interaction is also extensively used in the immobi lization of biotinylated molecular recognition elements. This noncovalent pro tein-ligand interaction is commonly used either via the physical adsorption o avidin onto the surface [67, 71, 73, 76, 101] or in the production of multi-layers often involving the use of both covalent and noncovalent interactions [34, 36 84, 102].

The study of receptor-ligand binding has meant the careful immobilizatior of the receptor protein, which is often membrane bound in its natural environment, such that the sensor surface mimics this environment ensuring that the receptor remains active [82]. Vogel's group [89, 90] successfully immobilized fully functional GFP and serotonin receptors onto planar waveguides in defined orientations. GFP and serotonin were genetically modified to contain poly-histidine (his) tags at the C- and N-termini, respectively. These tags bound specifically to waveguides coated with thiosilane and nitrilotriacetic acid (NTA)-bound metal ions; the his-tagged protein binding was reversible. Activity of the immobilized serotonin receptor was demonstrated through binding of a fluorescent ligand and competitive binding studies [90]. Pawlak et al. [82] created multi-layers on top of the planar waveguide. First, a long-chain alkyl phosphate monolayer was used to make the metal oxide surface hydrophobic. Lipid vesicles were then adsorbed onto the monolayer to form a lipid film. Membrane fragments containing the ATPase were finally adsorbed onto the lipid biolayer for use in a fluorescence quenching assay. Rowe-Taitt et al. [91] immobilized gangliosides as receptors for toxins by immobilizing a long-chain alkyl silane to create a hydrophobic surface for adsorption of the ganglioside; presumably the fatty acid tail of the ganglioside intercalated into the silane film. The resulting receptor immobilization was quite stable to rinsing, dehydration and rehydration. Both direct binding and sandwich assays using different gangliosides patterned on the waveguide were demonstrated.

2.4 Creation of Low Density Biomolecular Arrays

The evanescent wave excitation of a surface-bound fluorophore has been studied for two decades in the form of fiber optic technology. A number of the researchers currently involved in developing planar waveguide TIRF previously carried out much of their initial research in the field of fiber optics. Planar waveguides offer a number of advantages compared to fiber optic technology, including the relative ease of preparation and integration into fluidic systems. Researchers in the field have immobilized capture biomolecules uniformly over the planar surface and monitored the fluorescent signal intensity either as a function of time or the concentration of the labeled binding partner [73, 74, 81, 82, 101].

The most important advantage of using a planar waveguide is the possibility of creating patterns of immobilized biomolecules leading to multiple, parallel assays on a single waveguide. A number of techniques have been used in the creation of patterned biomolecular assemblies on planar surfaces as reviewed by Blawas and Reichert [103]. In terms of fluorescence studies, the production of these patterned surfaces has been investigated using either the fluorescence microscope or TIRF instrumentation. The patterns are typically created either by using photolithography or by depositing the recognition molecules in physically separate locations on the waveguide.

A method for the photolithographic patterning of proteins on surfaces was described by Bhatia et al. [104, 105]. Ultraviolet light was used to pattern (3-mercaptopropyl)trimethoxysilane on a glass surface. Exposed regions of the surface became protein resistant through the conversion of the thiol group to a sulfonate species, while the masked areas were subsequently used to bind the biomolecule. This proved to be a convenient method of creating high resolution patterns (less than 10 μm in width) of immobilized capture biomolecules. Unfortunately, this method had the disadvantage that only a single biomolecule could be patterned.

Photopatterning of the surface has also been utilized by Schwarz and co-workers [106]. Here photoablation of polymer substrates was used to produce avidin patterns on the exposed substrate. Likewise, Wadkins et al. [107] used glass slides coated with a photoactivated optical adhesive and a mask to create wells in the gel layer upon light exposure. A different biomolecule could then be covalently attached to the exposed glass in each well. A similar method was adopted by Guschin et al. [108] and Arenkov et al. [109] for the immobilization of oligonucleotides; however, the biomolecules were immobilized to the gel pads rather than the glass. Blawas and co-workers [110] used the caged-biotin-bovine serum albumin (BSA) compound, methyl α-nitropiperonyloxy-carbonyl-biotin-BSA, to pattern glass surfaces. First, caged-biotin-BSA was adsorbed onto the glass slide. Secondly, the slide was exposed to 353 nm light through a mask, which effectively removed the cage surrounding the biotin in unmasked areas. Thirdly, streptavidin was selectively bound to the irradiated regions of the surface. Finally, biotinylated capture antibody was bound to the streptavidin. The disadvantage of this method is that there was some cross-contamination of the first immobilized streptavidin with the biotinylated antibody intended for immobilization to a streptavidin bound in a second or third repetition of the process.

Two additional methods of photolithographically patterning proteins on planar waveguides use polyethylene glycol (PEG) moieties to prevent nonspecific protein adsorption. In the method developed by Conrad et al. [111, 112], the photochemically active silane, *o*-nitrobenzyl polyethylene glycol trichlorosilane, was attached to a glass waveguide. Photo-oxidation of the PEG-terminated silane through a mask cleaved the carbamate to yield an amino-terminal PEG and a surface-bound *o*-nitrosobenzaldehyde. The antibody reacted with the *o*-nitrosobenzaldehyde in a Schiff's base reaction, forming a stable amide linkage. The process was repeated for the addition of different antibodies to additional spots on the waveguide. Similarly, Liu et al. [113] tethered a benzophenone photophore through a PEG spacer to a maleimide group. After exposure to light, antibodies were immobilized and an assay conducted on the polystyrene waveguides. Spacer lengths of five ethylene glycol groups proved optimal for maximizing the signal-to-background ratio.

The use of ink jet printing is another popular choice for the production of patterned biomolecular surfaces. Silzel et al. [71] ink jet printed either the capture antibodies or avidin in 200 μm diameter zones on the surface of polystyrene waveguides. Biotinylated antibodies were later immobilized on the avidin spots. A checkerboard pattern of two different oligonucleotides was produced by Budach and co-workers [114] using the ink jet printing of capture biomolecules onto a Ta_2O_5 waveguide using (3-glycidoxypropyl)trimethoxysilane.

Physically isolated patterning using flow cells constructed from a variety of materials, including polydimethylsiloxane (PDMS) [70], a rubber gasket [72], a Teflon block fitted with O-rings [76] and a microfluidics network made of silicon [115], have been used. Typically the flow cell, containing a number of channels, was temporarily attached to the surface of the planar waveguide and each channel filled with a solution of the capture biomolecule, as shown in Fig. 4A. In this example [70], the resulting waveguide was patterned with stripes of immo-

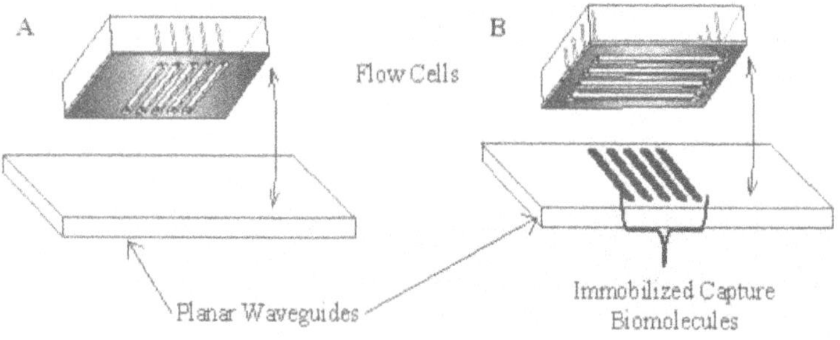

Fig. 4. The patterning of capture biomolecules using flow cells (adapted from [70]). A) A multichannel flow cell is pressed onto the planar waveguide and each channel filled with a solution of the capture biomolecule. B) Sample and fluorescent tracer antibody are passed over the waveguide surface perpendicular to the immobilized capture biomolecule channels using a second flow cell.

bilized biomolecules (Fig. 4B). The sample and fluorescently labeled antibody were then passed over the surface using a second flow cell orientated perpendicular to the immobilized capture biomolecule channels, as shown in Fig. 4B. The beauty of the physically isolated patterning technique is its ability to immobilize a number of different capture biomolecules, i.e. one type in each flow cell channel, onto a single surface, creating an array of recognition sites with no possibility of cross contamination. The main disadvantage of this approach is the difficultly in scaling up the process for manufacture of large numbers of waveguides. Yet it is easy to use for investigations with no special photolithographic or arraying equipment required.

3
State of the Art

As described in the previous sections, much of the initial investigation into the use of planar waveguides in TIRF biosensors has centered on both instrumentation development and reproducible immobilization of the capture biomolecules. However, once these systems have been optimized, the question of application becomes the driving force behind further development. Due to the vast number of biological systems which respond to a variety of analytes, the number of potential applications for biosensors in general is extensive, including medical diagnostics and healthcare, environmental monitoring, process monitoring in the chemical, food and beverage industries, and military defense. However, this chapter will deal with array biosensors developed specifically for environmental monitoring of pollutant species and BW agents.

A number of studies have investigated the use of a single capture biomolecule-analyte assay. Brecht et al. [74, 78] have developed TIRF immunoassays for the detection of atrazine, 2,4-dichlorophenoxyacetic acid and simazine in water; such a design has multi-analyte biosensing potential although these have not as yet been demonstrated. Schipper and co-workers [33, 35] have developed an atrazine biosensor for detection in water based on interferometric transduction. Schneider et al. [34, 36] have discussed multi-analyte detection using their interferometric instrumentation, but as of yet have only demonstrated single analyte detection for a number of species including hCG, influenza A virus, *Salmonella typhimurium* and DNA hybridization.

As previously stated, one of the major advantages of using a planar substrate is the ability to create arrays of different capture biomolecules for multi-analyte sensing. It is the development of this multi-analyte microarray technology that will give the resulting biosensor the edge over a number of current laboratory based measurements. There are to date at least five research groups involved in the immobilization of patterns of multiple capture biomolecules onto planar waveguides (Table 1), although only three of these groups have demonstrated multi-analyte measurements.

Zeller and coworkers [120] have developed a unique TIRF system in which the planar waveguide consists of multiple, single pad, sensing units. Each of these single pads has its own laser light input, background suppression, and coupling of the fluorescence emission to the detector. The authors demonstrat-

Table 1. Summary of research groups currently involved in multi-capture biomolecule patterning

Research group/ Technique	Analytes measured	Single- or multi-analyte detection	References
Schneider et al. Interferometry	hCG, influenza A virus, *S. typhimurium*, DNA oligonucleotides.	Single	34
Ligler et al. TIRF	Ricin, *Y. pestis* F1, staphylococcal enterotoxin B, ovalbumin, mouse and human IgG, D-dimer, *B. globigii*, MS2 bacteriophage, cholera toxin, botulinum toxoids A and B, *B. anthracis, F. tularensis, B. abortus*	Single and Multi	64, 70, 84, 85, 91, 116–119
Silzel et al. TIRF	Four different human IgG subclasses	Multi	71
Plowman et al. TIRF	Various IgG c. kinase MB c. troponin I myoglobin	Single and Multi	72
Budach et al. TIRF	16-Mer and 22-mer oligonucleotides	Single	114
Zeller et al. TIRF	Mouse and rabbit IgG	Single	120

ed a two-pad sensing device in which one pad was modified with mouse immunoglobulin (IgG) and the other with rabbit IgG. However, only one Cy5-labeled antibody directed against each immobilized antigen was assayed at a time. It was suggested that the laser light could be split into spatially different parts in multi-analyte measurements, using multiple single sensing pads. Such a process would probably involve a number of optical components, and therefore the robustness of the system for use outside the laboratory is still in question.

Experimental design is an important consideration when carrying out multi-analyte studies in terms of careful choice of capture biomolecule such that cross-reactivity is limited. The performance of single and multiple analyte assays was compared by Plowman et al. [72] using IgGs of different species and antibodies specific for cardiac proteins (creatin kinase MB, cardiac troponin I and myoglobin). Studies also investigated the effects of using polyclonal versus more specific monoclonal antibodies during assays. Results suggest that polyclonal antibodies are more prone to cross-reactivity in the multi-analyte assay format; therefore, monoclonal antibodies were the capture biomolecule of choice when available. Also, the single analyte assay format was found to have lower detection limits than its multi-analyte assay counterpart. Interestingly, however, a study by Rowe et al. [85] demonstrated the use of mixtures of polyclonal antibodies of different species with no cross-species interactions observed. Furthermore, the same sensitivity was achieved with mixtures of the fluorescent tracer antibodies as with single antibody assays.

The long-term aim of most biosensor research is the development of a fully-automated instrument geared towards portability and low cost, essential considerations for environmental monitoring applications. The Naval Research Laboratory (NRL) group including Ligler, Golden, Rowe-Taitt and coworkers first published papers on TIRF studies in 1998 for the detection of three analytes, ricin, *Yersinia pestis* F1 and staphylococcal enterotoxin B (SEB) [64]. In this particular study, the antigens and the Cy5 tracer antibodies were added sequentially and the slide imaged. Later the use of a PDMS flow cell for the patterning of capture biomolecules was developed and the immunoassay which was run prior to imaging was carried out using a fixed polymethylmethacrylate (PMMA) flow cell aligned perpendicular to the patterned antibody channels [116]. Not only was simultaneous detection of analytes demonstrated, but *Y. pestis* F1 was also detected in clinical fluids such as whole blood, plasma, urine, saliva and nasal secretions. This study was further extended in 1999 to the measurement of SEB and D-dimer in clinical fluids, all with detection limits suitable for clinical analysis requirements [84].

Studies which evaluate potential matrix interferents are essential not only for system development but are also a requirement if the instrumentation is to make the transition into commercial applications. The impact of potential environmental interferents was therefore also addressed [119]. Analyte samples of *Bacillus anthracis*, *Francisella tularensis* LVS, *Brucella abortus*, SEB, cholera toxin and ricin were assayed in the presence and absence of interferents such as sand, clay, pollen, and smoke extracts, and results were compared. No false positive or false negative responses were caused by the potential interferents; however, in some cases the signal amplitude was affected.

During multi-analyte assays for SEB, *Y. pestis* F1 and D-dimer, the room temperature CCD camera was replaced with a thermoelectrically cooled version, which resulted in a reduction of the background due to noise fluctuations [117] and an observed decrease in the limit of detection from 5 ng/mL [64] to 1 ng/mL for SEB [84]. Another variation on previous experiments was the use of a temporary, removable PDMS flow cell in the immunoassays as compared to the permanently mounted PMMA flow cell. In order to further develop the array technology, an automated image analysis program was developed and some of the optical and fluidic components were miniaturized [70]. To test the ability of the array biosensor to measure three diverse classes of analytes, assays were carried out using bacterial, viral and protein analytes [85]. Single or multi-analyte samples were run though the assay channels followed by either individual Cy5-labeled tracer antibodies or a mixture of polyclonal tracer antibodies. The array biosensor was used in the study of 126 blind samples and the automated image analysis proved reliable in the discrimination of fluorescent signals, with detection limits in the mid ng/mL range, equivalent to ELISA results using the same antibodies. Assays which used mixtures of fluorescent antibodies gave the same results as those in which the fluorescent antibodies were run individually. The approach using mixtures of tracer antibodies was later extended to monitor the six biohazardous analytes *B. anthracis*, *F. tularensis*, *B. abortus*, botulinum toxoids, cholera toxin and ricin, demonstrating simultaneous analysis of six samples for six analytes in 12 min. [118, 119].

4
Miniaturization and Automation of Array Biosensors

For environmental applications, it is important that the biosensor be fully automated, stand alone and be portable; therefore recent studies by the Ligler group have concentrated on realizing this goal. This has involved miniaturization of the optics and the combination of an automated fluidics system with an automated image analysis program [70, 79, 118]. The waveguides used for array biosensors are inherently small, generally the size of a 1 × 3 in. microscope slide or smaller. It is not the sensing component itself that limits the minimum size and weight of the biosensor, but the associated optics and automated fluidics system. To date, the smallest biosensor reported in the literature is a fully automated system that fits into a tackle box (Fig. 5) and weighs 16 kg [121]. Miniaturization of the fluidics made this portable device possible, but the electronics and optics are still a bit cumbersome for field use.

Two inventions made it possible to automate and miniaturize the fluidic system for the array biosensor. The first was a method for attaching flow channels to the waveguide without stripping evanescent light from the surface [70, 79]. The second invention was an integrated fluidics unit whereby fluid flow was controlled using small air valves [122, 123]. To solve the problem of attaching a flow cell to the waveguide without perturbing the confinement of the excitation light, but still allowing for optimization of the evanescent field in either air or water, a unique patterned reflective cladding was developed. This reflective cladding insulated the waveguide optically from the flow cell, yet allowed the evanescent excitation of fluorophores within the channels of the flow cell [70]. The

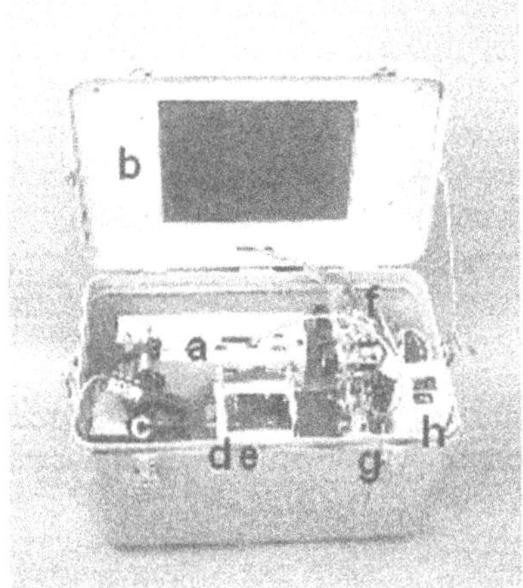

Fig. 5. Portable array biosensor. The fully automated biosensor includes: a. Pentium computer with screen (in lid) and removable keyboard (b., not shown); c. diode laser with line generator attached to output end; d. Peltier cooled CCD; e. scaffold containing emission filters and GRIN lens array and supporting the waveguide over which a flow channel and fluidics cube (not shown) are attached; f. three two-channel peristaltic pumps; g. small air valves; and h. laser power supply

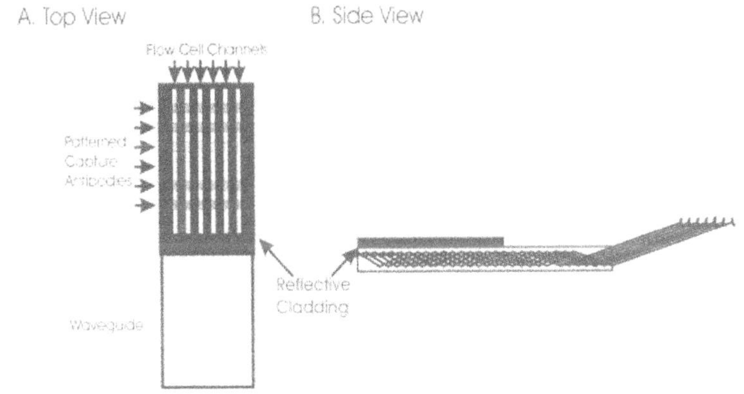

Fig. 6. Silver cladding stenciled onto the waveguide prevents stripping of evanescent light into the flow channels at the point of contact (adapted from [70]).

pattern of the reflective cladding, shown in Fig. 6, covered the area where a six-channel flow cell made contact with the waveguide. The rest of the waveguide surface was left unclad, and the array of biomolecules was subsequently immobilized on the exposed glass. The cladding was deposited using vacuum deposition through a metal mask, which acted like a stencil and allowed only selected areas to be coated. The silver-based cladding consisted of three layers: a thin, transparent dielectric material to promote adhesion, a silver film for reflectance, and a thin chromium film to protect the silver from dissolution in the saline buffers required for bioassays [79]. The comparison of the signal loss due to flow cell attachment for assay elements patterned onto the clad and unclad planar waveguides was impressive. The results from fluorescence immunoassays using 100 and 500 ng/mL of a protein toxin indicated a 90% decrease in signal after the flow cell was attached for the unclad waveguides. The silver-clad waveguides, by comparison, maintained 85–90% of their signal for the same assays after attachment [70].

In order to process fluids automatically over the waveguide, pumps and valves are required in addition to the flow channel component. Reservoirs may be needed to contain samples, buffers, and/or tracer reagents. In general, these pumps, valves, and reservoirs are cumbersome and the attachment points for the connecting tubing are susceptible to leakage. Furthermore, valves and other constriction points tend to clog when exposed repeatedly to complex, environmental samples. In order to reduce the size, weight, and power requirements of the fluidic components, the NRL group fabricated layered fluidic systems in plastic. Initial fluidic components could be designed, milled in polycarbonate sheets, assembled and tested. Once a design was optimized, the component could be produced inexpensively and in large quantities by injection molding. The most unique feature on these plastic fluidic cubes was that the fluid flow from multiple reservoirs through the flow channels could be regulated using a few tiny air valves. As depicted in Fig. 7A, groups of reservoirs (2 of 6 shown in this figure) were connected to a single air valve though a gasket across the top of the group

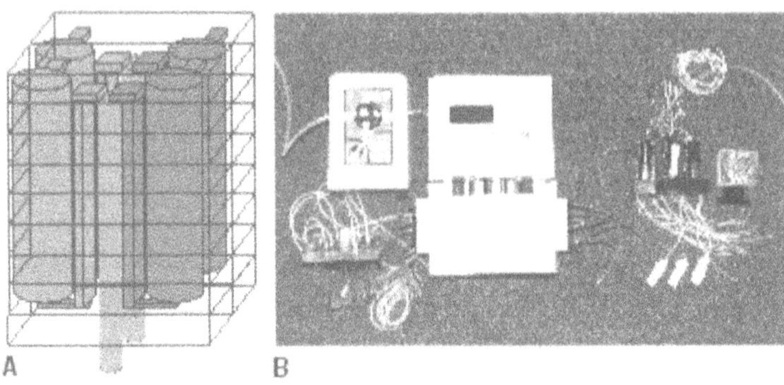

Fig. 7. A) Schematic of a portion of the fluidics cube. Groups of reservoirs (two reservoirs shown here out of a group of six actually fabricated) are connected at the top by conduit leading to a single air vent. At the bottom, each reservoir is connected to a J-tube which prevents release during the filling process. The J-tube empties into a conduit leading to a flow channel attached to the planar waveguide. B) The conventional automated fluidics system (left) with a 4-way valve, 6-channel pump and six sample reservoirs compared to the fluidics cube attached to the flow channel and waveguide and small valves and pumps (right)

of reservoirs. The exit from the reservoirs emptied through the flow channels across the waveguide. A negative pressure was exerted on the reservoirs using a finger-sized peristaltic pump placed after the waveguide. When one air valve was opened, the pump would pull the fluid from all reservoirs attached to that valve (e.g. all sample reservoirs) through the flow channels and out to waste. When the 1st (sample) valve was closed and the 2nd valve was opened, all the reservoirs containing fluorescent reagents would empty their contents into the flow channels.

Figure 7B shows two fluidic systems. The one on the left uses conventional pumps, valves, and reservoirs. The one on the right uses the fluidics cube, attached to the flow channels and waveguide, and associated small pumps and valves. The fluidics cube attaches directly to the flow channels and waveguide with no intervening tubing. Connections to the plastic block are limited to a buffer line, an air line, and two vents, all of which are inserted simultaneously through a gasket glued to the top of the cube.

In order to miniaturize the array biosensor further, Golden and Ligler [77] considered the possibility of replacing the CCD with the less expensive alternatives for image capture, a CMOS camera or a photodiode array. Both would provide smaller systems more amenable for portable sensors. The CMOS cameras, in particular, would have lower noise than the CCDs. However, careful comparison revealed that the Peltier-cooled CCDs still have an order of magnitude better signal-to-noise ratio than either of the other two imaging systems (Fig. 8).

The first tackle-box breadboard system included a small Pentium computer, which could be replaced by a chip and keyboard once the data analysis was automated. Thus in the next version, the Pentium was removed and the data recorded using a laptop computer. In addition to the Pentium with its keyboard and screen, the breadboard included a rather large and expensive (ca. $24,000) Pel-

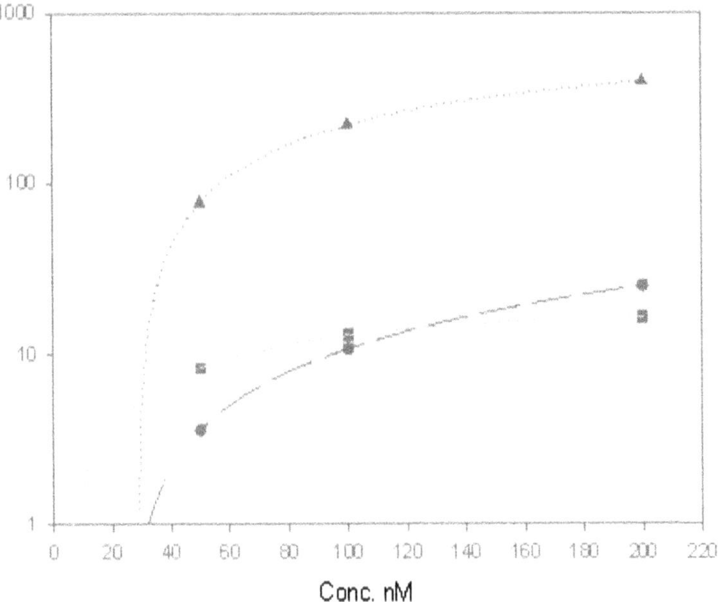

Fig. 8. Signal-to-noise ratio (*S*/*N*) for the three detection systems. Log plot shows the mean fluorescence signal (three data points) divided by the system noise (standard deviation of the background signal) for measurements of the fluorescent dye Cy5 in glycerol on a planer waveguide excited using evanescent excitation light at 635 nm. Shown are: photodiode (square, gray), CMOS (circle, dash) and CCD (triangle, dotted)

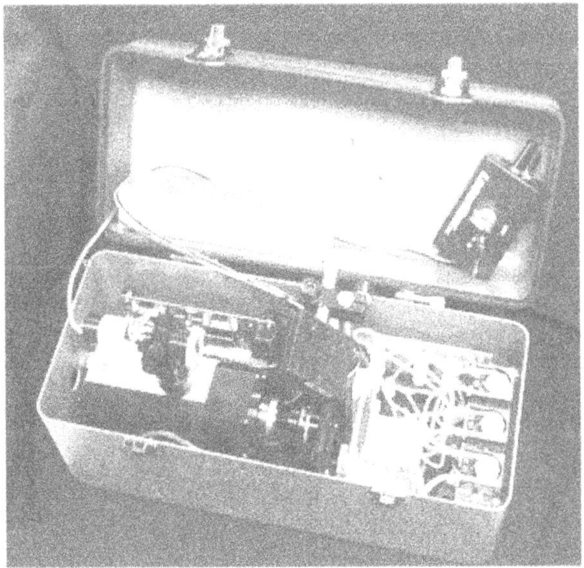

Fig. 9. The latest version of the portable array biosensor

tier-cooled CCD. While the photon collection capability of uncooled low light cameras was sufficient, the backgrounds were too variable for high sensitivity measurements. Recent advances in CCD technology have made it possible to replace the original CCD with a smaller, lighter, cheaper (ca. $7,000) CCD; for applications requiring less sensitivity, even smaller, cheaper, uncooled CCDs are now adequate. Furthermore, the "fire wire" technology used in the new CCD cameras eliminates the requirement for a specialized data interface board, simplifying the electronics.

Currently, the NRL group has assembled a smaller array biosensor for environmental monitoring and is in the process of optimizing its operation. It uses a relatively small, inexpensive CCD, the fluidics cube, and the small pumps and valves (Fig. 9). Image display and analysis are performed using a laptop computer (not shown). The device fits into a tackle box which is about 1/3 the size of the previous version and weighs only 5.5 kg.

5 The Future

The use of planar waveguide TIRF for the detection of multiple analytes has clearly been demonstrated as well as the ability to miniaturize the instrumentation making it the most promising of the devices mentioned for real-time field monitoring. The benefit of spatially distinct sensing regions is enabling these systems to gain advantage over single-analyte sensing systems, particularly in terms of environmental monitoring. The speed of signal transduction and relative resistance to matrix effects and other interfering influences, which are often problematic for interferometric and SPR transduction methods, are two key advantages of planar waveguide TIRF biosensors. There are a number of areas which can be exploited in the development of planar waveguide TIRF to further enhance its capabilities.

The published work using planar waveguides for fluorescence biosensors has used only limited types of recognition elements, primarily DNA, antibodies and avidin. Studies describing the immobilization and use of different kinds of recognition molecules, such as receptors, in planar waveguide TIRF systems are limited to a handful of papers [82, 90, 91]. A number of receptors have shown promise for use in fiber optic, SPR and resonant mirror sensors [124–128]. The methods used in these studies may also prove successful for the immobilization of functional receptors to planar waveguides for use in TIRF and interferometric biosensors. There are many other membrane receptors which could be studied, such as the nicotinic acetylcholine receptor for detection of organophosphate nerve agents.

The use of nucleic acids for the detection of non-nucleic acid ligands such as proteins or small organic molecules is another relatively unexplored field (although these systems have been studied with other biosensors). The use of DNA aptamers for analyte detection has recently been extended to use in planar waveguide TIRF biosensors [129]. Potyrailo and co-workers [129] studied the binding of thrombin to a fluorescently-labeled, immobilized DNA aptamer by monitoring changes in the evanescent-wave-induced fluorescence anisotropy. DNA and RNA aptamers can be synthesized for a number of target analytes with specificities comparable with the corresponding antibodies [130–132]. DNA and RNA aptamers, as well as catalytic DNAs [131], show promise for use in planar waveguide biosensors if appropriate capture molecules can be synthesized. The use of "molecular beacons" could be an interesting development in planar waveguide TIRF. These systems make use of fluorescent probe-fluorescent quencher pairs and could be utilized with either the probe and quencher species immobilized on different binding pairs or, as studied by Liu and Tan [133], on the same molecule. Liu and Tan synthesized an oligonucleotide such that it possessed a stem and loop structure with the 5' quencher-labeled and the fluorophore-labeled 3' ends in close proximity. Addition of the target DNA separated the fluorophore and quencher ends, resulting in an intensity increase of the fluorescent signal.

There are also a number of enzymes, such as oxidases and reductases, which react with a variety of clinically and environmentally important analytes. Enzymes are used for signal amplification in ELISAs, but have also been used as biological recognition elements in a wide variety of biosensors [134] although they have not as yet been utilized as capture biomolecules in planar waveguide TIRF studies. Though the majority of enzyme biosensors use an electrode as the signal transducer, some biosensors utilize enzymes that produce or consume an optically detectable species. Fiber optic systems using a number of luminescence-generating enzymes have been described [135]. Furthermore, enzymes have been used in a number of optical fiber measurements in which the target analyte is indirectly determined [136]. Here the enzyme is either labeled or co-immobilized with the fluorescent probe. Typically the enzyme activity, upon introduction of the analyte, produces changes in the concentration of another species, such as oxygen, water, ammonia, or pH. These changes can result in either the increase or decrease in the fluorescence intensity of the probe. For example, oxygen quenching of ruthenium fluorescence intensity was the basis of a fiber optic glucose biosensor [137]. In addition, a fiber optic sensor for acetylcholinesterase (AchE) inhibitors utilizes fluorescein-labeled AchE immobilized onto the surface. Enzyme activity, with its substrate acetylcholine results in the production of protons and a quenching of fluorescein fluorescence intensity [138]. However, to date enzyme-based recognition and detection systems have not been used widely in planar waveguide TIRF studies. The limitations of using enzyme-based systems in planar waveguide TIRF studies may lie in the requirement for a surface-associated fluorophore. Research by Pawlak and coworkers [82] may lead the way to immobilizing fully functional enzymatic recognition elements.

Another integral part of biosensor technology is the immobilization of biomolecules to the surfaces of planar waveguides such that they retain activity. Most of those mentioned earlier do not control the orientation of the molecule on the surface which can affect the sensitivity of the measurement. Biotinylation and attachment through an immobilized avidin does permit careful control of the number of sites though which the biomolecule is attached. While this helps maintain the mobility of a receptor molecule and minimizes steric hindrance, it does not control the location of the attachment site on the molecule.

Noncovalent attachment of antibodies via binding by Protein A or Protein G or covalent attachment through the carbohydrate side chain limit the attachment site to the Fc region of an antibody; thus there is no attachment in the binding site. Fab antibody fragments have been linked via the thiol group at the opposite end of the molecule from the antigen binding site [139, 140]. The methods for preparing antibodies to be immobilized through either the carbohydrate or the thiol groups result in a significant loss of antibody during the processing procedure but generally produce highly functional surfaces. An additional method for controlling the attachment site in a biomolecule is by genetically engineering unique attachment sites on the biomolecule's surface, a technique known as site directed mutagenesis [141, 142]. However, this is generally a complicated multistep procedure, not altogether favorable from a commercial viewpoint.

Another process which may reduce the sensitivity of TIRF and interferometry measurements is nonspecific binding of the target analyte species to non-sensing regions of the surface, increasing the background signal. Currently, researchers commonly use either BSA and/or polyoxyethylene-sorbitan monolaurate (Tween 20) to block the surface and minimize nonspecific adsorption after immobilization of the capture biomolecule. Another possibility is the introduction of monolayers resistant to nonspecific adsorption of proteins in the non-sensing regions of the planar waveguide. Monolayers consisting of oligoethylene oxide terminal groups have been found to be highly protein resistant in SPR [143, 144], TIRF studies [111] and interferometric studies [34, 36]. This approach, however, may not prevent nonspecific adsorption to the sensing regions of the planar waveguide [111].

The power of the larger scale arrays has been demonstrated using both DNA chips [13, 14] and antibody chips [15, 16, 145, 146]. Yang et al. recently developed a stacked microlaboratory for performing electric-field-driven immunoassays and DNA hybridization assays for target analytes SEB and *E. coli* [145]. The developed system consisted of miniaturizated flowcell and fluidics systems, with fluorescent images of the microelectrodes recorded using a cooled CCD camera coupled with surface illumination via a HeNe laser. However, the final fluorescence image of the electrodes was typically recorded separately using an epifluorescent microscope. The majority of larger scale arrays are currently being scanned using scanning confocal microscopy. TIRF biosensors such as that of Rowe-Taitt et al. [91] are also sufficiently sensitive to measure such assays (unpublished data), but the methods used to deposit the spots in high density arrays are currently being improved to attain reproducible surface concentrations of the capture biomolecules [146].

Asanov and co-workers [80] produced an IOW consisting of an ITO layer, a transparent conductor, and used it to investigate biotin-streptavidin binding. They found that electrochemical polarization of the waveguide dissociated the biotin-streptavidin complex effectively regenerating the surface. Regeneration of the surface could be useful for commercial applications as an alternative to disposable substrates, depending on the relative cost of the additional system components and the number of times regeneration can be carried out without loss of surface integrity. As mentioned previously, successful regeneration of DNA chips has been carried out using urea without apparent loss of activity [81, 88]. IOWs are also found to increase the sensitivity of the system. These waveguides are very promising for clinical and drug screening applications [19]. However, as pointed out by Feldstein et al. [70], the increased constraints and requirements of the optical components would reduce the robustness of the system from a commercial point of view, should the device be intended for field applications [74].

Herron, Christensen and their colleagues [68] developed a molded plastic device that included both the planar waveguide and the lens for distributing the light across the waveguide as a single piece. The array biosensor fluidics system and flow manifold described by Dodson et al. [122] and Leatzow et al. [147], respectively, could also be molded in a single plastic piece. Such plastic parts could be integrated along with a molded lens array for focusing the fluorescent signal onto the image camera and lyophilized fluorescence reagent into a disposable sensing element.

This chapter was concerned with the use of optical array biosensors for the detection of environmental pollutants and BW agents and has discussed the background, history and current developments in the field. The TIRF technique in particular has shown potential in this area. Prototypes exist for an automated, stand-alone device which is essential for environmental monitoring applications [121]. However, with the continuing development of smaller electronics and components, such as the CMOS camera currently used in the confocal microscopy studies of microarrays [148, 149], the potential for even smaller, handheld planar waveguide TIRF devices could become reality. Although the majority of studies have centered on antibody-antigen type systems, expansion of the work concerned with DNA/mRNA and receptor-ligand binding interactions, as well as expansion into the use of enzymes and other types of recognition biomolecules could lead to a much wider field of applications. In conclusion, the future looks bright for environmental monitoring using planar waveguide TIRF.

Acknowledgements. The authors would like to thank Caroline Schauer, Chris Rowe-Taitt and Yura Shubin for their assistance and advice. This work was supported by the Office of Naval Research and the National Aeronautics and Space Administration. The views are those of the authors and do not reflect opinion or policy of the U.S. Navy or Department of Defense.

List of Abbreviations

AchE	Acetylcholinesterase
ATR	Attenuated total reflectance
BOD	Biological oxygen demand
BSA	Bovine serum albumin
BW	Biological warfare
CCD	Charge coupled device
CMOS	Complementary metal oxide semiconductor
ELISA	Enzyme linked immunosorbent assay
GFP	Green fluorescent protein

GOPS	3-glycidoxypropyldimethylethoxysilane
GRIN	Graded index
hCG	Human chorionic gonadotropin
ITO	Indium tin oxide
IOW	Integrated optical waveguide
IRE	Internal reflection element
NRL	Naval Research Laboratory
NTA	Nitrilotriacetic acid
PMT	Photomultiplier tube
PDMS	Polydimethylsiloxane
PEG	Polyethylene glycol
PMMA	Polymethylmethacrylate
SAM	Self-assembled monolayer
SEB	Staphylococcal enterotoxin B
SPR	Surface plasmon resonance
TIRF	Total internal reflection fluorescence
TNT	2,4,6-trinitrotoluene

References

1. Hall EAH (1990) Biosensors, Open University Press, Milton Keynes
2. Eggins BR (1998) Biosensors: An Introduction, John Wiley & Sons, New York
3. Scheller FW, Schubert F, Renneberg R, Muller HG, Janchen M, Weise H (1985) Biosensors 1:135
4. Braguglia CM (1998) Chem Biochem Eng Q 12:183
5. Pearson JE, Gill A, Vadgama P (2000) Ann Clin Biochem 37:119
6. Weetall HH (1999) Biosen Bioelectron 14:237
7. Rogers K R (1995) Biosen Bioelectron 10:533
8. Alvarez-Icaza M, Bililewski U (1993) Anal Chem 65:525A
9. Preininger C, Klimant I, Wolfbeis OS (1994) Anal Chem 66:1841
10. Anon (2000) Sci Am 283:110
11. Vo-Dinh T, Cullum B (2000) Fresenius J Anal Chem 366:540
12. Mulchandani A, Sadik OA (2000) Chemical and Biological Sensors for Environmental Monitoring Biosensors, American Chemical Society, New York.
13. Brown PO, Botstein D (1999) Nat Genet Suppl 21:33
14. Lockhart DJ, Winzeler EA (2000) Nature 405:827
15. MacBeath G, Schreiber SL (2000) Science 289:1760
16. Haab BB, Dunham MJ, Brown PO (2001) Genome Biol 2-4.1
17. Gullans SR (2000) Nat Genet Suppl 26:4
18. Knight J (2001) Nature 410:860
19. Duveneck GL, Abel AP, Bopp MA, Kresbach GM, Ehrat M (2002) Anal Chim Acta 469:49
20. Wang J, Rivas G, Cai X, Palecek E, Nielsen P, Shiraishi H, Dontha N, Luo D, Parrado C, Chicharro M, Farias PAM, Valera FS, Grant DH, Ozsoz M, Flair MN (1997) Anal Chim Acta 347:1
21. Wu TZ (1999) Biosen Bioelectron 14:9
22. Kukla AL, Kanjuk NI, Starodub NF, Shirshov YM (1999) Sen Actuators B 57:213
23. Zhang S, Zhao H, John R (2000) Anal Chim Acta 421:175

24. Krantz-Rulcker C, Stenberg M, Winquist F, Lundstrom I (2001) Anal Chim Acta 426: 217
25. Pancrazio J (2001) Biosen Bioelectron 16:427
26. Young SJ, Hart JP, Dowman AA, Cowell DC (2001) Biosen Bioelectron 16:887
27. Gray SA, Kusel JK, Shaffer KM, Shubin YS, Stenger DA, Pancrazio JJ (2001) Biosen Bioelectron 16:535
28. Homola J, Yee SS, Myszka D (2002) Surface plasmon resonance biosensors. In: Ligler FS, Rowe-Taitt CA (eds) Optical Biosensors: Present and Future, Elsevier Science, Amsterdam, p207
29. Campbell DP, McCloskey CJ (2002) Interferometric biosensors. In: Ligler FS, Rowe-Taitt C.A (eds) Optical Biosensors: Present and Future, Elsevier Science, Amsterdam, p277
30. Sapsford K E, Rowe-Taitt CA, Ligler FS (2002) Planar waveguides for fluorescence biosensors. In: Ligler FS, Rowe-Taitt C.A (eds) Optical Biosensors: Present and Future, Elsevier Science, Amsterdam, p95
31. Plowman TE, Saavedra SS, Reichert WH (1998) Biomaterials 19:341
32. Schipper EF, Brugman AM, Dominguez C, Lechuga LM, Kooyman RPH, Greve J (1997) Sens Actuators B 40:147
33. Schipper EF, Bergevoet AJH, Kooyman RPH, Greve J (1997) Anal Chim Acta 341:171
34. Schneider BH, Edwards JG, Hartman NF (1997) Clin Chem 43:1757
35. Schipper, EF, Rauchalles S, Kooyman RPH, Hock B, Greve J (1998) Anal Chem 70: 1192
36. Schneider BH, Dickinson EL, Vach MD, Hoijer JV, Howard LV (2000) Biosen Bioelectron 15:13
37. Campbell DP, Hartman NF, Moore JL, Suggs JV, Cobb JM (1998) Proc SPIE 3253:20
38. Edwards JG, Campbell DP, Moore JL (1999) Proc SPIE 3534:614
39. Campbell DP, Moore JL, Cobb JM, Hartman NF, Schneider BH, Venugopal MG (1999) Proc SPIE 3540:153
40. Leidberg B, Nylander C, Lundstrom I (1983) Sens Actuators 4:299
41. Davies J (1994) Nanobiol 3:5
42. Leidberg B, Nylander C, Lundstrom I (1995) Biosens Bioelectron 10:i-ix
43. Homola J, Yee SS, Gauglitz G (1999) Sens Actuators B 54:3
44. Nikitin PI, Beloglazov AA, Kochergin VE, Valeiko MV, Ksenevich TI (1999) Sens Actuators B 54:43
45. Koubova V, Brynda E, Karasova L, Skvor J, Homola J, Dostalek J, Tobiska P, Rosicky J (2001) Sens Actuators B 74:100
46. Dostalek J, Ctyroky J, Homola J, Brynda E, Skalsky M, Nekvindova P, Spirkova J, Skvor J, Schrofel J (2001) Sens Actuators B 76:8
47. Homola J, Koudela I, Yee SS (1999) Sens Actuators B 54:16
48. Garland PB (1995) Q R Biophys 29:91
49. O'Brien MJ, Brueck SRJ, Perez-Luna VH, Tender LM, Lopez GP (1999) Biosens Bioelectron 14:145
50. Homola J, Dostalek J, Ctyroky J (2001) Proc SPIE 4416:86
51. Lu HB, Campbell CT, Nenninger GG, Yee SS, Ratner BD (2001) Sens Actuators B, 74: 91
52. Homola J, Lu HB, Nenninger GG, Dostalek J, Yee SS (2001) Sens Actuators B 76:403
53. O'Brien MJ, Perez-Luna VH, Brueck SRJ, Lopez GP (2001) Biosens Bioelectron 16:97
54. Jordan CE, Corn RM (1997) Anal Chem 69:1449
55. Jordan, CE, Frutos AG, Thiel AJ, Corn RM (1997) Anal Chem 69:4939
56. Thiel AJ, Frutos AG, Jordan CE, Corn RM, Smith LM (1997) Anal Chem 69:4948
57. Brockman JM, Frutos AG, Corn RM (1999) J Am Chem Soc 121:8044
58. Nelson BP, Frutos AG, Brockman JM, Corn RM (1999) Anal Chem 71:3928
59. Brockamn JM, Nelson BP, Corn RM (2000) Annu Rev Phys Chem 51:41
60. Lee HJ, Goodrich TT, Corn RM (2001) Anal Chem 73:5525

61. Nelson BP, Grimsrud TE, Liles MR, Goodman RM, Corn RM (2001) Anal Chem 73:1
62. Bradley RA, Drake RAL, Shanks IA, Smith AM, Stephenson PR (1987) Phil Trans R Soc Lond B 315:143
63. Lu B, Lu C, Wei Y (1992) Anal Lett 25:1
64. Wadkins RM, Golden JP, Pritsiolas LM, Ligler FS (1998) Biosens Bioelectron 13:407
65. Chittur KK (1998) Biomater 19:357
66. Duveneck GL, Neuschafer D, Ehrat M (1995) International Patent Go1N 21/77, 21/64
67. Herron JN, Christensen DA, Caldwell KD, Janatova V, Huang SC, Wang HK (1996) US Patent 5,512,492
68. Herron JN, Christensen DA, Wang HK, Caldwell KD (1997) US Patent 5,677,196
69. Golden JP (1998) US Patent 5,827,748
70. Feldstein MJ, Golden JP, Rowe CA, MacCraith BD, Ligler FS (1999) J Biomed Microdevices 1-2:139
71. Silzel JW, Cercek B, Dodson C, Tsay T, Obremski RJ (1998) Clin Chem 44:2036
72. Plowman TE, Durstchi JD, Wang HK, Christensen DA, Herron JN, Reichert WM (1999) Anal Chem 71:4344
73. Schult K, Katerkamp A, Trau D, Grawe F, Cammann K, Meusel M (1999) Anal Chem 71: 5430
74. Brecht A, Klotz A, Barzen C, Gauglitz G, Harris RD, Quigley GR, Wilkinson JS, Sztajnbok P, Abuknesha R, Gascon J, Oubina A, Barcelo D (1998) Anal Chim Acta 362:69
75. Lundgren JS, Watkins N, Racz D, Ligler FS (2000) Biosens Bioelectron 15:417
76. Schuderer J, Akkoyun A, Brandenburg A, Bilitewski U, Wagner E (2000) Anal Chem 72: 3942
77. Golden JP, Ligler FS (2002) Biosens Bioelectron 17:719
78. Koltz A, Brecht A, Barzen C, Gauglitz G, Harris RD, Quigley GR, Wilkinson JS, Abuknesha RA (1998) Sens Actuators B 51:181
79. Feldstein MJ, Golden JP, Ligler FS, Rowe CA (2000) US Patent 6,192,168
80. Asanov AN, Wilson WW, Oldham PB (1998) Anal Chem 70:1156
81. Duveneck GL, Pawlak M, Neuschafer D, Bar E, Budach W, Pieles U, Ehrat M (1997) Sens Actuators B 38–39:88
82. Pawlak M, Grell E, Schick E, Anselmetti D, Ehrat M (1998) Faraday Discuss 111:273
83. Iqbal SS, Mayo MW, Bruno JG, Bronk BV, Batt CA, Chambers JP (2000) Biosens Bioelectron 15:549
84. Rowe CA, Scruggs SB, Feldstein MJ, Golden JP, Ligler FS (1999) Anal Chem 71:433
85. Rowe CA, Tender LM, Feldstein MJ, Golden JP, Scruggs SB, MacCraith BD, Cras JJ, Ligler FS (1999) Anal Chem 71:3846
86. Rabbany SY, Donner BL, Ligler FS (1994) Crit Rev Biomed Eng 22:307
87. Sapsford KE, Charles PT, Patterson Jr. CH, Ligler FS (2002) Anal Chem 74:1061
88. Budach W, Abel AP, Bruno AP, Neuschafer D (1999) Anal Chem 71:3347
89. Schmid EL, Keller TA, Dienes Z, Vogel H (1997) Anal Chem 69:1979
90. Schmid EL, Tairi AP, Hovius R, Vogel H (1998) Anal Chem 70:1331
91. Rowe-Taitt CA, Cras JJ, Patterson CH, Golden JP, Ligler FS (2000) Anal Biochem 281: 123
92. Ulbrich R, Golbik R, Schellenberger A (1991) Biotechnol Bioeng 37:280
93. Bhatia SK, Shriver-Lake LC, Prior KJ, Georger JH, Calvert JM, Bredehorst R, Ligler FS (1989) Anal Biochem 178:408
94. Wink T, van Zuilen SJ, Bult A, Van Bennekom WP (1997) Analyst 122:43R
95. Mrksich M (2000) Chem Soc Rev 29:267
96. Ferretti S, Paynter S, Russell DA, Sapsford KE, Richardson DJ (2000) Trends Anal Chem 19:530
97. Brizzolara RA, Beard BC (1994) J Vac Sci Technol A 12:2981
98. Lee JE, Saavedra SS (1996) Langmuir 12:4025
99. Sojka B, Piunno PAE, Wust CC, Krull UJ (1999) Anal Chim Acta 395:273
100. Pirrung MC, Davis JD, Odenbaugh AL (2000) Langmuir 16:2185

101. Herron JN, Caldwell KD, Christensen DA, Dyer S, Hlady V, Huang P, Janatova V, Wang HK, Wei AP (1993) SPIE Adv Fluor Sen Tech 1885:28
102. Birkert O, Haake HM, Schutz A, Mack J, Brecht A, Jung G, Gauglitz G (2000) Anal Biochem 282:200
103. Blawas AS, Reichert WM (1998) Biomater 19:595
104. Bhatia SK, Hickman JJ, Ligler FS (1992) J Am Chem Soc 114:4432
105. Bhatia SK, Teixeira JL, Anderson M, Shriver-Lake LC, Calvert JM, Georger JH, Hickman JJ, Ducley CS, Schoen PE, Ligler FS (1993) Anal Biochem 208:197
106. Schwarz A, Rossier JS, Roulet E, Mermod N, Roberts MA, Girault HH (1998) Langmuir 14:5526
107. Wadkins RM, Golden JP, Ligler FS (1997) J Biomed Optics 2:74
108. Guschin D, Yershov G, Zaslavsky A, Gemmell A, Shick V, Proudnikov D, Arenkov P, Mirzabekov A (1997) Anal Biochem 250:203
109. Arenkov P, Kukhtin A, Gemmell A, Voloshchuk S, Chupeeva V, Mirzabekov A (2000) Anal Biochem 278:123
110. Blawas AS, Oliver TF, Pirrung MC, Reichert WH (1998) Langmuir 14:4243
111. Conrad DW, Davis AV, Golightley SK, Bart JC, Ligler FS (1997) SPIE 2978:12
112. Conrad DW, Golightley SK, Bart JC (1998) US Patent 5,773,308
113. Liu XH, Wang HK, Herron JN, Prestwich GD (2000) Bioconj Chem 11:755
114. Budach W, Abel AP, Bruno AP, Neuschafer D (1999) Anal Chem 71:3347
115. Bernard A, Michel B, Delamarche E (2001) Anal Chem 73:8
116. Ligler FS, Conrad DW, Golden JP, Feldstein MJ, MacCraith BD, Balderson SD, Czarnaski J, Rowe CA (1998) SPIE 3258:50
117. Golden JP, Rowe CA, Feldstein MJ, Ligler FS (1999) SPIE 3602:132
118. Rowe-Taitt CA, Golden JP, Feldstein MJ, Cas JJ, Hoffman KE, Ligler FS (2000) Biosens Bioelectron 14:785
119. Rowe-Taitt CA, Hazzard JW, Hoffman KE, Cras JJ, Golden JP, Ligler FS (2000) Biosens Bioelectron 15:579
120. Zeller PN, Voirin G, Kunz RE (2000) Biosens Bioelectron 15:591
121. Ligler FS, Golden JP, Rowe-Taitt CA, Dodson JP (2001) SPIE 4252:32
122. Dodson JM, Feldstein MJ, Leatzow DM, Flack LK, Golden JP, Ligler FS (2001) Anal Chem 73:3776
123. Feldstein MJ (2003) US Patient application No. 09/917,649, filed 08/07/01
124. Rogers KR, Valdes JJ, Eldefrawi ME (1989) Anal Biochem 182:353
125. Fisher MI, Tjarnhage T (2000) Biosens Bioelectron 15:463
126. Lang S, Xu J, Stuart F, Thomas RM, Vrijbloed JW, Robinson JA (2000) Biochem 39: 15674
127. Altin JG, White FAJ, Easton CJ (2001) Biochim Biophys Acta 1513:131
128. Puu G (2001) Anal Chem 73:72
129. Potyrailo RA, Conrad RC, Ellington AD, Hieftje GM (1998) Anal Chem 70:3419
130. Mauro JM, Cao LK, Kondracki LM, Walz SE, Campbell JR (2000) Anal Biochem 235: 61
131. Li J, Li Y (2000) J Am Chem Soc 122:10466
132. Wang L, Carrasco C, Kumar A, Stephens CE, Bailly C, Boykin DW, Wilson WD (2001) Biochem 40:2511
133. Liu X, Tan W (1999) Anal Chem 71:5054
134. Wilson GS, Hu Y (2000) Chem Rev 100:2693
135. Rowe-Taitt CA, Ligler FS (2001) In: Lopez-Higuera JM (ed) Handbook of Fiber Optic Sensing Technology. John Wiley & Sons, New York
136. Szurdoki F, Michael KL, Walt DR (2001) Anal Biochem 291:219
137. Moreno-Bondi, MC, Wolfbeis OS, Leiner MJP, Schaffer BHP (1999) Anal Chem 62: 2377
138. Rogers KR, Cao CJ, Valdes JJ, Eldefrawi AT, Eldefrawi ME (1991) Fundam Appl Toxicol 16:810

139. Lu EJ, Xie J, Lu C, Wu C, Wei Y (1995) Anal Chem 67:83
140. Huang, SC, Caldwell KD, Lin JN, Wang HK, Herron JN (1996) Langmuir 12:4292
141. McLean MA, Stayton PS, Sligar SG (1993) Anal Chem 65:2676
142. Vigmond SJ, Iwakura M, Mizutani F, Katsura T (1994) Langmuir 10:2860
143. Silin V, Weetall H, Vanderah DJ (1997) J Col Interface Sci 185:94
144. Lahiri J, Isaacs L, Tien J, Whitesides GM (1999) Anal Chem 71:777
145. Yang JM, Bell J, Huang Y, Tirado M, Thomas D, Forster AH, Haigis RW, Swanson PD, Wallace RB, Martinsons B, Krihak M (2002) Biosens Bioelectron 17:605
146. Delehanty JB, Ligler FS (2002) Anal Chem 74:5681
147. Leatzow DM, Dodson JM, Golden JP, Ligler FS (2002) Biosens Bioelectron 17:105
148. Vo-Dinh T, Alarie JP, Lsola N, Landis D, Wintenberg AL, Ericson MN (1999) Anal Chem 71:358
149. Askari M, Alarie JP, Moreno-Bondi M, Vo-Dinh T (2001) Biotechnol Prog 17:543

CHAPTER 15

Optical Techniques for Determination and Sensing of Hydrogen Peroxide in Industrial and Environmental Samples

HANNES VORABERGER

1 Introduction

Hydrogen peroxide (HP) has undergone somewhat of a rebirth in chemical industry circles over the last decade. The rather glib explanation for such a renaissance is that regulatory forces compelled the chemical industry to reduce, and in some instances to eliminate, environmental pollution. Since industry has learned to employ HP in a safer, more efficient and innovative manner, it can substitute many environmentally hazardous chemicals in an environmentally protective way. The most important applications of HP in industrial processes are summarized in Fig. 1 [1–4].

The production of paper, specifically the bleaching step of cellulose pulp, is the largest and fastest growing single application of HP. This growth has fundamentally been driven by environmental issues, and a large quantity of new and improved technologies have been devised in response to this demand.

HP, when supplied commercially, is usually stabilized with phosphates and tin(IV) compounds. The tin compounds are effective at the product's natural pH via hydro-colloid formation, which adsorb transition metals and reduce their catalytic activity in this way. Elevated temperatures and metal impurities tend to destabilize peroxygens. Where such conditions are unavoidable it may be necessary to employ additional stabilizers, added either to the HP or to the reaction mixture separately. For example a 30% HP solution from Merck (Darmstadt,

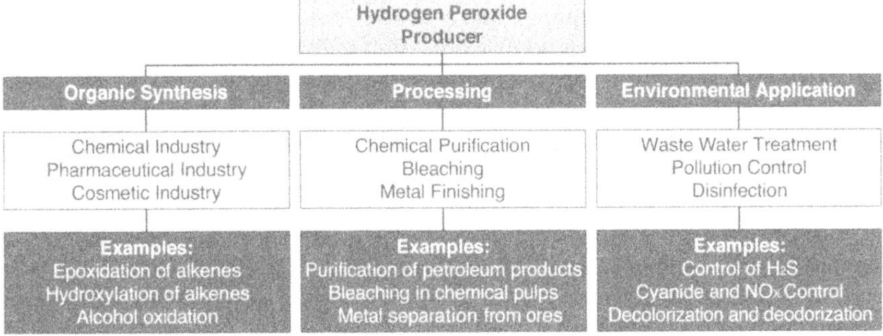

Fig. 1. Commercial uses of HP

Germany) contains 0.015% $Na_2H_2P_2O_7$, 0.01% phosphoric acid and 0.006% ammonium nitrate for stabilization [5].

HP is also important in many biological processes. It is the product of reactions catalyzed by a large number of oxidase enzymes. Therefore it is also essential in food, pharmaceutical, and environmental analysis.

Despite the large number of methods described in the literature for the measurement of HP, the method which is still most commonly used for the quantification of HP in industry is the titration of a sample solution with a reagent such as potassium permanganate [6]. Alternative methods can be divided into two main groups, namely into spectroscopic and electrochemical, while calorimetric methods also have been described [3].

Titrimetric analysis is the most commonly used and the only standardized method for quantification of HP in solution [6]. It is based on titration of HP with oxidizing agents such as potassium permanganate, cerium(IV) and potassium iodide/thiosulfate solutions. The endpoint of the titration can be determined amperometrically or by a change in color of the solution, detected either visually or photometrically. This method can be used over a wide concentration range [7].

Electrochemical methods for the determination of HP can be divided into amperometric [8–10], potentiometric [11] and conductometric methods [12]. Amperometric sensors are of particular interest, since they offer the possibility of direct on-line monitoring of HP. However, they suffer from interference from other substances, which may also react at the electrode surfaces. For this reason, special electrode materials and more complicated indirect measurement systems have been developed [12].

One such approach is based on a combination of a Clark-type oxygen electrode with a catalyst placed on the electrode surface [13]. The oxygen generated through decomposition of HP is measured and related to the HP concentration. In general it can be stated that the problem of measuring oxygen lies in the fact that oxygen is always present in the sample solution and its variable background concentration has to be compensated for (or measured separately).

In general, spectroscopic methods for the detection of HP are based on the change of optical parameters. Direct spectroscopy detects the absorption of light by the analyte itself. This can be done simply with a spectrometer and has the advantage that no preparation or chemical reaction is necessary. Indirect spectroscopy detects the change in the absorption, fluorescence or chemiluminescence of a substance caused by its reaction with HP. An alternative approach uses an optical oxygen sensor covered with a catalyst membrane the sensor detects the oxygen formed by HP decomposition in the catalyst membrane.

2
Direct Spectrometric Measurements of Hydrogen Peroxide

2.1
Hydrogen Peroxide in the Mid Infrared (Wavelength Range: 2.5–20 μm)

Recently, mid-infrared based optical sensor systems have gained increased importance due to the development of portable and compact instrumentation [14, 15]. The mid-infrared (MIR) spectral region considered from approximately 2.5–20 μm (4000–500 cm^{-1}) is particularly attractive for optical sensing, as molecule-specific information can be directly obtained due to the stimulation of fundamental vibrational modes. However, optical spectroscopic components in the mid infrared are restricted to a few materials like ZnSe, CaF_2 or Ge, because many materials are not transparent to mid infrared radiation (e.g. glass and quartz).

In the vapor state hydrogen peroxide (HP) shows some characteristic vibrations in the mid infrared region [16]. These can also be found in the spectrum of aqueous HP solutions in slightly different wavelength ranges [17]. In Fig. 2 transmission spectra of aqueous HP solutions using two CaF_2 disks as transmission windows display two additional absorption peaks compared to the spectrum of pure water. These two peaks are located at the shoulder of water bands, but are nevertheless clearly separated from the water peak.

Probing a liquid sample for quantitative evaluation is particularly facilitated using attenuated total reflection (ATR) and the resulting evanescent field, derived from the general principle of internal reflection spectroscopy (IRS) first

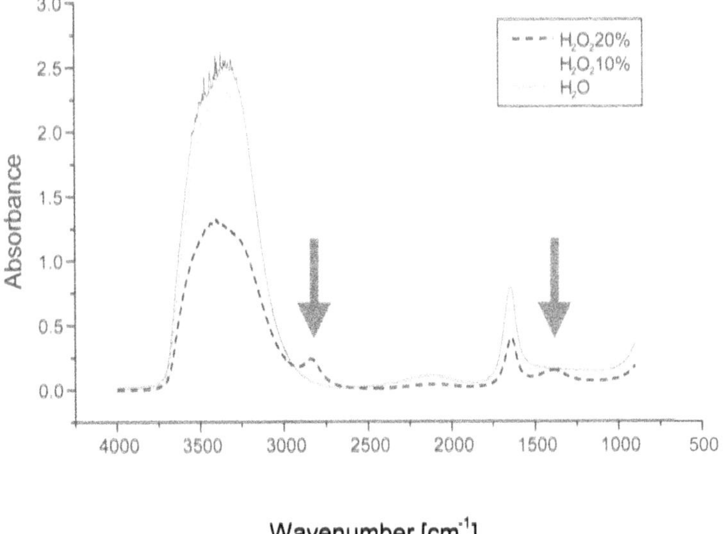

Fig. 2. Comparison of spectra of water (straight line) and aqueous HP solutions containing 10 % (dotted line) and 20 % (dashed line) (*w*/*w*) HP in the mid-infrared spectral region

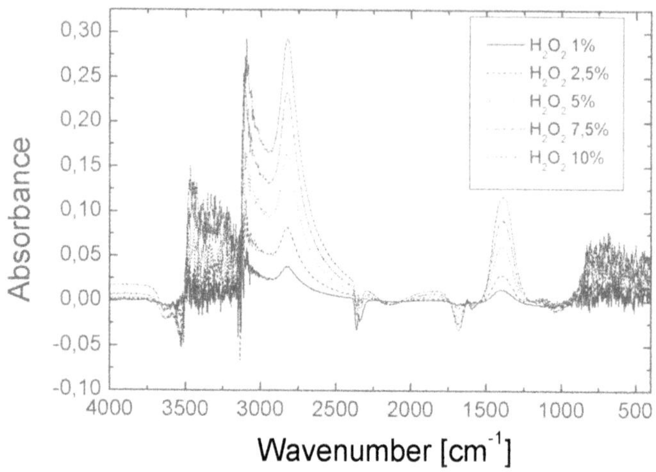

Fig. 3. Spectra of aqueous solutions of HP in varying concentrations after the subtracting water background (waveguide: ZnSe, temperature: 20–23 °C)

described by Fahrenfort and Harrick [18,19]. Interaction of incident and reflected light creates an exponentially decaying evanescent field, which penetrates, to a certain wavelength-dependent extent into the adjacent medium probing the analyte at the exterior of the waveguide.

Related spectra of HP recorded with ATR spectroscopy using a ZnSe crystal surface in contrast to transmission measurements are displayed in Fig. 3. As the water background spectrum is already subtracted from these spectra, the resulting bands are characteristic of HP. Figure 3 clearly shows a dependence of the area and the height of the bands on the HP concentration. Both figures show the O-H stretching vibration (2930–2680 cm^{-1}) and the O-H deformation vibration (1530–1260 cm^{-1}) of HP, which correspond to spectra of HP vapor reported in the literature [16].

Specific bands of HP evidently display a dependence on the concentration of aqueous HP solution. Therefore, evaluation of area and height of the absorption bands was performed in order to establish calibration functions. Figure 4 demonstrates linear correlation between the concentration of HP and both the area and the height of the O-H deformation vibration (1530–1260 cm^{-1}). Similar results were obtained for the O-H stretching vibration (2930–2680 cm^{-1}). The absolute values of the areas and heights, however, differ between these vibrational modes due to changes in their peak shape.

Aqueous solutions of HP used in industrial processes (e.g. for bleaching solutions) contain additives such as bases, acids and salts, as well as stabilizer compounds and particles from the product (e.g. fragments of bleached material). Data obtained for concentrations in the bleaching baths, however, fit the calibration graphs established with pure HP solutions quite well. In Fig. 5 the calibration graphs of two industrial samples (i.e. from a bleaching bath for cellulose

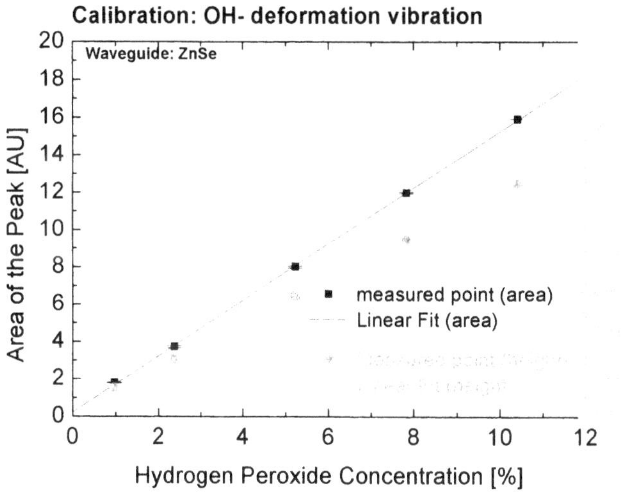

Fig. 4. Calibration curve for area and height of the O-H deformation vibration of aqueous HP solutions (waveguide: ZnSe, temperature: 20–23 °C)

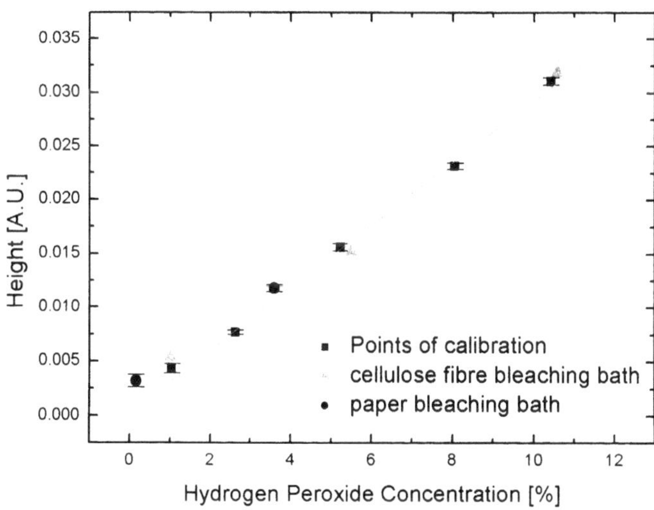

Fig. 5. Height at peak maximum of the O-H deformation vibrational band (1538–1215 cm^{-1}) of two different bleaching baths compared to the calibration graph of a pure HP solution (temperature: 20–23 °C)

fiber and a bleaching bath for paper pulp), spiked with HP solutions up to 11% by weight, are displayed. The area of the deformation vibration as well as the height and the area of the stretching vibration of aqueous HP solutions revealed similar results but are not presented here.

2.2 Near Infrared Spectroscopy of Hydrogen Peroxide

The absorption spectra in the near infrared display overtone and combination vibrations whose fundamental vibrations absorb in the mid-infrared region. The near-infrared region, 0.7 to 2.5 μm (wavenumber ca. 4000–14000 cm^{-1}), is more akin to the ultraviolet and visible regions with respect to pathlength than to the "normal" infrared region, in that cells with longer pathlength are employed. Such cells are easier to clean, more robust and (being made of glass with quartz windows or of silica) not attacked by water. Furthermore, immersion probes connected with fiber optics can be employed, allowing on-line monitoring of HP. Since the bands in this region are of the combination and overtone band type, the pathlength of the sample must be increased in order to examine successfully the higher frequency part of the range (pathlength: 0.1–10 cm depending on the part of the spectrum of interest).

The bands in the NIR are very broad and consequently overlap each other quite often. Hence, specific bands as found in mid-infrared spectra appear as shoulders in near-infrared spectra. As a result, quantitative evaluation is more difficult and requires more experience than for mid-infrared spectroscopy and is normally based on chemometric methods.

Two very weak bands occur in regions with very high wavenumbers for gaseous HP (10,283.7 and 10,291.1 cm^{-1}) [16]. Interestingly, the spectra of aqueous HP solutions seem to display three specific shoulders for HP at lower wavenumbers. The first is between 4500 and 5000 cm^{-1}, the second between 5400 and 6600 cm^{-1} (see Fig. 6; spectra recorded in a 0.1 cm cuvette) and the third between 8400 and 8700 cm^{-1} (not displayed).

Spectra are influenced not only by the varying HP concentration but also by temperature, ionic strength and other parameters of the sample matrix. Conse-

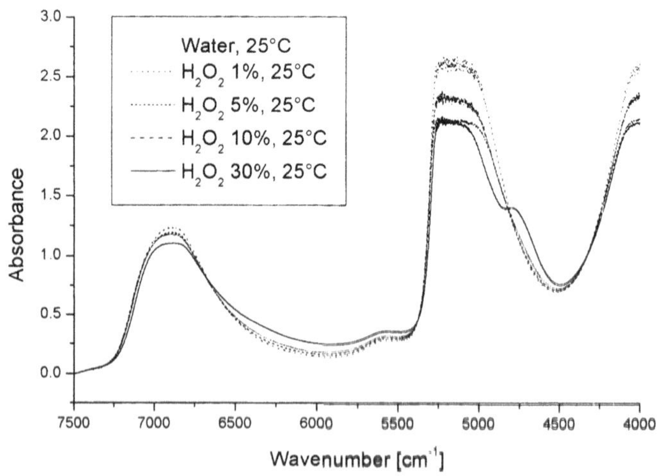

Fig. 6. Spectra of different HP solutions in the region of 4000–7500 cm^{-1} (temperature 25 °C)

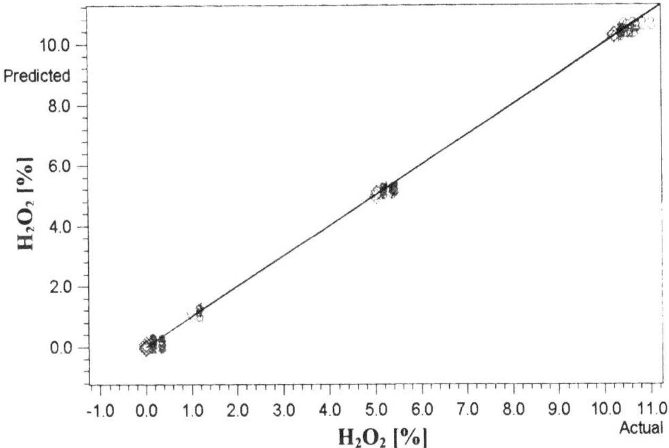

Fig. 7. Comparison of the values predicted with the chemometric model developed for samples from a paper mill with the actual values

quently, the development of a reliable method for the determination of HP in industrial sample solutions using near infrared spectroscopy requires the application of chemometric techniques, which has yielded very promising results for different industrial applications.

In Fig. 7 the effectiveness of a chemometric model developed for application in a paper mill is displayed. The values predicted with this model are plotted against the actual ones. The model was developed with a set of 41 spectra of pure HP solutions at different temperatures, ionic strength, pH and a number of spectra from a paper mill, spiked with different amounts of HP. The generated model has a correlation coefficient of 0.99997, a mean standard error of prediction of 0.0057 and a standard error of prediction of 0.076. This model, however, is only valid for this specific application (i.e. the specific production step in the paper mill where the sample originated). For other applications the basic model must be tailored by adding characteristic spectra of the specific application to the model.

2.3 Ultraviolet Spectroscopy of Hydrogen Peroxide

In the ultraviolet (UV) range HP shows a continual increase in absorption with decreasing wavelength. This is used for some biochemical applications where a change in HP concentration is followed by monitoring a certain absorption band in the UV (e.g. determination of activity of the enzyme catalase) [20]. There chemical reactions are monitored in a well defined matrix. However, the UV spectrum of HP displays a lack of selectivity for industrial or environmental samples, the scope of this report, because other species absorbing in the UV cannot be excluded.

3 Indirect Spectrometric Measurements of Hydrogen Peroxide

3.1 Introduction

Indirect spectroscopy detects the change in the absorption, fluorescence or chemiluminescence of a substance caused by its reaction with hydrogen peroxide (HP) or one of its derivatives. Normally, a dye or a precursor of a dye is mixed with a catalyst and added to the sample. The catalyst decomposes HP, forming a hydroxyl radical in the first step. This hydroxyl radical is the reactive species which reacts with the dye. Catalysts used include hemoproteins such as horseradish peroxidase and porphyrins. Other substances, which do not act as catalyst, can also be used for the formation of the hydroxyl radical, a typical example is iron(II) ions which are oxidized to iron(III) ions.

While direct spectroscopic assay is performed in cuvettes, flow-through cells or immersion probes in processes, additional measurement techniques, such as Flow Injection Analysis (FIA), test strips and sensors are employed for indirect spectroscopic detection of HP.

The simplest detection method is to mix the required substances in a cuvette, allow time for the reaction to occur and measure the absorption, fluorescence or chemiluminescence in a spectrometer. This procedure can be automated by using FIA. FIA uses the same principle as the cuvette method but the sample is introduced into a flow system where the reagents are added automatically. The solution flows through a reaction coil and finally passes a detector. In test strips reagents are immobilized on a carrier. Such test strips can only be used once.

A sensor system, in contrast, measures a reversible change in absorption or fluorescence. The reversible chemical reaction takes place in a so-called sensitive layer, consequently no addition of chemicals is necessary.

Indirect spectroscopic methods for analysis of HP depend strongly on the pH and the ionic strength of the solution. This results from the dependence of a dye's absorption on these parameters. Many of the dyes are acids which can be deprotonated by changing the pH, so it is often necessary that measurements are performed in buffered solutions. A problem reported to affect the analysis of HP arises from the presence of reductive substances such as ascorbic acid or bilirubin which cause negative errors [20].

Most of the published indirect spectrometric methods are based on one of the following main principles:

- Formation of a dye via an oxidative coupling reaction
- Formation of a dye by oxidation of leuco dyes
- Formation of a colored or fluorescent complex
- Destruction of a dye
- Chemiluminescence
- Indirect measurement via oxygen and chemiluminescence quenching

3.2 Formation of a Dye by Oxidative Coupling Reaction

Formation of a dye by oxidative coupling reaction is the most commonly used indirect spectroscopic method. Therefore, methods using different catalysts or reagents have been reported for a wide range of concentrations and applications. In all these methods, the main step is an oxidative coupling reaction similar to the one shown in Fig. 8.

These oxidative coupling reactions can be mediated either by light, as shown in Fig. 8, or by a catalytic reaction with horseradish peroxidase or related molecules. Since horseradish peroxidase is rather expensive and has a poor stability (it can be stored at 4 °C for one month approximately) attempts have been made to find a substitute. The most promising substitute is magnesium tetrakis(sulphophenyl) porphyrin ($MnTPPS_4$) which shows similar characteristics to horseradish peroxidase in HP decomposition but is stable and cheaper. In order to be able to reuse $MnTPPS_4$ it can be physically adsorbed on an anionic exchange resin [23].

Many reagents for this type of oxidative reaction have been reported. In general, however, they can be divided into three types:

- Aromatic compounds with nitrogen in the aromatic system
- Anthraquinone derivatives which are sulfonated [24]
- Hydroxyphenyl acids

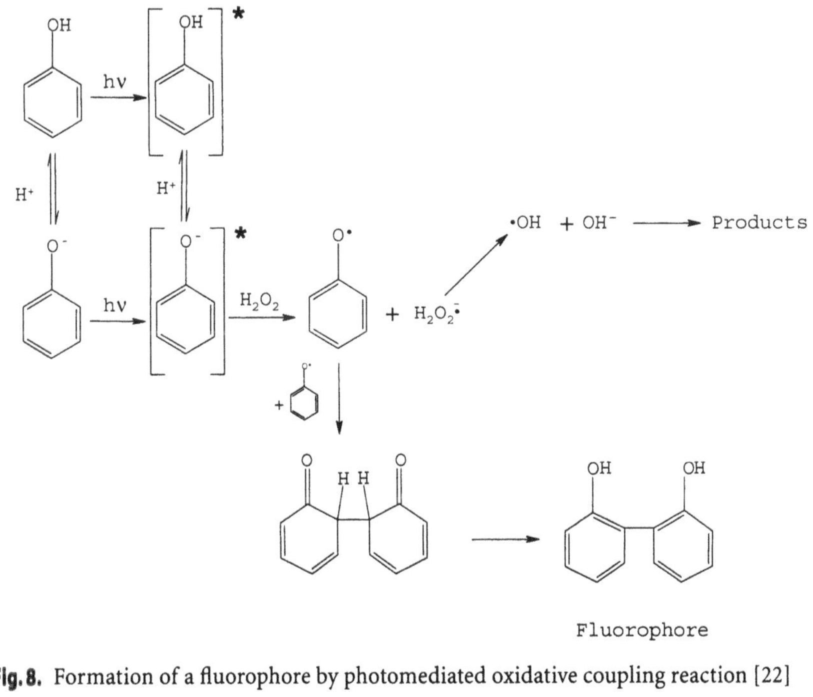

Fig. 8. Formation of a fluorophore by photomediated oxidative coupling reaction [22]

Table 1. Reactions to form fluorophores by oxidative coupling

Reactant	Catalyst	Linear range Detection limit	Remarks	Ref.
4-Hydroxyphenylacetic acid	Horseradish peroxidase	$0–1 \times 10^{-5}$ M 5×10^{-7} M (FIA)	pH: 10.5–11.5	[15]
4-Hydroxyquinoline	Horseradish peroxidase	$0–1 \times 10^{-3}$ M 5.1×10^{-6} M (FIA)	pH: 8.5–9.5	[15]
4-Hydroxyquinoline-2-carboxylic acid	Horseradish peroxidase	$0–1 \times 10^{-4}$ M 2.5×10^{-6} M (FIA)	pH: 6.0–6.5	[15]
4-Hydroxyproline	Horseradish peroxidase	$2.5 \times 10^{-5}–1 \times 10^{-4}$ M 1.84×10^{-5} M (FIA)	pH: 6.0–6.5	[15]
p-Hydroxyphenylpropionic acid	Iron-tetrasulfonato-phthalocyanine (FeTSPC)	$0.0 \times 10^{-6}–6 \times 10^{-6}$ M 3.1×10^{-8} M (cuvette)	Without catalyst very weak signal pH: 1–7 and 9.5–13 Interferences: Fe^{3+}, Sn^{2+}	[56]
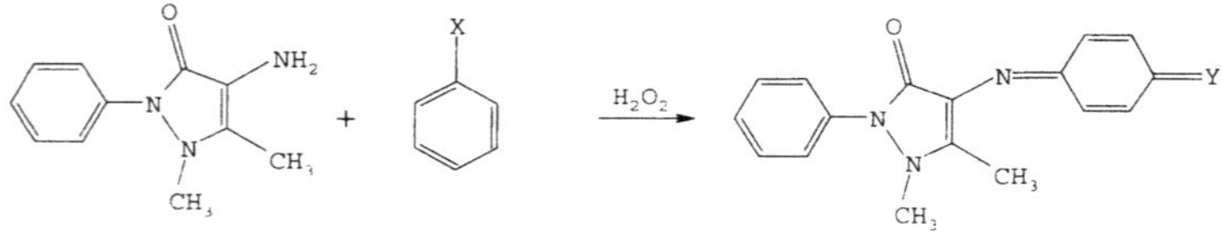 4-Aminoantipyrine + X-C6H5 → (H_2O_2) 4-Aminoantipyrine; X = OH, X = NEt_2 Y = O (λ_{max} = 505 nm) Y = N^+Et_2 (λ_{max} = 550 nm)	$Mn\text{-}TPPS_4$ Peroxidase	$0–5.9 \times 10^{-7}$ M (cuvette)	Interferences: Fe^{3+}	[57] [58]

Reactant	Catalyst	Linear range Detection limit	Remarks	Ref.
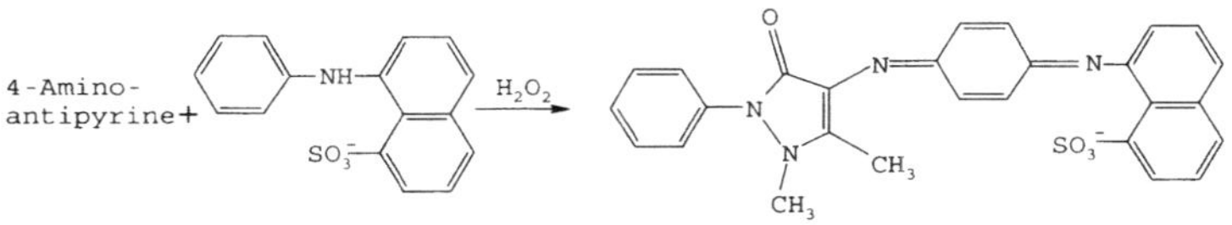 4-Amino-antipyrine + (anilinonaphthalenesulfonate, SO_3^-) $\xrightarrow{H_2O_2}$ (quinonimine dye, SO_3^-) Yellow	Hematin	0–5 × 10^{-5} M 3 × 10^{-7} M (FIA)	Reaction takes place in basic medium – immense bathochromic shift by making solution more acidic (blue)	[59]
Similar reaction with some derivatives of these reagents fixed on a support (reversible reaction)	peroxidase		Fixed via charge transfer Very similar reaction	[60]
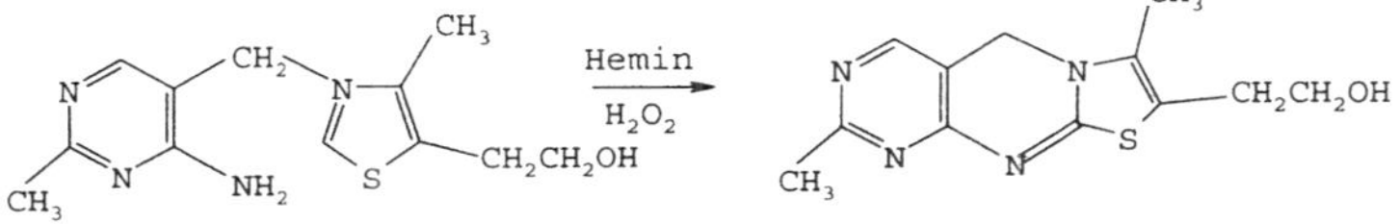 Thiamine $\xrightarrow{Hemin, H_2O_2}$ Thiochrome [strong fluorescence, λ (ex) 373 nm; λ (em) 440 nm]	Hemin	0–5 × 10^{-6} M 3 × 10^{-8} M (cuvette)	Interferences: Mn^{2+}, Cu^{2+}, S^{2-} pH:7.8–8.2	[61]
2 Homovanillic acid (HVA) $\xrightarrow{H_2O_2}$ [strong fluorescence, λ (ex) 315 nm; λ (em) 425nm]	$Mn\text{-}TPPS_4$	0–2.5 × 10^{-6} M 8.5 × 10^{-8} M (cuvette)	Interferences: Mn^{2+}, Cu^{2+}, S^{2-}, Pd^{2+}, Cd^{2+}	[62]
	Horseradish peroxidase	1.7 × 10^{-7}–1.7 × 10^{-5}M 1.7 × 10^{-7} M (cuvette)		
Reaction with 4-aminoantipyrine forming a quinonimine dye with λ_{max} at 497 nm	Horseradish peroxidase	0–1,2 × 10^{-4} M (cuvette)		[63]

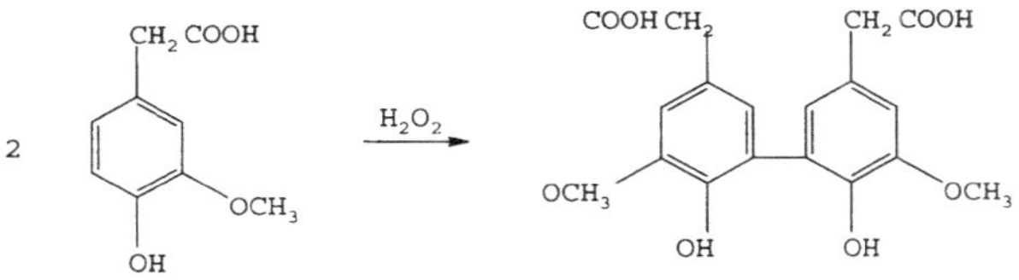

In Table 1 some methods and reaction schemes are summarized.

Many types of HP assays employ derivatives of these reactants. They have been revied by Aoyama [25].

3.3 Formation of a Dye by Oxidation of Leuco Dyes

The leucoform of a dye is usually colorless, sometimes weakly colored, or differently colored from its oxidized form. The dyes produced in this type of photometric determination generally have a very intense color which is quantified by absorption spectrometry. Detection methods based on these schemes have a lower detection limit, thus facilitating the determination of sub-μM concentrations of HP. Consequently these methods are used for very low HP concentrations such as those found in environmental samples of natural waters.

Examples of such reactions are shown in Figs. 9 and 10.

The first step in the reaction displayed in Fig. 9 is the so-called Fenton reaction which forms the reactive species. In the second step the fluorophore is

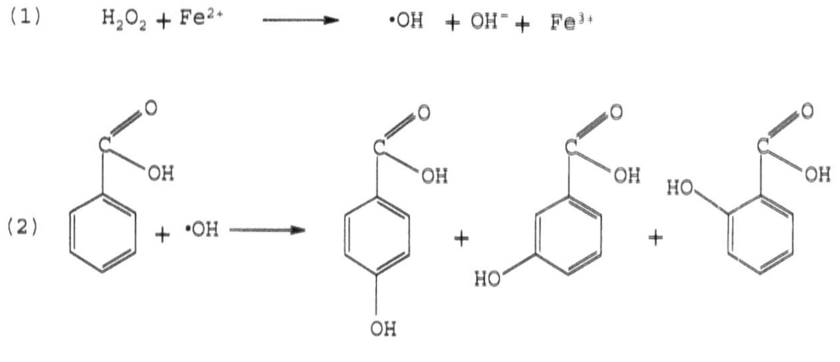

(1) $$H_2O_2 + Fe^{2+} \longrightarrow {\bullet}OH + OH^- + Fe^{3+}$$

(2) + •OH ⟶ + +

Fig. 9. Formation of a fluorophore by addition of a functional group

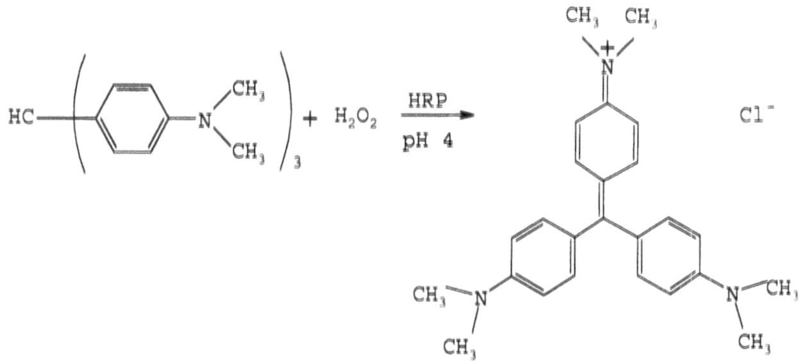

+ H_2O_2 $\xrightarrow[\text{pH 4}]{\text{HRP}}$ Cl^-

Fig. 10. Formation of a dye by oxidation of the leucoform

formed. The reported linear range of this method in a FIA is 10^{-8}–10^{-5} M HP with a detection limit of 2×10^{-8} M ($s/n = 1$) [26].

A similar method based on the use of 4-hydroxyphenylpropionic acid and horseradish peroxidase has been reported. Horseradish peroxidase is the catalyst for the first reaction. Here, the horseradish peroxidase is immobilized by physical adsorption on PTFE tubing which forms the reaction coil in the FIA. The adsorbed peroxidase has been reported to be stable for 7 hours. After this time a slow loss of activity took place (during use). The linear range of this assay is 1.2×10^{-7}–2.4×10^{-6} M with a limit of detection for 2×10^{-9} M HP [27].

With the method displayed in Fig. 10 a detection limit of 2×10^{-8} M has been achieved for this reaction in a cuvette. The oxidized form of the dye displays high stability with absorption remaining over several days; however, the dye is influenced by pH (different equilibria between protonated and deprotonated form) and salinity (molar absorption reduced if saline concentration is increased) [28].

A reaction of the leucoform of the dyestuff Patent Blue Violet has also been used for HP quantitative determination. This reaction works over a limited pH range 6.3 to 7.3 and displays linearity between 1×10^{-7}–1×10^{-5} M HP [29]. The highly fluorescent resorufin is formed by the oxidation of Amplex Red, a colorless and nonfluorescent derivative of dihydroresorufin [30].

The leuco form of some methine dyes have also been used for analysis of HP. Basically, there are three types of reaction which can result in the oxidized form of the methine dye. These are shown in Fig. 11. Methine dyes have the advantage that they can be fixed by simple adsorption on a carrier material and therefore can be used as probes [31].

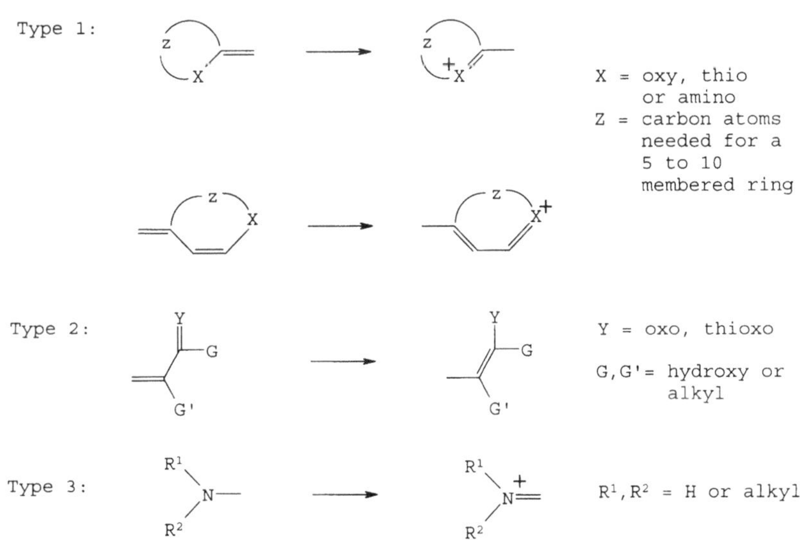

Fig. 11. Oxidation reactions with methine dyes

The most recent work reported on an optical sensor device for HP is based on this principle. A dye (Meldola Blue) is immobilized in a sol-gel layer placed in a flow-though cell (detection range 10^{-8} to 10^{-1} M HP) [32]. This technique was developed for industrial applications but some problems remain to be solved. The sensor did not show reversibility on simple change of HP concentration. Only on addition of reducing agents such as $Na_2S_2O_2$ the is response reversed. However, the relative signal change decreased as a result. Furthermore, the dye is leached (up to 4% within 4 h!) and the sensor suffers from pH sensitivity, so that solutions have to be adjusted to pH 7 to obtain reliable results. Consequently, the reported sensor is currently not industrially applicable.

A simple but not very sensitive method works on the principle of oxidation of iodide to give the colored iodine. This has been performed in a cuvette over a HP concentration range of 2.3×10^{-6} to 1.5×10^{-2} M. At higher concentrations there is the risk of precipitation of iodine. Interferents include oxidizing agents (chromate and permanganate) and substances reacting with iodine such as unsaturated organic compounds. Molybdates were used as catalysts [33].

3.4 Formation of a Colored or Fluorescent Complex

The first step in this method is a reduction of the central atom of the complex to be formed. In the second step, the reduced species forms a complex with another reactant. For example, the reaction of copper(II), HP and 2,9-dimethyl-1,10-phenanthroline (DMP) forms a copper(I)-DMP complex (Fig. 12). This copper(I)-DMP complex can be determined directly by spectrometric measurements. This method displays a linear range of 1×10^{-6}–1.2×10^{-4} M HP in a cuvette. The final absorbance remains constant over several days [34].

The formation of a Ti^{4+}– 4-(2-pyridylazo)resorcinol – H_2O_2 complex and a *N*-(4'-hydroxyphenyl)-*N*-(4-methylchinolinyl)amine- Co-H_2O_2 complex has also been reported to be useful for the analysis of HP [35, 36].

3.5 Destruction of a Dye

The first experiments based on this scheme were performed with the fluorophore scopoletin (7-hydroxy-6-methoxycoumarin). It is destroyed in the presence of HP with horseradish peroxidase as a catalyst. The decrease in fluorescence is measured and related to the concentration of HP in the solution. Cal-

$&,$# # %')+ # ($*$ ⟶ $&,!')+"$# # $(# # *$

Fig. 12. Formation of a copper(I)-DMP complex with absorption maximum at 454 nm

$)&,!+-./"%*%# # '$($ ⟶ $)&,!+-./"%*$# # ($ # $'#

Fig. 13. Reaction of HP with iron(III)tris(1,10-phenanthroline)

ibration is performed by standard addition. This method is very sensitive and was developed for detection of HP in ground water. Results are strongly influenced by reducing substances [37].

A similar reaction is based on iron(III)tris(1,10-phenanthroline), where a reduction takes place in contrast to the usual oxidation. The amount of HP is detected by the reduction of the absorption of the Fe^{3+} complex (Fig. 13). The assay is affected by the concentration of Fe(II), $CoSO_4$, Hg^{2+}, Ag^+, Sn^{2+} and by pH. The detection limit of this method in a cuvette is 4.5×10^{-7} M HP [38].

A similar method uses the fluorescence of thorium(III) in aqueous solution. The decrease of absorbance is caused by reduction of Th(III) to Th(I) with HP [39]. Methods have also been reported that correlate the reaction rate of dye destruction with the amount of HP [40].

3.6 Chemiluminescence

In chemiluminescence, the energy of a chemical reaction is released as light of a certain wavelength. The most popular reaction is the so-called luminol reaction (Fig. 14).

Luminol (5-amino-2,3-dihydro-1,4-phthalazindion) is oxidized by HP in the presence of a catalyst. The reaction has a pH optimum of 10 to 11 and is catalyzed by a variety of metal ions and heme-containing enzymes; cobalt(II) being the most efficient inorganic catalyst. After the components are mixed, blue light (λ_{max} = 440 nm) is emitted, the maximum chemiluminescence intensity being reached 2 s after mixing. This method is sensitive and consequently often used for environmental analysis (e.g. analysis of rain water, gaseous HP detection in atmospheric samples, sea water analysis). The detection limit of this method is 5×10^{-9} M and the linear range is between 5×10^{-9} and 5×10^{-7} M hydrogen peroxide (measured in sea water) [41–46].

Not only luminol but also certain aryl oxalates give a chemiluminescent reaction with HP. Some of them emit more reddish light. Therefore, interferences due to fluorescence of the biological matrix can be reduced [47].

Instead of using a catalyst, the luminol chemiluminescence can also be induced electrochemically by anodic oxidation of luminol or HP at electrodes made from platinum, gold or carbon. These methods display an extremely low detection limit of 6.6×10^{-11} M HP [48]. However, the strong passivation of the electrode surfaces and the high anodic potentials, which have to be applied

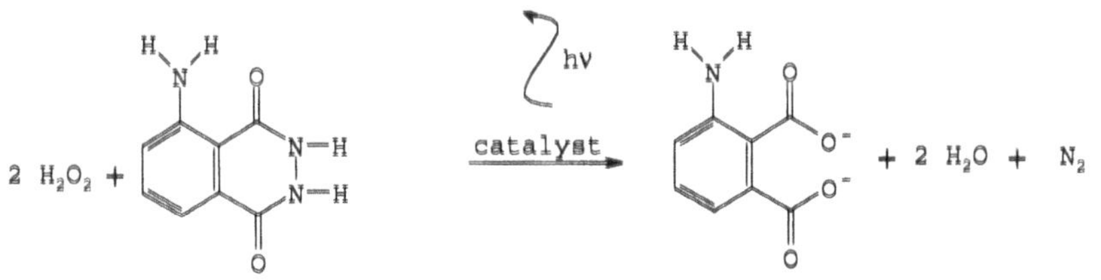

Fig. 14. Luminol reaction

(resulting in a decreased selectivity) are serious disadvantages. This problem was solved by covering the electrode surface with a HP-permeable membrane [49].

Another luminescent detection method employs electrochemically generated hydroxyl radicals, which interact with hydrated terbium(III) cations adsorbed on the oxide-covered surface of an electrode (e.g. zirconium, tantalum), resulting in a characteristic terbium(III) emission. This sensor shows a linear range between 10^{-6} and 10^{-3} M HP [50, 51].

Electroluminescent semiconductor optical sensors have also been reported. The electroluminescence arises from high energy $OH^{\bullet}$ radicals formed by the electroreduction of HP on the electrode surface [52, 53].

3.7 Indirect Measurement by Quenching of Fluorescence by Molecular Oxygen

This method is based on the catalytic decomposition of HP on a catalyst immobilized in a polymer membrane, mounted on an optical oxygen sensor. The oxygen sensor detects oxygen by luminescence quenching of a fluorophore. With this system and the catalysts catalase and silver powder, concentrations of HP from 1×10^{-4} to 1×10^{-2} M can be monitored [54].

For industrial applications more robust catalysts such as platinum or manganese dioxide are preferable. A sensor using manganese dioxide as catalyst is reported which quantifies reversibly HP concentrations from 0.1 to 2% (0.03–0.6 M, i.e. the range for many industrial applications). Determination of lower or higher concentrations of HP should be possible, but has not been tested to date. The sensor membrane can be covered with chemically resistant membranes, which enhance the oxygen concentration in the oxygen-sensitive layer up to five times, compared to an uncoated sensor. It also protects the underlying membranes from being chemically attacked by the sample medium. The sensor can be tailored for many applications by varying the cover membrane. The response times of these sensors depends on the type of covering layer, with the shortest response times being similar to that of uncoated sensors (t_{95} is approximately 1 min for a change from 0 to 0.2% by weight HP in water) [55].

4 Conclusions

While a vast number of methods exists for the determination of hydrogen peroxide (HP), only a limited number of sensor systems are commercially available, all of which are electrode systems which display several limitations. Parameters of the sample solution (e.g. sample conductivity and presence of other chemical species) may influence the sensor signals. Therefore for a given application it is necessary to tailor the sensor and the respective measurement conditions to allow precise analysis of HP.

Direct spectroscopic methods appear to be the most promising alternative technique for the determination of HP in industrial applications. The detection of HP in the near-infrared region offers a robust and easy-to-handle method for routine analysis in industrial solutions. The detection range of this method (0.1–>35% HP) covers the concentration range used in industrial processes in general. There are spectrometers available for this technique. Furthermore, immersion probes are available which may be connected to the spectrometer via glass fibers, allowing simple and effective on-line measurement. However, this method is restricted to a well defined matrix. Consequently, every new application or severe changes in the matrix demand a modification of the evaluation model.

The detection of HP in the mid infrared displays a technically more sophisticated method, compared to the near infrared. But the method is more sensitive (detection limit of this method is 0.01% of HP) and, as the peaks in the mid infrared are more separated, the interference by the matrix is weak. Moreover, mid-infrared spectroscopy offers the possibility of simple single spectrum multi-analyte detection.

Indirect spectroscopic methods, displaying high sensitivity, seem to provide the best solution for the analysis of environmental samples. Detection limits as low as 10^{-11} M HP have been reported. The majority of these methods are based on irreversible chemical reactions and consequently do not constitute real sensor systems. Test kits and test strips applying one of the described methods are commercially available, but do not fulfill many analytical needs, because they suffer from limited accuracy, inappropriate detection range, and matrix interferences. Therefore, an appropriate method needs to be selected from the vast number of methods reported here and tailored to the specific analytical need. The selected method may subsequently be automated by applying a Flow Injection Analysis (FIA) technique.

References

1. Jones CW (1999) Applications of HP and Derivatives, Royal Society of Chemistry, Cambridge
2. Raimo H (1987) IPPTA 24:31
3. Hoeoek JE, Aakerlund G (1988) PCT Int Appl WO 8802796
4. Ravensbergen DW, Holweg RBM (1993) J Soc Dyers Colour 109:72
5. Goeckeritz D, Friedrich F, Yahya M (1995) Pharmazie 50:437
6. Hydrogen peroxide for industrial use – Determination of hydrogen peroxide content – Titrimetric method., ISO/DIS standard No. 7157.
7. Spohn U, Ruettinger HH, Matschiner H (1987) Chem Tech (Leipzig) 39:284
8. Katayama M, Takeuchi H, Taniguchi H (1991) Anal Lett 24:1005
9. Westbroek P, Van Haute P, Temmerman E (1996) Fresenius' J Anal Chem 354:405
10. Heller A, Gregg BA (1993) PCT Int Appl WO 9323748
11. Ravensberger DW (1991) PCT Int Appl WO 9118296
12. Tay BT, Tat KP, Gunasingham H (1988) Analyst (Cambridge) 113:617
13. Gosgrove M, Moody GJ, Thomas JD (1989) Analyst (Cambridge) 114:1627
14. Mizaikoff B (1999) Proc SPIE 3849:7
15. Mizaikoff B, Lendl B (2002) Sensor Systems Based on Mid-Infrared Transparent Fibers. In Chalmers JM and Griffiths PR (eds) Handbook of Vibrational Spectroscopy, John Wiley & Sons, Ltd, p 1560
16. Herzberg G (1945) Molecular Spectra and Molecular Structure, van Nostrand, Princeton

17. Voraberger H, Ribitsch V, Janotta M, Mizaikoff B (2003) Appl Spectrosc 57:574
18. Fahrenfort J (1961) Spectrochim Acta17:698
19. Harrick NJ (1967) Internal Reflection Spectroscopy, Wiley, New York
20. Aebi HE (1983) Catalase. In: Bergmeyer HU, Bergmeyer J, Graßl M (eds) Methods of Enzymatic Analysis. VCH, Weinheim Deerfield Beach Basel, vol 3/ p 273
21. Hirai H, Hanada T, Yamanishi K, Nohara H (1987) Eur Pat Appl EP 223977
22. Genfa Z, Dasgupta PK, Edgemond WS, Marx JN (1991) Anal Chim Acta 243:207
23. Saito Y, Satouchi M, Mifune M (1987) Bull Chem Soc Jpn 60:2227
24. Klein C, Von der Eltz H, Herrmann R (1987) German Pat DE 3526566
25. Aoyama N, Okano M, Miike (1986) Eur Pat Appl EP 206316
26. Lee JH, Tang IN, Weinstein-Lloyd JB (1990) Anal Chem 62:2381
27. Li YZ, Townshend A (1998) Anal Chim Acta 359:149
28. Zhang L, Wong GTF 1994 Talanta 41:2137
29. Clapp PA, Evans DF (1991) Anal Chim Acta 243:217
30. Zhou M, Diwu Z, Panchuk-Voloshina N, Haugland (1997) Anal Biochem 253:162
31. Norton GE (1987) Eur Pat Appl EP 247845
32. Lobnik A, Čajlaković M (2001) Sens Actuators B 74:194
33. Jadesjoe G, Magnusson B, Sundstrand S (1992) Eur Pat Appl EP 485000
34. Baga AN, Johnson GRA, Nazhat B (1988) Anal Chim Acta 204:349
35. Mori I, Fujita Y, Toyoda M, Kato K, Yoshida N, Akagi M (1991) Talanta 38:683
36. Possanzini M, Di Palo V (1995) Anal Chim Acta 315:225
37. Holm TR, George GK, Barcelona MJ (1987) Anal Chem 59:582
38. Afsar H, Apak R, Tor I (1990) Analyst (London) 115:99
39. Rao MSP, Rao ARM, Ramana KV (1990) Talanta 37:753
40. Perez-Ruiz T, Martinez-Lozano C, Tomas V, Val O (1992) Analyst (Cambridge) 117: 1771
41. Price D, Worsfold PJ, Mantoura RFC (1994) Anal Chim Acta 298:121
42. Li J, Dasgupta GK (2001) Anal Chim Acta 442:63
43. Hanaoka S, Lin J, Yamada M (2001) Anal Chim Acta 426:57
44. Li B, Zhang Z, Jin Y (2001) Sens Actuators B 72:115
45. Diaz AN, Peinado MCR., Minguez MCT (1998) Anal Chim Acta 363:221
46. Hara T, Tsukagoshi K Kimoto H (1990) Bull Chem Soc Jpn 63:2423
47. Katayama M, Takeuchi H, Taniguchi H (1991) Anal Lett 24:1005
48. Sakura S (1992) Anal Chim Acta 262:49
49. Janasek D, Spohn U, Beckmann D (1998) Sens Actuators B 51:107
50. Lipiainen K, Kulmala S, Haapakka K (1992) Anal Chim Acta 266:51
51. Haapakka K, Kulmala S (1988) Anal Chim Acta 208:69
52. Poznyak SK, Kulak AI (1994) Sens Actuators B B22:97
53. Poznyak SK, Kulak AI (1996) Talanta 43:1607
54. Posch HE, Wolfbeis OS (1989) Microchim Acta 1:41
55. Voraberger HS, Trettnak W, Ribitsch V (2003) Sens Actuators B 90:324
56. Chen QY, Li DH, Zhu QZ, Zheng H, Xu JG (1998) Anal Chim Acta 373:261
57. Saito Y, Mifune M, Nakashima S (1987) Talanta 34:667
58. Saito Y, Nakashima S, Mifune M, Odo J Tanaka Y, Chikuma M, Tanaka H (1985) Anal Chim Acta 172:285
59. Chung HK, Dasgupta PK, Marx JN (1993) Talanta 40:981
60. Mohr P, Kahrig E, Mueller A (1986) German Pat (East) DD 236654
61. Zhu Q, Li Q, Lu J (1996) Anal Lett 29:1729
62. Ci Y, Wang F (1990) Anal Chim Acta 233:299
63. Goedicke W, Goedicke I (1986) German Pat (East) DD236554

Subject Index

D

T

www.ingramcontent.com/pod-product-compliance
Ingram Content Group UK Ltd.
Pitfield, Milton Keynes, MK11 3LW, UK
UKHW021859190726
13853UKWH00003B/1340

* 9 7 8 3 6 6 2 0 9 1 1 2 8 *